MW00844297

Spectroscopic Ellipsometr

Spectroscopic Ellipsometry
Principles and Applications

Hiroyuki Fujiwara
Department of Electrical and Electronic Engineering, Gifu University, Gifu, Japan

John Wiley & Sons, Ltd

Japanese Edition, Copyright 2003, Hiroyuki Fujiwara, ISBN 4 621 07253 6
Published by Maruzen Co. Ltd, Tokyo, Japan

Telephone (+44) 1243 779777

Email (for orders and customer service enquiries): cs-books@wiley.co.uk
Visit our Home Page on www.wiley.com

Printed with corrections April 2009

Other Wiley Editorial Offices

John Wiley & Sons Inc., 111 River Street, Hoboken, NJ 07030, USA
Jossey-Bass, 989 Market Street, San Francisco, CA 94103-1741, USA
Wiley-VCH Verlag GmbH, Boschstr. 12, D-69469 Weinheim, Germany
John Wiley & Sons Australia Ltd, 42 McDougall Street, Milton, Queensland 4064, Australia
John Wiley & Sons (Asia) Pte Ltd, 2 Clementi Loop #02-01, Jin Xing Distripark, Singapore 129809
John Wiley & Sons Ltd, 6045 Freemont Blvd, Mississauga, Ontario L5R 4J3, Canada
Wiley also publishes its books in a variety of electronic formats. Some content that appears in print may not be available in electronic books.

Library of Congress Cataloging in Publication Data

Fujiwara, Hiroyuki.
Spectroscopic ellipsometry : principles and applications / Hiroyuki Fujiwara.
p. cm.
Includes bibliographical references and index.
ISBN-13: 978-0-470-01608-4 (cloth : alk. paper)
ISBN-10: 0-470-01608-6 (cloth : alk. paper)
1. Ellipsometry. 2. Spectrum analysis. 3. Materials—Optical properties. I. Title.
QC443.F85 2007
620.1′1295—dc22

2006030741

British Library Cataloguing in Publication Data

A catalogue record for this book is available from the British Library

ISBN 978-0-470-01608-4

Typeset in 10/12pt Times by Integra Software Services Pvt. Ltd, Pondicherry, India

This book is printed on acid-free paper responsibly manufactured from sustainable forestry in which at least two trees are planted for each one used for paper production.

Dedicated to my father, Sadao Fujiwara

Contents

Appendices

Foreword

It is a pleasure and an honor to comment on this outstanding book, *Spectroscopic Ellipsometry: Principles and Applications* by Dr H. Fujiwara. It is a tutorial introduction, yet offers considerable depth into advanced topics such as generalized ellipsometry and advanced dispersion and oscillator models for analysis of complex materials systems. Each chapter is extremely well referenced, with over 400 literature citations in total, providing the reader rapid access to considerable published literature from fundamentals to recent advances. It is also well illustrated, with over 200 figures, making this an excellent possible textbook for teaching ellipsometry at both the beginning and intermediate to advanced levels. The book will be appropriate as a text in an educational institution. Equally it will be excellent to help educate and train researchers in institutes and industrial laboratories to learn practical applications of the technique.

For decades the book, *Ellipsometry and Polarized Light*, by R. M. A. Azzam and N. M. Bashara, (North-Holland, New York, 1977), has probably been the most widely cited general reference on ellipsometry. However, this book is now 30 years old, and out of print. Fujiwara-san's book offers the reader a modern, up-to-date, clear discussion of many of the same topics: fundamentals of optics, polarization, ellipsometry and instrumentation, in the first few chapters. This follows naturally into more advanced and well referenced chapters on data analysis, anisotropy, experimental examples, and *in situ* ellipsometry.

A perspective of the role of Fujiwara-san's book in the context of existing literature on ellipsometry might be helpful. Often cited references on ellipsometry are:

- *Infrared Spectroscopic Ellipsometry*, by A. Röseler, (Akademie-Verlag, Berlin 1990);
- *Selected Papers on Ellipsometry*, R. M. A. Azzam, Ed., SPIE Milestone Series, MS 27, (SPIE, Bellingham 1990);
- R. Muller, Ellipsometry as an *in situ* probe for the study of electrode processes, in *Techniques of Characterization of Electrodes and Electrochemical Processes*, R. Varma and J. Selman, Eds, (John Wiley & Sons, Inc., New York, 1991);
- H. G. Tompkins and W. A. McGahan, *Spectroscopic Ellipsometry and Reflectometry: A User's Guide*, (John Wiley & Sons, Inc., New York, 1999).

There are also recent books for specialists in ellipsometry. These include M. Schubert, *Infrared Ellipsometry on Semiconductor Layer Structures: Phonons, Plasmons and Polaritons* (Springer, Berlin, 2004); and *Handbook of Ellipsometry*,

H. G. Tompkins and E. A. Irene, Eds, (Andrew, Norwich, 2005). One can also find contributed papers and brief reviews in proceedings of the International Conferences on Spectroscopic Ellipsometry. However, Fujiwara-san's book covers the topic in unique and valuable ways, subsequently allowing advanced literature, such as found in the conference proceedings and books named above, to be comfortably read and understood.

Dr Fujiwara's *Spectroscopic Ellipsometry: Principles and Applications* offers a welcome new contribution as both a tutorial text and an introduction to advanced topics and applications. This book will become a 'must have' for every new user and university student, as well as a specialist wishing for a greater depth of understanding of this technique. It also contains complete and up-to-date references to a wealth of published information on spectroscopic ellipsometry and its applications.

John A. Woollam
J.A. Woollam Company (Founder and President)
University of Nebraska (George Holmes University Professor)
10 July 2006

Preface

Historically, the development of a new measurement technique implies advance in science. Unless clear scientific evidence is presented, scientific facts are often treated as merely experimental knowledge. In other words, advances in scientific fields can be viewed as a consequence of various measurements used to confirm scientific significance. In the last 50 years alone, a variety of characterization techniques have been established, and some scientific fields owe their progress to the innovation of such measurement techniques. The development of scanning tunneling microscope (STM), for example, has revolutionized surface science and contributed greatly to the rapid progress of surface science.

Ironically, basic principles of ellipsometry were established more than 100 years ago, but ellipsometry had been perceived as an 'unproductive instrument' until recently. During the 1990s, however, this situation changed drastically due to rapid advances in computer technology that allowed the automation of ellipsometry instruments as well as ellipsometry data analyses. With the commercialization of such spectroscopic ellipsometry instruments in the mid-1990s, the ellipsometry technique became quite popular, and now is applied to wide research areas from semiconductors to organic materials. Recent developments in spectroscopic ellipsometry have further allowed the real-time characterization of film growth and evaluation of optical anisotropy. Consequently, spectroscopic ellipsometry has established its position as a high precision optical characterization technique, and more researchers in universities and companies have started using this technique. Nevertheless, principles of ellipsometry are often said to be difficult, partly due to a lack of proper knowledge of polarized light used as a probe in ellipsometry. Besides, the meaning of (ψ, Δ) obtained from ellipsometry measurements is not straightforward, and procedures of ellipsometry data analysis are rather unique.

The key objective of this book is to provide a fundamental understanding for spectroscopic ellipsometry particularly for researchers who are not familiar with the ellipsometry technique. Although some aspects are complicated, the understanding of the ellipsometry technique is not essentially difficult, if one comprehends the principles in order. Based on this point of view, this book provides general descriptions for measurement and data analysis methods employed widely in spectroscopic ellipsometry. Since ellipsometry is quite a geometrical measurement method, various illustrations are included to help readers. To simplify descriptions, unnecessary equations for electromagnetics and quantum mechanics have been eliminated. Instead, the derivations of important formulae used in spectroscopic ellipsometry are shown in this book.

In order to comprehend spectroscopic ellipsometry, however, a fundamental knowledge of optics is required. In the book, therefore, 'Principles of optics' and 'Polarization of light' are described in Chapter 2 and Chapter 3, respectively. From these two chapters, 'Principles of spectroscopic ellipsometry' (Chapter 4) can be understood more easily. We focus on data analysis of spectroscopic ellipsometry in Chapters 5–8. In particular, principles and physical backgrounds of ellipsometry analysis are discussed in detail in Chapter 5. Since there is growing interest in optical anisotropy, the data analysis of anisotropic materials is explained in Chapter 6. In 'Data analysis examples' (Chapter 7), examples of ellipsometry analyses for various materials used in different fields are described. In 'Real-time monitoring by spectroscopic ellipsometry' (Chapter 8), the applications of spectroscopic ellipsometry for growth monitoring and feedback control of processing are addressed.

Most of the content in this book is a translation from the Japanese book *Spectroscopic Ellipsometry*, published in 2003 by Maruzen. In this English edition, the overall content is expanded and thc description for anisotropic materials (Chapter 6) has been added. For the English edition, I am especially grateful to Prof. John A. Woollam (University of Nebraska, Lincoln and J. A. Woollam Co.) who kindly reviewed this book and gave me very thoughtful comments. The author gratefully acknowledges Prof. Isamu Shimizu for his continued support. I would also like to thank Dr Michio Kondo (AIST), Dr Akihisa Matsuda (AIST), and Prof. Christopher R. Wronski (Pennsylvania State University) for their kind advice. I am grateful to Mr Michio Suzuki (J. A. Woollam Co., Japan), Mr Teruaki Kuwahara (Maruzen) and Miss Jenny Cossham (John Wiley & Sons, Ltd) who have supported the publication of this book. Finally, I wish to express my sincere gratitude to Prof. Robert W. Collins (University of Toledo) who has taught me everything concerning real-time spectroscopic ellipsometry.

Hiroyuki Fujiwara

Acknowledgments

The author wishes to thank the authors and publishers for permission to reproduce the following figures and tables used in this book:

Fig. 1.4 and Table 1.3, K. Vedam, *Thin Solid Films*, **313–314** (1998) 1. Fig. 1.5, I. An, Y. M. Li, H. V. Nguyen, and R. W. Collins, *Rev. Sci. Instrum.*, **63** (1992) 3842. Fig. 1.6, D. E. Aspnes, *Thin Solid Films*, **455–456** (2004) 3. Fig. 2.12, S. Adachi, *J. Appl. Phys.*, **53** (1982) 8775. Fig. 3.9(a), S. N. Jasperson and S. E. Schnatterly, *Rev. Sci. Instrum.*, **40** (1969) 761. Fig. 3.9(b), J. C. Canit and J. Badoz, *Appl. Opt.*, **22** (1983) 592. Fig. 4.11, W. M. Duncan and S. A. Henck, *Appl. Surf. Sci.*, **63** (1993) 9. Fig. 4.13, A. Röseler, *Thin Solid Films*, **234** (1993) 307. Fig. 4.14, P. S. Hauge, *Surf. Sci.*, **96** (1980) 108. Figs. 4.16(a) and 4.16(c), G. Jin, R. Jansson, and H. Arwin, *Rev. Sci. Instrum.*, **67** (1996) 2930. Fig. 4.22(a), D. E. Aspnes, *J. Opt. Soc. Am.*, **64** (1974) 812. Fig. 4.23, B. Johs, *Thin Solid Films*, **234** (1993) 395. Fig. 4.24, J. Lee, P. I. Rovira, I. An, and R. W. Collins, *J. Opt. Soc. Am. A*, **18** (2001) 1980. Fig. 4.26, S. Kawabata, *OYO BUTURI*, **57** (1988) 1868. Fig. 4.27, J. Lee, P. I. Rovira, I. An, and R. W. Collins, *Rev. Sci. Instrum.*, **69** (1998) 1800. Fig. 4.28, D. E. Aspnes and A. A. Studna, *Appl. Opt.*, **14** (1975) 220. Fig. 4.29, R. W. Collins and K. Vedam, 'Optical properties of solids,' in *Encyclopedia of Applied Physics, Vol. 12*, Wiley-VCH (1995) 285. Fig. 5.6(b), R. J. Archer and G. W. Gobeli, *J. Phys. Chem. Solids*, **26** (1965) 343. Figs. 5.15(b), 5.19 and 5.20(a), K. Kobayashi, *Physics of Light: Why Light Refracts, Reflects and Transmits*, in Japanese, Tokyo University Publisher (2002). Fig. 5.21(b) G. E. Jellison, Jr and F. A. Modine, *Appl. Phys. Lett.*, **69** (1996) 371. Fig. 5.28, H. Fujiwara, J. Koh, P. I. Rovira, and R. W. Collins, *Phys. Rev. B*, **61** (2000) 10832. Fig. 5.30, J. Koh, Y. Lu, C. R. Wronski, Y. Kuang, R. W. Collins, T. T. Tsong, and Y. E. Strausser, *Appl. Phys. Lett.*, **69** (1996) 1297. Fig. 5.31(b), R. H. Muller and J. C. Farmer, *Surf. Sci.*, **135** (1983) 521. Fig. 5.37, M. Kildemo, R. Ossikovski, and M. Stchakovsky, *Thin Solid Films*, **313–314** (1998) 108. Fig. 5.38, G. E. Jellison, Jr and J. W. McCamy, *Appl. Phys. Lett.*, **61** (1992) 512. Figs. 5.43 and 7.17, D. E. Aspnes, A. A. Studna, and E. Kinsbron, *Phys. Rev. B*, **29** (1984) 768. Fig. 7.5, C. M. Herzinger, B. Johs, W. A. McGahan, J. A. Woollam, and W. Paulson, *J. Appl. Phys.*, **83** (1998) 3323. Fig. 7.7(a), J. R. Chelikowsky, and M. L. Cohen, *Phys. Rev. B*, **14** (1976) 556. Fig. 7.7(b), U. Schmid, N. E. Christensen, and M. Cardona, *Phys. Rev. B*, **41** (1990) 5919. Figs. 7.10(b) and 7.11, T. Yang, S. Goto, M. Kawata, K. Uchida, A. Niwa, and J. Gotoh, *Jpn. J. Appl. Phys.*, **37** (1998) L1105-1108. Fig. 7.12(b), P. Petrik, M. Fried, T. Lohner, R. Berger, L. P. Bíro, C. Schneider, J. Gyulai, H. Ryssel, *Thin Solid Films*, **313–314** (1998) 259. Figs. 7.13(a),

7.14 and Table 7.3, F. L. Terry, Jr, *J. Appl. Phys.*, **70** (1991) 409. Fig. 7.13(b), D. E. Aspnes, S. M. Kelso, R. A. Logan and R. Bhat, *J. Appl. Phys.*, **60** (1986) 754. Figs. 7.15 and 7.16, P. Lautenschlager, M. Garriga, L. Viña, and M. Cardona, *Phys. Rev. B*, **36** (1987) 4821. Fig. 7.18(a), C. Pickering and R. T. Carline, *J. Appl. Phys.*, **75** (1994) 4642. Fig. 7.18(b), R. T. Carline, C. Pickering, D. J. Robbins, W. Y. Leong, A. D. Pitt, and A. G. Cullis, *Appl. Phys. Lett.*, **64** (1994) 1114. Fig. 7.19(b), S. Boultadakis, S. Logothetidis, S. Ves, and J. Kircher, *J. Appl. Phys.*, **73** (1993) 914. Fig. 7.20(a), H. Ehrenreich and H. R. Philipp, *Phys. Rev.*, **128** (1962) 1622. Fig. 7.20(b), Frederick Wooten, *Optical Properties of Solids*, Academic Press (1972). Figs. 7.22(b), 7.23, 7.24 and 7.25, H. Fujiwara and M. Kondo, *Phys. Rev. B*, **71** (2005) 075109. Fig. 7.26, T. E. Tiwald, D. W. Thompson, J. A. Woollam, W. Paulson, and R. Hance, *Thin Solid Films*, **313–314** (1998) 661. Fig. 7.27, Y.-T. Kim, D. L. Allara, R. W. Collins, K. Vedam, *Thin Solid Films*, **193/194** (1990) 350. Fig. 7.28, D. Tsankov, K. Hinrichs, A. Röseler, and E. H. Korte, *Phys. Stat. Sol. A*, **188** (2001) 1319. Fig. 7.29, K. Postava, T. Yamaguchi, and M. Horie, *Appl. Phys. Lett.*, **79** (2001) 2231. Fig. 7.30(b), A. C. Zeppenfeld, S. L. Fiddler, W. K. Ham, B. J. Klopfenstein, and C. J. Page, *J. Am. Chem. Soc.*, **116** (1994) 9158. Fig. 7.31, H. Arwin, *Thin Solid Films*, **313–314** (1998) 764. Fig. 7.32, U. Jönsson, M. Malmqvist, and I. Rönnberg, *J. Colloid. Interface Sci.*, **103** (1985) 360. Fig. 7.33, D. E. Gray, S. C. Case-Green, T. S. Fell, P. J. Dobson, and E. M. Southern, *Langmuir*, **13** (1997) 2833. Figs. 7.34(b) and 7.35, J. Humlíček, and A. Röseler, *Thin Solid Films*, **234** (1993) 332. Figs. 7.37 and 7.38, M. Schubert, B. Rheinländer, J. A. Woollam, B. Johs and C. M. Herzinger, *J. Opt. Soc. Am. A*, **13** (1996) 875. Fig. 7.39, T. Wagner, J. N. Hilfiker, T. E. Tiwald, C. L. Bungay, and S. Zollner, *Phys. Stat. Sol. A*, **188** (2001) 1553-1562. Figs. 7.41 and 7.42, C. M. Ramsdale and N. C. Greenham, *Adv. Mater.*, **14** (2002) 212. Fig. 8.2, H. Z. Massoud, J. D. Plummer, and E. A. Irene, *J. Electrochem. Soc.*, **132** (1985) 2685. Fig. 8.4, M. Wakagi, H. Fujiwara, and R. W. Collins, *Thin Solid Films*, **313–314** (1998) 464. Fig. 8.10, H. Fujiwara, Y. Toyoshima, M. Kondo, and A. Matsuda, *Phys. Rev. B*, **60** (1999) 13598. Fig. 8.14, H. Fujiwara, J. Koh, C. R. Wronski, and R. W. Collins, *Appl. Phys. Lett.*, **70** (1997) 2150. Fig. 8.15, H. Fujiwara, J. Koh, C. R. Wronski, R. W. Collins, and J. S. Burnham, *Appl. Phys. Lett.*, **72** (1998) 2993. Fig. 8.16, Y. M. Li, I. An, H. V. Nguyen, C. R. Wronski, and R. W. Collins, *J. Non-Cryst. Solids*, **137&138** (1991) 787. Figs. 8.17 and 8.18, H. Fujiwara, M. Kondo, and A. Matsuda, *Phys. Rev. B*, **63** (2001) 115306. Figs. 8.19 and 8.20, H. Fujiwara, M. Kondo, and A. Matsuda, *J. Appl. Phys.*, **93** (2003) 2400. Fig. 8.21, E. A. Irene, *Thin Solid Films*, **233** (1993) 96. Fig. 8.22, H. Fujiwara, and M. Kondo, *Appl. Phys. Lett.*, **86** (2005) 032112. Figs. 8.23, 8.24, and 8.25, H. L. Maynard, N. Layadi, J. T. C. Lce, *Thin Solid Films*, **313–314** (1998) 398. Fig. 8.26, D. E. Aspnes, W. E. Quinn, M. C. Tamargo, M. A. A. Pudensi, S. A. Schwarz, M. J. S. P. Brasil, R. E. Nahory, and S. Gregory, *Appl. Phys. Lett.*, **60** (1992) 1244. Fig. 8.27, B. Johs, D. Doerr, S. Pittal, I. B. Bhat, and S. Dakshinamurthy, *Thin Solid Films*, **233** (1993) 293.

1 Introduction to Spectroscopic Ellipsometry

Because of recent advances in computer technology, the spectroscopic ellipsometry technique has developed rapidly. As a result, the application area of spectroscopic ellipsometry has also expanded drastically. In spectroscopic ellipsometry, process diagnosis including thin-film growth can be performed in real time by employing light as a measurement probe. More recently, 'feedback control,' in which complicated device structure is controlled in real time, has been carried out using spectroscopic ellipsometry. In this chapter, we review the features and applications of spectroscopic ellipsometry. This chapter will provide an overview of measurement techniques and data analysis procedures in spectroscopic ellipsometry.

1.1 FEATURES OF SPECTROSCOPIC ELLIPSOMETRY

Ellipsometry is an optical measurement technique that characterizes light reflection (or transmission) from samples [1–4]. The key feature of ellipsometry is that it measures the change in polarized light upon light reflection on a sample (or light transmission by a sample). The name 'ellipsometry' comes from the fact that polarized light often becomes 'elliptical' upon light reflection. As shown in Table 1.1, ellipsometry measures the two values (ψ, Δ). These represent the amplitude ratio ψ and phase difference Δ between light waves known as p- and s-polarized light waves (see Fig. 4.1). In spectroscopic ellipsometry, (ψ, Δ) spectra are measured by changing the wavelength of light. In general, the spectroscopic ellipsometry measurement is carried out in the ultraviolet/visible region, but measurement in the infrared region has also been performed widely.

The application area of spectroscopic ellipsometry is quite wide (Chapter 7). For real-time monitoring, not only characterization of thin-film growth but also process diagnoses including etching and thermal oxidation can be performed (Chapter 8). In particular, spectroscopic ellipsometry allows characterization of thin films formed in

Spectroscopic Ellipsometry: Principles and Applications H. Fujiwara

Table 1.1 Features of spectroscopic ellipsometry

Measurement probe:	Light
Measurement value:	(ψ, Δ) Amplitude ratio ψ and phase difference Δ between p- and s-polarized light waves
Measurement region:	Mainly in the infrared–visible/ultraviolet region
Application area:	
Semiconductor	Substrates, thin films, gate dielectrics, lithography films
Chemistry	Polymer films, self-assembled monolayers, proteins, DNA
Display	TFT films, transparent conductive oxides, organic LED
Optical coating	High and low dielectrics for anti-reflection coating
Data storage	Phase change media for CD and DVD, magneto-optic layers
Real-time monitoring:	Chemical vapor deposition (CVD), molecular beam epitaxy (MBE), etching, oxidation, thermal annealing, liquid phase processing etc.
General restrictions:	i) Surface roughness of samples has to be small ii) Measurement has to be performed at oblique incidence

solution (Section 7.4), because light is employed as the probe. However, there are two general restrictions on the ellipsometry measurement; specifically: (1) surface roughness of samples has to be rather small, and (2) the measurement must be performed at oblique incidence. When light scattering by surface roughness reduces the reflected light intensity severely, the ellipsometry measurement becomes difficult as ellipsometry determines a polarization state from its light intensity. If the size of surface roughness exceeds $\sim 30\%$ of a measurement wavelength, measurement errors generally increase, although this effect depends completely on the type of instrument (Section 4.4).

In ellipsometry, an incidence angle is chosen so that the sensitivity for the measurement is maximized. The choice of the incidence angle, however, varies according to the optical constants of samples. For semiconductor characterization, the incidence angle is typically 70–80° (Section 2.3.4). It should be noted that, at normal incidence, the ellipsometry measurement becomes impossible, since p- and s-polarizations cannot be distinguished anymore at this angle (Section 2.3.2). One exception is the characterization of in-plane optical anisotropy. In this case, the ellipsometry measurement is often performed at normal incidence to determine the variation of optical constants with the rotation of a sample (Chapter 6).

Table 1.2 summarizes the advantages and disadvantages of the spectroscopic ellipsometry technique. One of the remarkable features of spectroscopic ellipsometry is the high precision of the measurement, and very high thickness sensitivity (~ 0.1 Å) can be obtained even for conventional instruments (Section 4.4.3). As we will see in the next section, spectroscopic ellipsometry allows various characterizations including optical constants and thin-film structures. Moreover, as the ellipsometry measurement takes only a few seconds, real-time observation and feedback control of processing can be performed relatively easily (Chapter 8).

The one inherent drawback of the ellipsometry technique is the indirect nature of this characterization method. Specifically, ellipsometry data analysis requires an

Table 1.2 Advantages and disadvantages of spectroscopic ellipsometry

Advantages:	High precision (thickness sensitivity: ~0.1 Å)
	Nondestructive measurement
	Fast measurement
	Wide application area
	Various characterizations including optical constants and film thicknesses are possible
	Real-time monitoring (feedback control) is possible
Disadvantages:	Necessity of an optical model in data analysis (indirect characterization)
	Data analysis tends to be complicated
	Low spatial resolution (spot size: several mm)
	Difficulty in the characterization of low absorption coefficients ($\alpha < 100\,\mathrm{cm}^{-1}$)

optical model defined by the optical constants and layer thicknesses of a sample (see Fig. 5.39). In an extreme case, one has to construct an optical model even when the sample structure is not clear at all. In addition, this ellipsometry analysis using an optical model tends to become complicated, which can be considered as another disadvantage of the technique. The spot size of a light beam used for spectroscopic ellipsometry is typically several millimeters, leading to the low spatial resolution of the measurement. However, it is possible to determine the surface area ratio of different materials that cover the sample surface (see Fig. 5.31). Recently, in order to improve spatial resolution, imaging ellipsometry has been developed (Section 4.2.8). As shown in Table 1.2, in ellipsometry, characterization of small absorption coefficients ($\alpha < 100\,\mathrm{cm}^{-1}$) is rather difficult (Section 4.4.3).

1.2 APPLICATIONS OF SPECTROSCOPIC ELLIPSOMETRY

Spectroscopic ellipsometry has been applied to evaluate optical constants and thin-film thicknesses of samples. However, the application area of spectroscopic ellipsometry has been expanded recently, as it allows process diagnosis on the atomic scale from real-time observation. Figure 1.1 shows various physical properties that can be determined from spectroscopic ellipsometry. In particular, this figure summarizes the characterization by *ex situ* measurement. Here, *ex situ* measurement means a measurement performed after finishing sample preparation (processing).

As shown in Fig. 1.1, spectroscopic ellipsometry measures (ψ, Δ) spectra for photon energy $h\nu$ or wavelength λ. In general, the interpretation of measurement results is rather difficult from the absolute values of (ψ, Δ). Thus, construction of an optical model is required for data analysis. From this data analysis, physical properties including the optical constants and film thicknesses of the sample can be extracted. Unlike reflectance/transmittance measurement, ellipsometry allows the direct measurement of the refractive index n and extinction coefficient k, which are also referred to as optical constants. From the two values (n, k), the complex refractive index defined by $N \equiv n - ik(i = \sqrt{-1})$ is determined.

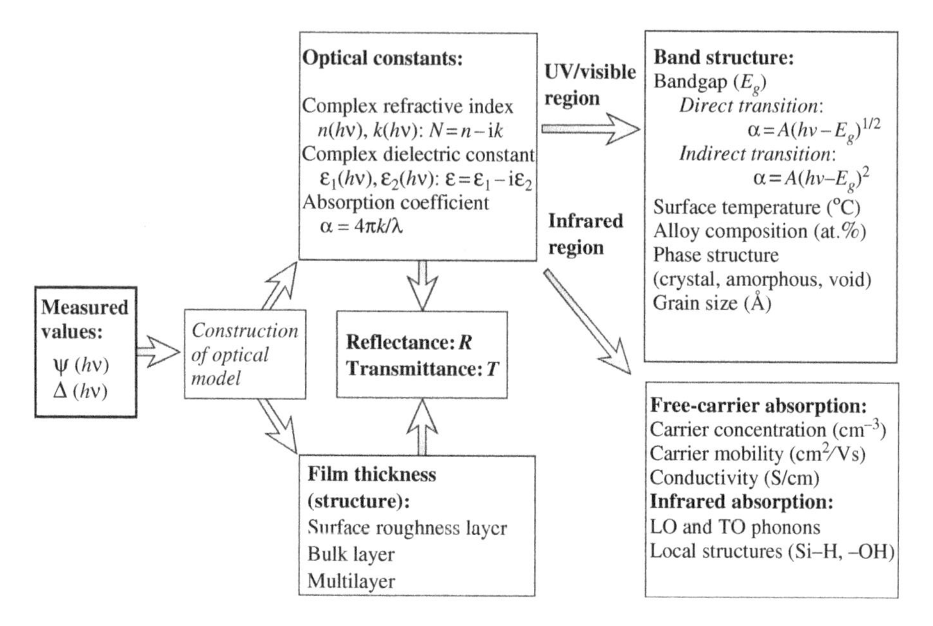

Figure 1.1 Characterization of physical properties by spectroscopic ellipsometry.

The complex dielectric constant ε and absorption coefficient α can also be obtained from the simple relations expressed by $\varepsilon = N^2$ and $\alpha = 4\pi k/\lambda$, respectively (Chapter 2). Moreover, from optical constants and film thicknesses obtained, the reflectance R and transmittance T at a different angle of incidence can be calculated.

From the measurements in the ultraviolet/visible region, interband transitions (band structures) are characterized. In particular, the bandgap E_g can be deduced from the variation of α with $h\nu$ (Section 7.2.1). Since band structure generally varies according to surface temperature, alloy composition, phase structure, and crystal grain size, these properties can also be determined from the spectral analysis of optical constants (Section 7.2.4). In the infrared region, on the other hand, there exists free carrier absorption induced by free electrons (or holes) in solids. When carrier concentration is high enough ($>10^{18}\,\mathrm{cm}^{-3}$), electrical properties including carrier mobility, carrier concentration, and conductivity can be obtained (Section 7.3.2). Moreover, in the infrared region, lattice vibration modes (LO and TO phonons) as well as local atomic structures, such as Si–H and –OH, can also be studied (Sections 7.5.1 and 7.4).

In real-time spectroscopic ellipsometry, (ψ, Δ) spectra are measured continuously during processing. This technique further allows a number of characterizations illustrated in Fig. 1.2 (Chapter 8). From real-time monitoring, for example, initial growth processes or interface structures can be studied in detail (Section 8.2). In a compositionally modulated layer in which alloy composition varies continuously in the growth direction, the alloy compositions of each layer are determined. In particular, the real-time measurement enables us to characterize reaction rate during

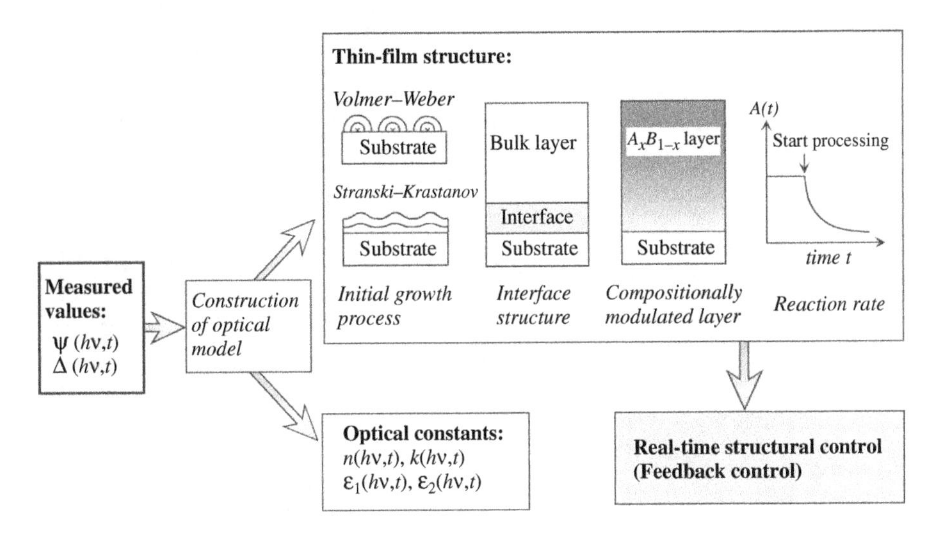

Figure 1.2 Characterization of thin film structures by real-time spectroscopic ellipsometry.

processing. Real-time spectroscopic ellipsometry can be applied further to perform process control. From real-time observation, the feedback control of semiconductor alloy composition has already been performed (Section 8.3.3). Accordingly, the ability of spectroscopic ellipsometry has opened up a new way for more advanced process control.

1.3 DATA ANALYSIS

Figure 1.3 shows (a) optical model consisting of an air/thin film/substrate structure and (b) (ψ, Δ) spectra obtained from a hydrogenated amorphous silicon (a-Si:H) thin film formed on a crystalline Si (c-Si) substrate. As mentioned earlier, an optical model is represented by the complex refractive index and layer thickness of each layer. In Fig. 1.3(a), N_0, N_1 and N_2 denote the complex refractive indices of air, thin film, and substrate, respectively. The transmission angles (θ_1 and θ_2) can be calculated from the angle of incidence θ_0 by applying Snell's law (Section 2.3.1). As shown in Fig. 1.3(a), when light absorption in a thin film is small, optical interference occurs by multiple light reflections within the thin film. In particular, this figure illustrates the optical interference in which each optical wave is superimposed destructively. Of course, the total intensity of the reflected light becomes smaller in this case.

In ellipsometry, the two ellipsometry parameters (ψ, Δ) are defined by $\rho \equiv \tan\psi \exp(i\Delta)$ (Section 4.1.1). In the optical model shown in Fig. 1.3(a), ρ is expressed by the following equation (Section 5.1):

$$\tan\psi \exp(i\Delta) = \rho(N_0, N_1, N_2, d, \theta_0) \tag{1.1}$$

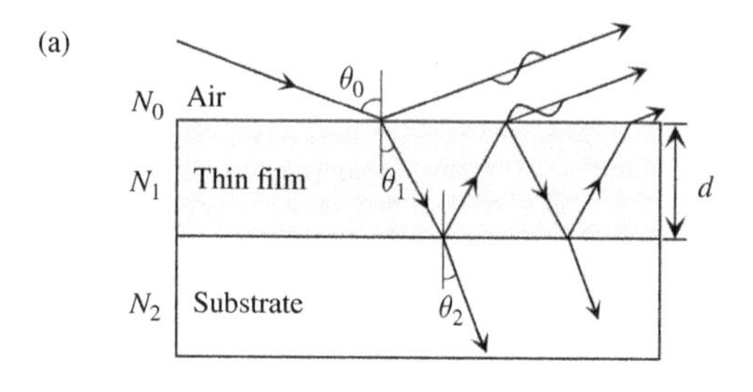

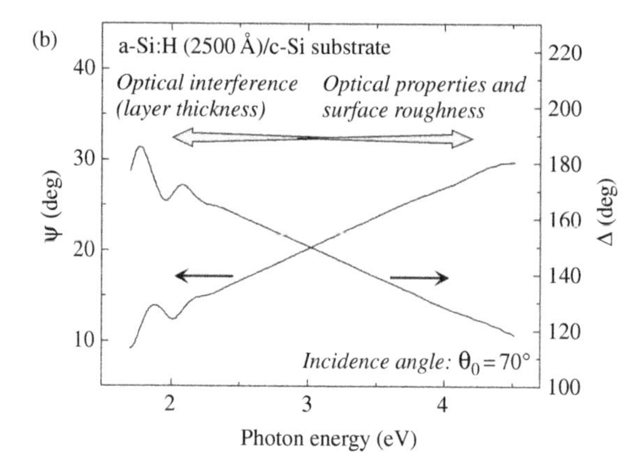

Figure 1.3 (a) Optical model consisting of an air/thin film/substrate structure and (b) (ψ, Δ) spectra obtained from an a-Si:H thin film (2500 Å) formed on a c-Si substrate.

Notice that the above equation shows only variables used in the calculation. The complex refractive index of air is given by $N_0 = 1$, and the values of N_2 and θ_0 are usually known in advance. In the (ψ, Δ) spectra shown in Fig. 1.3(b), the optical interference effect appears in the energy region where optical light absorption is relatively small (<2.5 eV). From the analysis of this interference pattern, the thin-film thickness d can be estimated. If d is determined from this analysis, the unknown parameters in Eq. (1.1) are only $N_1 = n_1 - ik_1$. In this condition, these two values (n_1, k_1) can be obtained directly from the two measured values (ψ, Δ) (Section 5.5.3). In spectroscopic ellipsometry, the optical constants and thickness of the thin film are determined in this manner. In the high-energy region, on the other hand, light absorption in samples generally increases and penetration depth of light becomes smaller. Thus, optical interference is negligible in this region. From the analysis of this energy region, band structure and effect of surface roughness can be studied. In spectroscopic ellipsometry, therefore, from (ψ, Δ) spectra measured in a wide energy range, characterization of various physical properties becomes possible.

Figure 1.4 shows the data analysis example of a multilayer structure by spectroscopic ellipsometry [5]. In this figure, (a) the cross-sectional image obtained

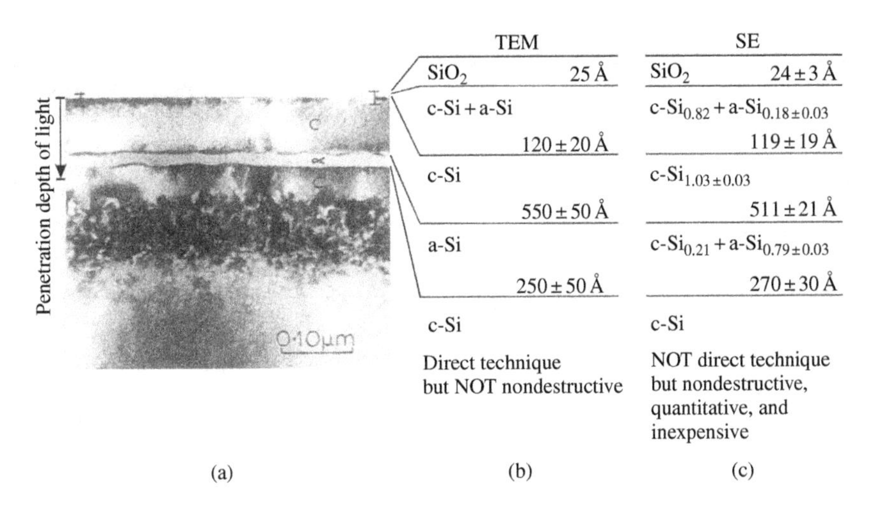

Figure 1.4 (a) Cross-sectional TEM image of a Si(100) wafer implanted with Si ions, (b) structure obtained from TEM, and (c) structure estimated from spectroscopic ellipsometry (SE). Reprinted from *Thin Solid Films*, **313–314**, K. Vedam, Spectroscopic ellipsometry: a historical overview, 1–9. Copyright (1998), with permission from Elsevier.

from transmission electron microscope (TEM), (b) the structure obtained from TEM, and (c) the structure estimated from spectroscopic ellipsometry (SE) are shown. The sample is a Si(100) wafer implanted with Si ions and, by this Si ion implantation, a partial phase change from c-Si to a-Si occurs. As confirmed from Fig. 1.4, the results obtained from TEM and spectroscopic ellipsometry show excellent agreement. Nevertheless, spectroscopic ellipsometry further allows the characterization of the volume fractions for the c-Si and a-Si components. As shown in Fig. 1.4(a), structural characterization by TEM is very reliable since TEM is a direct measurement technique. In TEM, however, difficulties in sample preparation as well as measurement itself generally limit the number of samples for the measurement. In contrast, although spectroscopic ellipsometry is an indirect measurement technique, highly quantitative results can be obtained. Moreover, spectroscopic ellipsometry provides fast and easy measurement, which permits characterization of many samples. Accordingly, for samples that allow proper data analysis (see Fig. 5.32), spectroscopic ellipsometry is a quite effective characterization tool.

1.4 HISTORY OF DEVELOPMENT

Table 1.3 summarizes the history of development for ellipsometry instruments (ellipsometers) [5]. As shown in Table 1.3, ellipsometry was developed first by Drude in 1887. He also derived the equations of ellipsometry, which are used even today. Drude is well known from 'the Drude model' which expresses the optical properties of metals

Table 1.3 History of ellipsometry development

Year	Technique[a]	Parameters determined[b]	Number of data	Time taken (s)	Precision (deg)	Author and reference
1887	E	Δ, ψ	2	Theory and first experiment		Drude [6]
1945	E	Δ, ψ	2	3600	$\Delta = 0.02$ $\psi = 0.01$	Rothen [7]
1971	E	Δ, ψ, R	3	3600	$\Delta = 0.02$ $\psi = 0.01$	Paik, Bockris [8]
1975	SE	$(\Delta, \psi)\lambda$	200	3600	$\Delta = 0.001$ $\psi = 0.0005$	Aspnes, Studna [9]
1984	RTSE	$\{(\Delta, \psi)\lambda\}t$	80 000	3–600	$\Delta = 0.02$ $\psi = 0.01$	Muller, Farmer [10]
1990	RTSE (PDA)[c]	$\{(\Delta, \psi)\lambda\}t$	2×10^5[d]	0.8–600	$\Delta = 0.02$ $\psi = 0.01$	Kim, Collins, Vedam [11]
1994	RTSE (PDA)[c]	$\{(\Delta, \psi, R)\lambda\}t$	3×10^5[d]	0.8–600	$\Delta = 0.007$ $\psi = 0.003$	An, Collins *et al.*[12]

[a] ellipsometry (E), spectroscopic ellipsometry (SE), real-time spectroscopic ellipsometry (RTSE)
[b] reflectance (R), wavelength (λ), time (t)
[c] photodiode array (PDA),
[d] maximum capacity. Reprinted from *Thin Solid Films*, **313–314**, K. Vedam, Spectroscopic ellipsometry: a historical overview, 1–9. Copyright (1998), with permission from Elsevier.

(Section 5.2.5). Until the early 1970s, most ellipsometers were operated manually and the ellipsometry measurement was very time consuming. In 1975, however, Aspnes *et al.* realized the complete automation of spectroscopic ellipsometry measurements [9] (Section 4.2). As shown in Table 1.3, the development of this instrument improved not only the measurement time but also the measurement precision significantly. A spectroscopic ellipsometry instrument for real-time monitoring was reported first by Muller and Farmer in 1984 [10], and this instrument increased the number of measurement data drastically. In 1990, a group from the Pennsylvania State University developed a real-time instrument that has been used widely up to now [11]. In particular, this instrument unitizes a photodiode array (PDA) detector that allows the simultaneous measurement of light intensities at multiwavelengths (Section 4.2). Figure 1.5 shows real-time spectra obtained from this instrument [13]. In this figure, $\langle\varepsilon_1\rangle$ and $\langle\varepsilon_2\rangle$ represent pseudo-dielectric function that can be calculated from (ψ, Δ) spectra (Section 5.4.2). In this measurement, the total of 250 spectra were measured in 16 seconds with a repetition time of 64 ms during the a-Si:H growth on a c-Si substrate. From analysis of the real-time data set, the initial growth process of the thin film can be characterized on the atomic scale (Section 8.2).

Up to now, spectroscopic ellipsometry instruments have been improved continuously and four different types of instruments are mainly used. Nevertheless, ranges and errors for the (ψ, Δ) measurement vary significantly depending on the type of instrument (see Tables 4.2 and 4.3). In order to perform accurate data analysis, therefore, understanding of the ellipsometry measurement is necessary.

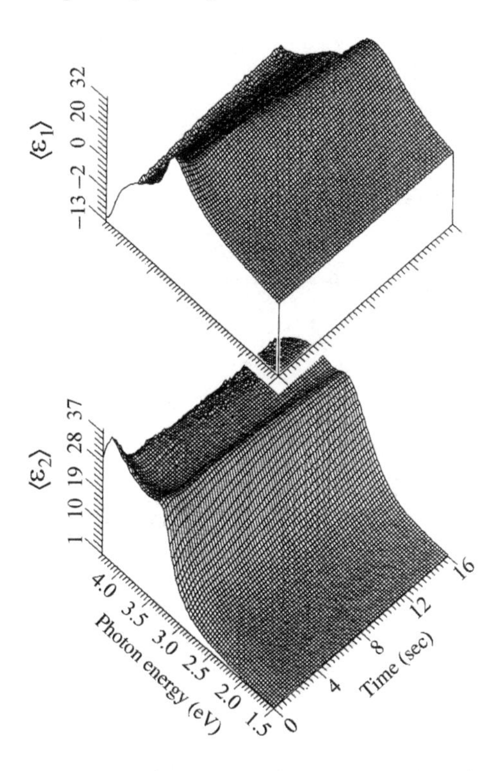

Figure 1.5 Real-time spectra obtained from the spectroscopic ellipsometry measurement performed during the a-Si:H growth. Reprinted with permission from *Review of Scientific Instruments*, **63**, I. An, Y. M. Li, H. V. Nguyen, and R. W. Collins, Spectroscopic ellipsometry on the millisecond time scale for real-time investigations of thin-film and surface phenomena, 3842–3848 (1992). Copyright 1992, American Institute of Physics.

1.5 FUTURE PROSPECTS

Recently, optically anisotropic materials have been studied extensively by applying Mueller matrix ellipsometry that allows the complete characterization of optical behavior in anisotropic materials (Section 4.2.7). For the characterization of conventional isotropic samples, current spectroscopic ellipsometry instruments are highly satisfactory. Thus, most of recent ellipsometry studies have been made on material characterization, rather than the development of ellipsometry instruments.

Figure 1.6 shows the number of papers published each year with ‘ellipsometry’ in the title [14]. The two large peaks at 1993 and 1997 are due to publications of the ellipsometry conference proceedings [15–17]. Since the early 1990s, research that applies spectroscopic ellipsometry has increased drastically due to the commercialization of spectroscopic ellipsometry instruments. During the 1990s, spectroscopic ellipsometry was mainly employed to characterize semiconductor materials. Now, from advances in instruments as well as data analysis methods,

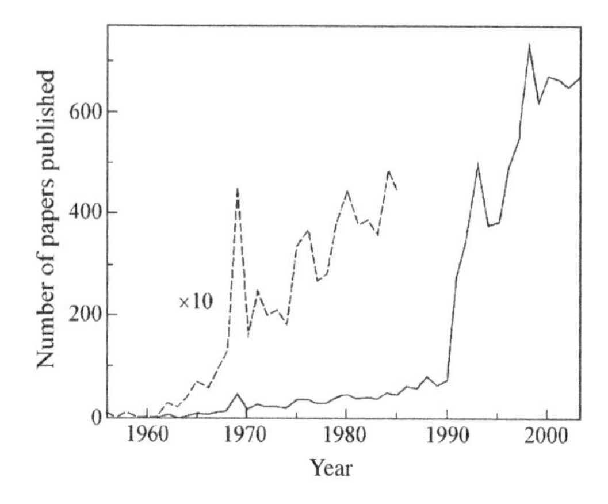

Figure 1.6 Number of papers published with 'ellipsometry' in the title versus year. Reprinted from *Thin Solid Films*, **455–456**, D. E. Aspnes, Expanding horizons: new developments in ellipsometry and polarimetry, 3–13. Copyright (2004), with permission from Elsevier.

the application of the spectroscopic ellipsometry technique has become quite common in wider scientific fields from semiconductors to biomaterials (Chapters 7 and 8). Moreover, some characterizations including the feedback control of alloy composition can be performed only using spectroscopic ellipsometry. Therefore, the application of spectroscopic ellipsometry is expected to expand further in the future. For some materials, however, no optical data is available. Thus, the construction of a larger optical database has been required in this field. As mentioned earlier, ellipsometry data analysis requires the construction of an optical model. In Chapters 5–8, we will see examples that will explain how data analyses are performed using various optical models and when data analyses are difficult.

REFERENCES

[1] R. M. A. Azzam and N. M. Bashara, *Ellipsometry and Polarized Light*, North-Holland, Amsterdam (1977).
[2] H. G. Tompkins and W. A. McGahan, *Spectroscopic Ellipsometry and Reflectometry: A User's Guide*, John Wiley & Sons, Inc., New York (1999).
[3] H. G. Tompkins and E. A. Irene, Eds, *Handbook of Ellipsometry*, William Andrew, New York (2005).
[4] M. Schubert, *Infrared Ellipsometry on Semiconductor Layer Structures: Phonons, Plasmons, and Polaritons*, Springer, Heidelberg (2004).
[5] For a review, see K. Vedam, Spectroscopic ellipsometry: a historical overview, *Thin Solid Films*, **313–314** (1998) 1–9.
[6] P. Drude, *Ann. Phys.*, **32** (1887) 584; *Ann. Phys.*, **34** (1888) 489.
[7] A. Rothen, The ellipsometer, an apparatus to measure thicknesses of thin surface films, *Rev. Sci. Instrum.*, **16** (1945) 26–30.
[8] W. Paik and J. O'M. Bockris, Exact ellipsometric measurement of thickness and optical properties of a thin light-absorbing film without auxiliary measurements, *Surf. Sci.*, **28** (1971) 61–68.

[9] D. E. Aspnes and A. A. Studna, High precision scanning ellipsometer, *Appl. Opt.*, **14** (1975) 220–228.
[10] R. H. Muller and J. C. Farmer, Fast, self-compensating spectral-scanning ellipsometer, *Rev. Sci. Instrum.*, **55** (1984) 371–374.
[11] Y.-T. Kim, R. W. Collins and K. Vedam, Fast scanning spectroelectrochemical ellipsometry: in-situ characterization of gold oxide, *Surf. Sci.*, **233** (1990) 341–350.
[12] I. An, H. V. Nguyen, A. R. Heyd, and R. W. Collins, Simultaneous real-time spectroscopic ellipsometry and reflectance for monitoring thin-film preparation, *Rev. Sci. Instrum.*, **65** (1994) 3489–3500.
[13] I. An, Y. M. Li, H. V. Nguyen, and R. W. Collins, Spectroscopic ellipsometry on the millisecond time scale for real-time investigations of thin-film and surface phenomena, *Rev. Sci. Instrum.*, **63** (1992) 3842–3848.
[14] D. E. Aspnes, Expanding horizons: new developments in ellipsometry and polarimetry, *Thin Solid Films*, **455–456** (2004) 3–13.
[15] A. C. Boccara, C. Pickering, and J. Rivory, Eds, *The Proceedings of the 1st International Conference on Spectroscopic Ellipsometry*; *Thin Solid Films*, **233** (1993) 1–306; *Thin Solid Films*, **234** (1993) 307–572.
[16] R. W. Collins, D. E. Aspnes, and E. A. Irene, Eds, *The Proceedings of the 2nd International Conference on Spectroscopic Ellipsometry*; *Thin Solid Films* **313–314** (1998) 1–835.
[17] For the latest proceedings, see M. Fried, K. Hingerl and J. Humlíček, Eds, *The Proceedings of the 3rd International Conference on Spectroscopic Ellipsometry*; *Thin Solid Films*, **455–456** (2004) 1–836.

2 Principles of Optics

Spectroscopic ellipsometry is an optical measurement technique. Accordingly, to understand spectroscopic ellipsometry, accurate knowledge of optics is required. In this chapter, we will look at basic principles of optics, including optical constants of materials as well as reflection and transmission of light. In optics, optical constants and amplitude reflection coefficients, for example, are expressed by using complex numbers. Thus, it is generally difficult to comprehend optical phenomena by intuition. In this chapter, in order to help understanding, behavior of light will be explained by visualizing the propagation processes of each light wave.

2.1 PROPAGATION OF LIGHT

As is known widely, light has the character of waves. Maxwell showed that light also has electromagnetic properties, and thus light is called an electromagnetic wave in scientific fields. Mathematically, light waves are described easily by using sinusoidal waves, such as $y = \sin x$ or $y = \cos x$. When light enters into media, however, light shows rather complicated behavior due to refraction or absorption of light. The propagation of light in media can be expressed by the complex refractive index or by two values known as the refractive index and extinction coefficient. In this section, we will examine the propagation of light and complex refractive index.

2.1.1 PROPAGATION OF ONE-DIMENSIONAL WAVES

The behavior of a light wave that advances with time can be expressed from the simplest one-dimensional wave. Now consider that a sinusoidal wave is traveling in the positive direction of the x axis with a constant speed s (Fig. 2.1). In this case, the wave at the position x before the propagation is given by

$$\varphi = A \sin Kx \tag{2.1}$$

Spectroscopic Ellipsometry: Principles and Applications H. Fujiwara

Here, A is the wave amplitude and K is called the propagation number. From the wavelength λ, K is defined as

$$K \equiv 2\pi/\lambda \tag{2.2}$$

Thus, K shows the number of sinusoidal waves present in the distance from 0 to 2π. In general, the propagation number is denoted as k. However, as the complex refractive index is also defined by using k, the propagation number K will be used throughout this book.

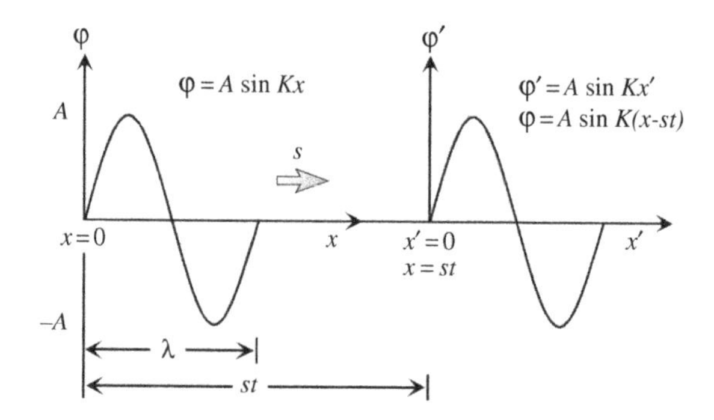

Figure 2.1 One-dimensional wave traveling along the x axis with a speed s.

As shown in Fig. 2.1, after a time t, the wave traveling with the speed of s moves a distance st along the x axis. Now we introduce a new coordinate system x'-φ' so that the wave becomes identical to the one before the propagation. In this new coordinate system, the wave is expressed by $\varphi' = A \sin Kx'$. From Fig. 2.1, it is obvious that $x' = x - st$ and $\varphi' = \varphi$. Therefore, the one-dimensional wave traveling at a speed s is described by

$$\varphi = A \sin K(x - st) \tag{2.3}$$

On the other hand, the time necessary to travel the distance λ is given by

$$\tau = \lambda/s \tag{2.4}$$

where τ is referred to as the temporal period. The frequency ν and angular frequency ω are defined from τ as

$$\nu \equiv 1/\tau \tag{2.5}$$

$$\omega \equiv 2\pi/\tau = 2\pi\nu \tag{2.6}$$

Similarly to K, ω shows the number of waves included in the elapsed time 2π. By rewriting Eq. (2.3) using Eqs. (2.2), (2.4), and (2.6), we get the familiar form for the propagation of one-dimensional waves:

$$\varphi = A\sin(Kx - \omega t) \tag{2.7}$$

From Eq. (2.7), it can be seen that the one-dimensional wave is a function of x and t. In general, the term $(Kx - \omega t)$ in Eq. (2.7) is referred to as the phase.

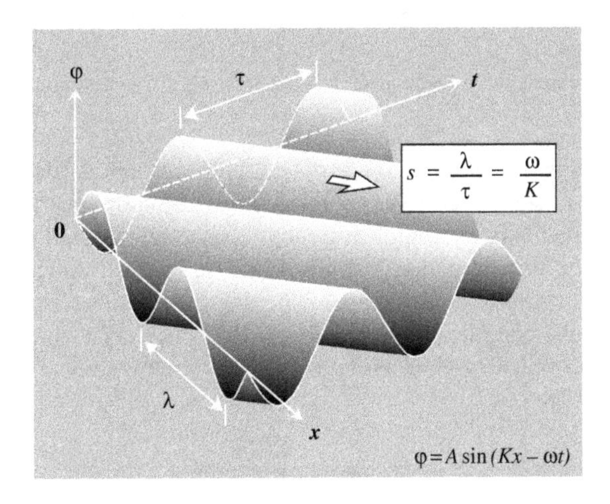

Figure 2.2 Propagation of a one-dimensional wave, plotted as a function of the position x and time t. In this figure, λ and τ represent the wavelength and temporal period, respectively.

Figure 2.2 shows the propagation of a one-dimensional wave calculated from Eq. (2.7). As shown in this figure, the sine wave moves along the positive direction of the x axis with increasing t. In particular, the propagating wave is represented as a periodic function of x and t in the x-t plane. If $x = 0$ in Eq. (2.7), it follows that $A\sin(-\omega t) = -A\sin(\omega t)$ (see Appendix 1). In Fig. 2.2, therefore, the shape of the sine wave along the t axis is reversed. The velocity of the wave is represented by the direction of the arrow in Fig. 2.2 and is obtained from Eq. (2.2) and Eqs. (2.4)–(2.6):

$$s = \frac{\lambda}{\tau} = \lambda\nu = \frac{\omega}{K} \tag{2.8}$$

When light waves are reflected by media, the phase $(Kx - \omega t)$ generally shows a large change. In order to express this phenomenon, we introduce the initial phase δ into Eq. (2.7):

$$\varphi = A\sin(Kx - \omega t + \delta) \tag{2.9}$$

In Eq. (2.9), when $\delta = \pi/2$, a sine wave becomes a cosine wave since $\sin(x + \pi/2) = \cos(x)$ (see Appendix 1).

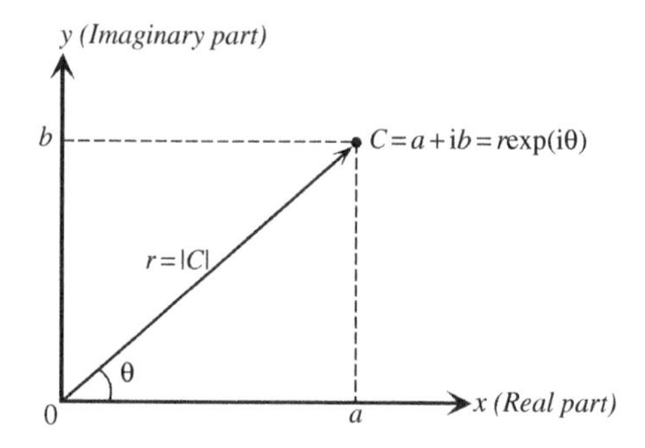

Figure 2.3 Representation of a complex number ($C = a + ib$) in the complex plane.

As will be shown in Section 2.1.3, optical constants are defined using complex numbers. Thus, it is essential to express the propagation of one-dimensional waves using complex numbers. A complex number can be represented by a point on the complex plane or Argand plane shown in Fig. 2.3. In this figure, the x and y axes show real and imaginary components of the complex number, respectively. In the complex plane, the complex number $C = a + ib$ ($i = \sqrt{-1}$) is shown by point C in Fig. 2.3. The absolute value or modulus of the complex number is obtained from the length r in Fig. 2.3:

$$|C| = r = \sqrt{a^2 + b^2} \tag{2.10}$$

The complex conjugate, indicated by an asterisk, is expressed by replacing i with −i:

$$C^* = (a + ib)^* = a - ib \tag{2.11}$$

If we use the complex conjugate, the real part (Re) and imaginary part (Im) can be written as

$$\mathrm{Re}(C) = a = \frac{C + C^*}{2} \quad \text{and} \quad \mathrm{Im}(C) = b = \frac{C - C^*}{2i} \tag{2.12}$$

The absolute value of C can also be calculated using C^*:

$$|C| = \sqrt{CC^*} \tag{2.13}$$

In the polar coordinate system, point C in the complex plane is represented by

$$\mathrm{Re}(C) = r\cos\theta \quad \text{and} \quad \mathrm{Im}(C) = r\sin\theta. \tag{2.14}$$

Thus,

$$C = a + ib = r(\cos\theta + i\sin\theta) \tag{2.15}$$

Here, θ is referred to as the argument of C and is denoted by $\theta = \arg C$. Since $\tan\theta = b/a$, we get

$$\theta = \arg C = \begin{cases} \tan^{-1}(b/a) & \text{for } a > 0, \\ \tan^{-1}(b/a) + 180^\circ & \text{for } a < 0, b \geq 0, \\ \tan^{-1}(b/a) - 180^\circ & \text{for } a < 0, b < 0. \end{cases} \tag{2.16}$$

In Eq. (2.16), the conditions for a and b were used to convert the range of θ from $-90^\circ \leq \theta \leq 90^\circ$ to $-180^\circ \leq \theta \leq 180^\circ$. It can be seen from Fig. 2.3 that, when $a = 0$ in Eq. (2.16), $\theta = 90^\circ (b > 0)$ and $\theta = -90^\circ (b < 0)$. From Euler's formula,

$$\exp(\mathrm{i}\theta) = \cos\theta + \mathrm{i}\sin\theta \tag{2.17}$$

Thus, the complex number shown in Eq. (2.15) can be expressed by

$$C = r\exp(\mathrm{i}\theta) = |C|\exp(\mathrm{i}\theta) \tag{2.18}$$

If we use Eqs. (2.13) and (2.17), the absolute value of $\exp(\mathrm{i}\theta)$ is given by

$$|\exp(\mathrm{i}\theta)| = \sqrt{\cos^2\theta + \sin^2\theta} = 1. \tag{2.19}$$

As shown in Eq. (2.14), in the polar coordinate system, the real and imaginary parts are expressed by cosine and sine functions, respectively. Accordingly, if we choose either the real part or the imaginary part, we can express a sinusoidal wave as follows:

$$\varphi = A\cos(\omega t - Kx + \delta) = \mathrm{Re}\{A\exp[\mathrm{i}(\omega t - Kx + \delta)]\} \tag{2.20}$$

In general, to simplify the expression, the above equation is written as

$$\varphi = A\exp[\mathrm{i}(\omega t - Kx + \delta)] \tag{2.21}$$

In Eq. (2.21), however, it is given that we take the real part whenever actual waveforms are necessary. From Eq. (2.19), the absolute value of Eq. (2.21) can be obtained as follows:

$$|\varphi| = |A\exp[\mathrm{i}(\omega t - Kx + \delta)]| = |A| \tag{2.22}$$

In Eqs. (2.20)–(2.22), the phase $(Kx - \omega t)$ in Eq. (2.9) is rewritten as $(\omega t - Kx)$. Since $\sin(-x) = \cos(x + \pi/2)$ (see Appendix 1), the phase conversion from $(Kx - \omega t)$ to $(\omega t - Kx)$ does not alter the propagation of waves, although the initial phase δ changes. Nevertheless, the definition of this phase is extremely important, since the signs of many important formulas vary according to the definition of the phase (see Appendix 2). In this book, whenever possible, the phase of light waves will be expressed by $(\omega t - Kx)$, instead of $(Kx - \omega t)$, because the propagation of light can be understood more easily with this definition.

2.1.2 ELECTROMAGNETIC WAVES

In 1849, Maxwell found out that light waves are electromagnetic waves that follow electromagnetic theory. Figure 2.4 shows the propagation of an electromagnetic wave derived from the well known Maxwell's equations (see Appendix 3). In this figure, $\boldsymbol{E}$ and $\boldsymbol{B}$ show the electric field and magnetic induction (or magnetic flux density), respectively. In the electromagnetic wave, $\boldsymbol{E}$ and $\boldsymbol{B}$ are perpendicular to each other, and $\boldsymbol{E}=0$ when $\boldsymbol{B}=0$. These behaviors are consistent with the finding that a time-varying $\boldsymbol{B}$ field generates $\boldsymbol{E}$ perpendicular to $\boldsymbol{B}$ (Faraday's induction law) and, conversely, a time-varying $\boldsymbol{E}$ field generates $\boldsymbol{B}$ perpendicular to $\boldsymbol{E}$ (Ampère's law). The direction of light propagation is perpendicular to both $\boldsymbol{E}$ and $\boldsymbol{B}$. Thus, in the traveling direction, $\boldsymbol{E}=\boldsymbol{B}=0$. In general, such a wave is called the transverse wave. Quite interestingly, the speed of light waves (electromagnetic waves) does not depend on the wavelength of light and shows a constant value of

$$c = 2.99792 \times 10^8 \text{ m/s} \tag{2.23}$$

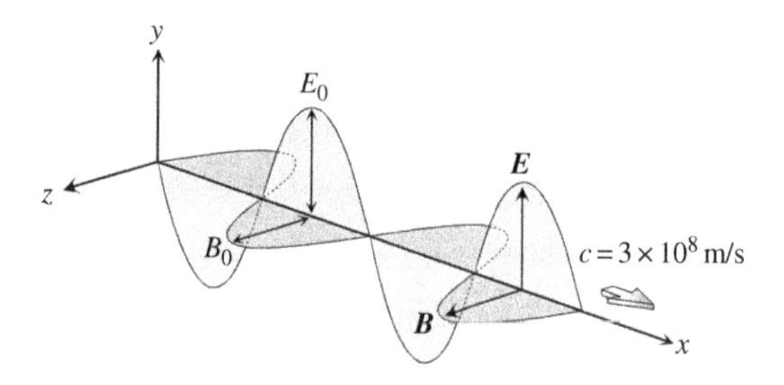

Figure 2.4 Propagation of an electromagnetic wave. In this figure, $\boldsymbol{E}$ and $\boldsymbol{B}$ denote the electric field and magnetic induction, respectively.

The electromagnetic wave can be treated as the one-dimensional wave described in the previous section:

$$E = E_0 \exp[\mathrm{i}(\omega t - Kx + \delta)] \tag{2.24a}$$

$$B = B_0 \exp[\mathrm{i}(\omega t - Kx + \delta)] \tag{2.24b}$$

In Eq. (2.24), there is the relation given by $E_0 = cB_0$ [1]. Thus,

$$E = cB \tag{2.25}$$

In 1905, on the other hand, Einstein proposed a new form of theory that treats light as energy particles, now known as photons. Later, it was confirmed that light *also* has the properties of particles, and light is emitted or absorbed as a photon. When light is treated as a photon, the photon energy En is given by

$$\mathrm{En} = h\nu = \hbar\omega \tag{2.26}$$

where h is referred to as Planck's constant and $\hbar = h/2\pi$. In general, the photon energy is denoted by using E. In this book, however, photon energy will be written as En, since E has already been used to represent the electric field.

Table 2.1 shows physical constants used in optics fields. If we set $s = c$ in Eq. (2.8), the relation $c = \lambda\nu$ is obtained. Thus, by substituting $\nu = c/\lambda$ into Eq. (2.26) and using a unit of electron volts [eV], we can express En as follows:

$$\mathrm{En} = \frac{1}{1.60218 \times 10^{-19}} \frac{hc}{\lambda} = \frac{1.23984 \times 10^{-6}}{\lambda} \mathrm{eV} \tag{2.27}$$

Here, the unit of λ is m. Table 2.2 summarizes conversion of various units used in optics. In theoretical expressions, ω is frequently used, while ellipsometry spectra in the ultraviolet/visible region are generally expressed using En or λ. In the infrared region, the wave number W defined by $W = 1/\lambda$ is mainly used.

Table 2.1 Physical constants used in optics fields

Speed of light	c	2.99792×10^{8} m/s
Planck's constant	h	6.62607×10^{-34} J·s
	$\hbar = h/2\pi$	1.05457×10^{-34} J·s
Free-space permittivity	ε_0	8.85419×10^{-12} F/m
Electron charge[a]	e	1.60218×10^{-19} C
Free-electron mass	m_0	9.10938×10^{-31} kg

[a] 1 electron volt is given by 1 eV $= 1.60218 \times 10^{-19}$ J.

2.1.3 REFRACTIVE INDEX

The refraction of light occurs when light advances into optically different media. We can observe this phenomenon easily when we look at a water surface from an oblique direction. The refraction of light is determined from the refractive index n and, classically, n is defined by

$$n \equiv c/s \tag{2.28}$$

where s represents the speed of light in a medium. Accordingly, the propagation of light waves becomes slower in a medium with high n. The refractive index of air is $n = 1.0003$ [1,2] and is almost the same as $n = 1$ in vacuum.

Table 2.2 Conversion of units for light

	Photon energy En (eV)	Wavelength		Wave number $W(\mathrm{cm}^{-1})$	Angular frequency ω(rad/s)	Frequency ν(Hz)
		λ(Å)	λ(μm)			
En(eV) =		$12398/\lambda$	$1.240/\lambda$	$1.240 \times 10^{-4} W$	$6.582 \times 10^{-16}\omega$	$4.136 \times 10^{-15}\nu$
λ(Å) =	12398/En		$10^4\lambda$	$10^8/W$	$1.884 \times 10^{19}/\omega$	$2.998 \times 10^{18}/\nu$
λ(μm) =	1.240/En	$10^{-4}\lambda$		$10^4/W$	$1.884 \times 10^{15}/\omega$	$2.998 \times 10^{14}/\nu$
$W(\mathrm{cm}^{-1})$ =	8066En	$10^8/\lambda$	$10^4/\lambda$		$5.309 \times 10^{-12}\omega$	$3.336 \times 10^{-11}\nu$
ω(rad/s) =	1.519×10^{15} En	$1.884 \times 10^{19}/\lambda$	$1.884 \times 10^{15}/\lambda$	$1.884 \times 10^{11} W$		6.283ν
ν(Hz) =	2.418×10^{14} En	$2.998 \times 10^{18}/\lambda$	$2.998 \times 10^{14}/\lambda$	$2.998 \times 10^{10} W$	0.159ω	

When there is no light absorption in media, the propagation number K can be obtained by inserting Eq. (2.28) into Eq. (2.8):

$$K = \frac{\omega n}{c} = \frac{2\pi n}{\lambda} \tag{2.29}$$

Here, λ is the wavelength of light in vacuum. In particular, when $n = 1$, it follows that $\omega/c = 2\pi/\lambda$. By substituting Eq. (2.29) into Eq. (2.24a), we get

$$E = E_{t0} \exp[i(\omega t - Kx + \delta)] = E_{t0} \exp\left[i\left(\omega t - \frac{2\pi n}{\lambda} x + \delta\right)\right] \tag{2.30}$$

where E_{t0} corresponds to E_0 in the transparent medium. Eq. (2.30) represents an electromagnetic wave traveling in a transparent medium. Figure 2.5(a) illustrates the waveform when an electromagnetic wave advances into a transparent medium. It can be seen from Eq. (2.30) that the wavelength of light becomes λ/n in the medium due to the reduction in s. Moreover, $E_0 > E_{t0}$, since light reflection occurs at the interface.

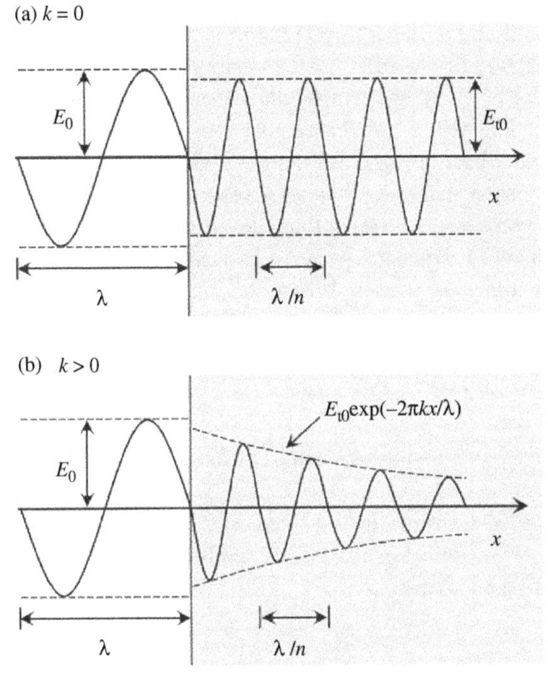

Figure 2.5 Propagation of electromagnetic waves in (a) a transparent medium ($k = 0$) and (b) a light-absorbing medium ($k > 0$).

As mentioned above, in transparent media, n determines the propagation of electromagnetic waves completely. Nevertheless, there are media that show strong light absorption, and such a phenomenon cannot be expressed only with n. Thus, in order to describe light absorption by media, we introduce the extinction coefficient k and define the complex refractive index N as

$$N \equiv n - \mathrm{i}k \tag{2.31}$$

If we replace n in Eq. (2.30) with N, we obtain the following equation, which represents an electromagnetic wave traveling in a light-absorbing medium:

$$\begin{aligned} E &= E_{t0} \exp\left[\mathrm{i}\left(\omega t - \frac{2\pi N}{\lambda}x + \delta\right)\right] \\ &= E_{t0} \exp\left(-\frac{2\pi k}{\lambda}x\right) \exp\left[\mathrm{i}\left(\omega t - \frac{2\pi n}{\lambda}x + \delta\right)\right] \end{aligned} \tag{2.32}$$

As shown in Eq. (2.32), by defining k using an imaginary number, we can now treat the absorption of the electromagnetic wave, expressed by $\exp(-2\pi kx/\lambda)$, as a real number. Figure 2.5(b) shows the electromagnetic wave described by Eq. (2.32). The wavelength of light in the absorbing medium is λ/n, as confirmed from Eq. (2.32). Thus, light absorption has no effects on wavelengths in media. However, when light absorption occurs ($k > 0$), the amplitude of the electromagnetic wave decreases along the x direction with $\exp(-2\pi kx/\lambda)$.

As mentioned earlier, in Eqs. (2.30) and (2.32), the phase of the electromagnetic waves was expressed using $(\omega t - Kx)$. Nevertheless, if Eq. (2.32) is expanded using the phase $(Kx - \omega t)$, the term that shows the decay of electromagnetic waves becomes $\exp(2\pi kx/\lambda)$. Thus, this leads to a wrong expression that shows an increase in the amplitude along the x axis. When the phase $(Kx - \omega t)$ is applied, therefore, the complex refractive index is defined by $N \equiv n + \mathrm{i}k$ (see Appendix 2).

If we neglect a proportional constant, the light intensity of electromagnetic waves is given by

$$I = |E|^2 = EE^* \tag{2.33}$$

In the calculation of transmittance, however, $I = n|E|^2$ should be used [1,2]. By substituting Eq. (2.32) into Eq. (2.33) and applying Eq. (2.22), we obtain the light intensity in media as follows:

$$I = \left| E_{t0} \exp\left(-\frac{2\pi k}{\lambda}x\right)\right|^2 = |E_{t0}|^2 \exp\left(-\frac{4\pi k}{\lambda}x\right) \tag{2.34}$$

In optical measurements, light intensity in media is characterized by applying an empirical relation, referred to as Beer's law:

$$I = I_0 \exp(-\alpha d) \tag{2.35}$$

Here, α is the absorption coefficient of media and d is a distance from the surface. By comparing Eqs. (2.34) and (2.35), it follows that $I_0 = |E_{t0}|^2$ and more importantly

$$\alpha = \frac{4\pi k}{\lambda} \tag{2.36}$$

If there is no light absorption in media, it can be seen that $\alpha = k = 0$.

Figure 2.6(a) shows the light intensity ratio I/I_0 calculated from Eq. (2.35), plotted as a function of the depth from the surface. In a light-absorbing medium, the light intensity decreases with increasing d and the decay of the light intensity is more pronounced at high α. When $I/I_0 = 1/\mathrm{e} \sim 37\%$ in Eq. (2.35), it follows that $\alpha d = 1$. This depth d_p, known as the penetration depth, is defined by

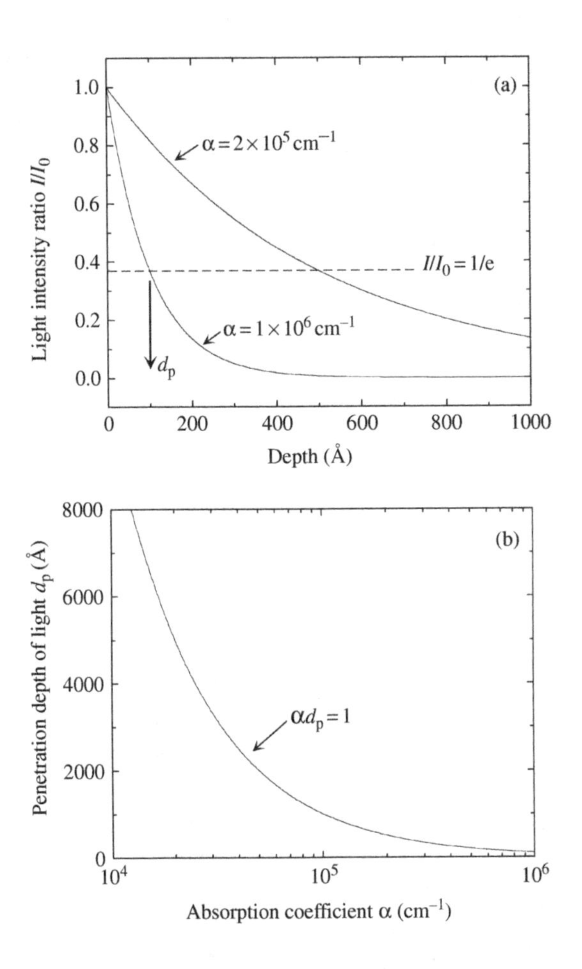

Figure 2.6 (a) Light intensity ratio I/I_0, plotted as a function of the depth from the surface and (b) penetration depth of light d_p, plotted as a function of the absorption coefficient α of media. In (a), the depth when $I/I_0 = 1/\mathrm{e}$ defines the penetration depth of light d_p.

$$d_p \equiv 1/\alpha. \tag{2.37}$$

Figure 2.6(b) shows d_p calculated from Eq. (2.37), plotted as a function of α. As shown in this figure, d_p reduces rapidly with increasing α. As will be mentioned in Section 5.1.3, spectroscopic ellipsometry has very high precision and measurement up to a thickness of $\sim 5d_p$ is possible [3]. Thus, even when a film thickness is thicker than d_p, measured spectra generally include the effect of underlying films or substrates. In ellipsometry data analysis, therefore, the effect of d_p should be taken into account.

2.2 DIELECTRICS

In the previous section, we have seen that light propagation in media can be described by the complex refractive index. However, the complex refractive index itself is determined from the dielectric polarization generated in a medium. The dielectric polarization represents the phenomenon that external electric fields separate electric charges in a medium spatially. In general, the media that show large dielectric polarization are called dielectrics. On the other hand, metals show quite weak dielectric polarization because electric fields applied to metals lead to current flows. In this section, we will review polarization phenomena in dielectrics and will discuss light propagation processes in more detail. Furthermore, this section will relate the complex refractive index with the dielectric constant.

2.2.1 DIELECTRIC POLARIZATION

When an electric field is applied to a medium, positive and negative charges in the medium receive electric forces in the opposite direction. In dielectrics, however, electric charges cannot move freely since atoms, for example, are bound together by strong chemical bonding. Nevertheless, in the presence of the electric field, the spatial distributions of positive and negative charges are modified slightly and are separated into regions that are more electrically positive and negative. This phenomenon is referred to as dielectric polarization. As shown in Fig. 2.7, several types of dielectric polarization exist. The most important polarization for semiconductor characterization is electric polarization shown in Fig. 2.7(a). In the classical view, negatively charged electrons in an atom are bound strongly to a positively charged atomic nucleus with springs. The electric polarization occurs when electric fields distort the positions of the electrons and the nucleus in opposite directions. On the other hand, ionic crystals including NaCl consist of electrically positive and negative ions. Accordingly, charge distortion occurs in ionic crystals by external electric fields, similar to electric polarization. This is called the atomic polarization or ionic polarization [Fig. 2.7(b)]. As shown in Fig. 2.7(c), hydrogen and oxygen atoms in H_2O molecules are charged due to the difference in electronegativity. When an electric field is applied to such a molecule, the orientation

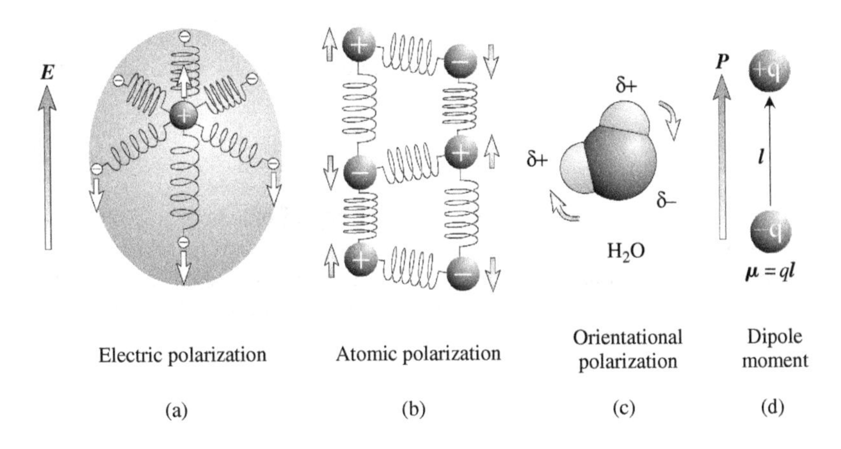

Figure 2.7 Dielectric polarization in dielectrics.

of the molecule is aligned along the direction of the electric field. The dielectric polarization generated by this effect is known as the orientational polarization. When H is incorporated into a solid as Si–H, no orientational polarization occurs and only atomic polarization takes place since the Si atom cannot move freely. A pair of electric charges generated by dielectric polarization is referred to as the electric dipole. As shown in Fig. 2.7(d), let q and l be the electric charge of an electric dipole and the distance between the charged pair, respectively. In this case, the dipole moment is given by

$$\boldsymbol{\mu} = q\boldsymbol{l} \tag{2.38}$$

From the sum of the dipole moment per unit volume, the dielectric polarization $\boldsymbol{P}$ is written as

$$\boldsymbol{P} = \sum_i \boldsymbol{\mu}_i \tag{2.39}$$

The electric field $\boldsymbol{E}$ is defined by a vector whose direction is from a positive charge to a negative charge. In contrast, the dielectric polarization $\boldsymbol{P}$ is defined by a vector whose direction is from the negative charge to the positive charge, as shown in Fig. 2.7(d).

2.2.2 DIELECTRIC CONSTANT

The magnitude of the polarization generated within a dielectric is expressed by the permittivity or dielectric constant. In order to define the permittivity physically, consider a parallel plate capacitor shown in Fig. 2.8. In this figure, a dielectric medium is inserted between the two electrodes of the capacitor and an ac electric field is applied to the capacitor. When the medium between the electrodes is vacuum, the electric field E in the capacitor is given by

$$E = D/\varepsilon_0 \tag{2.40}$$

The above equation represents Gauss's law for capacitors. In Eq. (2.40), ε_0 is the free-space permittivity shown in Table 2.1 and D is the surface charge density of the capacitor. In electromagnetics, D is called the electric displacement (D has the same unit of C/m^2). Eq. (2.40) shows that E is proportional to D on the electrode.

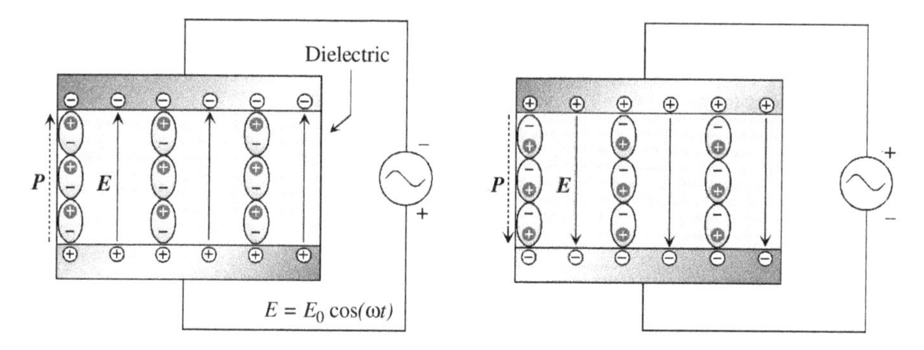

Figure 2.8 Dielectric introduced into a parallel plate capacitor.

When a dielectric is present between the electrodes, the external electric field induces dielectric polarization within the dielectric by electrostatic induction. In this case, the electric field E within the capacitor is expressed by

$$E = D/\varepsilon_p \tag{2.41}$$

where ε_p is the permittivity of the dielectric medium. By the insertion of the dielectric, the electric field inside the capacitor reduces since generally $\varepsilon_p > \varepsilon_0$. For example, the permittivity of glass is $\varepsilon_p \sim 2\varepsilon_0$. Thus, when the glass is introduced into the capacitor, the density of the electric field becomes half, compared with the case of vacuum, as shown in Fig. 2.8. As confirmed from Fig. 2.8, the surface charges on the electrodes contribute to generate either $\boldsymbol{E}$ or $\boldsymbol{P}$. Accordingly, if we use the dielectric polarization P, the electric field within the dielectric can be written as

$$E = (D - P)/\varepsilon_0 \tag{2.42}$$

In general, the relative permittivity or relative dielectric constant expressed by the following equation is used:

$$\varepsilon = \varepsilon_p/\varepsilon_0 \tag{2.43}$$

In optics fields, however, ε is simply called the dielectric constant and we will also follow this in this book. From Eqs. (2.41)–(2.43), the dielectric constant ε is given by

$$\varepsilon = 1 + \frac{P}{\varepsilon_0 E} = 1 + \chi \tag{2.44}$$

where χ is referred to as the dielectric susceptibility $[\chi \equiv P/(\varepsilon_0 E)]$. It can be seen from Eq. (2.44) that ε increases as the dielectric polarization and susceptibility increase.

As shown in Fig. 2.8, when the ac electric field $E = E_0 \cos(\omega t)$ is applied to the capacitor, the polarity of the surface charges on the electrodes varies with time. Thus, the electric dipoles induced within the dielectric also vary with time and the time-varying external field accelerates the charges of the electric dipoles continuously. When the charges of the electric dipoles are accelerated, the electric dipoles radiate electromagnetic waves. This phenomenon is identical to the principle of synchrotron radiation, in which electromagnetic waves are produced by accelerating charged particles [1]. In particular, the radiation of electromagnetic waves from electric dipoles is known as electric dipole radiation. Figure 2.9 shows the radiation process of an electromagnetic wave by electric dipole radiation. Now suppose that the ac electric field shown in Fig. 2.9(a) is applied externally to the capacitor. As shown in Figs. 2.9(b) and (c), the positive and negative charges are attracted in opposite directions and electric fields are generated between these charges. As shown in Fig. 2.9(e), the electric dipole disappears when the external electric field is zero and the electric fields created by the electric dipole become closed circles. Consequently, continuous acceleration of the electric dipole leads to radiation of the electromagnetic wave shown in Fig. 2.9(g). It can be seen from Fig. 2.9 that the frequency of the electromagnetic wave radiated from the electric dipole is identical to that of the external ac field.

As we discussed in Fig. 2.4, the electric field of light waves (electromagnetic waves) is an ac electric field with a sinusoidal shape. Thus, when the light advances into a medium, the same electric dipole radiation occurs. Figure 2.10(a) illustrates the propagation process of light when the light radiation from the electric dipole is taken into account. The electric dipole radiation in this figure corresponds to the one observed when we look down Fig. 2.9(g) from the top. As shown in Fig. 2.10(a), the electric field of light accelerates the electric dipoles of the medium and light waves that have the same frequency as the incident light are radiated from the electric dipoles. Since wavelengths of light are quite large, compared with atomic spacing, we can always find the same atom at a distance equal to λ from the surface. Therefore, light waves emitted from each atom overlap constructively and propagate in the medium as the single transmitted wave. This is consistent with the well-known Huygens's principle in which a propagating wave is expressed by superimposing spherical waves radiated from the surface acting as point light source [Fig. 2.10(b)]. In Fig. 2.10(a), the radiated light waves are represented by semi-circles, similar to

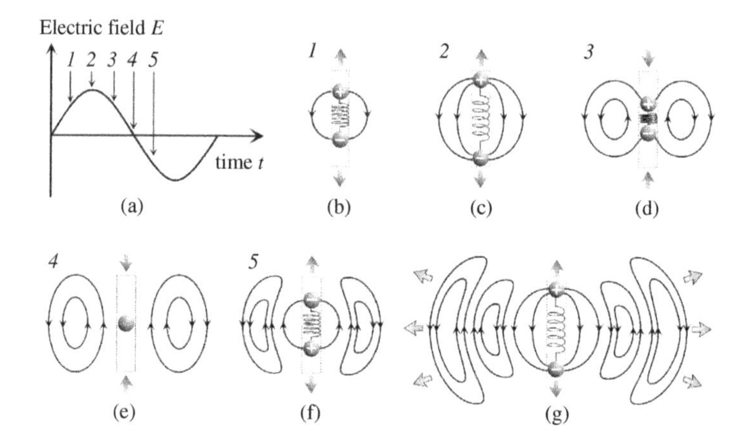

Figure 2.9 Electric dipole radiation in a dielectric: (a) an external ac electric field applied to the dielectric and (b)–(f) the behavior of an electric dipole when the external electric field of 1–5 in (a) is applied.

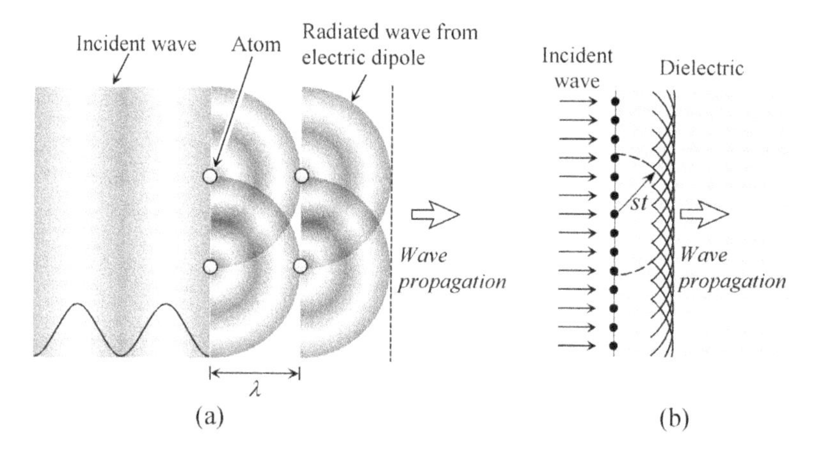

Figure 2.10 (a) Propagation process of light in a dielectric and (b) propagation of light according to Huygens's principle. In (a), atoms are shown with the interval of the light wavelength λ and other atoms are not illustrated. In (b), st shows a distance to which light travels with a speed s during a time t.

Huygens's principle, since the radiated waves also propagate backward if they are completely spherical.

As we have discussed in the previous section, the propagation speed of light decreases in a medium that has a high refractive index. We can understand this effect more easily if we consider the phenomenon that the sequential light radiation from the electric dipole slows the propagation speed of light (see Section 5.2.2

for more details). This implies that the propagation speed becomes slower in dielectrics with high ε values since electric dipoles are created more easily in these dielectrics. Accordingly, there is a close relation between the dielectric constant and the refractive index. From Maxwell's equations for conductors (see Appendix 3), the complex refractive index $N \equiv n - \mathrm{i}k$ is defined as

$$N^2 \equiv \varepsilon \tag{2.45}$$

In Eq. (2.45), ε is a complex number and the complex dielectric constant is defined by

$$\varepsilon \equiv \varepsilon_1 - \mathrm{i}\varepsilon_2 \tag{2.46}$$

For the definition of $N \equiv n + \mathrm{i}k$, the complex dielectric constant is defined by $\varepsilon \equiv \varepsilon_1 + \mathrm{i}\varepsilon_2$ (see Appendix 2). From Eqs. (2.45) and (2.46), we get

$$\varepsilon_1 = n^2 - k^2 \tag{2.47a}$$

$$\varepsilon_2 = 2nk \tag{2.47b}$$

It can be seen from Eq. (2.47) that $\varepsilon_1 = n^2$ and $\varepsilon_2 = 0$ when there is no light absorption ($k = 0$). In this case, ε_1 values become larger with increasing n. On the other hand, if we use the complex dielectric constant, the complex refractive index is given by

$$n = \left\{ \left[\varepsilon_1 + \left(\varepsilon_1^2 + \varepsilon_2^2 \right)^{1/2} \right] / 2 \right\}^{1/2} \tag{2.48a}$$

$$k = \left\{ \left[-\varepsilon_1 + \left(\varepsilon_1^2 + \varepsilon_2^2 \right)^{1/2} \right] / 2 \right\}^{1/2} \tag{2.48b}$$

From k in Eq. (2.48b), the absorption coefficient α can also be obtained using Eq. (2.36).

2.2.3 DIELECTRIC FUNCTION

So far, we have treated ε as a single complex number. However, ε observed in actual measurements varies significantly according to the angular frequency ω of incident light. Figure 2.11 shows the real part ε_1 and imaginary part ε_2 of the complex dielectric constant, plotted as a function of the angular frequency of light ($\log \omega$). When the angular frequency is quite low, the value of ε_1 is represented by the static dielectric constant ε_s. This ε_s includes the contributions of both atomic and electric polarizations. As shown in Fig. 2.7, in the classical model, the dielectric polarization is described by oscillation of springs. Thus, when the angular frequency of incident

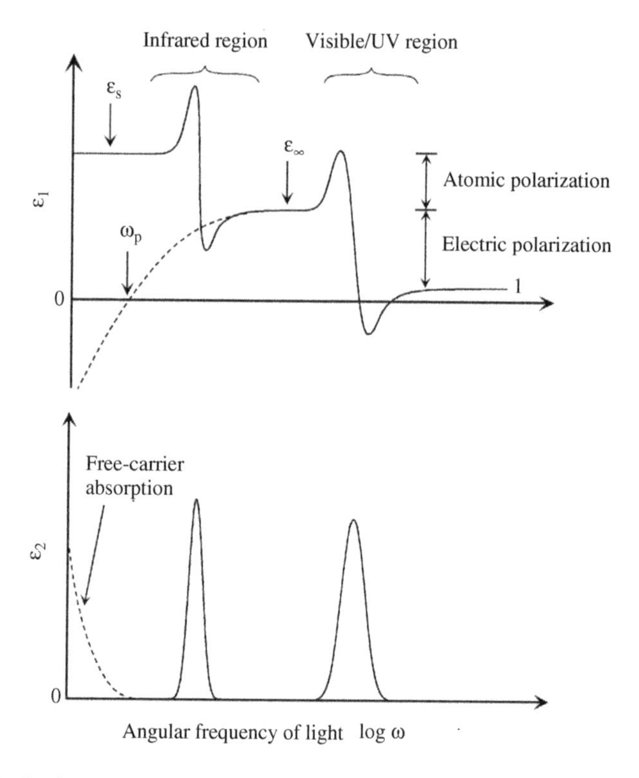

Figure 2.11 Real part ε_1 and imaginary part ε_2 of the complex dielectric constant, plotted as a function of the angular frequency of light log ω.

light coincides with an oscillatory frequency of the spring, resonant oscillation occurs and light is absorbed by the medium. It is obvious from Eq. (2.47b) that ε_2 is proportional to k, which represents light absorption. Accordingly, the ε_2 peaks in Fig. 2.11 show that light is absorbed in specific regions corresponding to the resonant frequencies of the springs. The resonant oscillation for atomic polarization is generally observed in the infrared region, while the resonant oscillation for electric polarization occurs in the UV/visible region. On the other hand, orientational polarization absorbs electromagnetic waves in the microwave region. In microwave ovens, the microwave absorption of H_2O molecules is utilized as the heating system.

At angular frequencies higher than the infrared region, the spring oscillation of atomic polarization cannot follow the oscillation of incoming light, and the atomic polarization disappears. Consequently, the ε_1 value reduces to the value of the high-frequency dielectric constant ε_∞. When the angular frequency is increased further, the electric polarization cannot follow the oscillation of light either, and finally ε_1 reaches the value in a vacuum ($\varepsilon_1 = 1$). Accordingly, the complex dielectric constant of media varies depending on the frequency response of the dielectric polarization. This dielectric response for angular frequency or photon energy is referred to as the dielectric function or dielectric dispersion. In Fig. 2.11, the dielectric function near the resonant frequency has been shown using the Lorentz model (see Section 5.2.1). However, the dielectric functions observed in experiments generally exhibit rather complicated structures in which several resonant oscillators are overlapped (see Fig. 5.11). In Si and Ge crystals that have no ionicity, we observe $\varepsilon_s = \varepsilon_\infty$ since there is no atomic polarization. In such cases, only the value of ε_s is used.

When there are free electrons and holes in a medium, light absorption occurs by these free carriers. In metals or semiconductors that have high carrier concentration, we observe the dielectric function indicated by the dotted lines in Fig. 2.11. The free-carrier absorption increases at lower angular frequencies and ε_1 values become negative at frequencies lower than the plasma angular frequency ω_p. We will discuss free-carrier absorption in greater detail in Section 5.2.5.

Figure 2.12 shows ε_∞ plotted as a function of the bandgap E_g of semiconductor crystals [4]. As shown in this figure, ε_∞ reduces almost linearly with increasing bandgap. In addition, ε_s also shows a similar trend [4]. Classically, the observed trend can be explained as follows. In general, the bandgap of materials becomes larger when constituent atoms have fewer electrons and thus have higher bonding energies. In III-V compound semiconductors, for example, the bandgap increases in the order GaSb< GaAs< GaP. However, the electric polarization becomes smaller in a medium having fewer electrons since the electric polarization is essentially induced by the distortion of electrons in atoms. From these effects, we can understand the reduction in ε_∞ with increasing bandgap qualitatively.

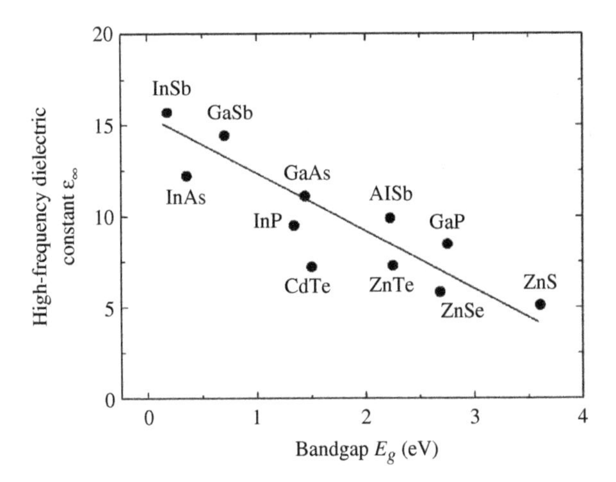

Figure 2.12 High-frequency dielectric constant ε_∞, plotted as a function of the bandgap E_g of semiconductor crystals. Reprinted with permission from *Journal of Applied Physics*, **53**, S. Adachi, Material parameters of $In_{1-x}Ga_xAs_yP_{1-y}$ and related binaries, 8775–8792 (1982). Copyright 1982, American Institute of Physics.

2.3 REFLECTION AND TRANSMISSION OF LIGHT

The reflection and transmission of light are determined by the complex refractive indeces (or complex dielectric constants) of media. In this section, we review basic principles of light reflection and transmission, including Snell's law and Fresnel equations. In particular, p- and s-polarized light waves discussed in this section are of significant importance for the understanding of spectroscopic ellipsometry. This section will also explain the Brewster angle since ellipsometry measurements are generally performed at the Brewster angle. In addition, we will see the phenomenon called total reflection. Total reflection has been utilized in polarizers, which are a vital component of ellipsometry instruments. This section will also introduce optical interference in thin-film and multilayer structures.

2.3.1 REFRACTION OF LIGHT

When light moves into a medium at oblique incidence, the propagation direction of light generally changes by the refraction of light. We can interpret this phenomenon from the variation of light speed at an interface. Figure 2.13(a) illustrates light propagation in an optically dense medium at oblique incidence. In this figure, the incident light is a plane wave that has the same in-plane phase, and the constant phase of the electric field is indicated by lines. When light advances into the medium, light is emitted from the atoms present on the surface. In Fig. 2.13(a), the electric dipole radiation occurs sequentially from the atom located at the left-hand side. As we have seen in Fig. 2.10, the light waves emitted from the atoms overlap constructively and propagate as a transmitted wave.

Now suppose that the incident wave travels from point B to D in Fig. 2.13(a) during a time t. In the meantime, the transmitted wave advances from point A to E, and the reflected wave moves to point C. Let θ_i, θ_r and θ_t be the angles of the incidence, reflection, and transmission, respectively. From Fig. 2.13(a), it can be seen that

$$\frac{\sin\theta_i}{\overline{BD}} = \frac{\sin\theta_r}{\overline{AC}} = \frac{\sin\theta_t}{\overline{AE}} = \frac{1}{\overline{AD}} \tag{2.49}$$

Recall that the speed of light in a medium is given by $s = c/n$ [Eq. (2.28)] and light advances by $st = ct/n$ during the time t. Therefore,

$$\overline{BD} = ct/n_i \quad \overline{AC} = ct/n_r \quad \overline{AE} = ct/n_t \tag{2.50}$$

where n_i, n_r and n_t show the refractive indices for the incident, reflected, and transmitted light, respectively. It is obvious that $\overline{BD} = \overline{AC}$ since $n_i = n_r$. From this, we obtain the law of reflection:

$$\theta_i = \theta_r \tag{2.51}$$

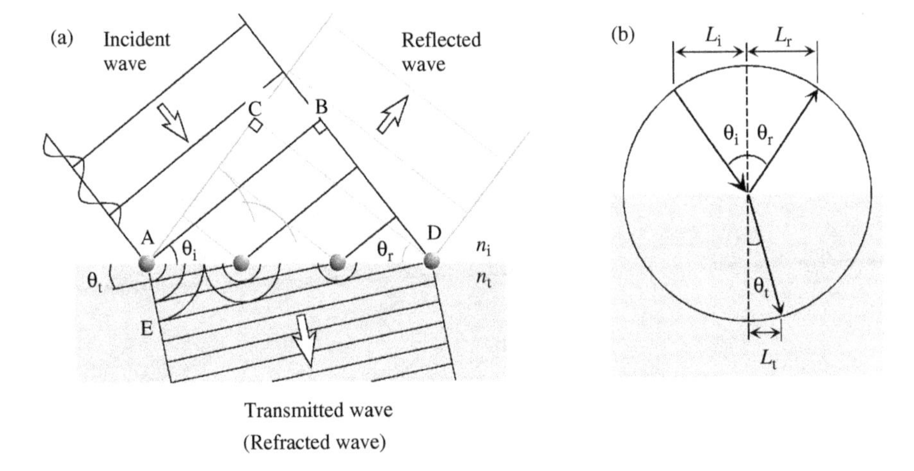

Figure 2.13 Light reflection and transmission at oblique incidence. This figure shows the light reflection and transmission when $n_i < n_t$.

By combining Eqs. (2.49) and (2.50), we can derive the well-know Snell's law:

$$n_\mathrm{i} \sin \theta_\mathrm{i} = n_\mathrm{t} \sin \theta_\mathrm{t} \tag{2.52}$$

As shown in Fig. 2.13(b), if we assume that there is a circle at an interface, we obtain $L_\mathrm{i} = L_\mathrm{r}$ and $n_\mathrm{t}/n_\mathrm{i} = L_\mathrm{i}/L_\mathrm{t}$ from law of reflection and Snell's law, respectively.

Snell's law can also be applied for the complex refractive index. In this case, Snell's law is expressed by

$$N_\mathrm{i} \sin \theta_\mathrm{i} = N_\mathrm{t} \sin \theta_\mathrm{t} \tag{2.53}$$

If we use the complex refractive index, θ_i and θ_t become complex numbers. In this case, however, visualization of light propagation processes becomes very difficult.

2.3.2 p- AND s-POLARIZED LIGHT WAVES

Figure 2.14 shows light reflection on a sample surface. When light is reflected or transmitted by samples at oblique incidence, the light is classified into p- and s-polarized light waves depending on the oscillatory direction of its electric field and each light wave shows quite different behavior. In p-polarization, the electric fields of incident and reflected light waves oscillate within the same plane. This particular plane is called the plane of incidence. Figure 2.15 shows reflection and transmission of p- and s-polarized waves. In this figure, the light reflection and transmission are represented by the vectors of the electric field ***E*** and magnetic

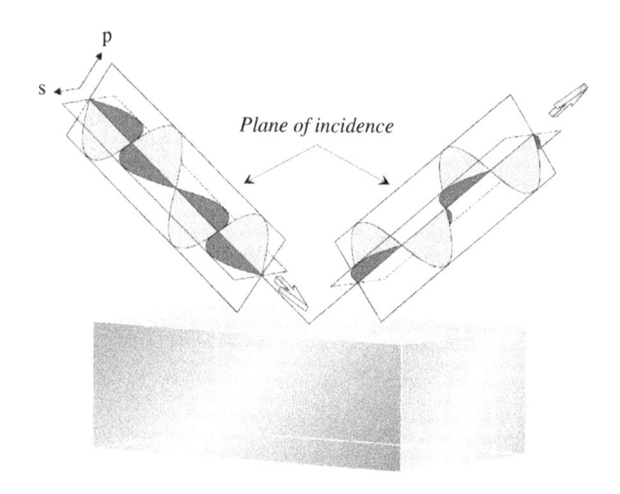

Figure 2.14 Reflection of p- and s-polarized light waves.

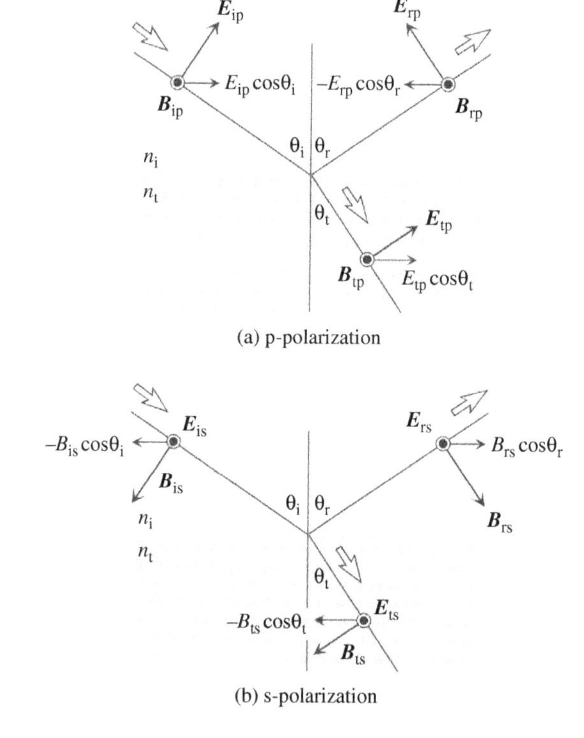

Figure 2.15 Electric field $\boldsymbol{E}$ and magnetic induction $\boldsymbol{B}$ for (a) p-polarization and (b) s-polarization. In these figures, $\boldsymbol{B}$ in (a) and $\boldsymbol{E}$ in (b) are perpendicular to the plane of the paper and are pointing to the reader.

induction $\boldsymbol{B}$. The boundary conditions for electromagnetic waves require that E and B components parallel to an interface are continuous at the interface. In other words, the parallel components on the incident side must be equal to that on the transmission side.

In the case of p-polarized light, the boundary conditions for E and B are given by

$$E_{\mathrm{ip}} \cos\theta_{\mathrm{i}} - E_{\mathrm{rp}} \cos\theta_{\mathrm{r}} = E_{\mathrm{tp}} \cos\theta_{\mathrm{t}} \tag{2.54}$$

$$B_{\mathrm{ip}} + B_{\mathrm{rp}} = B_{\mathrm{tp}} \tag{2.55}$$

where the subscripts ip, rp, and tp represent the incidence, reflection, and transmission of p-polarized light, respectively. In a medium with a refractive index of n, it follows that $E = sB(s = c/n)$ since $E = cB$ [Eq. (2.25)]. If we use $E = sB$, we can rewrite Eq. (2.55) as follows:

$$n_{\mathrm{i}}\left(E_{\mathrm{ip}} + E_{\mathrm{rp}}\right) = n_{\mathrm{t}} E_{\mathrm{tp}} \tag{2.56}$$

By eliminating E_{tp} from Eqs. (2.54) and (2.56) and using $\theta_{\mathrm{i}} = \theta_{\mathrm{r}}$, we obtain the amplitude reflection coefficient for p-polarized light, defined by $r_{\mathrm{p}} \equiv E_{\mathrm{rp}}/E_{\mathrm{ip}}$:

$$r_{\mathrm{p}} \equiv \frac{E_{\mathrm{rp}}}{E_{\mathrm{ip}}} = \frac{n_{\mathrm{t}} \cos\theta_{\mathrm{i}} - n_{\mathrm{i}} \cos\theta_{\mathrm{t}}}{n_{\mathrm{t}} \cos\theta_{\mathrm{i}} + n_{\mathrm{i}} \cos\theta_{\mathrm{t}}} \tag{2.57}$$

If we eliminate E_{rp} from Eqs. (2.54) and (2.56), we get the amplitude transmission coefficient for p-polarized light:

$$t_{\mathrm{p}} \equiv \frac{E_{\mathrm{tp}}}{E_{\mathrm{ip}}} = \frac{2n_{\mathrm{i}} \cos\theta_{\mathrm{i}}}{n_{\mathrm{t}} \cos\theta_{\mathrm{i}} + n_{\mathrm{i}} \cos\theta_{\mathrm{t}}} \tag{2.58}$$

On the other hand, the boundary conditions for s-polarized light are given by

$$E_{\mathrm{is}} + E_{\mathrm{rs}} = E_{\mathrm{ts}} \tag{2.59}$$

$$-B_{\mathrm{is}} \cos\theta_{\mathrm{i}} + B_{\mathrm{rs}} \cos\theta_{\mathrm{r}} = -B_{\mathrm{ts}} \cos\theta_{\mathrm{t}} \tag{2.60}$$

where the subscripts is, rs, and ts represent the incidence, reflection, and transmission for s-polarized light, respectively. Similarly, the amplitude reflection (transmission) coefficient for s-polarized light is expressed by

$$r_{\mathrm{s}} \equiv \frac{E_{\mathrm{rs}}}{E_{\mathrm{is}}} = \frac{n_{\mathrm{i}} \cos\theta_{\mathrm{i}} - n_{\mathrm{t}} \cos\theta_{\mathrm{t}}}{n_{\mathrm{i}} \cos\theta_{\mathrm{i}} + n_{\mathrm{t}} \cos\theta_{\mathrm{t}}} \tag{2.61}$$

$$t_{\mathrm{s}} \equiv \frac{E_{\mathrm{ts}}}{E_{\mathrm{is}}} = \frac{2n_{\mathrm{i}} \cos\theta_{\mathrm{i}}}{n_{\mathrm{i}} \cos\theta_{\mathrm{i}} + n_{\mathrm{t}} \cos\theta_{\mathrm{t}}} \tag{2.62}$$

The above equations for r_{p}, r_{s}, t_{p}, and t_{s} are known as Fresnel equations. These Fresnel equations still hold if the refractive index n is replaced with the complex refractive index N. In this case, amplitude coefficients become complex numbers. We can rewrite the above Fresnel equations using Snell's law as described below. From the relations $N_i \sin\theta_i = N_t \sin\theta_t$, $\sin^2\theta + \cos^2\theta = 1$ and $N^2 = \varepsilon$, we obtain

$$N_t \cos\theta_t = \left(N_t^2 - N_i^2 \sin^2\theta_i\right)^{1/2} = \left(\varepsilon_t - \varepsilon_i \sin^2\theta_i\right)^{1/2} \quad (2.63)$$

If we apply this equation, Fresnel equations for light reflection are written as

$$r_p = \frac{N_{ti}^2 \cos\theta_i - \left(N_{ti}^2 - \sin^2\theta_i\right)^{1/2}}{N_{ti}^2 \cos\theta_i + \left(N_{ti}^2 - \sin^2\theta_i\right)^{1/2}} \quad r_s = \frac{\cos\theta_i - \left(N_{ti}^2 - \sin^2\theta_i\right)^{1/2}}{\cos\theta_i + \left(N_{ti}^2 - \sin^2\theta_i\right)^{1/2}} \quad (2.64)$$

where $N_{ti} = N_t/N_i$. The following equations, which can be obtained by rearranging the above equations, have also been used widely [5,6]:

$$r_p = \frac{\varepsilon_t N_{ii} - \varepsilon_i N_{tt}}{\varepsilon_t N_{ii} + \varepsilon_i N_{tt}} \quad r_s = \frac{N_{ii} - N_{tt}}{N_{ii} + N_{tt}} \quad (2.65)$$

where

$$N_{ii} \equiv N_i \cos\theta_i \quad N_{tt} \equiv \left(\varepsilon_t - \varepsilon_i \sin^2\theta_i\right)^{1/2} \quad (2.66)$$

In the polar coordinate representation [Eq. (2.18)], amplitude coefficients are expressed by

$$r_p = |r_p| \exp(i\delta_{rp}) \quad r_s = |r_s| \exp(i\delta_{rs}) \quad (2.67)$$

$$t_p = |t_p| \exp(i\delta_{tp}) \quad t_s = |t_s| \exp(i\delta_{ts}) \quad (2.68)$$

If we apply the above equations, we can interpret light reflection and transmission in terms of variations in amplitude and phase. Figure 2.16(a) shows the amplitude coefficients at an air/glass interface, plotted as a function of the angle of incidence for p- and s-polarized light waves. The amplitude coefficients were obtained from Fresnel equations using the refractive indices of air ($n_i = 1$) and glass ($n_t = 1.49$ at En = 4 eV [7]). As shown in Fig. 2.16(a), the amplitude reflection coefficient r_s shows negative values and its amplitude is -1 at $\theta_i = 90°$. On the other hand, the r_p values become negative when the angle of incidence is larger than the Brewster angle θ_B (see Section 2.3.4). t_p and t_s show positive values and their amplitudes are zero at $\theta_i = 90°$. Figures 2.16(b) and (c) show the absolute value of the amplitude coefficient and phase change, obtained from Eqs. (2.67) and (2.68). We can apply Eqs. (2.10) and (2.16) to determine the amplitude and phase of complex numbers. The result shown in Fig. 2.16(b) can be obtained directly from Fig. 2.16(a) by simply reversing the negative sign to positive. The absolute value of the amplitude coefficient represents the amplitude ratio between incident and

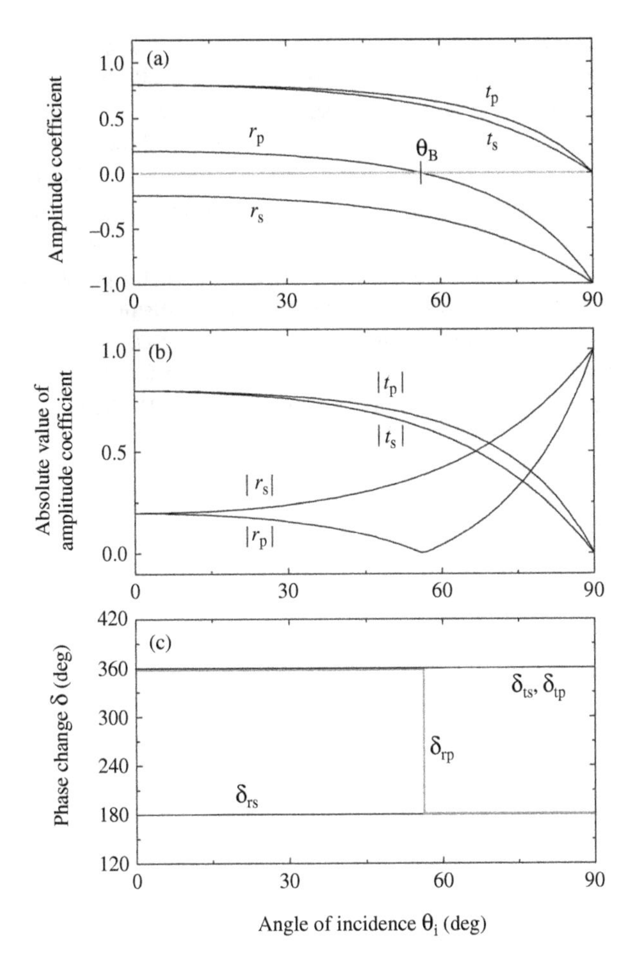

Figure 2.16 (a) Amplitude coefficient, (b) absolute value of the amplitude coefficient and (c) phase change at an air/glass interface, plotted as a function of the angle of incidence θ_i.

reflected (transmitted) waves. Thus, it can be seen that all the light waves are reflected at $\theta_i = 90°$.

The signs of the amplitude reflection coefficients in Fig. 2.16(a) are determined from the phase variation that occurs when polarized light is reflected at the air/glass interface. By comparing Fig. 2.16(a) with Fig. 2.16(c), it can be understood that $\delta = 180°$ for the negative amplitude coefficients and $\delta = 0°(360°)$ for the positive amplitude coefficients. This phase variation can be interpreted from wave reflections at fixed and free ends shown in Fig. 2.17. The wave reflection at a fixed end can be seen when a wave propagates on a string whose end is fixed by hook. In this case, when the wave reaches the reflection surface, the wave receives a force from the hook in the opposite direction, which reverses the phase completely ($\delta = 180°$).

On the other hand, we can see the wave reflection at a free end when a wave propagating on the water surface is reflected by a wall. In this case, since the wave can move freely at the reflection surface, the wave propagates without phase variation ($\delta = 0°$). As shown in Fig. 2.16(c), r_s always shows reflection by a fixed end, while r_p shows reflection by fixed and free ends depending on θ_i. It can be seen from Eq. (2.61) that the r_s values are always negative since generally $n_i < n_t$ and $\cos\theta_i < \cos\theta_t$. This is the reason why s-polarized light shows a constant phase change of $\delta = 180°$. In the case of light transmission, no phase variation occurs since the values of t_p and t_s are always positive, as confirmed from Eqs. (2.58) and (2.62). When there is no light absorption in a medium ($k = 0$), the phase change upon light reflection is either $0°(360°)$ or $180°$. When $k > 0$, however, the phase shows rather complicated variations (see Sections 5.2.2). As confirmed from the complex plane shown in Fig. 2.3, amplitude coefficients become real numbers only when $\theta = 0°(360°)$ or $180°$.

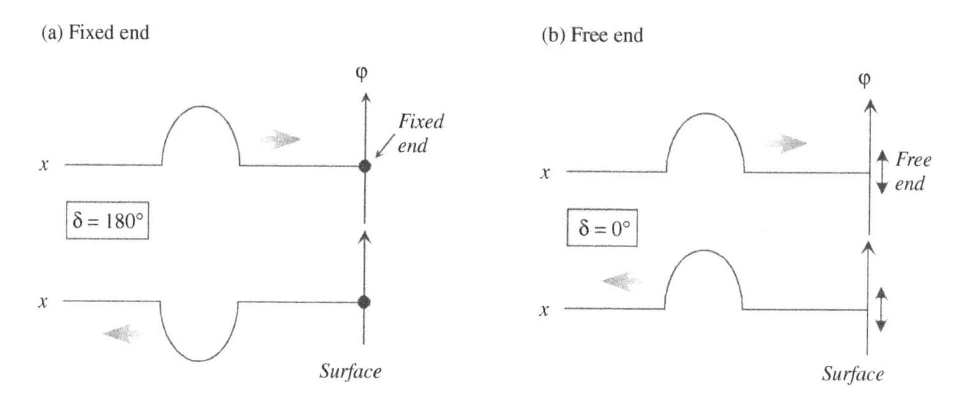

Figure 2.17 Reflection of waves at (a) a fixed end and (b) a free end.

At $\theta_i = 0°$ in Fig. 2.16(c), the phase of p-polarized light is different from that of s-polarized light, although p- and s-polarizations cannot be distinguished anymore at this angle. This obvious contradiction arises from the definition of the vectors shown in Fig. 2.15. In the case of s-polarization, the vectors $\boldsymbol{E}_{is}$ and $\boldsymbol{E}_{rs}$ overlap completely when $\theta_i = \theta_r = 0°$. In p-polarization, however, the vectors $\boldsymbol{E}_{ip}$ and $\boldsymbol{E}_{rp}$ do not overlap at $\theta_i = \theta_r = 0°$ and their directions are completely opposite to each other. Accordingly, although the phase change for p-polarized light (δ_{rp}) shows $0°(360°)$ at $\theta_i < \theta_B$, this is an artifact caused by the definition of the vectors and the actual phase change is 180°, which is identical to that of s-polarized light. Thus, when we consider the actual shape of reflected light, the phase of p-polarized light has to be reversed [8].

2.3.3 REFLECTANCE AND TRANSMITTANCE

The reflectance R obtained in conventional measurements is defined by the ratio of reflected light intensity I_r to incident light intensity I_i ($R \equiv I_r/I_i$). If we use Eq. (2.33), the reflectances for p- and s-polarized waves are expressed by

$$R_p \equiv \frac{I_{rp}}{I_{ip}} = \left|\frac{E_{rp}}{E_{ip}}\right|^2 = |r_p|^2 \quad R_s \equiv \frac{I_{rs}}{I_{is}} = \left|\frac{E_{rs}}{E_{is}}\right|^2 = |r_s|^2 \tag{2.69}$$

As mentioned earlier, for the light transmittance, we use $I = n|E|^2$. In this calculation, the ratio of the cross-sectional areas for transmitted and incident beams ($\cos\theta_t/\cos\theta_i$) should be taken into account. Consequently, the transmittances for p- and s-polarized waves are given by the following equations [1,2]:

$$T_p \equiv \frac{I_{tp}\cos\theta_t}{I_{ip}\cos\theta_i} = \left(\frac{n_t\cos\theta_t}{n_i\cos\theta_i}\right)\left|\frac{E_{tp}}{E_{ip}}\right|^2 = \left(\frac{n_t\cos\theta_t}{n_i\cos\theta_i}\right)|t_p|^2 \tag{2.70a}$$

$$T_s \equiv \frac{I_{ts}\cos\theta_t}{I_{is}\cos\theta_i} = \left(\frac{n_t\cos\theta_t}{n_i\cos\theta_i}\right)\left|\frac{E_{ts}}{E_{is}}\right|^2 = \left(\frac{n_t\cos\theta_t}{n_i\cos\theta_i}\right)|t_s|^2 \tag{2.70b}$$

When $k = 0$, it follows that $R_p + T_p = 1$ and $R_s + T_s = 1$. In the case of $k > 0$, we observe $R_p + T_p < 1$ and $R_s + T_s < 1$. The reflectance for natural light or unpolarized light is given by

$$R_n = (R_p + R_s)/2 \tag{2.71}$$

Figure 2.18 shows the reflectance at (a) an air/glass interface and (b) an air/c-Si (crystalline Si) interface, plotted as a function of the angle of incidence θ_i. In the calculation, $N_t = 1.49 - i0$ for glass [7] and $N_t = 5 - i3.7$ for c-Si [9] were used. In both cases, R_s increases as θ_i increases. At the air/glass interface, only s-polarized light is reflected at θ_B since $R_p = 0$. Therefore, θ_B is also called the polarizing angle. At the air/c-Si interface, R_p shows a minimum value at the pseudo-Brewster angle $\theta_{B'}$. As confirmed from Fig. 2.18, reflectances at a low incidence angle increase as (n, k) values of samples increase. If we substitute $\theta_i = \theta_t = 0°$ and $N_i = 1 - i0$ into Fresnel equations, we obtain the reflectance at normal incidence:

$$R = R_p = R_s = \frac{(n-1)^2 + k^2}{(n+1)^2 + k^2} \tag{2.72}$$

It is evident from Eq. (2.72) that the reflectances increase with increasing (n, k) values of samples. From Eq. (2.72), the reflectance at the air/glass interface ($R =$ 4%) and air/c-Si interface ($R = 60\%$) can be obtained easily.

2.3.4 BREWSTER ANGLE

As we will see in Chapter 4, ellipsometry measures the ratio of the amplitude reflection coefficients (r_p/r_s). Since the difference between r_p and r_s is maximized at the Brewster angle θ_B (see Fig. 2.16), sensitivity for the measurement also increases at this angle. Thus, ellipsometry measurement is generally performed at the Brewster angle. As we have seen in Fig. 2.18(a), when $k = 0$, $R_p = 0$ at θ_B. This phenomenon can be explained from electric dipole radiation at a media interface [1]. Now consider that p-polarized light is irradiated at $\theta_i < \theta_B$ and the light transmits with an angle of θ_t, as shown in Fig. 2.19(a). In this case, the electric dipole radiation occurs from the atoms located near the interface, and only light waves that satisfy law of reflection

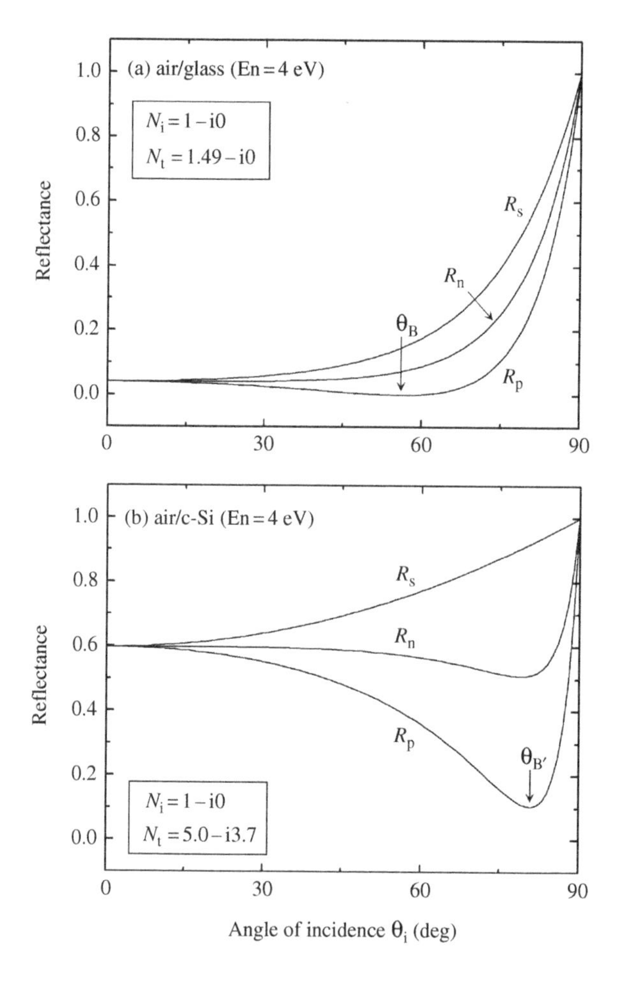

Figure 2.18 Reflectance at (a) an air/glass interface and (b) an air/c-Si (crystalline Si) interface, plotted as a function of the angle of incidence θ_i. In these figures, θ_B and $\theta_{B'}$ represent the Brewster and pseudo-Brewster angles, respectively.

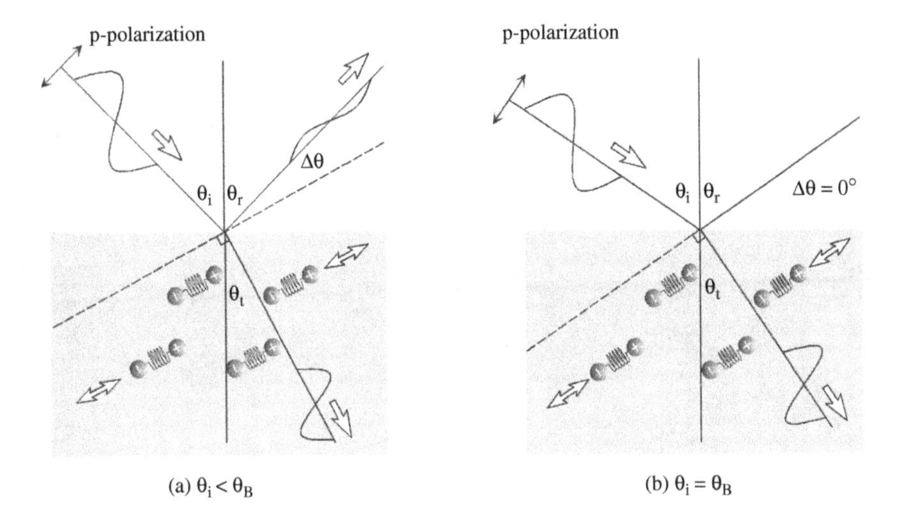

Figure 2.19 Electric dipole radiation at (a) $\theta_i < \theta_B$ and (b) $\theta_i = \theta_B$.

($\theta_i = \theta_r$) are extracted as the reflected light. When $\theta_i \neq \theta_B$, the angle between the propagation direction of the reflected light and the oscillatory direction of the electric dipoles is $\Delta\theta \neq 0°$. However, $\Delta\theta$ becomes zero at $\theta_i = \theta_B$ [Fig. 2.19(b)]. As confirmed from Fig. 2.9, in electric dipole radiation, no light radiation occurs toward the oscillating direction of electric dipoles. In other words, when the oscillatory direction of electric dipoles is perpendicular to the vibrational direction of reflected light, light reflection disappears. This is the reason why $R_p = 0$ at θ_B. In s-polarization, on the other hand, the oscillatory direction of electric dipoles is always parallel to the vibrational direction of reflected light. Thus, R_s increases gradually with increasing θ_i. As mentioned above, electric dipole radiation at an interface differs significantly in p- and s-polarized waves. As a result, p- and s-polarized light waves show quite different reflectances. When there is light absorption in a medium ($k > 0$), $R_p \neq 0$ at $\theta_{B'}$, as shown in Fig. 2.18(b). This implies that light reflection still occurs at $\theta_{B'}$ by the contribution of k, as confirmed from Eq. (2.72).

It can be seen from Fig. 2.19(b) that $\theta_B + \theta_t = 90°$ at $\theta_B(= \theta_i = \theta_r)$. If we use Snell's law ($n_i \sin\theta_i = n_t \sin\theta_t$), $\theta_t = 90° - \theta_B$ and $\sin(90° - \theta) = \cos\theta$ (see Appendix 1), we get

$$\tan\theta_B = n_t/n_i \tag{2.73}$$

This equation is known as the Brewster's law. At the air/glass interface ($n_t/n_i = 1.49$), $\theta_B = \tan^{-1}(1.49) = 56.1°$ is obtained using Eq. (2.73). The θ_B increases as n_t becomes larger and θ_B increases to 73.7° in c-Si ($\varepsilon_s = 11.6$ and $n_t = 3.41$). As we have seen in Section 2.2.3, n_t generally shows strong wavelength dependence. Thus, θ_B varies according to the wavelength or photon energy of measurement light. In semiconductor characterization, θ_i used in spectroscopic ellipsometry measurement is typically 70–80°.

2.3.5 TOTAL REFLECTION

So far, we have considered the case when $n_i < n_t$. If $n_i > n_t$, we observe the total reflection when the incidence angle is larger than the critical angle θ_c. Figure 2.20 shows the light reflection at a glass/air interface ($n_t/n_i = 1/1.49$)

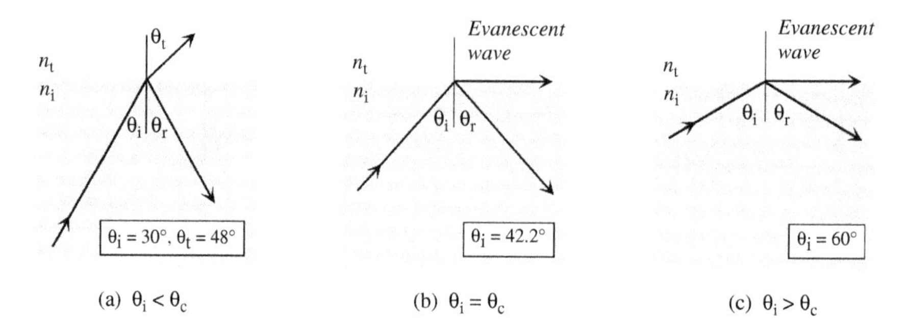

Figure 2.20 Light reflection at (a) $\theta_i < \theta_c$, (b) $\theta_i = \theta_c$, and (c) $\theta_i > \theta_c$ at a glass/air interface ($n_t/n_i = 1/1.49$). In these figures, θ_i and θ_c represent the incidence and critical angles, respectively.

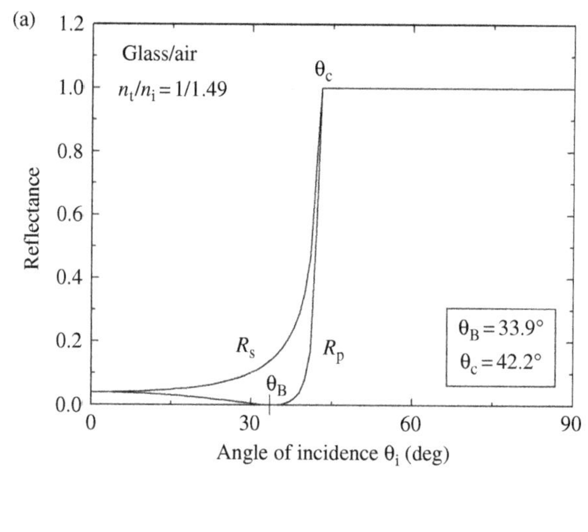

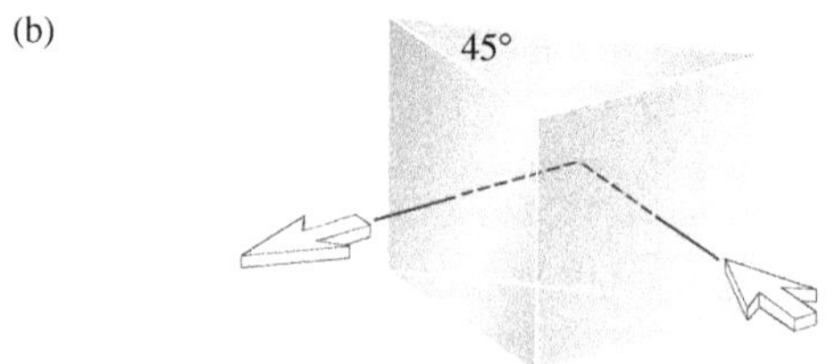

Figure 2.21 (a) Reflectance for p- and s-polarizations at a glass/air interface ($n_t/n_i = 1/1.49$) and (b) total reflection of light in a prism.

when the light is illuminated from the glass side. It can be seen from Fig. 2.20(a) that $\theta_i < \theta_t$ when $n_i > n_t$. If we increase the incidence angle further, we observe $\theta_t = 90°$ at $\theta_i = \theta_c$ [Fig. 2.20(b)]. In this case, the light cannot advance into the medium of n_t and all the light waves are reflected at the interface (total reflection). By substituting $\theta_t = 90°$ into Snell's law, we obtain

$$\sin\theta_c = n_t/n_i \tag{2.74}$$

When total reflection occurs at $\theta_i \geq \theta_c$, a wave called an evanescent wave is generated at an interface. If there is a medium ($k > 0$) at an interface, light absorption by the evanescent wave leads to the attenuated total reflection.

Figure 2.21(a) shows the reflectance when $n_t/n_i = 1/1.49$, plotted as a function of the angle of incidence θ_i. As shown in Fig. 2.21(a), total reflection occurs at $\theta_c = 42.2°$ for both p- and s-polarizations and $R_p = R_s = 1$ at $\theta_i \geq \theta_c$. Observe that $R_p = 0$ at θ_B even if $n_i > n_t$. In the prism shown in Fig. 2.21(b), total reflection occurs since $\theta_i(45°) > \theta_c(42°)$. The prisms of this shape have been utilized in various polarizers.

2.4 OPTICAL INTERFERENCE

When light waves generated at different positions overlap, optical interference occurs by the superposition of the light waves. This optical interference is also observed in a thin film formed on a substrate. In the analysis of spectroscopic ellipsometry, thicknesses of thin films and multilayers are determined from this interference effect. This section will review optical interference in thin-film and multilayer structures, which plays an important role in ellipsometry data analysis.

2.4.1 OPTICAL INTERFERENCE IN THIN FILMS

Figure 2.22 shows an optical model constructed for a thin film formed on a substrate (ambient/thin film/substrate). As shown in this figure, an optical model is represented by the complex refractive index and thickness of each layer. Let N_0, N_1, and N_2 be the complex refractive indices of air, the thin film and the substrate, respectively. When light absorption within the thin film is weak, an incident wave is reflected at the film surface and film/substrate interface. In this case, the light wave reflected first on the surface (primary beam) overlaps with the light wave reflected at the film/substrate interface (secondary beam) and optical interference occurs. In Fig. 2.22, the wave amplitude becomes larger by this interference effect. If the primary and secondary beams are out of phase ($\delta = 180°$), the amplitude of reflected light becomes smaller. The phase difference between the primary and secondary beams can be obtained as follows [1]. Consider that the primary beam is reflected at point A and reaches point D. Mathematically, the primary beam is expressed by

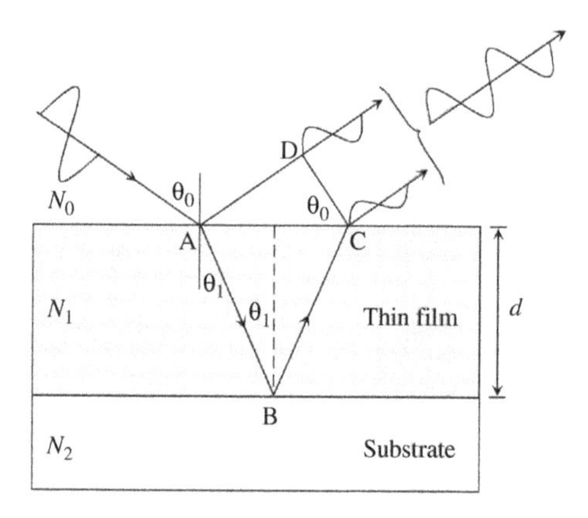

Figure 2.22 Optical interference in a thin film formed on a substrate.

$E = E_0 \exp[\mathrm{i}(\omega t - K_0 x + \delta)]$, where K_0 is the propagation number in air. If we consider the phase variation only for the position x, the phase change induced by the traveling distance $\overline{\mathrm{AD}}$ is expressed by $K_0\overline{\mathrm{AD}}$. On the other hand, the optical pass length of the secondary beam is $\overline{\mathrm{AB}}+\overline{\mathrm{BC}}$. Thus, the phase variation for the secondary beam is given by $K_1(\overline{\mathrm{AB}}+\overline{\mathrm{BC}})$, where K_1 shows the propagation number in the thin film. As shown in Fig. 2.22, the difference in the optical path length between the primary and secondary beams is given by $\overline{\mathrm{AB}}+\overline{\mathrm{BC}}-\overline{\mathrm{AD}}$. Therefore, using $K = 2\pi N/\lambda$ [Eq. (2.29)], we obtain the total phase difference between the two beams as follows:

$$\alpha = \frac{2\pi N_1}{\lambda}\left(\overline{\mathrm{AB}}+\overline{\mathrm{BC}}\right) - \frac{2\pi N_0}{\lambda}\overline{\mathrm{AD}} \tag{2.75}$$

It can be seen from Fig. 2.22 that $\overline{\mathrm{AD}} = \overline{\mathrm{AC}}\sin\theta_0$ and $\overline{\mathrm{AC}} = 2d\tan\theta_1$. If we transform these using Snell's law, we get

$$\overline{\mathrm{AD}} = 2d\frac{\sin^2\theta_1}{\cos\theta_1}\cdot\frac{N_1}{N_0} \tag{2.76}$$

By substituting Eq. (2.76) and $\overline{\mathrm{AB}} = \overline{\mathrm{BC}} = d/\cos\theta_1$ into Eq. (2.75), we obtain

$$\alpha = \frac{4\pi d N_1}{\lambda}\left(\frac{1-\sin^2\theta_1}{\cos\theta_1}\right) = \frac{4\pi d N_1}{\lambda}\cos\theta_1 \tag{2.77}$$

This α shows the total phase variation for the secondary beam. In general, to express the phase difference between the surface and interface, the phase variation β defined by $\alpha = 2\beta$ is used:

$$\beta = \frac{2\pi d}{\lambda}N_1\cos\theta_1 = \frac{2\pi d}{\lambda}\left(N_1^2 - N_0^2\sin^2\theta_0\right)^{1/2} \tag{2.78}$$

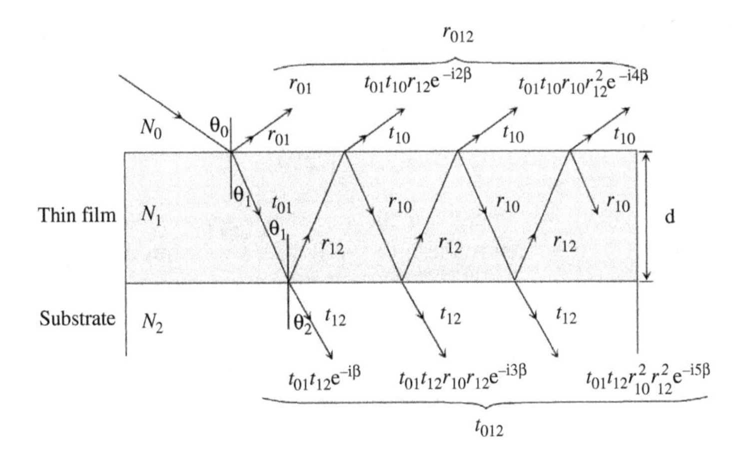

Figure 2.23 Optical model for an ambient/thin film/substrate structure. In this figure, $r_{jk}(t_{jk})$ represents the amplitude reflection (transmission) coefficient.

In the transformation of Eq. (2.78), Eq. (2.63) was used. The above β is also called the film phase thickness.

Figure 2.23 illustrates optical interference in an ambient/thin film/substrate structure. In this figure, $r_{jk}(t_{jk})$ shows the amplitude reflection (transmission) coefficient at each interface. From Fresnel equations for p- and s-polarized waves (see Section 2.3.2), we obtain

$$r_{jk,\mathrm{p}} = \frac{N_k \cos\theta_j - N_j \cos\theta_k}{N_k \cos\theta_j + N_j \cos\theta_k} \qquad r_{jk,\mathrm{s}} = \frac{N_j \cos\theta_j - N_k \cos\theta_k}{N_j \cos\theta_j + N_k \cos\theta_k} \tag{2.79}$$

$$t_{jk,\mathrm{p}} = \frac{2N_j \cos\theta_j}{N_k \cos\theta_j + N_j \cos\theta_k} \qquad t_{jk,\mathrm{s}} = \frac{2N_j \cos\theta_j}{N_j \cos\theta_j + N_k \cos\theta_k} \tag{2.80}$$

As shown in Fig. 2.23, the amplitude reflection coefficient for the primary beam is r_{01}. On the other hand, the phase variation caused by the difference in the optical pass length is given by $\exp(-i2\beta)$ since $\exp\{i[\omega t - (Kx + 2\beta) + \delta]\} = \exp[i(\omega t - Kx + \delta)]\exp(-i2\beta)$. By multiplying the phase variation and amplitude coefficients at each interface, we obtain $t_{01}t_{10}r_{12}\exp(-i2\beta)$ for the secondary beam. It should be noted that, when the phase of electromagnetic waves is expressed by $(Kx - \omega t)$, the phase variation for the secondary beam becomes $\exp(i2\beta)$ (see Appendix 2).

The amplitude reflection coefficient for the ambient/thin film/substrate structure is expressed from the sum of all the reflected waves:

$$r_{012} = r_{01} + t_{01}t_{10}r_{12}e^{-i2\beta} + t_{01}t_{10}r_{10}r_{12}^2e^{-i4\beta} + t_{01}t_{10}r_{10}^2r_{12}^3e^{-i6\beta} + \cdots \tag{2.81}$$

Since the infinite series $y = a + ar + ar^2 + ar^3 + \cdots$ is reduced to $y = a/(1-r)$, we get

$$r_{012} = r_{01} + \frac{t_{01}t_{10}r_{12}\exp(-i2\beta)}{1 - r_{10}r_{12}\exp(-i2\beta)} \tag{2.82}$$

In Eqs. (2.79) and (2.80), there are the relations expressed by $r_{10} = -r_{01}$ and $t_{01}t_{10} = 1 - r_{01}^2$ [10]. If we apply these to Eq. (2.82), we obtain the familiar form:

$$r_{012} = \frac{r_{01} + r_{12}\exp(-i2\beta)}{1 + r_{01}r_{12}\exp(-i2\beta)} \tag{2.83}$$

Similarly, the amplitude transmission coefficient t_{012} is given by

$$t_{012} = t_{01}t_{12}e^{-i\beta} + t_{01}t_{12}r_{10}r_{12}e^{-i3\beta} + t_{01}t_{12}r_{10}^2r_{12}^2e^{-i5\beta} + \cdots \tag{2.84}$$

and, after the rearrangement, we obtain

$$t_{012} = \frac{t_{01}t_{12}\exp(-i\beta)}{1 + r_{01}r_{12}\exp(-i2\beta)} \tag{2.85}$$

From Eqs. (2.83) and (2.85), the amplitude reflection (transmission) coefficients for p- and s-polarized waves are expressed by

$$r_{012,\mathrm{p}} = \frac{r_{01,\mathrm{p}} + r_{12,\mathrm{p}}\exp(-i2\beta)}{1 + r_{01,\mathrm{p}}r_{12,\mathrm{p}}\exp(-i2\beta)} \quad r_{012,\mathrm{s}} = \frac{r_{01,\mathrm{s}} + r_{12,\mathrm{s}}\exp(-i2\beta)}{1 + r_{01,\mathrm{s}}r_{12,\mathrm{s}}\exp(-i2\beta)} \tag{2.86}$$

$$t_{012,\mathrm{p}} = \frac{t_{01,\mathrm{p}}t_{12,\mathrm{p}}\exp(-i\beta)}{1 + r_{01,\mathrm{p}}r_{12,\mathrm{p}}\exp(-i2\beta)} \quad t_{012,\mathrm{s}} = \frac{t_{01,\mathrm{s}}t_{12,\mathrm{s}}\exp(-i\beta)}{1 + r_{01,\mathrm{s}}r_{12,\mathrm{s}}\exp(-i2\beta)} \tag{2.87}$$

By using Eq. (2.69), we obtain the reflectances for p- and s-polarized waves as follows:

$$R_\mathrm{p} = \left|r_{012,\mathrm{p}}\right|^2 \quad R_\mathrm{s} = \left|r_{012,\mathrm{s}}\right|^2 \tag{2.88}$$

The incidence and transmission angles at each interface can be obtained by applying Snell's law:

$$N_0 \sin\theta_0 = N_1 \sin\theta_1 = N_2 \sin\theta_2 \tag{2.89}$$

2.4.2 MULTILAYERS

Optical interference in multilayer structures can also be calculated from the procedure described in the previous section. Figure 2.24(a) shows an optical model

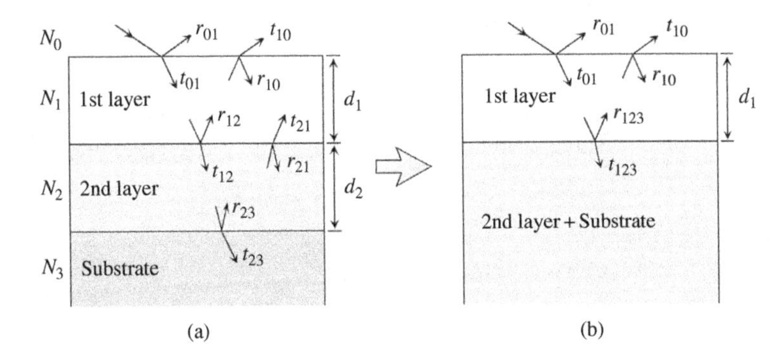

Figure 2.24 Calculation method for optical interference in a multilayer structure.

in which two thin layers are formed on a substrate. As shown in Fig. 2.24(b), we first calculate the amplitude coefficients for the second layer and substrate by applying Eqs. (2.83) and (2.85):

$$r_{123} = \frac{r_{12} + r_{23}\exp(-i2\beta_2)}{1 + r_{12}r_{23}\exp(-i2\beta_2)} \tag{2.90}$$

$$t_{123} = \frac{t_{12}t_{23}\exp(-i\beta_2)}{1 + r_{12}r_{23}\exp(-i2\beta_2)} \tag{2.91}$$

The phase variation β_2 is given by $\beta_2 = 2\pi d_2 N_2 \cos\theta_2/\lambda$, where d_2 is the thickness of the second layer. From these r_{123} and t_{123}, we obtain the amplitude coefficients for the multilayer as follows:

$$r_{0123} = \frac{r_{01} + r_{123}\exp(-i2\beta_1)}{1 + r_{01}r_{123}\exp(-i2\beta_1)} \tag{2.92}$$

$$t_{0123} = \frac{t_{01}t_{123}\exp(-i\beta_1)}{1 + r_{01}r_{123}\exp(-i2\beta_1)} \tag{2.93}$$

In these equations, $\beta_1 = 2\pi d_1 N_1 \cos\theta_1/\lambda$, where d_1 is the thickness of the first layer. In this manner, we can perform the calculation upward from the substrate, even if there are many layers in the multilayer structure. In this method, however, the calculation cannot be performed from the top layer toward the substrate. If we apply virtual substrate approximation (see Section 8.1.4), we can extract optical constants of the top layer without detailed knowledge of underlying structures [6,11].

The calculation process for multilayers can be simplified if we apply Eq. (2.64) in which $\cos\theta_t$ is eliminated from the Fresnel equations. Finally, we show the results obtained from the substitution of Eqs. (2.90) and (2.91) into Eqs. (2.92) and (2.93) [10]:

$$r_{0123} = \frac{r_{01} + r_{12}\exp(-i2\beta_1) + [r_{01}r_{12} + \exp(-i2\beta_1)]\, r_{23}\exp(-i2\beta_2)}{1 + r_{01}r_{12}\exp(-i2\beta_1) + [r_{12} + r_{01}\exp(-i2\beta_1)]\, r_{23}\exp(-i2\beta_2)} \tag{2.94}$$

$$t_{0123} = \frac{t_{01}t_{12}t_{23}\exp[-i(\beta_1 + \beta_2)]}{1 + r_{01}r_{12}\exp(-i2\beta_1) + [r_{12} + r_{01}\exp(-i2\beta_1)]\, r_{23}\exp(-i2\beta_2)} \tag{2.95}$$

REFERENCES

[1] E. Hecht, *Optics*, 4th edition, Addison Wesley, San Francisco (2002).

[2] M. Born and E. Wolf, *Principles of Optics*, 7th edition, Cambridge University Press, Cambridge (1999).

[3] M. Wakagi, H. Fujiwara, and R. W. Collins, Real time spectroscopic ellipsometry for characterization of the crystallization of amorphous silicon by thermal annealing, *Thin Solid Films*, **313–314** (1998) 464–468.

[4] S. Adachi, Material parameters of $In_{1-x}Ga_xAs_yP_{1-y}$ and related binaries, *J. Appl. Phys.*, **53** (1982) 8775–8792.

[5] D. E. Aspnes, Spectroscopic ellipsometry of solids, in *Optical Properties of Solids: New Developments*, edited by B. O. Seraphin, Chapter 15, 801–846, North-Holland, Amsterdam (1976).

[6] D. E. Aspnes, Minimal-data approaches for determining outer-layer dielectric responses of films from kinetic reflectometric and ellipsometric measurements, *J. Opt. Soc. Am. A*, **10** (1993) 974–983.

[7] E. D. Palik (Ed.), *Handbook of Optical Constants of Solids*, Academic Press, San Diego (1985).

[8] R. T. Holm, Convention confusions, in *Handbook of Optical Constants of Solids II*, edited by E. D. Palik, Chapter 2, 21–55, Academic Press, San Diego (1991).

[9] S. Adachi, *Optical Constants of Crystalline and Amorphous Semiconductors: Numerical Data and Graphical Information*, Kluwer Academic Publishers, Norwell (1999).

[10] R. M. A. Azzam and N. M. Bashara, *Ellipsometry and Polarized Light*, North-Holland, Amsterdam (1977).

[11] D. E. Aspnes, Optical approaches to determine near-surface compositions during epitaxy, *J. Vac. Sci. Technol. A*, **14** (1996) 960–966.

3 Polarization of Light

In spectroscopic ellipsometry, optical constants of samples are determined from the change in the polarization state by light reflection (or transmission). In this chapter, we will review polarization of light, which provides a strong basis for the ellipsometry technique. This chapter will also examine various optical elements, including a polarizer and compensator, which have been used widely in ellipsometry measurement. In order to understand the principles of ellipsometry measurement, it is essential to comprehend the Jones vector and Stokes parameters that enable us to describe polarization of light mathematically. In this chapter, we will define the states of polarization and will address mathematical methods for their descriptions.

3.1 REPRESENTATION OF POLARIZED LIGHT

When electric fields of light waves are oriented in specific directions, such light is referred to as polarized light. As we have seen in the previous chapter, p- and s-polarizations represent polarized waves whose orientations are defined by Fig. 2.14. If the oscillating direction of light waves is completely random, the light is called unpolarized light (or natural light). When light waves propagate in the same direction, the polarization is expressed by superimposing each electric field. In this case, a phase difference between the light waves has to be taken into account, in order to describe the state of polarization properly. According to this phase difference, the state of polarization changes into various states from linear polarization to circular polarization. In this section, we will address the basic principles of polarized light, including the phase of light and types of polarization.

3.1.1 PHASE OF LIGHT

The initial phase of light waves (or electromagnetic waves) is of particular importance in describing the state of polarization. As we have seen in the previous

Spectroscopic Ellipsometry: Principles and Applications H. Fujiwara

chapter, the propagation of light is described by $E = E_0 \exp[i(\omega t - Kx + \delta)]$ [Eq. (2.24a)]. If we transform this using Eq. (2.20), we obtain $E = E_0 \cos(\omega t - Kx + \delta)$. Now consider that the initial phase δ of a light wave varies from 0 to $\pi/2$. Figure 3.1 shows the waveforms of each light wave traveling in the positive direction of the x axis at $t = 0$. When δ is increased to $\pi/2$, the waveform shifts toward the right-hand side, as shown in Fig. 3.1. This simply shows that the cosine wave starts from the position where $(-Kx + \delta) = 0$ ($\omega t = 0$ at $t = 0$). Thus, the shift of the waveform by δ is expressed as δ/K. In particular, when $K = 1$, the waveform shifts to the right-hand side by $\delta = \pi/2$ and the wave advances forward with increasing δ. This effect can be understood more easily if we consider the phase as $(\omega t + \delta)$, because, in this case, the increase in δ corresponds to an increase in time of $\omega t = \delta$. Conversely, if the sign of δ is minus, the wave lags by $\omega t = \delta$. When the phase is defined by $(Kx - \omega t + \delta)$, however, the positive sign of δ implies phase lag since the ωt term has the minus sign (see Appendix 2).

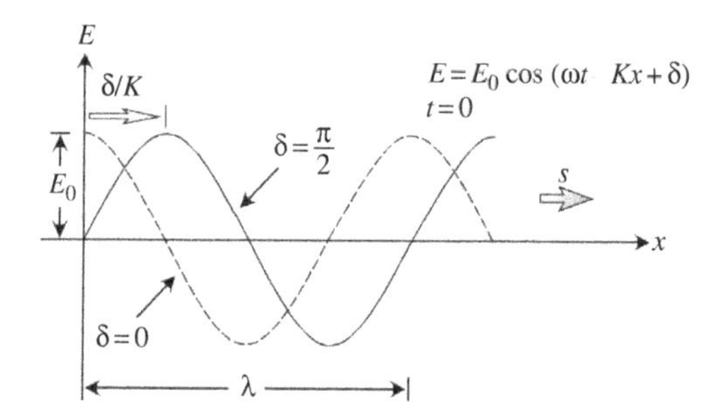

Figure 3.1 Variation of a light wave with the initial phase δ.

3.1.2 POLARIZATION STATES OF LIGHT WAVES

The polarization state of light traveling along the z axis, for example, is described by superimposing two electric fields whose directions are parallel to the x and y axes. In this case, we can express the electromagnetic wave traveling along the z axis as the vector sum of the electric fields $\boldsymbol{E}_x$ and $\boldsymbol{E}_y$:

$$\begin{aligned} \boldsymbol{E}(z,t) &= \boldsymbol{E}_x(z,t) + \boldsymbol{E}_y(z,t) \\ &= \{E_{x0} \exp[i(\omega t - Kz + \delta_x)]\}\boldsymbol{x} + \{E_{y0} \exp[i(\omega t - Kz + \delta_y)]\}\boldsymbol{y} \end{aligned} \tag{3.1}$$

where $\boldsymbol{x}$ and $\boldsymbol{y}$ are unit vectors along the coordinate axes. When we describe the states of polarization, the absolute values of the initial phases (δ_x, δ_y) are not required, and only the relative phase difference $\delta_y - \delta_x$ (or $\delta_x - \delta_y$) is taken into account. Similarly, only the phase difference is accounted for in spectroscopic ellipsometry.

Figure 3.2 shows the variation of the polarization state with the phase difference $\delta_y - \delta_x$. In this figure, $E_{x0} = E_{y0}$ and $K = 1$ are assumed. As shown in Fig. 3.2(a), when $\delta_y - \delta_x = 0$, there is no phase difference between $\boldsymbol{E}_x$ and $\boldsymbol{E}_y$, and the orientation of the synthesized vector $(\boldsymbol{E}_x + \boldsymbol{E}_y)$ is always 45° in the $x-y$ plane. In other words, an electromagnetic wave oriented at 45° can be resolved into two electromagnetic waves vibrating parallel to the x and y axes. However, the amplitude of the synthesized vector is $\sqrt{2}$ times larger than that of $E_{x0}(= E_{y0})$. The polarization state shown in Fig. 3.2(a) is referred to as linear polarization. When the phase difference between $\boldsymbol{E}_x$ and $\boldsymbol{E}_y$ is $90°(\delta_y - \delta_x = \pi/2)$, the synthesized vector rotates in the $x-y$ plane as the light propagates. This polarization state shown in Fig. 3.2(b) is

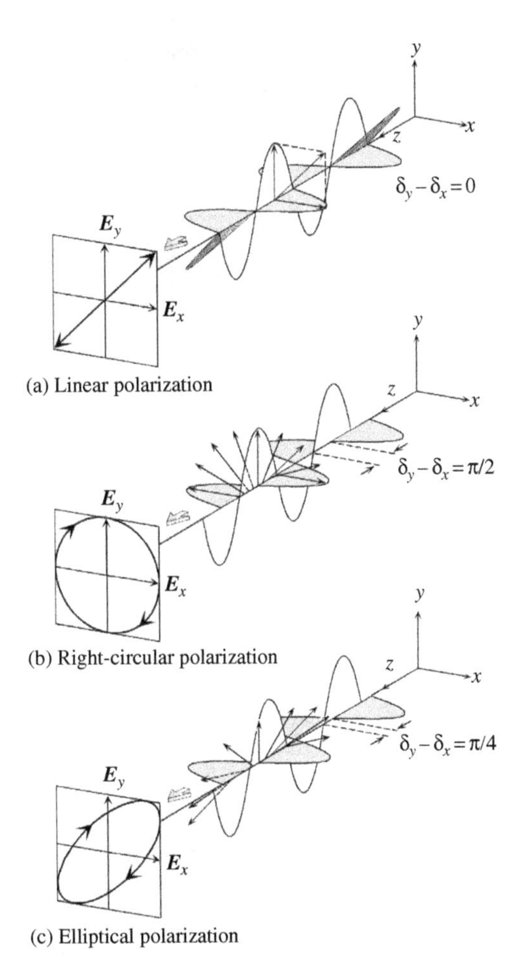

Figure 3.2 Representations of (a) linear polarization, (b) right-circular polarization and (c) elliptical polarization. Phase differences between the electric fields parallel to the x and y axes $(\delta_y - \delta_x)$ are (a) 0, (b) $\pi/2$, and (c) $\pi/4$.

known as circular polarization. As confirmed from Fig. 3.1, since $\delta_y - \delta_x = \pi/2$ and $K = 1$ in this case, the wave oscillating along the y axis advances forward by $\pi/2$, compared with the wave oscillating along the x axis. In Fig. 3.2(b), if we choose a point on the z axis, the synthesized vector on the point rotates toward the right (clockwise) as the light propagates with time (the rotation is counterclockwise in the positive direction of the z axis). This particular polarization is called right-circular polarization. The polarization shown in Fig. 3.2(c) is referred to as elliptical polarization and rotates toward the right (clockwise) when $\delta_y - \delta_x = \pi/4$.

Figure 3.3 shows the variation in the state of polarization when $\delta_y - \delta_x$ and $\delta_x - \delta_y$ are changed sequentially from 0 to 2π. It can be seen from this figure that the relative phase difference changes the state of polarization significantly. When the phase difference is either 0 or π, a light wave is linearly polarized. If $E_{x0} \neq E_{y0}$, the state of polarization becomes elliptical polarization even at $\delta_y - \delta_x = \pi/2$. As we will see in the next chapter, the phase difference in ellipsometry is basically expressed by $\delta_x - \delta_y$, instead of $\delta_y - \delta_x$. Accordingly, we will use the phase difference $\delta_x - \delta_y$ in the following part.

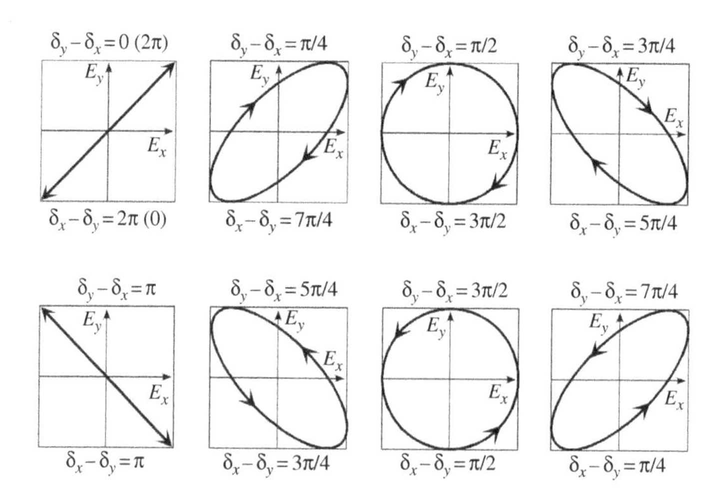

Figure 3.3 Variation of the state of polarization with phase difference ($\delta_y - \delta_x$ and $\delta_x - \delta_y$). In this figure, the amplitudes of the waves in the x and y directions are the same ($E_{x0} = E_{y0}$).

3.2 OPTICAL ELEMENTS

In order to determine the state of polarization, an optical instrument composed of various optical elements is required. Optical elements used in ellipsometry measurement can be classified into polarizers (analyzers), compensators (retarders), and depolarizers. The polarizer is employed to extract linearly polarized light from unpolarized light, while the compensator (retarder) is used when we convert linear polarization to circular polarization. The depolarizer is utilized when polarized light is changed to unpolarized light. This section will provide an overview of these optical elements.

3.2.1 POLARIZER (ANALYZER)

A polarizer is generally placed in front of the light source and is utilized to extract linearly polarized light from unpolarized light. On the other hand, an analyzer is placed in front of a light detector and the state of polarization is determined from the intensity of light transmitted through the analyzer. Although a polarizer and an analyzer are the same optical element, they are named separately due to the difference in their roles. In general, a polarizer (analyzer) consists of prisms mainly made from a $CaCO_3$ crystal called calcite. Figure 3.4(a) shows the crystal structure of calcite. As we have seen in Section 2.2.3, the refractive index (or dielectric constant) in the UV/visible region is determined from the electric polarization of materials. Accordingly, if electron concentration in a material is distributed inhomogeneously toward a specific direction, the refractive index also varies according to its direction and the material shows optical anisotropy. As shown in Fig. 3.4(a), calcite has a multilayer structure in which Ca and CO_3 layers are formed alternately. Thus, the electron concentration of CO_3 is quite high in the plane parallel to the paper and dielectric polarization occurs more easily in this plane. As shown in Eq. (2.44), high dielectric polarization leads to an increase in refractive index. Consequently, the refractive index for the electric field parallel to the paper (E_o) becomes larger than that for the electric field perpendicular to the paper (E_e). If the refractive indices for E_o and E_e are n_o and n_e, respectively, we observe $n_o > n_e$. In particular, E_o and E_e are known as the ordinary ray and extraordinary ray, respectively. As confirmed from Fig. 3.4(a), calcite shows optical anisotropy only in the plane perpendicular to the paper. Such a crystal is known as a uniaxial crystal. Moreover, the direction of E_e in Fig. 3.4(a) is called the optical axis. Quartz also shows optical anisotropy, but $n_o < n_e$ in quartz. In general, optical anisotropy resulting from refractive index differences is referred to as birefringence. Optical anisotropy will be reviewed in Chapter 6 in more detail.

Figure 3.4(b) shows the propagation of the ordinary and extraordinary rays in calcite. As shown in Fig. 2.5(a), since the wavelength of light in a medium is given by λ/n, the wavelengths for the ordinary and extraordinary rays become λ/n_o and λ/n_e, respectively ($\lambda/n_o < \lambda/n_e$). Recall that the speed of light in a transparent medium is expressed by $s = c/n$ [Eq. (2.28)]. Thus, in calcite ($n_o > n_e$), the speed of the extraordinary ray is faster than that of the ordinary ray. In the case of calcite, the oscillatory direction of the extraordinary ray is called the fast axis since the wave propagation is faster in this direction, while the direction of the ordinary ray is called the slow axis. In quartz, however, the direction of the ordinary ray becomes the fast axis since $n_o < n_e$. At normal incidence, the light propagates as illustrated in Fig. 3.4(b). At oblique incidence, however, the ordinary and extraordinary rays

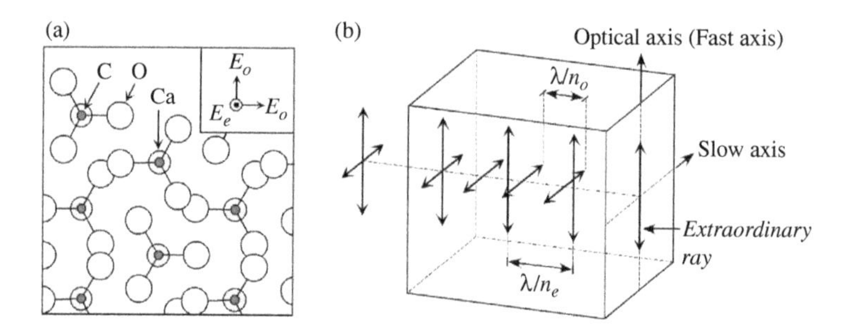

Figure 3.4 (a) Crystal structure of calcite ($CaCO_3$) looking down the optical axis and (b) propagation of light waves in calcite. In (a), E_o and E_e show the electric fields parallel to the paper and perpendicular to the paper, respectively. The refractive indices for E_o and E_e are n_o and n_e, respectively.

propagate at different transmission angles due to the difference in the refractive index (see Section 6.1.1).

Figure 3.5 shows the structure of a polarizer (analyzer) known as a Glan–Taylor prism. A Glan–Taylor prism consists of two prisms and extracts only linearly polarized light from unpolarized light. In Fig. 3.5(a), only the light whose direction is parallel to the x axis passes though the polarizer. Thus, the x axis of this polarizer is called the transmission axis. In ellipsometry measurement, Glan–Taylor prisms made from calcite have been used widely. In a Glan–Taylor prism, linear polarization is extracted by utilizing the total reflection (see Section 2.3.5). Figure 3.5(b) shows the top view of Fig. 3.5(a). In this polarizer, the optical axes of two calcite prisms are aligned in parallel with the paper and the ambient between these prisms is air. Now consider that incident light propagates in the first prism. When the light propagating in this prism is reflected at the prism/air interface, the critical angle for the total reflection is given by $\sin\theta_c = 1/n_i$ [Eq. (2.74)]. In calcite, however, θ_c varies depending on the direction of electric fields due to the anisotropy of the refractive index. As mentioned earlier, θ_c increases with reducing n_i and, in Fig. 3.5(b), θ_c becomes larger in the direction of E_e (refractive index n_e). Thus, if

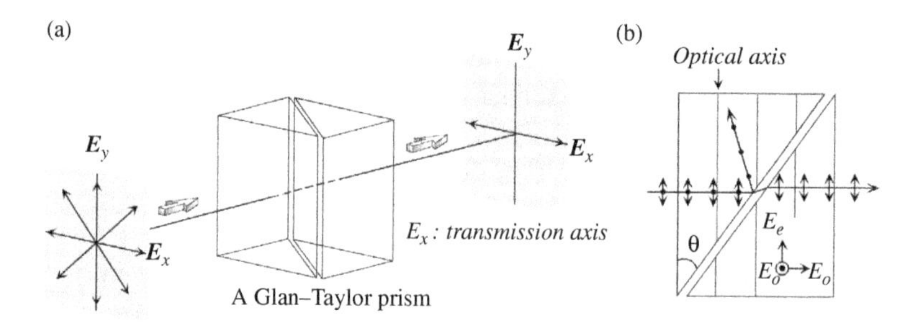

Figure 3.5 (a) A Glan–Taylor prism used as a polarizer (analyzer) and (b) its structure.

we adjust the incidence angle θ shown in Fig. 3.5(b), it becomes possible to remove the ordinary ray by total reflection and to extract only the extraordinary ray. It can be seen from Eq. (2.74) that this condition is given by $1/n_o < \sin\theta < 1/n_e$. Since the refractive indices of calcite at a photon energy of En = 2.1 eV are $n_o = 1.6584$ and $n_e = 1.4864$ [1], only the extraordinary ray transmits at incidence angles of $37.1° < \theta < 42.3°$. The second calcite prism is installed so that the extraordinary ray emerging from the second prism becomes parallel with the incident light. The performance of polarizers (analyzers) is represented by extinction ratio and, in the configuration of Fig. 3.5(a), the extinction ratio is defined as

$$\kappa \equiv |E_{x0}|^2/|E_{y0}|^2 = I_x/I_y \tag{3.2}$$

where I_x and I_y denote light intensities for the x and y directions. In a Glan–Taylor prism using calcite, a high performance of $\kappa \sim 10^5$ can be obtained. Since the light transmission region is $\lambda = 0.21$–$5\,\mu$m in calcite, a Glan–Taylor prism can also be applied in this region.

Figure 3.6 shows the structures of (a) a Glan–Thompson prism and (b) a Rochon prism. A Glan–Thompson prism has a structure similar to a Glan–Taylor prsim, but the directions of the optical axes are different. In a Glan–Thompson prism, two calcite prisms are cemented together using optically transparent glue. One drawback of this polarizer is low light transmittance in the UV region due to light absorption in the glue. Accordingly, this polarizer has not been used in spectroscopic ellipsometry instruments. In addition, since the refractive index of the glue is $n > 1$, θ in a Glan–Thompson prism becomes larger than that in a Glan–Taylor prism.

Although calcite is used for Glan-Thompson prisms, quartz and MgF_2 are employed for Rochon prisms. As shown in Fig. 3.6(b), a Rochon prism is composed of two prisms whose optical axes are orthogonal. In a Rochon prism, the light oscillating parallel to the paper (E_o) transmits the two prisms without any disturbance since the refractive indices of the first and second prisms are the same. However, with respect to the light oscillating perpendicular to the paper, the

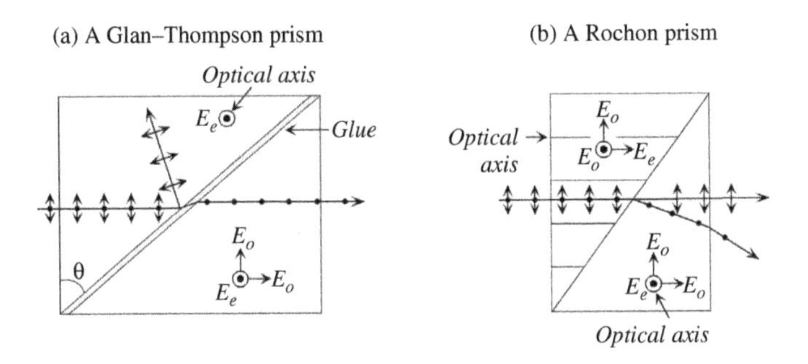

Figure 3.6 Structures of (a) a Glan–Thompson prism and (b) a Rochon prism used as a polarizer (analyzer).

refractive indices of the first and second prisms are different ($n_o \neq n_e$). In particular, since $n_o < n_e$ in quartz ($n_o = 1.5443$, $n_e = 1.5534$ at En = 2.1 eV [1]), the light refracts toward the direction shown in Fig. 3.6(b). In a Rochon prism, the light path does not deviate even when the prism is rotated since the light path is completely straight in this prism. This is quite advantageous for ellipsometry instruments that use rotating optical elements. However, quartz has a character known as optical activity [1,2]. Optical activity is the phenomenon that the oscillating plane of linearly polarized light undergoes a continuous rotation as the light propagates along an optical axis. When a quartz Rochon prism is employed for ellipsometry measurement, calibration for the optical activity is required [3–5]. On the other hand, MgF_2 shows no optical activity and has superior transmittance characteristics in the UV region. In recent years, Rochon prisms made from MgF_2 have been applied in spectroscopic ellipsometry instruments allowing measurement up to the deep UV region (~6 eV) [6,7].

The polarizer (analyzer) mentioned above cannot be employed in the infrared region ($\lambda \sim 10\,\mu$m or $\sim 1000\,cm^{-1}$) due to low light transmittance in that region. In infrared spectroscopic ellipsometry, therefore, a wire-grid polarizer shown in Fig. 3.7 has been used in the region around 100–4000 cm^{-1} [8,9]. In this polarizer, narrow metal wires are formed on a substrate that has no light absorption in the infrared region. In some wire-grid polarizers, very narrow metal wires are formed by applying the photolithography technique. Now suppose that light waves oscillating parallel to the x and y axes advance into this polarizer at normal incidence. When the incident wave in the y direction impinges on the metal wires, the electric field of the light moves electrons in the wires along the y direction. This leads to joule heating and the light is absorbed by the metal wires. In addition, light waves that are not absorbed will be reflected by the metal wires. On the other hand, when the light in the x direction enters into the wire-grid polarizer, the traveling distance of electrons inside the metal wires is limited, since the oscillatory direction of the light is perpendicular to the direction of the metal wires. Consequently, the light absorption for the x component becomes smaller. In the wire-grid polarizer shown in Fig. 3.7, therefore, only the light oscillating in the x direction transmits

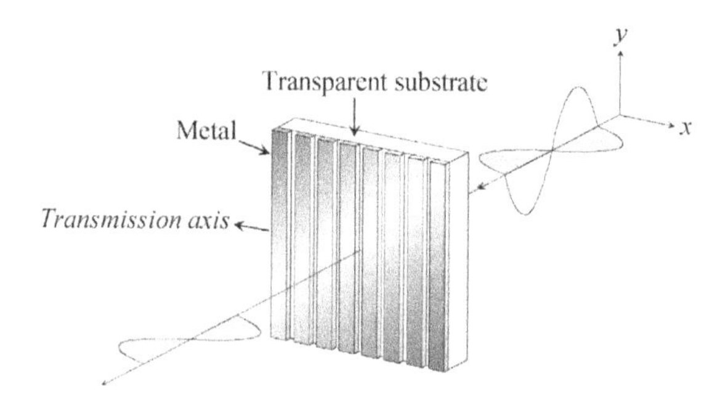

Figure 3.7 Structure of a wire-grid polarizer.

though the polarizer. Notice that the transmission axis of wire-grid polarizers is perpendicular to the direction of the metal wires. Such optical anisotropy for light absorption (or extinction coefficient) is referred to as dichroism and is distinguished from birefringence, which represents optical anisotropy for refractive index. The extinction ratio of a wire-grid polarizer is typically $\kappa \sim 10^3$ and its performance is relatively lower than that of a Glan–Taylor polarizer.

3.2.2 COMPENSATOR (RETARDER)

A compensator (or retarder) is generally placed behind a polarizer or in front of an analyzer and is often employed to convert linear polarization to circular polarization and vice versa. A compensator also utilizes optical anisotropy for refractive index and is composed of a birefringent crystal only. Figure 3.8 shows the propagation of light in a compensator. In particular, this figure illustrates the case when the linear polarization oriented at 45° is converted to left-circular polarization. This figure corresponds to Fig. 3.4(b) in which the light wave was represented by arrows. As mentioned earlier, the light wave parallel to the fast axis propagates faster than that parallel to the slow axis. Therefore, the compensator generates a phase difference between the electric field vectors $\boldsymbol{E}_x$ and $\boldsymbol{E}_y$. If we transform Eq. (2.30), this phase difference is given by

$$\delta = \frac{2\pi}{\lambda} |n_e - n_o| d \tag{3.3}$$

where d denotes the thickness of the compensator. It can be seen from Eq. (3.3) that δ varies depending on λ. In Fig. 3.8, the phase difference generated by the compensator is $\delta = \pi/2$. In particular, when a phase difference corresponds to a wavelength of $\lambda/4$, as in the case shown in Fig. 3.8, the compensator is also called a quarter-wave plate. Since the fast axis is parallel to the x axis in Fig. 3.8,

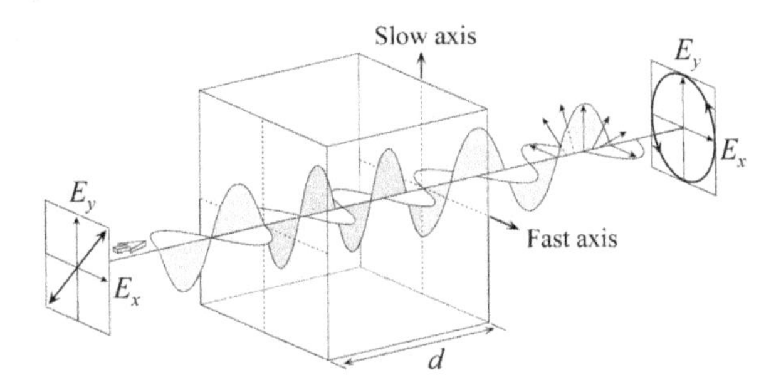

Figure 3.8 Change in the state of polarization by a compensator (retarder). In this figure, d denotes the thickness of the compensator.

the circular polarization represents the one shown in Fig. 3.3 at $\delta_x - \delta_y = \pi/2$. Traditionally, the optical element shown in Fig. 3.8 has been called a retarder and an optical element that allows control of phase shift has been called a compensator. However, if the retarder in Fig. 3.8 is rotated in the x–y plane, the relative phase difference varies and thus phase control is possible. In this case, it is referred to as a rotating compensator. Recently, such a rotating compensator has been applied widely to spectroscopic ellipsometry, and the distinction between a retarder and a compensator has become rather ambiguous.

For compensators, calcite described in the previous section is rarely used. This originates from the fact that, since the value of $|n_e - n_o|$ is quite large in calcite, the thickness required for making a compensator becomes too thin. Thus, in spectroscopic ellipsometry measurement, compensators made from MgF_2 [10] and mica [11,12] have been employed. In particular, MgF_2 (transmission wavelength $>0.12\,\mu m$) shows superior light transmittance in the UV region, compared with mica (transmission wavelength $>0.29\,\mu m$). In recent years, therefore, MgF_2 compensators have been used widely.

3.2.3 PHOTOELASTIC MODULATOR

When stress is applied to an optically isotropic material, the electron density of the material varies in the direction of the stress and, consequently, the material shows optically anisotropic character. This phenomenon is known as photoelasticity. In particular, the birefringence induced by photoelasticity is proportional to the stress applied and the direction of an optical axis coincides with that of the stress. A photoelastic modulator is a different type of compensator that utilizes the photoelasticity. Figure 3.9 shows structures of photoelastic modulators. The photoelastic modulator shown in Fig. 3.9(a) was developed first by Jasperson *et al.* in 1969 [13] and Fig. 3.9(b) represents a modulator that was modified later [14]. In recent spectroscopic ellipsometry instruments that employ a photoelastic modulator, the photoelastic element shown in Fig. 3.9(b) is mainly used.

The modulator shown in Fig. 3.9(a) has a structure in which quartz and fused quartz are cemented together, and the modulator is supported by knife edges. The quartz crystal in Fig. 3.9(a) is used as a piezo transducer. When a 50 kHz electric field, corresponding to the resonant frequency of the piezo transducer, is applied to an electrode pair formed on the crystal quartz, a periodic stress is applied to the fused quartz. If linear light oriented at 45° enters the photoelastic modulator, as shown in Fig. 3.9(a), a phase difference δ relative to the x and y directions will be generated by the photoelasticity. In photoelastic modulators, δ varies continuously with time and is expressed as follows [13]:

$$\delta = F \sin(\omega t) \tag{3.4}$$

where $\omega = 2\pi\nu$ and $\nu = 50\,\text{kHz}$. In Eq. (3.4), F shows the phase amplitude and is proportional to V/λ, where V and λ represent the voltage applied to the piezo

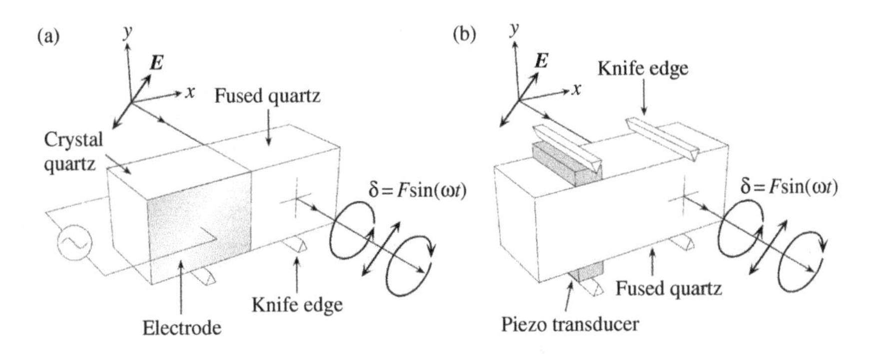

Figure 3.9 Structures of photoelastic modulators. In (a), the structure of the photoelastic modulator developed first is shown, while a modified structure of (a) is shown in (b). Structure (a): Reprinted with permission from *Review of Scientific Instruments*, **40**, S. N. Jasperson and S. E. Schnatterly, An improved method for high reflectivity ellipsometry based on a new polarization modulation technique, 761–767 (1969). Copyright 1969, American Institute of Physics. Structure (b): From *Applied Optics*, **22**, J. C. Canit and J. Badoz, New design for a photoelastic modulator, 592–594 (1983). Reproduced with permission of the Optical Society of America.

transducer and the wavelength of incident light, respectively [13]. Thus, in order to keep δ constant for different wavelengths, the applied voltage has to be adjusted.

In the photoelastic modulator shown in Fig. 3.9(b), a pair of piezo transducers are attached to the fused quartz and the structure of this device is simpler, compared with the one shown in Fig. 3.9(a). One transducer in Fig. 3.9(b) is employed to apply stress to the fused quartz and another one is used to monitor the stress. The phase difference generated by this modulator is also expressed by Eq. (3.4) and the resonant frequency is also 50 kHz. Since both of the photoelastic modulators shown in Fig. 3.9 are highly sensitive to temperature variation, a precise temperature adjustment is necessary for accurate control of δ.

3.2.4 DEPOLARIZER

A depolarizer is used when we convert polarized light to unpolarized light. Although spectroscopic ellipsometry requires a light source for the measurement, the light emitted from the light source is not perfectly random (unpolarized) and has slightly polarized components. This phenomenon is referred to as source polarization [5]. A depolarizer is often employed to eliminate such source polarization. In addition, diffraction efficiencies of grating spectrometers generally show polarization dependence [15] and sensitivity of light detectors may also change according to the state of polarization [4]. If we install a depolarizer in front of a spectrometer or light detector, such polarization dependence can be eliminated.

Figure 3.10 shows the structure of a depolarizer made from a birefringent crystal. This depolarizer has a wedge shape structure and light transmits though

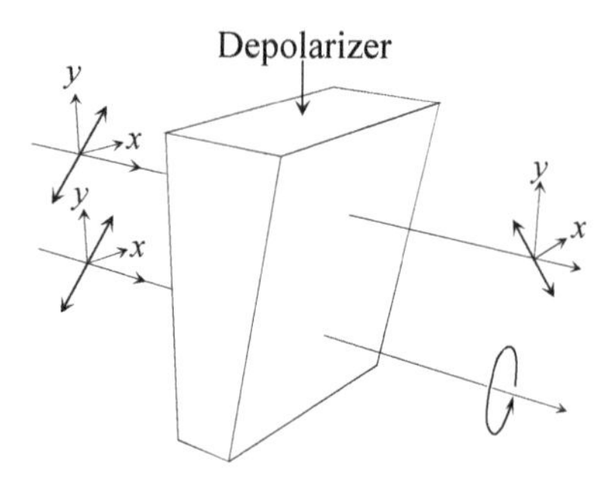

Figure 3.10 Structure of a depolarizer.

the depolarizer at different thicknesses. As shown in Eq. (3.3), a phase difference induced by a birefringent crystal changes with its thickness. Thus, this depolarizer introduces a phase difference that varies continuously along the vertical direction of Fig. 3.10. As a result, the light emerging from the depolarizer becomes unpolarized light. There is another type of a depolarizer called a Cornu prism in which the optical activity of quartz is utilized [16].

3.3 JONES MATRIX

A matrix representation called the Jones matrix allows the mathematical description of optical measurements. If we apply the Jones matrix, we can express variations in polarized light from matrix calculation, even when there are many optical elements in a measurement. The Jones matrix is also utilized when we describe ellipsometry measurement mathematically. On the other hand, the Jones vector is used when we express states of polarization including linear and elliptical polarizations. In this section, we will address the Jones vector and matrix, which provide a basis for the understanding of ellipsometry measurement.

3.3.1 JONES VECTOR

As mentioned earlier, the polarization state of light is represented by superimposing two waves oscillating parallel to the x and y axes (see Section 3.1.2). The Jones vector is defined by the electric field vectors in the x and y directions [17]. If we use Eq. (3.1), the Jones vector is given by

$$\boldsymbol{E}(z, t) = \begin{bmatrix} E_{x0} \exp\{\mathrm{i}(\omega t - Kz + \delta_x)\} \\ E_{y0} \exp\{\mathrm{i}(\omega t - Kz + \delta_y)\} \end{bmatrix} = \exp\{\mathrm{i}(\omega t - Kz)\} \begin{bmatrix} E_{x0} \exp(\mathrm{i}\delta_x) \\ E_{y0} \exp(\mathrm{i}\delta_y) \end{bmatrix} \quad (3.5)$$

In general, the above equation is expressed by omitting the term $\exp\{i(\omega t - Kz)\}$:

$$\boldsymbol{E}(z, t) = \begin{bmatrix} E_{x0}\exp(i\delta_x) \\ E_{y0}\exp(i\delta_y) \end{bmatrix} \tag{3.6}$$

This equation can be simplified further to

$$\boldsymbol{E}(z, t) = \begin{bmatrix} E_x \\ E_y \end{bmatrix} \tag{3.7}$$

where

$$E_x = E_{x0}\exp(i\delta_x) = |E_x|\exp(i\delta_x) \tag{3.8a}$$

$$E_y = E_{y0}\exp(i\delta_y) = |E_y|\exp(i\delta_y) \tag{3.8b}$$

In Eq. (3.8), the transformation given by Eq. (2.18) was used, and we assume positive values for E_{x0} and E_{y0}. If we use the phase difference $(\delta_x - \delta_y)$, Eq. (3.8) can be rewritten as

$$E_x = E_{x0}\exp[i(\delta_x - \delta_y)] = |E_x|\exp[i(\delta_x - \delta_y)] \tag{3.9a}$$

$$E_y = E_{y0} = |E_y| \tag{3.9b}$$

From Eq. (2.33), light intensity is given by

$$I = I_x + I_y = E_{x0}^2 + E_{y0}^2 = |E_x|^2 + |E_y|^2 = E_x E_x^* + E_y E_y^* \tag{3.10}$$

In conventional optical measurements including ellipsometry, only relative changes in amplitude and phase are taken into account. Accordingly, the Jones vector is generally expressed by the normalized light intensity ($I = 1$). In this case, linearly polarized waves parallel to the x and y directions are expressed by

$$\boldsymbol{E}_{\text{linear},x} = \begin{bmatrix} 1 \\ 0 \end{bmatrix} \quad \boldsymbol{E}_{\text{linear},y} = \begin{bmatrix} 0 \\ 1 \end{bmatrix} \tag{3.11}$$

If we normalize light intensity, linearly polarized light oriented at 45° is written as

$$\boldsymbol{E}_{+45^\circ} = \frac{1}{\sqrt{2}}\begin{bmatrix} 1 \\ 1 \end{bmatrix} \tag{3.12}$$

On the other hand, right-circular polarization ($\boldsymbol{E}_\text{R}$) and left-circular polarization ($\boldsymbol{E}_\text{L}$) are given by

$$\boldsymbol{E}_\text{R} = \frac{1}{\sqrt{2}}\begin{bmatrix} 1 \\ i \end{bmatrix} \quad \boldsymbol{E}_\text{L} = \frac{1}{\sqrt{2}}\begin{bmatrix} 1 \\ -i \end{bmatrix} \tag{3.13}$$

The above $\boldsymbol{E}_{\mathrm{R}}$, for example, can be obtained by simply substituting $\delta_x = 0$, $\delta_y = \pi/2$ and $E_{x0} = E_{y0} = 1$ into Eq. (3.6). In this case, from Euler's formula [Eq. (2.17)], we obtain $\exp(\mathrm{i}\pi/2) = \mathrm{i}$. Recall from Fig. 3.2 that $\delta_y - \delta_x = \pi/2$ in $\boldsymbol{E}_{\mathrm{R}}$. It should be noted that, when the phase of electromagnetic waves is expressed by $(Kz - \omega t + \delta)$, the signs of i in Eq. (3.13) should be reversed (see Appendix 2). If we assume that $\delta_x = \pi/4$, $\delta_y = \pi/2$, and $E_{x0} = E_{y0} = 1$, we can express the elliptical polarization by

$$\boldsymbol{E}_{\mathrm{elli}} = \frac{1}{\sqrt{2}}\begin{bmatrix} 0.707 + \mathrm{i}0.707 \\ \mathrm{i} \end{bmatrix} \tag{3.14}$$

This polarization corresponds to the one shown in Fig. 3.3 at $\delta_x - \delta_y = 7\pi/4$. Table 3.1 summarizes the Jones vectors for various polarization states. The elliptical polarization and the Stokes vectors in Table 3.1 will be explained in Section 3.4.

3.3.2 TRANSFORMATION OF COORDINATE SYSTEMS

In ellipsometry measurement, a polarizer or compensator is generally installed with a certain rotation angle relative to the x or y axis. In this case, if we rotate the $x - y$ coordinate system itself using mathematical transformation, we can simplify equations, as will be shown in Section 3.3.4. Figure 3.11 shows the transformation of the $x - y$ coordinates into the $x' - y'$ coordinates by coordinate rotation. In this figure, we assume that the positive direction for the rotation of the $x' - y'$ coordinate system is counterclockwise. It can be seen from Fig. 3.11 that, in the $x' - y'$ coordinate system, point P (E_x, E_y) is represented by

$$E_{x'} = E_x \cos\alpha + E_y \sin\alpha \tag{3.15a}$$

$$E_{y'} = -E_x \sin\alpha + E_y \cos\alpha \tag{3.15b}$$

where α is the rotation angle of the $x' - y'$ coordinate system. In matrix form, Eq. (3.15) is expressed as

$$\begin{bmatrix} E_{x'} \\ E_{y'} \end{bmatrix} = \begin{bmatrix} \cos\alpha & \sin\alpha \\ -\sin\alpha & \cos\alpha \end{bmatrix}\begin{bmatrix} E_x \\ E_y \end{bmatrix} \tag{3.16}$$

From Eq. (3.16), we obtain a matrix that represents the coordinate rotation:

$$\boldsymbol{R}(\alpha) = \begin{bmatrix} \cos\alpha & \sin\alpha \\ -\sin\alpha & \cos\alpha \end{bmatrix} \quad \boldsymbol{R}(-\alpha) = \begin{bmatrix} \cos\alpha & -\sin\alpha \\ \sin\alpha & \cos\alpha \end{bmatrix} \tag{3.17}$$

The above $\boldsymbol{R}(-\alpha)$ corresponds to the matrix when the coordinate system is rotated clockwise and is obtained easily from $\boldsymbol{R}(\alpha)$ using the relation $\sin(-A) = -\sin(A)$. Now suppose that (E_x, E_y) in Fig. 3.11 is the Jones vector. In this case, the Jones

Table 3.1 Representations of the states of polarization by the Jones and Stokes vectors

Polarization	Polarization state	Jones vector	Stokes vector
Linear polarization parallel to x axis	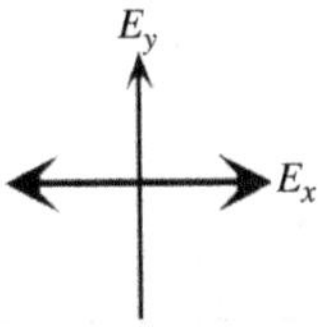	$\begin{bmatrix}1\\0\end{bmatrix}$	$\begin{bmatrix}1\\1\\0\\0\end{bmatrix}$
Linear polarization parallel to y axis	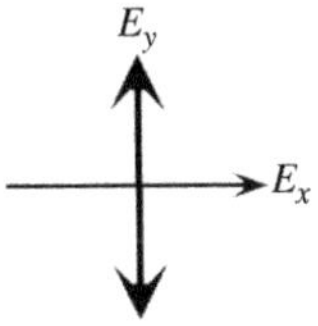	$\begin{bmatrix}0\\1\end{bmatrix}$	$\begin{bmatrix}1\\-1\\0\\0\end{bmatrix}$
Linear polarization oriented at 45°	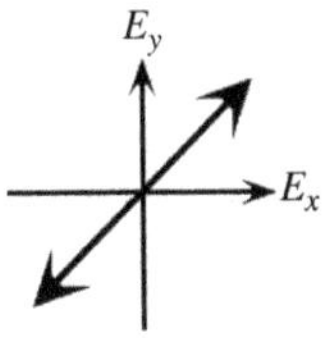	$\frac{1}{\sqrt{2}}\begin{bmatrix}1\\1\end{bmatrix}$	$\begin{bmatrix}1\\0\\1\\0\end{bmatrix}$
Right-circular polarization	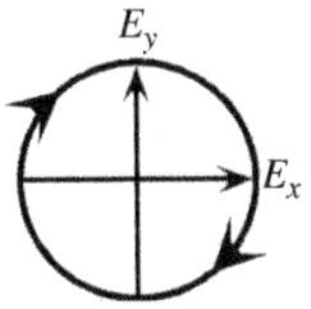	$\frac{1}{\sqrt{2}}\begin{bmatrix}1\\i\end{bmatrix}$	$\begin{bmatrix}1\\0\\0\\1\end{bmatrix}$
Left-circular polarization	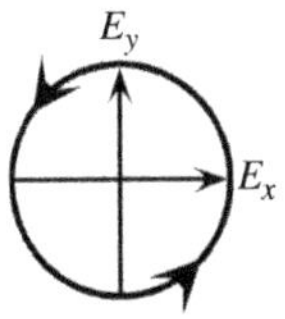	$\frac{1}{\sqrt{2}}\begin{bmatrix}1\\-i\end{bmatrix}$	$\begin{bmatrix}1\\0\\0\\-1\end{bmatrix}$
Elliptical polarization	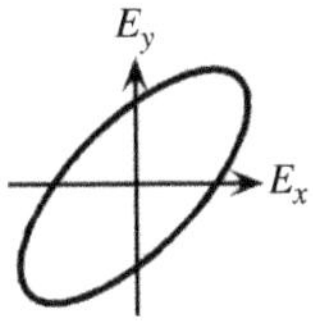	$\begin{bmatrix}\sin\psi\exp(i\Delta)\\\cos\psi\end{bmatrix}$	$\begin{bmatrix}1\\-\cos 2\psi\\\sin 2\psi\cos\Delta\\-\sin 2\psi\sin\Delta\end{bmatrix}$
Natural light (unpolarized light)			$\begin{bmatrix}1\\0\\0\\0\end{bmatrix}$

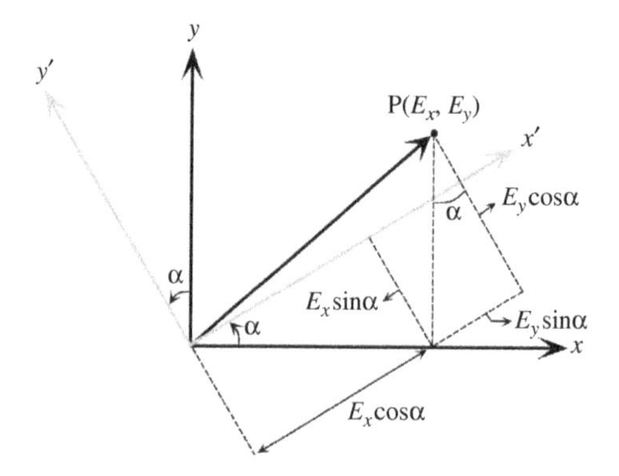

Figure 3.11 Transformation of the $x-y$ coordinates into the $x'-y'$ coordinates by coordinate rotation.

vector (E_x, E_y) is transformed into ($E_{x'}$, $E_{y'}$) by coordinate rotation. Such a 2×2 matrix that transforms a Jones vector is referred to as a Jones matrix [17]. When we perform rotations of two coordinate systems with different rotation angles of α and −β, we get

$$\boldsymbol{R}(\alpha)\boldsymbol{R}(-\beta) = \boldsymbol{R}(\alpha - \beta) \tag{3.18}$$

Eq. (3.18) can be obtained easily by applying the addition theorem (see Appendix 1).

The states of polarization can also be expressed from the vector sum of left- and right-circular polarizations. Figure 3.12 shows the left- and right-circular polarizations expressed using polar coordinates. As shown in Fig. 3.12, the left-circular polarization E_L and right-circular polarization E_R are given by

$$E_L = |E_L| \exp(i\delta_L) \quad E_R = |E_R| \exp(i\delta_R) \tag{3.19}$$

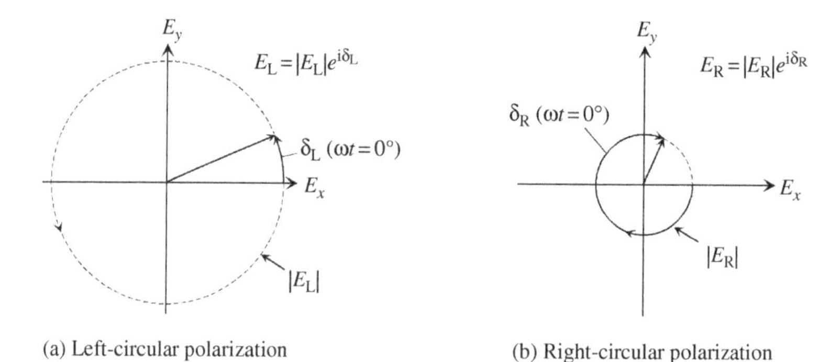

Figure 3.12 Representations of (a) left-circular polarization and (b) right-circular polarization expressed using polar coordinates.

where $|E|$ and δ represent the size of the circle and phase when $\omega t = 0°$, respectively. In this case, the Jones vector is expressed by

$$\boldsymbol{E} = \begin{bmatrix} E_{\mathrm{L}} \\ E_{\mathrm{R}} \end{bmatrix} \tag{3.20}$$

The polar coordinate system used to express these polarizations can be transformed into the Cartesian coordinate system shown in Fig. 3.11 by the following equation [2]:

$$\begin{bmatrix} E_x \\ E_y \end{bmatrix} = \frac{1}{\sqrt{2}} \begin{bmatrix} 1 & 1 \\ -\mathrm{i} & \mathrm{i} \end{bmatrix} \begin{bmatrix} E_{\mathrm{L}} \\ E_{\mathrm{R}} \end{bmatrix} \tag{3.21}$$

It is clear from Eq. (3.21) that the coordinate transformation from polar coordinates to Cartesian coordinates is written as

$$\boldsymbol{T} = \frac{1}{\sqrt{2}} \begin{bmatrix} 1 & 1 \\ -\mathrm{i} & \mathrm{i} \end{bmatrix} \tag{3.22}$$

The above matrix can be obtained easily by combining the Jones vectors of E_{L} and E_{R} shown in Table 3.1, since a matrix basically represents the projection of a vector onto a new coordinate system. By substituting $E_{\mathrm{L}} = 1$ and $E_{\mathrm{R}} = 0$ into Eq. (3.22), we obtain

$$\begin{bmatrix} E_x \\ E_y \end{bmatrix} = \frac{1}{\sqrt{2}} \begin{bmatrix} 1 \\ -\mathrm{i} \end{bmatrix} \tag{3.23}$$

Naturally, Eq. (3.23) shows left-circular polarization in the Cartesian coordinate system. If we use the inverse transformation matrix $\boldsymbol{T}^{-1}$, defined by $\boldsymbol{T}\boldsymbol{T}^{-1} = 1$, the Cartesian coordinates are transformed into the original polar coordinates as follows:

$$\begin{bmatrix} E_{\mathrm{L}} \\ E_{\mathrm{R}} \end{bmatrix} = \frac{1}{\sqrt{2}} \begin{bmatrix} 1 & \mathrm{i} \\ 1 & -\mathrm{i} \end{bmatrix} \begin{bmatrix} E_x \\ E_y \end{bmatrix} \tag{3.24}$$

Figure 3.13 shows the representation of the elliptical polarization by the superposition of left- and right-circular polarizations. The example shown in Fig. 3.13 can be obtained by substituting $E_x = 1$ and $E_y = \exp(-\mathrm{i}\pi/4)(\delta_x - \delta_y = \pi/4)$ into Eq. (3.24). As shown in Eq. (3.21), when we try to match the polar coordinates with the Cartesian coordinates, there is a need to multiply the amplitudes by $1/\sqrt{2}$. In Fig. 3.13, therefore, the amplitudes of the circular polarizations are illustrated using a factor of $1/\sqrt{2}$. As shown in Fig. 3.13(a), the synthesized vector locates at $(E_x, E_y) = (1, 0.71)$ when $\omega t = 0°$. Each circular polarization shown in this figure has been indicated in Fig. 3.12. Figure 3.13(b) shows the state of polarization when each circular polarization shown in Fig. 3.13(a) rotates by 90°. It is confirmed from Fig. 3.13 that elliptical polarization can be described as the vector sum of left- and right-circular polarizations. The linear polarization parallel to the x axis, for example, is also expressed by simply assuming $E_{\mathrm{L}} = E_{\mathrm{R}}$.

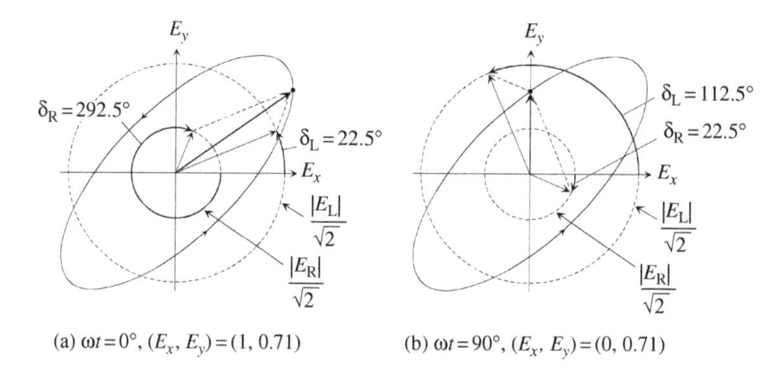

Figure 3.13 Representation of elliptical polarization by the superposition of left- and right-circular polarizations: (a) $\omega t = 0°$ and (b) $\omega t = 90°$.

3.3.3 JONES MATRICES OF OPTICAL ELEMENTS

The transformation of polarization states by optical elements can also be described from the Jones matrix. The Jones matrices for a polarizer (**P**) and an analyzer (**A**) are expressed by

$$\boldsymbol{P} = \boldsymbol{A} = \begin{bmatrix} 1 & 0 \\ 0 & 0 \end{bmatrix} \tag{3.25}$$

The above matrix represents the case when the transmission axis is parallel to the x axis. As shown in Eq. (3.16), the calculation of the Jones matrix is performed by multiplying the Jones vector (incident light) by the Jones matrix (an optical element) from the left side. For example, if we assume that linear polarization oriented at 45° enters a polarizer, the light emerging from the polarizer is calculated from

$$\begin{bmatrix} E_x \\ E_y \end{bmatrix} = \frac{1}{\sqrt{2}} \begin{bmatrix} 1 & 0 \\ 0 & 0 \end{bmatrix} \begin{bmatrix} 1 \\ 1 \end{bmatrix} = \frac{1}{\sqrt{2}} \begin{bmatrix} 1 \\ 0 \end{bmatrix} \tag{3.26}$$

It can be seen from Eq. (3.26) that $E_y = 0$ since the polarizer transmits only the light whose direction is parallel to the x axis. Furthermore, the total light intensity ($I = |E_x|^2$) becomes 1/2 after the light passes though the polarizer. Accordingly, if we apply the Jones matrix, the final state of polarization can be determined easily from matrix calculation. When there are many optical elements in an optical system, we can introduce additional Jones matrices sequentially to the left side of the Jones vector.

On the other hand, the Jones matrix corresponding to a compensator (retarder) is expressed by

$$\boldsymbol{C} = \begin{bmatrix} 1 & 0 \\ 0 & \exp(-\mathrm{i}\delta) \end{bmatrix} \tag{3.27}$$

Table 3.2 Jones and Mueller matrices for optical elements and coordinate rotation

Optical element	Jones matrix	Mueller matrix
Polarizer[a] (Analyzer) ***P(A)***	$\begin{bmatrix} 1 & 0 \\ 0 & 0 \end{bmatrix}$	$\frac{1}{2}\begin{bmatrix} 1 & 1 & 0 & 0 \\ 1 & 1 & 0 & 0 \\ 0 & 0 & 0 & 0 \\ 0 & 0 & 0 & 0 \end{bmatrix}$
Compensator[b] (Retarder) ***C***	$\begin{bmatrix} 1 & 0 \\ 0 & \exp(-i\delta) \end{bmatrix}$	$\begin{bmatrix} 1 & 0 & 0 & 0 \\ 0 & 1 & 0 & 0 \\ 0 & 0 & \cos\delta & \sin\delta \\ 0 & 0 & -\sin\delta & \cos\delta \end{bmatrix}$
Photoelastic modulator[c] ***M***	$\begin{bmatrix} 1 & 0 \\ 0 & \exp(i\delta) \end{bmatrix}$	$\begin{bmatrix} 1 & 0 & 0 & 0 \\ 0 & 1 & 0 & 0 \\ 0 & 0 & \cos\delta & -\sin\delta \\ 0 & 0 & \sin\delta & \cos\delta \end{bmatrix}$
Coordinate rotation[d] ***R(α)***	$\begin{bmatrix} \cos\alpha & \sin\alpha \\ -\sin\alpha & \cos\alpha \end{bmatrix}$	$\begin{bmatrix} 1 & 0 & 0 & 0 \\ 0 & \cos 2\alpha & \sin 2\alpha & 0 \\ 0 & -\sin 2\alpha & \cos 2\alpha & 0 \\ 0 & 0 & 0 & 1 \end{bmatrix}$
Sample[e] ***S***	$\begin{bmatrix} \sin\psi\exp(i\Delta) & 0 \\ 0 & \cos\psi \end{bmatrix}$	$A\begin{bmatrix} 1 & -\cos 2\psi & 0 & 0 \\ -\cos 2\psi & 1 & 0 & 0 \\ 0 & 0 & \sin 2\psi\cos\Delta & \sin 2\psi\sin\Delta \\ 0 & 0 & -\sin 2\psi\sin\Delta & \sin 2\psi\cos\Delta \end{bmatrix}$
Depolarizer ***D***		$\begin{bmatrix} 1 & 0 & 0 & 0 \\ 0 & 0 & 0 & 0 \\ 0 & 0 & 0 & 0 \\ 0 & 0 & 0 & 0 \end{bmatrix}$

[a] Transmission axis is parallel to the x axis
[b] fast axis is parallel to the x axis and δ is given by Eq. (3.3)
[c] δ is given by Eq. (3.4)
[d] coordinate rotation is counterclockwise (see Fig. 3.11)
[e] $A = (r_p r_p^* + r_s r_s^*)/2$.

The above equation represents a compensator whose fast axis is parallel to the x axis. In Eq. (3.27), δ shows the phase difference given by Eq. (3.3). Notice that the value of δ is always positive since we expressed the refractive index difference using $|n_e - n_o|$ in Eq. (3.3). As we have seen in Section 3.1.1, when δ values are positive, light waves advance forward. Eq. (3.27) shows that light oscillating in the x direction (fast axis) propagates without any change and light oscillating in the y direction (slow axis) lags with a relative phase difference of $-\delta$. Using Eq. (3.27), we can express the conversion of polarized light by the compensator shown in Fig. 3.8 as follows:

$$\begin{bmatrix} E_x \\ E_y \end{bmatrix} = \frac{1}{\sqrt{2}}\begin{bmatrix} 1 & 0 \\ 0 & \exp(-i\pi/2) \end{bmatrix}\begin{bmatrix} 1 \\ 1 \end{bmatrix} = \frac{1}{\sqrt{2}}\begin{bmatrix} 1 \\ -i \end{bmatrix} \tag{3.28}$$

From this calculation, we can confirm that linear polarization oriented at 45° is transformed into left-circular polarization by the compensator with $\delta = \pi/2$.

Similarly, the Jones matrix for the photoelastic modulator is given by

$$\boldsymbol{M} = \begin{bmatrix} 1 & 0 \\ 0 & \exp(\mathrm{i}\delta) \end{bmatrix} \tag{3.29}$$

where δ shows the phase shift given by Eq. (3.4). Table 3.2 summarizes the Jones matrices corresponding to optical elements and coordinate rotation. The matrix for samples will be treated in Section 4.1.3 and the Mueller matrix will be explained in Section 3.4.4.

3.3.4 REPRESENTATION OF OPTICAL MEASUREMENT BY JONES MATRICES

Figure 3.14 shows a simple optical instrument expressed by Jones matrices. In this instrument, light emitted from a light source is transformed into linear polarization by a polarizer and a light detector measures the intensity of light transmitted though an analyzer. Here, we assume that the transmission axis of the polarizer is rotated by an angle α in the $x - y$ plane and that of the analyzer is parallel to the x direction. In the polarizer placed behind the light source, light transmission occurs only in the direction of the transmission axis. Accordingly, with respect to the light emitted from the light source, only the light transmitted though the polarizer (E_P) is taken into account. If we choose the $x' - y'$ coordinates so that the transmission axis of the polarizer is parallel to the x' axis, the x' axis is rotated by α relative to the x axis. In this case, in order to perform matrix calculation, coordinate rotation from (x', y') to (x, y) is required. The coordinate rotation in Fig. 3.14 is expressed by $\boldsymbol{R}(-\alpha)$ since the coordinate rotation is clockwise, as confirmed from Fig. 3.11. If we apply Jones matrices to this instrument, we get

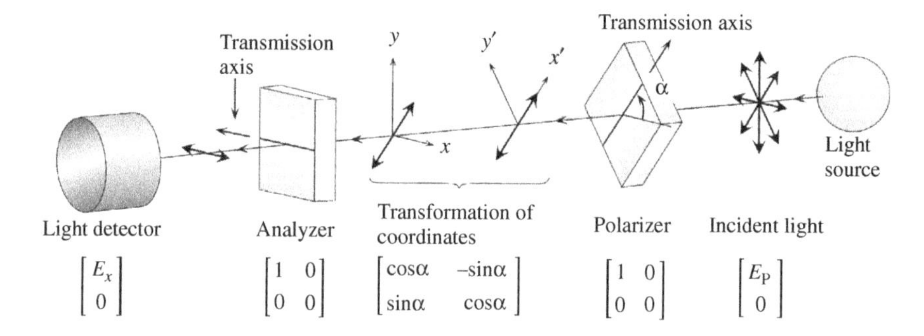

Figure 3.14 Representation of an optical instrument by Jones matrices. This optical instrument is composed of light source/polarizer/analyzer/light detector.

$$\begin{aligned}\begin{bmatrix}E_x\\E_y\end{bmatrix}&=\begin{bmatrix}1&0\\0&0\end{bmatrix}\begin{bmatrix}\cos\alpha&-\sin\alpha\\\sin\alpha&\cos\alpha\end{bmatrix}\begin{bmatrix}1&0\\0&0\end{bmatrix}\begin{bmatrix}E_P\\0\end{bmatrix}\\&=\begin{bmatrix}E_P\cos\alpha\\0\end{bmatrix}\end{aligned}\tag{3.30}$$

It is clear from Eq. (3.30) that $E_x = E_P \cos\alpha$ and $E_y = 0$. By using Eq. (3.10), we obtain the light intensity measured by the light detector:

$$I = |E_x|^2 = |E_P|^2 \cos^2\alpha \tag{3.31}$$

The above result is known as Malus's law. Figure 3.15 shows light intensity plotted as a function of the rotation angle α, obtained from Eq. (3.31). In Fig. 3.15, the light intensity is normalized using $E_P = 1$. As shown in Fig. 3.15, we find that $I = 1$ at $\alpha = 0°$. This implies that light advances without any disturbance when the transmission axes of the polarizer and analyzer are parallel. On the other hand, when $\alpha = 90°$, light intensity becomes zero because the transmission axes of the polarizer and analyzer are perpendicular with each other. Furthermore, with 180° rotation of the polarizer, the light intensity is restored to $I = 1$. This shows the simple fact that 180° rotation of the polarizer is equivalent to 0° rotation, since there is no distinction between the upper and lower sides of the transmission axis. Thus, one complete rotation of the polarizer varies the light intensity with two periods. In this example, the transmission axis of the analyzer was fixed in parallel to the x axis. If

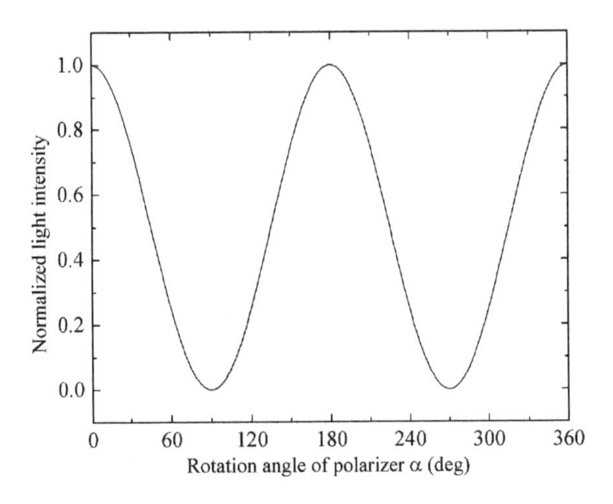

Figure 3.15 Normalized light intensity plotted as a function of the rotation angle α of a polarizer. This result can be obtained by changing the rotation angle of the polarizer in the optical instrument shown in Fig. 3.14.

the analyzer also rotates by an angle of α, we obtain the rotation matrix $\boldsymbol{R}(\alpha-\beta)$ by applying Eq. (3.18). Here, $-\beta$ is the rotation angle of the polarizer (its rotation is clockwise). Accordingly, in the optical configuration shown in Fig. 3.14, the light intensity is determined by the relative angle difference between the polarizer and analyzer $[I=|E_{\mathrm{P}}|^2\cos^2(\alpha-\beta)]$.

3.4 STOKES PARAMETERS

Although the Jones vector provides an elegant method for describing polarized light, unpolarized light cannot be expressed by using the Jones vector. Furthermore, there is a polarization state referred to as partial polarization. In order to describe unpolarized or partially polarized light, Stokes parameters (vectors) are used. The Stokes parameters enable us to describe all types of polarization. In actual ellipsometry measurement, these Stokes parameters are measured. In the Stokes vector representation, optical elements are described by the Mueller matrix. In this section, we will address the Stokes parameters (vectors), the Mueller matrix and the Poincaré sphere and will discuss the states of polarization in greater detail.

3.4.1 DEFINITION OF STOKES PARAMETERS

There are several ways to describe the Stokes parameters. If we use the light intensity of polarized light, the four parameters that define the Stokes parameters (S_{0-3}) are expressed by

$$S_0=I_x+I_y \tag{3.32a}$$

$$S_1=I_x-I_y \tag{3.32b}$$

$$S_2=I_{+45^\circ}-I_{-45^\circ} \tag{3.32c}$$

$$S_3=I_{\mathrm{R}}-I_{\mathrm{L}} \tag{3.32d}$$

Here, S_0 represents the total light intensity and S_1 shows the light intensity determined by subtracting the light intensity of linear polarization in the y direction (I_y) from that in the x direction (I_x) [Fig. 3.16(a)]. On the other hand, S_2 represents the light intensity obtained by subtracting the light intensity of linear polarization at $-45^\circ(I_{-45^\circ})$ from that at $+45^\circ(I_{+45^\circ})$ [Fig. 3.16(b)]. With respect to the parameter S_3, the light intensity of left-circular polarization (I_{L}) is subtracted from that of right-circular polarization (I_{R}) [Fig. 3.16(c)]. Thus, the parameters S_{1-3} represent the relative difference in light intensity between each state of polarization. When $S_1>0$, for example, the light is polarized toward the x direction, while polarization of light is oriented in the y direction when $S_1<0$.

The Stokes parameters shown in Eq. (3.32) can also be expressed by using electric fields as described below. The Stokes parameters described by electric fields

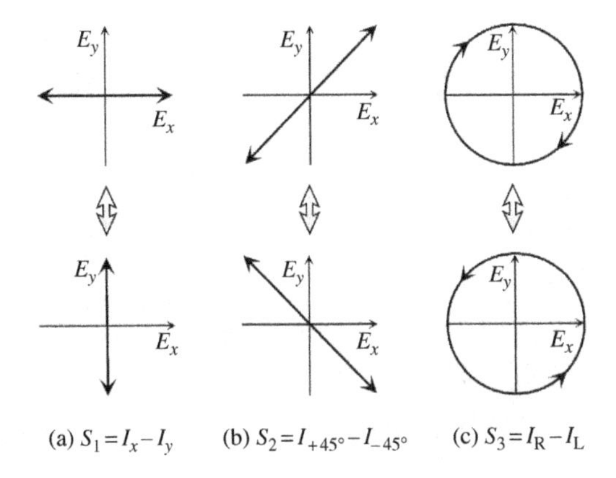

Figure 3.16 Definitions of Stokes parameters (S_{1-3}) based on light intensity.

are quite important for the interpretation of measured values in ellipsometry. If we apply Eq. (3.10), S_0 and S_1 are given by

$$S_0 = I_x + I_y = E_{x0}^2 + E_{y0}^2 = E_x E_x^* + E_y E_y^* \tag{3.33}$$

$$S_1 = I_x - I_y = E_{x0}^2 - E_{y0}^2 = E_x E_x^* - E_y E_y^* \tag{3.34}$$

In order to express S_2 using electric fields, we first determine the electric fields of the linear polarizations oriented at +45° and −45°. We can obtain these from the coordinate transformation of the Jones vector [2]:

$$\begin{bmatrix} E_{-45°} \\ E_{+45°} \end{bmatrix} = \begin{bmatrix} \cos(-45°) & \sin(-45°) \\ -\sin(-45°) & \cos(-45°) \end{bmatrix} \begin{bmatrix} E_x \\ E_y \end{bmatrix} = \frac{1}{\sqrt{2}} \begin{bmatrix} E_x - E_y \\ E_x + E_y \end{bmatrix} \tag{3.35}$$

It can be seen from Eq. (3.35) that $E_{-45°}$ and $E_{+45°}$ are calculated from −45° rotation of the $x-y$ coordinates. From Eqs. (3.32c) and (3.35), we get

$$\begin{aligned} S_2 &= E_{+45°} E_{+45°}^* - E_{-45°} E_{-45°}^* \\ &= 1/2\left[(E_x + E_y)(E_x^* + E_y^*) - (E_x - E_y)(E_x^* - E_y^*)\right] \\ &= E_x E_y^* + E_x^* E_y \end{aligned} \tag{3.36}$$

If we use $(E_x E_y^*)^* = E_x^* E_y$ and Eq. (2.12), Eq. (3.36) is rewritten as

$$S_2 = 2\,\mathrm{Re}(E_x E_y^*) = 2\,\mathrm{Re}(E_x^* E_y) \tag{3.37}$$

Recall from Section 2.1.1 that $\mathrm{Re}(C^*) = \mathrm{Re}(C)$ and $\mathrm{Im}(C^*) = -\mathrm{Im}(C)$ in complex conjugate numbers. It follows from Eq. (3.9) that $E_x^* = E_{x0}\exp[-\mathrm{i}(\delta_x - \delta_y)]$ and $E_y = E_{y0}$. By substituting these into Eq. (3.37), we obtain

$$S_2 = 2E_{x0}E_{y0}\mathrm{Re}\{\exp[-\mathrm{i}(\delta_x - \delta_y)]\} = 2E_{x0}E_{y0}\cos\Delta \tag{3.38}$$

where $\Delta = \delta_x - \delta_y$. Notice that $\exp(-i\Delta) = \cos\Delta - i\sin\Delta$.

Similarly, in order to express S_3 from electric fields, we calculate E_L and E_R using Eq. (3.24) [2]:

$$\begin{bmatrix} E_L \\ E_R \end{bmatrix} = \frac{1}{\sqrt{2}} \begin{bmatrix} 1 & i \\ 1 & -i \end{bmatrix} \begin{bmatrix} E_x \\ E_y \end{bmatrix} = \frac{1}{\sqrt{2}} \begin{bmatrix} E_x + iE_y \\ E_x - iE_y \end{bmatrix} \tag{3.39}$$

Thus, S_3 is given by

$$\begin{aligned} S_3 &= E_R E_R^* - E_L E_L^* \\ &= 1/2\left[(E_x - iE_y)(E_x^* + iE_y^*) - (E_x + iE_y)(E_x^* - iE_y^*)\right] \\ &= i(E_x E_y^* - E_x^* E_y) \end{aligned} \tag{3.40}$$

By applying Eq. (2.12), we get

$$S_3 = -2\,\mathrm{Im}(E_x E_y^*) = 2\,\mathrm{Im}(E_x^* E_y) \tag{3.41}$$

Finally, substituting $E_x^* = E_{x0}\exp[-i(\delta_x - \delta_y)]$ and $E_y = E_{y0}$ into Eq. (3.41) yields

$$S_3 = 2E_{x0}E_{y0}\mathrm{Im}\{\exp[-i(\delta_x - \delta_y)]\} = -2E_{x0}E_{y0}\sin\Delta. \tag{3.42}$$

3.4.2 POINCARÉ SPHERE

If we choose the Stokes parameters (S_{1-3}) as the axes of the three-dimensional coordinates, the state of polarization can be represented as a point on the surface of a sphere. This sphere shown in Fig. 3.17 is known as the Poincaré sphere. The size of the Poincaré sphere itself shows the total light intensity S_0. If the Poincaré sphere is compared to the earth, all polarizations on the equator are linear polarizations and the direction of linear polarization varies according to a position on the equator. As shown in Fig. 3.17, polarization is oriented in the x axis when $S_1 > 0$ and, conversely, polarization is oriented in the y axis when $S_1 < 0$. On the other hand, on the S_2 axis, the polarization is $+45°$ when $S_2 > 0$ and is $-45°$ when $S_2 < 0$. The positions corresponding to the North Pole and the South Pole represent right- and left-circular polarizations, respectively. It can be seen from Fig. 3.17 that all of the elliptical and circular polarizations in the northern hemisphere rotate in the right direction (clockwise), while the rotations are counterclockwise in the southern hemisphere.

A point on the surface of the Poincaré sphere can be described by using the (ε, θ) and (ψ, Δ) coordinate systems shown in Fig. 3.18. In the (ε, θ) system, we choose the major axis (length $2a$) and minor axis (length $2b$) as the coordinates of

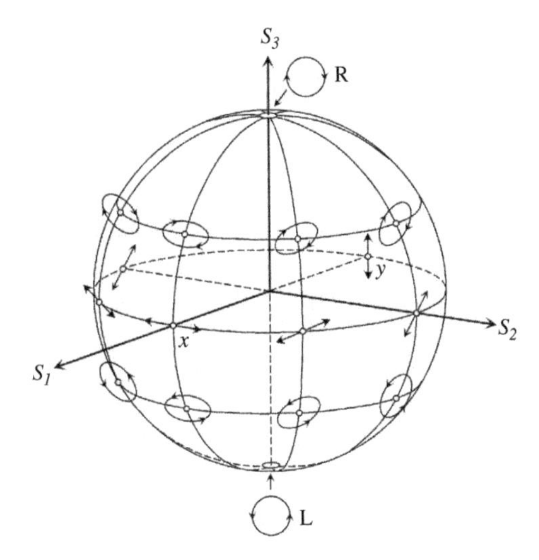

Figure 3.17 Representation of polarization by the Poincaré sphere.

the elliptical polarization. The angle of the major axis relative to the E_x direction is referred to as the azimuth θ, while ε represents the ellipticity angle given by $\tan\varepsilon = b/a$. When $\tan\varepsilon = 0$, the state of polarization becomes linear polarization. On the equator of the Poincaré sphere, only θ changes with $\varepsilon = 0$. On the other hand, if we vary the latitude from a fixed position on the equator, only ε changes with a constant θ. In the (ε, θ) coordinate system, therefore, the state of polarization is represented by the two angles (ε, θ) illustrated in Fig. 3.19. It is clear from Fig. 3.19 that point $P(S_1, S_2, S_3)$ on the surface of the Poincaré sphere is expressed by

$$S_1 = \cos 2\varepsilon \cos 2\theta \tag{3.43a}$$

$$S_2 = \cos 2\varepsilon \sin 2\theta \tag{3.43b}$$

$$S_3 = \sin 2\varepsilon \tag{3.43c}$$

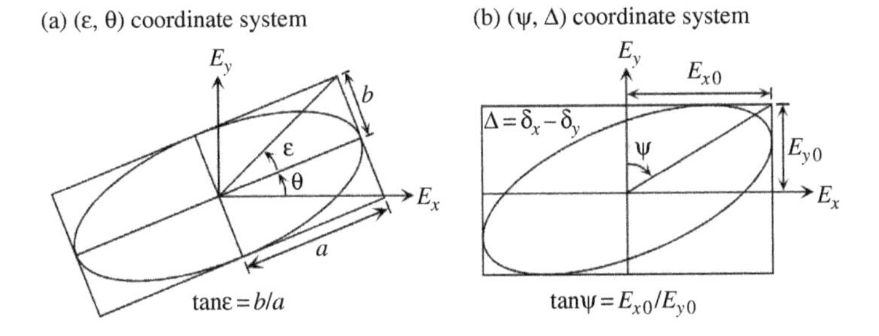

Figure 3.18 Representations of the elliptical polarization by (a) the (ε, θ) coordinate system and (b) the (ψ, Δ) coordinate system.

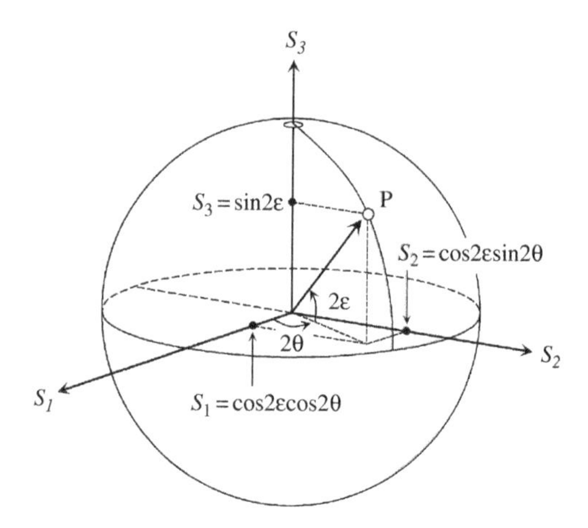

Figure 3.19 Representation of a point P on the surface of the Poincaré sphere using the (ε, θ) coordinates.

As shown in Eq. (3.43), the values of S_1 and S_2 are calculated from 2θ. This implies that there is no distinction between the upper and lower sides of polarized light and the 180° rotation corresponds to one complete optical rotation. Moreover, 2ε in Eq. (3.43) originates from the fact that $a = b$ at $\varepsilon = 45°$. If we use Eq. (3.43), we can calculate (ε, θ) values from the Stokes parameters:

$$\theta = \frac{1}{2}\tan^{-1}\left(\frac{S_2}{S_1}\right) \qquad \varepsilon = \frac{1}{2}\sin^{-1}(S_3) \tag{3.44}$$

As shown in Fig. 3.18(b), the (ψ, Δ) coordinate system is expressed using the (E_x, E_y) coordinates. Here, ψ and Δ represent the angles of the amplitude ratio ($\tan\psi = E_{x0}/E_{y0}$) and phase difference ($\Delta = \delta_x - \delta_y$), respectively. It can be seen from Fig. 3.18(b) that

$$E_{x0} = \sin\psi \quad E_{y0} = \cos\psi \tag{3.45}$$

Thus, using Eq. (3.9) and $\Delta = \delta_x - \delta_y$, we can rewrite the Jones vector as

$$\begin{bmatrix} E_x \\ E_y \end{bmatrix} = \begin{bmatrix} E_{x0}\exp(i\Delta) \\ E_{y0} \end{bmatrix} = \begin{bmatrix} \sin\psi\exp(i\Delta) \\ \cos\psi \end{bmatrix} = \cos\psi \begin{bmatrix} \tan\psi\exp(i\Delta) \\ 1 \end{bmatrix} \tag{3.46}$$

The elliptical polarization shown in Table 3.1 represents the above equation. By substituting Eq. (3.45) into Eqs. (3.34), (3.38), and (3.42), we get

$$S_1 = \sin^2\psi - \cos^2\psi = -\cos 2\psi \tag{3.47a}$$

$$S_2 = 2\sin\psi\cos\psi\cos\Delta = \sin 2\psi\cos\Delta \tag{3.47b}$$

$$S_3 = -2\sin\psi\cos\psi\sin\Delta = -\sin 2\psi\sin\Delta \tag{3.47c}$$

In Eq. (3.47), double-angle formulas were used for the transformation (see Appendix 1). Table 3.3 summarizes the Stokes parameters expressed by various physical parameters. As shown in this table, we obtain the normalized value for S_0 when the Stokes parameters are described by (ε, θ) and (ψ, Δ).

Table 3.3 Stokes parameters (S_{0-3})

	S_0	S_1	S_2	S_3
Light intensity	$I_x + I_y$	$I_x - I_y$	$I_{+45°} - I_{-45°}$	$I_R - I_L$
Electric field *A*	$E_x E_x^* + E_y E_y^*$	$E_x E_x^* - E_y E_y^*$	$E_x E_y^* + E_x^* E_y$	$i(E_x E_y^* - E_x^* E_y)$
Electric field *B*	$E_x E_x^* + E_y E_y^*$	$E_x E_x^* - E_y E_y^*$	$2\mathrm{Re}(E_x^* E_y)$	$2\mathrm{Im}(E_x^* E_y)$
Electric field *C*	$E_{x0}^2 + E_{y0}^2$	$E_{x0}^2 - E_{y0}^2$	$2E_{x0}E_{y0}\cos\Delta$	$-2E_{x0}E_{y0}\sin\Delta$
(ε, θ) system	1	$\cos 2\varepsilon\cos 2\theta$	$\cos 2\varepsilon\sin 2\theta$	$\sin 2\varepsilon$
(ψ, Δ) system	1	$-\cos 2\psi$	$\sin 2\psi\cos\Delta$	$-\sin 2\psi\sin\Delta$

3.4.3 PARTIALLY POLARIZED LIGHT

In totally polarized light, the light is represented by one specific state of polarization, while polarization of light is completely random in unpolarized light. In partially

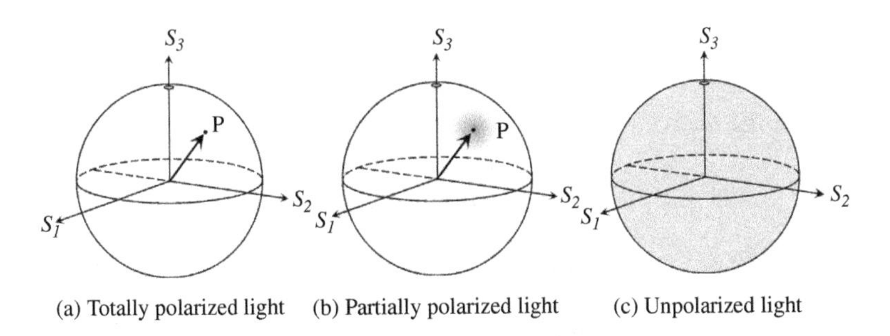

Figure 3.20 Representations of (a) totally polarized light, (b) partially polarized light and (c) unpolarized light using the Poincaré sphere.

polarized light, on the other hand, the light waves consist of a mixture of polarized and unpolarized waves. Figure 3.20 shows each state of polarization represented by the Poincaré sphere. In total polarization, the state of polarization is indicated by a point P on the surface of the Poincaré sphere. In contrast to totally polarized light, the polarization states of partially polarized and unpolarized waves vary with time. Thus, points on the Poincaré sphere are distributed around point P in partially polarized light and all the points are completely scattered in unpolarized light. The state of polarization for partially polarized and unpolarized waves is described statistically as the average of all the points on the Poincaré sphere over a certain time [2]. In the case of unpolarized light, therefore, the state of polarization is represented by the point $S_1 = S_2 = S_3 = 0$. In the depolarizer shown in Fig. 3.10, however, unpolarized light is generated by using the spatial variation of the polarized light, rather than the time variation. Accordingly, such depolarization is generally referred to as quasi-depolarization.

In totally polarized light, a distance from the point $S_1 = S_2 = S_3 = 0$ to P shows the total light intensity (S_0) and the following equation holds:

$$S_0^2 = S_1^2 + S_2^2 + S_3^2 \tag{3.48}$$

Using the values of the electric field C in Table 3.3, we can confirm the above relation as

$$S_0^2 = S_1^2 + S_2^2 + S_3^2 = E_{x0}^4 + E_{y0}^4 + 2E_{x0}^2 E_{y0}^2 \tag{3.49}$$

If we normalize the light intensity ($S_0 = 1$), we get

$$S_1^2 + S_2^2 + S_3^2 = 1 \tag{3.50}$$

This equation represents a sphere with a radius of 1. Accordingly, the polarization state of totally polarized light is expressed by a point on the Poincaré sphere with a radius of 1.

For partially polarized light, we find

$$S_0^2 > S_1^2 + S_2^2 + S_3^2 \tag{3.51}$$

In this case, the degree of polarization is given by

$$p = \left(S_1^2 + S_2^2 + S_3^2\right)^{1/2} / S_0 \tag{3.52}$$

It is obvious from Eqs. (3.50) and (3.52) that $p = 1$ for totally polarized light and $p = 0$ for unpolarized light ($S_1 = S_2 = S_3 = 0$). In actual ellipsometry measurement, we often assume that reflected light from a sample is totally polarized. Nevertheless, if light scattering occurs on a sample surface, totally polarized incident light is often transformed into partially polarized light. In some cases, depolarization of incident light affects the measurement seriously [12,18], although this effect depends completely on the types of instruments. We will discuss depolarization effects in greater detail in Section 4.4.4.

3.4.4 MUELLER MATRIX

We can describe the Stokes parameters using a vector representation, known as the Stokes vector. The Stokes vector is given by

$$\boldsymbol{S} = \begin{bmatrix} S_0 \\ S_1 \\ S_2 \\ S_3 \end{bmatrix} \tag{3.53}$$

Table 3.1 shows the normalized Stokes vectors ($S_0 = 1$) for various polarizations. For example, the linear polarization in the x direction is represented by $S_0 = S_1 = 1$, while the linear polarization in the y direction is denoted by $S_0 = 1$ and $S_1 = -1$. In Table 3.1, the Stokes vector for elliptical polarization is expressed by using the (ψ, Δ) coordinate system shown in Table 3.3. The Stokes vector representation allows the description of natural (unpolarized) light, as shown in Table 3.1. Futhermore, we can express partially polarized light by

$$\boldsymbol{S} = \begin{bmatrix} 1 \\ -p\cos 2\psi \\ p\sin 2\psi\cos\Delta \\ -p\sin 2\psi\sin\Delta \end{bmatrix} \tag{3.54}$$

where p is the degree of polarization. From Eq. (3.54), we find $\left(S_1^2 + S_2^2 + S_3^2\right)^{1/2} = p$. Thus, Eq. (3.52) holds since $S_0 = 1$.

We can describe the transformation of a Stokes vector by a 4×4 matrix representation, referred to as a Mueller matrix. When we treat totally polarized light, the Jones matrix can be converted to the Mueller matrix (see Appendix 4). Therefore, it can be seen from Table 3.2 that the Mueller matrix consists of matrix elements similar to those of the Jones matrix. However, the Mueller matrix enables us to describe a depolarizer mathematically. The Mueller matrix calculation is performed in a manner similar to the Jones matrix calculation. For example, when linear polarization oriented at 45° passes through a polarizer whose transmission axis is in the x direction, the light emerging from the polarizer is calculated by

$$\frac{1}{2}\begin{bmatrix} 1 & 1 & 0 & 0 \\ 1 & 1 & 0 & 0 \\ 0 & 0 & 0 & 0 \\ 0 & 0 & 0 & 0 \end{bmatrix}\begin{bmatrix} 1 \\ 0 \\ 1 \\ 0 \end{bmatrix} = \begin{bmatrix} 1/2 \\ 1/2 \\ 0 \\ 0 \end{bmatrix} \tag{3.55}$$

This simple calculation shows clearly that only light in the x direction transmits the polarizer and the total light intensity becomes 1/2. On the other hand, when linear light oriented at 45° transmits through a compensator whose phase shift is $\pi/2$, we get

$$\begin{bmatrix} 1 & 0 & 0 & 0 \\ 0 & 1 & 0 & 0 \\ 0 & 0 & \cos(\pi/2) & \sin(\pi/2) \\ 0 & 0 & -\sin(\pi/2) & \cos(\pi/2) \end{bmatrix} \begin{bmatrix} 1 \\ 0 \\ 1 \\ 0 \end{bmatrix} = \begin{bmatrix} 1 \\ 0 \\ 0 \\ -1 \end{bmatrix} \tag{3.56}$$

Thus, in this case, the light is transformed into left-circular polarization, as shown in Fig. 3.8. If we apply the Muller matrix, Malus's law described in Section 3.3.4 is expressed as

$$\boldsymbol{AR}(-\alpha)\boldsymbol{P} \begin{bmatrix} 1 \\ 0 \\ 0 \\ 0 \end{bmatrix} = \frac{1}{4} \begin{bmatrix} 1+\cos(2\alpha) \\ 1+\cos(2\alpha) \\ 0 \\ 0 \end{bmatrix} \tag{3.57}$$

In Eq. (3.57), the Mueller matrices were described using the symbols shown in Table 3.2. In addition, the Stokes vector of natural light shown in Table 3.1 was used as the incident light. Using the equation shown in Appendix 1(f), we obtain

$$S_0 = S_1 = \frac{1}{4}[1+\cos(2\alpha)] = \frac{1}{2}\cos^2\alpha \tag{3.58}$$

which describes the same result shown in Eq. (3.31). The coefficient 1/2 in Eq. (3.58) represents the fact that the light intensity of natural light ($S_0 = 1$) becomes 1/2 after the light passes through the polarizer, similar to the case of Eq. (3.55).

REFERENCES

[1] E. Hecht, *Optics*, 4th edition, Addison Wesley, San Francisco (2002).
[2] R. M. A. Azzam and N. M. Bashara, *Ellipsometry and Polarized Light*, North–Holland, Amsterdam (1977).
[3] D. E. Aspnes, Effects of component optical activity in data reduction and calibration of rotating-analyzer ellipsometers, *J. Opt. Soc. Am.*, **64** (1974) 812–819.
[4] For a review, see R. W. Collins, Automatic rotating element ellipsometers: calibration, operation, and real-time applications, *Rev. Sci. Instrum.*, **61** (1990) 2029–2062.
[5] N. V. Nguyen, B. S. Pudliner, I. An, and R. W. Collins, Error correction for calibration and data reduction in rotating-polarizer ellipsometry: applications to a novel multichannel ellipsometer, *J. Opt. Soc. Am. A*, **8** (1991) 919–931.
[6] J. A. Zapien, R. W. Collins, and R. Messier, Multichannel ellipsometer for real time spectroscopy of thin film deposition from 1.5 to 6.5 eV, *Rev. Sci. Instrum.*, **71** (2000) 3451–3460.

[7] J. N. Hilfiker, C. L. Bungay, R. A. Synowicki, T. E. Tiwald, C. M. Herzinger, B. Johs, G. K. Pribil and J. A. Woollam, Progress in spectroscopic ellipsometry: applications from vacuum ultraviolet to infrared, *J. Vac. Sci. Technol. A*, **21** (2003) 1103–1108, and references therein.
[8] A. Röseler, IR spectroscopic ellipsometry: instrumentation and results, *Thin Solid Films*, **234** (1993) 307–313.
[9] A. Canillas, E. Pascual, and B. Drévillon, Phase-modulated ellipsometer using a Fourier transform infrared spectrometer for real time applications, *Rev. Sci. Instrum.*, **64** (1993) 2153–2159.
[10] J. Lee, P. I. Rovira, I. An, and R. W. Collins, Alignment and calibration of the MgF_2 biplate compensator for applications in rotating-compensator multichannel ellipsometry, *J. Opt. Soc. Am. A*, **18** (2001) 1980–1985.
[11] P. S. Hauge, Generalized rotating-compensator ellipsometry, *Surf. Sci.*, **56** (1976) 148–160.
[12] J. Lee, P. I. Rovira, I. An, and R. W. Collins, Rotating-compensator multichannel ellipsometry: applications for real time Stokes vector spectroscopy of thin film growth, *Rev. Sci. Instrum.*, **69** (1998) 1800–1810.
[13] S. N. Jasperson and S. E. Schnatterly, An improved method for high reflectivity ellipsometry based on a new polarization modulation technique, *Rev. Sci. Instrum.*, **40** (1969) 761–767.
[14] J. C. Canit and J. Badoz, New design for a photoelastic modulator, *Appl. Opt.*, **22** (1983) 592–594.
[15] B. Johs, Regression calibration method for rotating element ellipsometers, *Thin Solid Films*, **234** (1993) 395–398.
[16] S. Huard, *Polarization of Light*, John Wiley & Sons, Ltd, Chichester (1997).
[17] R. C. Jones, A new calculus for the treatment of optical systems: I. Description and discussion of the calculus, *J. Opt. Soc. Am.*, **31** (1941) 488–493.
[18] U. Rossow, Depolarization/mixed polarization corrections of ellipsometry spectra, *Thin Solid Films*, **313–314** (1998) 97–101.

4 Principles of Spectroscopic Ellipsometry

In this chapter, we will review principles and measurement methods of spectroscopic ellipsometry. There are several types of spectroscopic ellipsometry instruments and, depending on the type of instruments, precision and error in measurements vary. Accordingly, in order to perform appropriate ellipsometry data analysis, it is essential to understand the characteristics of measurement methods as well. In actual spectroscopic ellipsometry instruments, we need to correct instrument imperfections to minimize measurement errors. Moreover, extra care is required when we install spectroscopic ellipsometry instruments to various processing systems. In this chapter, we will address the installation of ellipsometry instruments, various calibration methods, and measurement errors, in addition to the principles and measurement methods of spectroscopic ellipsometry.

4.1 PRINCIPLES OF ELLIPSOMETRY MEASUREMENT

In this section, we will see what quantity ellipsometry measures. Furthermore, we will define the coordinate system in ellipsometry, which is of significant importance for the interpretation of ellipsometry measurements. This section will also introduce the Jones and Mueller matrices corresponding to a measurement sample. As we will see in next section, these matrices enable us to describe ellipsometry measurements mathematically.

4.1.1 MEASURED VALUES IN ELLIPSOMETRY

In ellipsometry, p- and s-polarized light waves are irradiated onto a sample at the Brewster angle (see Section 2.3.4), and the optical constants and film thickness of the sample is measured from the change in the polarization state by light reflection or transmission. Figure 4.1 illustrates the measurement principle of ellipsometry. As

Spectroscopic Ellipsometry: Principles and Applications H. Fujiwara

we have seen in Chapter 3, the state of polarization is expressed by superimposing waves propagating along two orthogonal axes. In ellipsometry measurement, the polarization states of incident and reflected light waves are described by the coordinates of p- and s-polarizations. The incident vectors $\boldsymbol{E}_{ip}$ and $\boldsymbol{E}_{is}$ in Fig. 4.1 are identical to those defined in Fig. 2.15. From comparison with Fig. 2.15, it can be seen that the directions of the electric field vectors for p- and s-polarizations are reversed on both incident and reflection sides in Fig. 4.1, in order to make the understanding of ellipsometry easier. When the vectors are defined by the directions shown in Fig. 4.1, all the equations described in Chapter 2 remain the same. Notice that the vectors on the incident and reflection sides overlap completely when $\theta = 90°$ (straight-through configuration). In Fig. 4.1, the incident light is linear polarization oriented at $+45°$ relative to the E_{ip} axis. In particular, $E_{ip} = E_{is}$ holds for this polarization since the amplitudes of p- and s-polarizations are the same and the phase difference between the polarizations is zero.

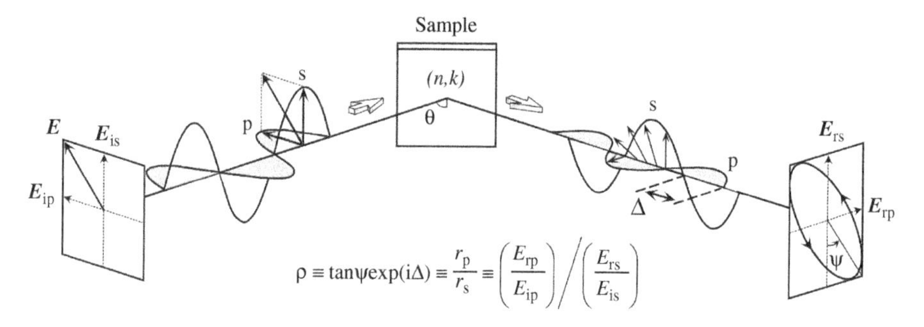

Figure 4.1 Measurement principle of ellipsometry.

As mentioned earlier, the amplitude reflection coefficients for p- and s-polarizations differ significantly due to the difference in electric dipole radiation (see Section 2.3). Thus, upon light reflection on a sample, p- and s-polarizations show different changes in amplitude and phase. As shown in Fig. 4.1, ellipsometry measures the two values (ψ, Δ) that express the amplitude ratio and phase difference between p- and s-polarizations, respectively. In ellipsometry, therefore, *the variation of light reflection with p- and s-polarizations is measured as the change in polarization state*. In particular, when a sample structure is simple, the amplitude ratio ψ is characterized by the refractive index n, while Δ represents light absorption described by the extinction coefficient k (see Section 5.1.1). In this case, the two values (n, k) can be determined directly from the two ellipsometry parameters (ψ, Δ) obtained from a measurement by applying the Fresnel equations. This is the basic principle of ellipsometry measurement.

The (ψ, Δ) measured from ellipsometry are defined from the ratio of the amplitude reflection coefficients for p- and s-polarizations:

$$\rho \equiv \tan\psi \exp(i\Delta) \equiv \frac{r_p}{r_s} \tag{4.1}$$

When we measure light transmission, instead of light reflection, (ψ, Δ) are defined as

$$\rho \equiv \tan\psi \exp(\mathrm{i}\Delta) \equiv \frac{t_\mathrm{p}}{t_\mathrm{s}} \tag{4.2}$$

If we apply the definitions of the amplitude reflection coefficients r_p and r_s [Eqs. (2.57) and (2.61)], we can rewrite Eq. (4.1) as follows:

$$\rho \equiv \tan\psi \exp(\mathrm{i}\Delta) \equiv \frac{r_\mathrm{p}}{r_\mathrm{s}} \equiv \left(\frac{E_\mathrm{rp}}{E_\mathrm{ip}}\right) \Big/ \left(\frac{E_\mathrm{rs}}{E_\mathrm{is}}\right) \tag{4.3}$$

As confirmed from Eq. (4.3), r_p and r_s are originally defined by the ratios of reflected electric fields to incident electric fields, and $\tan\psi \exp(\mathrm{i}\Delta)$ is defined further by the ratio of r_p to r_s. In the case of Fig. 4.1, Eq. (4.3) can be simplified to $\tan\psi \exp(\mathrm{i}\Delta) = E_\mathrm{rp}/E_\mathrm{rs}$ since $E_\mathrm{ip} = E_\mathrm{is}$. In Fig. 4.1, therefore, ψ represents the angle determined from the amplitude ratio between reflected p- and s-polarizations, while Δ expresses the phase difference between reflected p- and s-polarizations. Although ψ is determined from the 4th quadrant in Fig. 4.1, other quadrants also provide the same ψ, as ψ is defined from the absolute value of the amplitude ratio $(0° \le \psi \le 90°)$.

If we use polar coordinates to represent the amplitude reflection coefficients [Eq. (2.67)], it follows from Eq. (4.1) that

$$\tan\psi = |r_\mathrm{p}|/|r_\mathrm{s}| \quad \Delta = \delta_\mathrm{rp} - \delta_\mathrm{rs} \tag{4.4}$$

Recall from Eq. (2.69) that $R_\mathrm{p} = |r_\mathrm{p}|^2$ and $R_\mathrm{s} = |r_\mathrm{s}|^2$. By applying these and Eq. (2.22), we get the following equation [1]:

$$\psi = \tan^{-1}(|\rho|) = \tan^{-1}\left(\frac{|r_\mathrm{p}|}{|r_\mathrm{s}|}\right) = \tan^{-1}\left[\left(\frac{R_\mathrm{p}}{R_\mathrm{s}}\right)^{1/2}\right] \tag{4.5}$$

Using Eq. (2.16), we obtain Δ from ρ as follows:

$$\Delta = \arg(\rho) = \begin{cases} \tan^{-1}[\mathrm{Im}(\rho)/\mathrm{Re}(\rho)] & \text{for } \mathrm{Re}(\rho) > 0, \\ \tan^{-1}[\mathrm{Im}(\rho)/\mathrm{Re}(\rho)] + 180° & \text{for } \mathrm{Re}(\rho) < 0, \quad \mathrm{Im}(\rho) \ge 0, \\ \tan^{-1}[\mathrm{Im}(\rho)/\mathrm{Re}(\rho)] - 180° & \text{for } \mathrm{Re}(\rho) < 0, \quad \mathrm{Im}(\rho) < 0 \end{cases} \tag{4.6}$$

In Eq. (4.6), if $\mathrm{Re}(\rho) = 0$, $\Delta = 90°[\mathrm{Im}(\rho) > 0]$ and $\Delta = -90°[\mathrm{Im}(\rho) < 0]$. In general, the range of Δ in ellipsometry is expressed by either $-180° \le \Delta \le 180°$ or $0° \le \Delta \le 360°$. We can convert the range of Δ from $-180° \le \Delta \le 180°$ to $0° \le \Delta \le 360°$ by simply adding 360° to the region of $-180° \le \Delta < 0°$.

The above equations correspond to the ones when the definition of $N \equiv n - \mathrm{i}k$ is used. For the definition of $N \equiv n + \mathrm{i}k$, we need to rewrite Eq. (4.1) as $\rho \equiv \tan\psi \exp(-i\Delta)$ (see Appendix 2). Traditionally, Eq. (4.1) has also been applied to the convention of $N \equiv n + ik$. In this case, however, we have to reverse the signs of Δ. Unfortunately, this procedure is highly confusing and we will use $\rho \equiv \tan\psi \exp(-i\Delta)$ for the definition of $N \equiv n + ik$ throughout this book.

4.1.2 COORDINATE SYSTEM IN ELLIPSOMETRY

When the incident light is linear polarization oriented at $45°(E_{ip} = E_{is})$, the (E_x, E_y) coordinates described in Chapter 3 can be transformed easily into the (E_{rp}, E_{rs}) coordinates used in ellipsometry (Fig. 4.2). In this case, the representation of the polarization states described in Chapter 3 also holds without any change, except for the coordinate axis conversion of $E_x \rightarrow E_{rp}$ and $E_y \rightarrow E_{rs}$. In this conversion, the amplitudes are transformed using $E_{x0} \rightarrow |r_p|$ and $E_{y0} \rightarrow |r_s|$ and the initial phases are rewritten as $\delta_x \rightarrow \delta_{rp}$ and $\delta_y \rightarrow \delta_{rs}$. Table 4.1 represents the polarization states of reflected light waves obtained from this coordinate transformation. The polarization states in ellipsometry measurement can be expressed not only by (ψ, Δ) but also by the Stokes vector and (ε, θ) coordinates shown in Fig. 3.18(a). If we replace $\delta_x - \delta_y$ in Fig. 3.3 with Δ, we can understand easily that reflected light becomes linear polarization when $\Delta = 0°$ and $180°$. On the other hand, reflected light waves become right- and left-circular polarizations when $\Delta = -90°(270°)$ and $90°$, respectively. In the case of the (ε, θ) coordinates, the states of polarization can be expressed easily from the Poincaré sphere representation (see Fig. 3.19). As shown in Fig. 3.17, in the southern hemisphere of the Poincaré sphere, all the elliptical and circular polarizations show counterclockwise rotations. Thus, the ellipticity angle ε shows a negative value in the southern hemisphere. The values of the Stokes vectors shown in Table 4.1 can be obtained easily by substituting the values of (ψ, Δ) and (ε, θ) into the terms shown in Table 3.3.

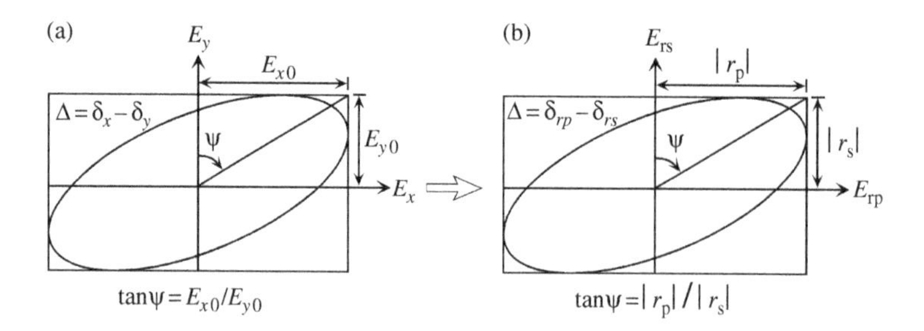

Figure 4.2 Coordinate transformation from (a) the (E_x, E_y) coordinates to (b) the (E_{rp}, E_{rs}) coordinates. In (a), the same figure as Figure 3.18(b) is shown.

Table 4.1 Representations of the polarization states of reflected light in ellipsometry measurement[a]

Polarization	Polarization state	Stokes vector	ψ (deg) Δ (deg)	ε (deg) θ (deg)
Linear polarization oriented at −45°	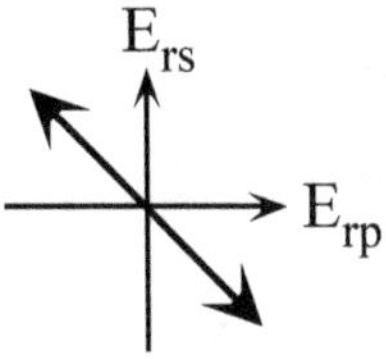 E_{rs}, E_{rp}	$\begin{bmatrix} 1 \\ 0 \\ -1 \\ 0 \end{bmatrix}$	$\psi = 45°$ $\Delta = 180°$	$\varepsilon = 0°$ $\theta = -45°$
Linear polarization oriented at −65°	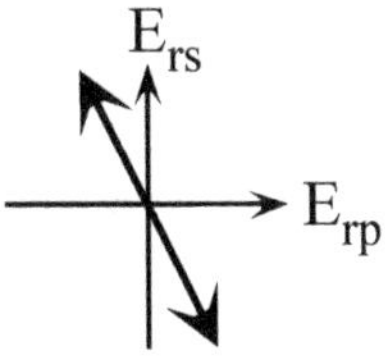 E_{rs}, E_{rp}	$\begin{bmatrix} 1 \\ -0.643 \\ -0.766 \\ 0 \end{bmatrix}$	$\psi = 25°$ $\Delta = 180°$	$\varepsilon = 0°$ $\theta = -65°$
Right-circular polarization	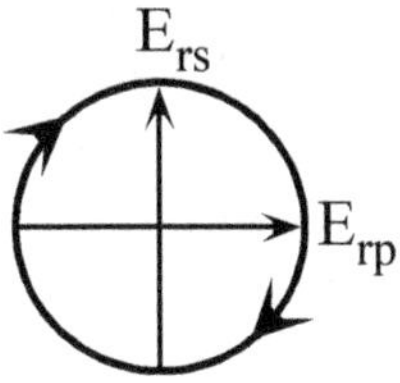 E_{rs}, E_{rp}	$\begin{bmatrix} 1 \\ 0 \\ 0 \\ 1 \end{bmatrix}$	$\psi = 45°$ $\Delta = -90°$	$\varepsilon = 45°$ $\theta = 0°$
Left-circular polarization	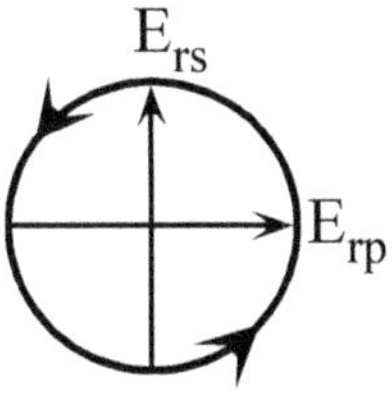 E_{rs}, E_{rp}	$\begin{bmatrix} 1 \\ 0 \\ 0 \\ -1 \end{bmatrix}$	$\psi = 45°$ $\Delta = 90°$	$\varepsilon = -45°$ $\theta = 0°$
Elliptical polarization $(\psi, \Delta) = (45°, 135°)$	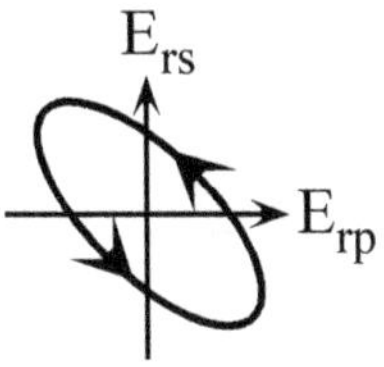 E_{rs}, E_{rp}	$\begin{bmatrix} 1 \\ 0 \\ -0.707 \\ -0.707 \end{bmatrix}$	$\psi = 45°$ $\Delta = 135°$	$\varepsilon = -22.5°$ $\theta = -45°$

[a] $E_{ip} = E_{is}$ for incident light.

4.1.3 JONES AND MUELLER MATRICES OF SAMPLES

In ellipsometry measurement, the Jones matrix that corresponds to light reflection by a sample is given by

$$\boldsymbol{S} = \begin{bmatrix} r_{\mathrm{p}} & 0 \\ 0 & r_{\mathrm{s}} \end{bmatrix} \tag{4.7}$$

For example, when linearly polarized light oriented at 45° is reflected by a sample, this light reflection is expressed as

$$\begin{bmatrix} E_{\mathrm{rp}} \\ E_{\mathrm{rs}} \end{bmatrix} = \begin{bmatrix} r_{\mathrm{p}} & 0 \\ 0 & r_{\mathrm{s}} \end{bmatrix} \begin{bmatrix} E_{\mathrm{ip}} \\ E_{\mathrm{is}} \end{bmatrix} = \begin{bmatrix} r_{\mathrm{p}} \\ r_{\mathrm{s}} \end{bmatrix} \tag{4.8}$$

In the above calculation, $E_{\mathrm{ip}} = E_{\mathrm{is}} = 1$ is assumed. Eq. (4.8) shows the straightforward fact that the incident waves E_{ip} and E_{is} are reflected by a sample with the coefficients of r_{p} and r_{s}, respectively. Notice that Eq. (4.8) represents the light reflection illustrated in Fig. 4.1. A similar calculation can also be performed using the Mueller matrix. If we use Eq. (4.1), Eq. (4.7) can be rewritten as

$$\boldsymbol{S} = r_{\mathrm{s}} \begin{bmatrix} r_{\mathrm{p}}/r_{\mathrm{s}} & 0 \\ 0 & 1 \end{bmatrix} = r_{\mathrm{s}} \begin{bmatrix} \tan\psi\exp(\mathrm{i}\Delta) & 0 \\ 0 & 1 \end{bmatrix} = \frac{r_{\mathrm{s}}}{\cos\psi} \begin{bmatrix} \sin\psi\exp(\mathrm{i}\Delta) & 0 \\ 0 & \cos\psi \end{bmatrix} \tag{4.9}$$

The Jones matrix in Table 3.2 represents the above equation. In Table 3.2, however, the proportional constant of Eq. (4.9) was neglected since only relative changes are taken into account in ellipsometry measurement.

From Eq. (4.8), it is obvious that

$$E_{\mathrm{rp}} = r_{\mathrm{p}} \quad E_{\mathrm{rs}} = r_{\mathrm{s}} \tag{4.10}$$

If we substitute $E_x = E_{\mathrm{rp}} = r_{\mathrm{p}}$ and $E_y = E_{\mathrm{rs}} = r_{\mathrm{s}}$ into Table 3.3 (Electric field B), we can express the normalized Stokes parameters using the amplitude reflection coefficients as follows:

$$\frac{S_1}{S_0} = \frac{r_{\mathrm{p}}r_{\mathrm{p}}^* - r_{\mathrm{s}}r_{\mathrm{s}}^*}{r_{\mathrm{p}}r_{\mathrm{p}}^* + r_{\mathrm{s}}r_{\mathrm{s}}^*} \tag{4.11a}$$

$$\frac{S_2}{S_0} = \frac{2\mathrm{Re}\left(r_{\mathrm{p}}^* r_{\mathrm{s}}\right)}{r_{\mathrm{p}}r_{\mathrm{p}}^* + r_{\mathrm{s}}r_{\mathrm{s}}^*} \tag{4.11b}$$

$$\frac{S_3}{S_0} = \frac{2\mathrm{Im}\left(r_{\mathrm{p}}^* r_{\mathrm{s}}\right)}{r_{\mathrm{p}}r_{\mathrm{p}}^* + r_{\mathrm{s}}r_{\mathrm{s}}^*} \tag{4.11c}$$

Conversely, using Table 3.3 [(ψ, Δ) system], we can calculate (ψ, Δ) values from the Stokes parameters:

$$\psi = \frac{1}{2}\cos^{-1}\left(\frac{-S_1}{S_0}\right) \tag{4.12a}$$

$$\Delta = \begin{cases} \tan^{-1}(-S_3/S_2) & \text{for} \quad \cos\Delta > 0, \\ \tan^{-1}(-S_3/S_2) + 180° & \text{for} \quad \cos\Delta < 0, \quad \sin\Delta \geq 0, \\ \tan^{-1}(-S_3/S_2) - 180° & \text{for} \quad \cos\Delta < 0, \quad \sin\Delta < 0 \end{cases} \tag{4.12b}$$

In Eq. (4.12b), when $\cos\Delta = 0$, it follows that $\Delta = 90°(\sin\Delta > 0)$ and $\Delta = -90°(\sin\Delta < 0)$.

The Jones matrix shown in Eq. (4.7) represents the light reflection by an optically isotropic sample. When a sample shows optical anisotropy including birefringence and dichroism, the Jones matrix corresponding to a sample is described by the following matrix [2–4]:

$$\boldsymbol{S}_{\text{ani}} = \begin{bmatrix} r_{\text{pp}} & r_{\text{ps}} \\ r_{\text{sp}} & r_{\text{ss}} \end{bmatrix} \tag{4.13}$$

It can be seen from Eq. (4.13) that the off-diagonal elements of the Jones matrix are no longer zero in samples that show optical anisotropy. Accordingly, the characterization of anisotropic samples generally becomes complicated. The Jones matrix for anisotropic samples will be explained in Section 6.1.4 in greater detail.

4.2 ELLIPSOMETRY MEASUREMENT

Until the early 1970s, only an ellipsometry instrument called the null ellipsometry had been used for measurements [1]. However, this ellipsometer is now seldom used, except for imaging ellipsometry in which measurement is performed in a two-dimensional plane [5–7]. Spectroscopic ellipsometry instruments that are used now can be classified into two major categories: instruments that use rotating optical elements [8–22] and instruments that use a photoelastic modulator [21–30]. The rotating-element ellipsometers can further be separated into rotating-analyzer ellipsometry (RAE) [8–16,20–22] and rotating-compensator ellipsometry (RCE) [17–21]. Up to now, the capability of spectroscopic ellipsometry measurement has been extended to the infrared region [31–40]. Furthermore, Mueller matrix ellipsometry, which allows the complete analysis of optically anisotropic samples, has been developed [21,41–51]. In this section, we will examine the characteristics and measurement principles of various ellipsometry instruments.

4.2.1 MEASUREMENT METHODS OF ELLIPSOMETRY

Figure 4.3 shows the schematic diagrams of ellipsometry instruments that have been used widely up to now. In general, these instruments are expressed using symbols of optical elements (see Table 3.2); the rotating-analyzer ellipsometer, for example, is described by PSA_R. Here, P, S, and A represent the polarizer, sample, and

(a) Rotating-analyzer ellipsometry (PSA_R)

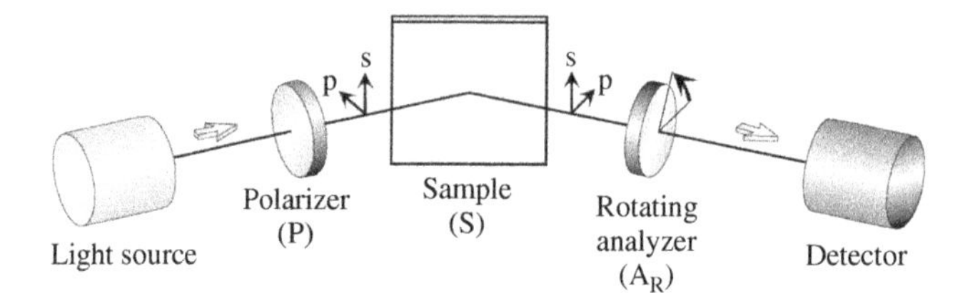

(b) Rotating-analyzer ellipsometry with compensator ($PSCA_R$)

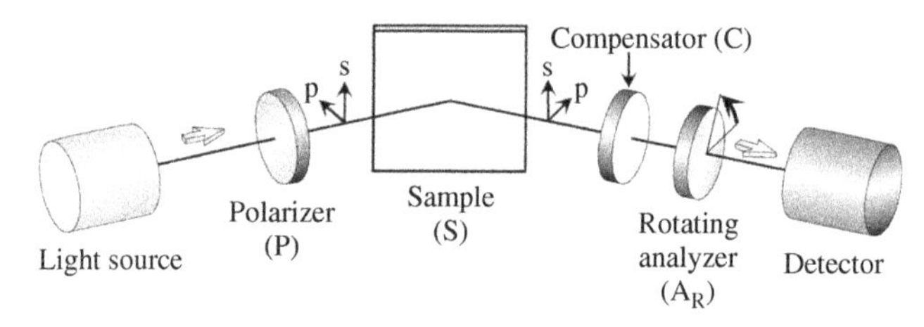

(c) Rotating-compensator ellipsometry (PSC_RA)

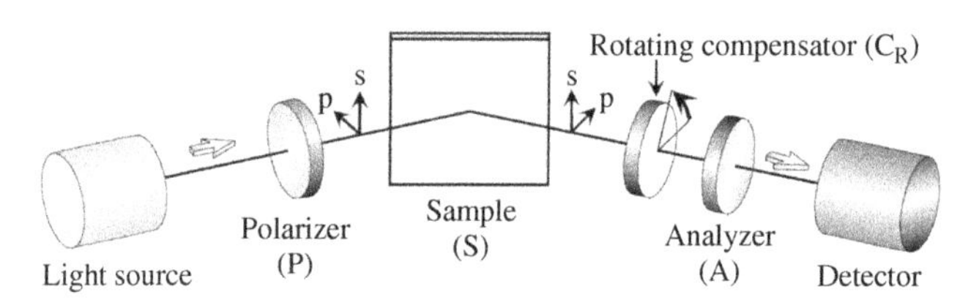

(d) Phase-modulation ellipsometry (PSMA)

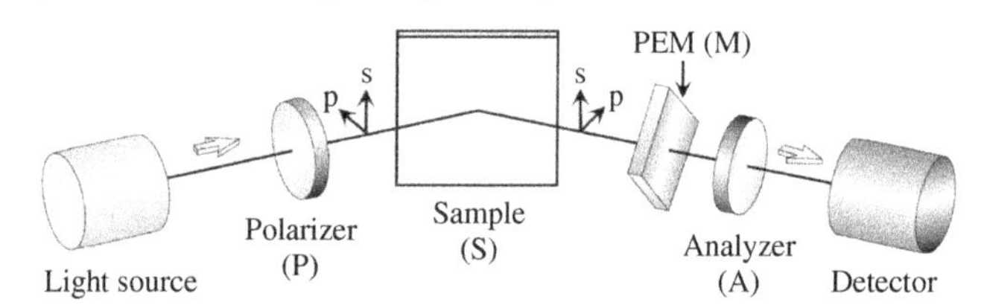

Figure 4.3 Optical configurations of ellipsometry instruments: (a) the rotating-analyzer ellipsometry (RAE), (b) the rotating-analyzer ellipsometry with compensator, (c) the rotating-compensator ellipsometry (RCE), and (d) the phase-modulation ellipsometry (PME).

analyzer, respectively, and the subscript R of A indicates that the analyzer rotates continuously. All the instruments in Fig. 4.3 have a light source and polarizer on the incident side and an analyzer and a detector on the reflection side. The coordinates of the optical elements in Fig. 4.3 follow the definition shown in Fig. 4.1. When looking against the direction of the beam (looking down the propagation axis toward the origin), counterclockwise rotation is the positive direction for the rotation of optical elements (see Section 3.3.2).

The rotating-analyzer ellipsometry (RAE) shown in Fig. 4.3(a) was perfected by Aspnes *et al.* in 1975 (see Table 1.3) [11]. In 1990, a group from The Pennsylvania State University first developed a real-time instrument that utilized a photodiode array as a light detector [15]. In this instrument, a polarizer is rotated (P_RSA), instead of an analyzer. Rotating-analyzer ellipsometry with compensator shown in Fig. 4.3(b) was developed to overcome the disadvantages of RAE (see Table 4.3) [32,33]. The rotating-compensator ellipsometry (RCE) in Fig. 4.3(c) is the latest ellipsometer that is commercially available now. Notice that RAE with compensator and RCE have identical optical configurations. The RCE instrument was reported first by Hauge *et al.* in 1975 [17]. With respect to a real-time spectroscopic instrument, the first result was reported in 1998 [19] and the commercialization of this instrument was also made around the same period. The phase-modulation ellipsometry (PME) shown in Fig. 4.3(d) has relatively old history and Jasperson *et al.* succeeded in developing PME in 1969 [23].

In conventional single-wavelength ellipsometry, a He–Ne laser and photomultiplier tube are employed as a light source and detector, respectively [10]. In spectroscopic ellipsometry, the wavelength of incident light is changed using a monochromator and the monochromatic light is detected by a photomultiplier tube [11,14]. In this case, however, a spectral measurement takes a long time ($\sim$ 10 min). In spectroscopic ellipsometry instruments that allow real-time monitoring, white light illuminates a sample, and all the light waves of different wavelengths are detected simultaneously by a photodiode array [15,16,19]. Figure 4.4 shows the schematic diagram of this instrument (PSA$_R$ configuration). In this instrument, light is detected continuously during the measurement, except for a short time during signal reading,

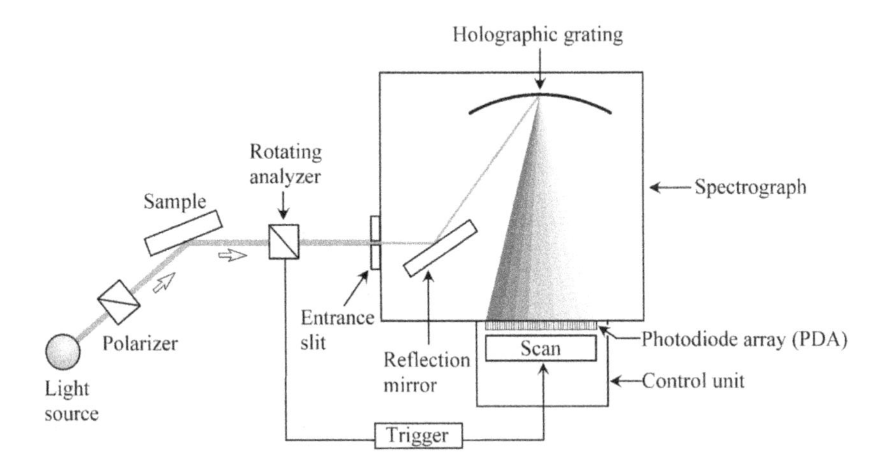

Figure 4.4 Schematic diagram of a real-time spectroscopic ellipsometry instrument (PSA$_R$ configuration).

and the scanning of light signals starts when the rotating analyzer reaches a rotation position determined in advance (see Section 4.3.3). This instrument allows real-time monitoring of thin-film growth with a repetition time of 64 ms (see Fig. 1.5) [16]. Although a holographic grating is used for the spectrograph in Fig. 4.4, we can also employ a more conventional grating-type or prism-type spectrograph. In grating-type spectrographs, spacing of data points is constant versus measurement wavelength. In this case, the number of data points becomes smaller at high energy, as confirmed from Eq. (2.27). In prism-type spectrographs, on the other hand, data spacing is constant for photon energy, but wavelength calibration becomes more difficult.

Table 4.2 summarizes the characteristics of each spectroscopic ellipsometry instrument. As we will find later, ellipsometry basically measures the Stokes parameters (vector) [21]. As shown in Table 4.2, however, the Stokes parameters measured in each ellipsometer differ. The variation of measurement error with the instrument types is basically based on this difference (see Section 4.4.1). As shown in Table 4.2, S_3 cannot be measured in PSA_R and P_RSA (see Section 4.2.2). In RAE with compensator ($PSCA_R$ or $PCSA_R$ configurations), at least two measurements have to be performed with different compensator settings to obtain all the Stokes parameters (see Section 4.2.3). In RCE, on the other hand, all the Stokes parameters can be obtained from a single measurement (see Section 4.2.4). For PME measurement, there are two optical configurations [21,25–27], and two measurements are necessary to obtain all the parameters (see Section 4.2.5). It should be emphasized that the measurement ranges for (ψ, Δ) are restricted when all the Stokes parameters are not measured. Specifically, in RAE, S_3 is not measured and thus only $\cos\Delta$ is determined with respect to Δ, as confirmed from Table 3.3. In this case, when $\cos\Delta = 0.707$, for example, we cannot distinguish $\Delta = +45°$ from $\Delta = -45°$. Consequently, the measurement range of RAE becomes half $(0° \leq \Delta \leq 180°)$ of the full range $(-180° \leq \Delta \leq 180°)$. Similarly, the

Table 4.2 Characteristics of spectroscopic ellipsometry instruments

Instrument type[a]	Measurable Stokes parameters	Measurable region	Minimum measurement time	Number of wavelengths measured[b]
PSA_R (P_RSA)	S_0, S_1, S_2	$0° \leq \psi \leq 90°$ $0° \leq \Delta \leq 180°$	~ 10 ms	~ 200
$PSCA_R$ ($PCSA_R$)	S_0, S_1, S_2, S_3^c	$0° \leq \psi \leq 90°$ $-180° \leq \Delta \leq 180°$	~ 10 ms[d]	~ 200
PSC_RA (PC_RSA)	S_0, S_1, S_2, S_3	$0° \leq \psi \leq 90°$ $-180° \leq \Delta \leq 180°$	~ 10 ms	~ 200
$PSMA^e$ (PMSA)	S_0, S_1, S_3 S_0, S_2, S_3	$0° \leq \psi \leq 90°$ $0° \leq \Delta \leq 180°$ $0° \leq \psi \leq 45°$ $-180° \leq \Delta \leq 180°$	20 μs	~ 10

[a] Polarizer (P), sample (S), analyzer (A), compensator (C), and photoelastic modulator (M). The subscript R indicates the rotation of the optical element [b] capabilities for measurements in the visible/UV region [c] two measurements with different angle setting of a compensator are necessary to obtain all the Stokes parameters [d] when a compensator position is fixed [e] there exist two measurement configurations.

measurement ranges of (ψ, Δ) are restricted in PME and the ranges vary depending on the measurement configuration of PME. In RCE, on the other hand, the (ψ, Δ) measurement can be performed over the full range since all the Stokes parameters are measured in this method.

It can be seen from Table 4.2 that a minimum measurement time is around 10 ms in the ellipsometers that use rotation of the optical elements. In general, electrically powered motors are employed to rotate these optical elements [11]. In these rotating-element ellipsometers, a minimum data acquisition time is determined by a rotation speed of the optical elements (10–100 Hz). In PME, on the other hand, a minimum measurement time is determined from the resonant frequency of piezo transducers (50 kHz) [23–30]. Accordingly, a fast measurement of 20 μs is possible in PME [26]. By applying this technique, the responses of liquid crystal molecules to electric fields (< 1 ms) have been characterized [52].

The number of measurement wavelengths in Table 4.2 shows the capability for real-time measurement performed in the visible/UV region. In conventional ellipsometry measurement (*ex situ* measurement), the number of wavelengths in a measurement is determined by the performance of the monochromator and light detector. As shown in Table 4.2, the number of wavelengths measured in PME is one order of magnitude lower than that in the rotating-element ellipsometers. This originates from the wavelength dependence of the phase shift in the photoelastic modulator. Specifically, in PME, a voltage applied to the photoelastic modulator has to be changed according to the wavelength of probe light, in order to obtain a constant phase shift (see Section 3.2.3). So far, real-time spectroscopic measurements using PME have been performed by employing a photodiode array [28,29] or several photomultiplier tubes [30]. When a photodiode array is used for PME, a measurement time of ~5 ms is required for each wavelength to obtain enough light signals [28,29]. In RAE, on the other hand, the maximum number of measurement wavelengths is determined by the number of pixels in the photodiode array since a polarizer (analyzer) shows no wavelength dependence over a wide range. In particular, light intensities of different wavelengths are measured simultaneously in RAE, as shown in Fig. 4.4. However, if there are too many pixels, measurement accuracy degrades due to the reduction in light intensity per pixel. In general, when an optical element shows no wavelength dependence, this element is said to be achromatic, while an optical element is said to be chromatic when there is wavelength dependence. Thus, RAE is an achromatic instrument, whereas the RCE instrument is chromatic since the compensator in RCE shows wavelength dependence [Eq. (3.3)]. However, if the wavelength dependence of a compensator is known, multiwavelength measurement can be performed simultaneously using a photodiode array [19,20].

Table 4.3 summarizes the advantages and disadvantages of each measurement method. As shown in this table, the rotating-analyzer (polarizer) ellipsometry has the advantages that the optical configuration is simple and the instrument is achromatic. However, this instrument has the disadvantages that S_3 is not measured and measurement error increases at $\Delta \cong 0°$ and 180° (see Section 4.4.1). The RAE

Table 4.3 Advantages and disadvantages of spectroscopic ellipsometry instruments

Instrument type[a]	Advantage	Disadvantage
$PSA_R(P_RSA)$	• Optical configuration is simple • Instrument is achromatic	• $S_3(-180° \leq \Delta < 0°)$ cannot be measured • Measurement error increases at $\Delta \cong 0°$ and 180°
$PSCA_R(PCSA_R)$	• All the range of (ψ, Δ) can be measured • Depolarization spectrum can be measured • Uniform measurement sensitivity for (ψ, Δ)	• Longer data acquisition time, compared with PSA_R and PSC_RA • Optical configuration is complicated, compared with PSA_R • Instrument is chromatic
$PSC_RA(PC_RSA)$	• All the range of (ψ, Δ) can be measured • Depolarization spectrum can be measured • Uniform measurement sensitivity for (ψ, Δ)	• Optical configuration is complicated, compared with PSA_R • Instrument is chromatic
PSMA (PMSA)	• Fast measurement • Capability for real-time spectroscopic measurement in the infrared region • Depolarization spectrum can be measured	• S_1 or S_2 cannot be measured in a single measurement • Increases in measurement error in specific regions of (ψ, Δ) • Instrument is chromatic

[a] Polarizer (P), sample (S), analyzer (A), compensator (C), and photoelastic modulator (M). The subscript R indicates the rotation of the optical element.

with compensator has been developed to improve these characteristics [32,33]. In this method, however, several measurements are necessary to suppress the measurement errors at $\Delta \cong 0°$ and 180° (see Section 4.2.3) and consequently actual measurement time becomes longer, compared with RAE. The main advantage of RCE and RAE with compensator is the capability of (ψ, Δ) measurement over the full range. Moreover, these instruments enable us to measure the degree of polarization versus wavelength or photon energy, which is often referred to as the depolarization spectrum [see Fig. 4.27(a)]. In particular, these instruments allow accurate measurements even when samples depolarize incident light (see Section 4.4.4). Furthermore, measurement sensitivity for (ψ, Δ) is uniform in all the (ψ, Δ) ranges in these instruments. However, optical configuration and calibration become complicated in RCE [19] and RAE with compensator, compared with RAE. As mentioned earlier, the major advantage of PME over the rotating-element ellipsometers is a fast ellipsometry measurement of 20 μs [26]. Another advantage of PME is the capability for real-time measurement in the infrared region [39,40]. In contrast to PME, application of the rotating-element ellipsometers to real-time spectroscopic measurement is rather difficult in the infrared region, as we will see in Section 4.2.6. From PME, a depolarization spectrum can also be measured, although two measurements are necessary [53], similar to RAE with compensator [33]. On the other hand, PME has the disadvantages that the number of wavelengths in real-time measurement is limited (visible/UV region) and S_1 or S_2 cannot be measured from a single measurement. In PME, measurement error increases in specific regions of (ψ, Δ), similar to RAE (see Section 4.4.1).

4.2.2 ROTATING-ANALYZER ELLIPSOMETRY (RAE)

Ellipsometry measurement can be expressed from the Jones and Mueller matrices described in the previous chapter. Here, we will examine the RAE instrument using simple Jones matrices. If we apply the symbols of the Jones matrices shown in Table 3.2, the ellipsometry instrument with $\mathrm{PSA_R}$ configuration is expressed as

$$\boldsymbol{L}_{\mathrm{out}} = \boldsymbol{A}\boldsymbol{R}(A)\boldsymbol{S}\boldsymbol{R}(-P)\boldsymbol{P}\boldsymbol{L}_{\mathrm{in}} \tag{4.14}$$

where $\boldsymbol{L}_{\mathrm{out}}$ represents the Jones vector of the light detected by a light detector and is given by $\boldsymbol{L}_{\mathrm{out}} = [E_{\mathrm{A}}, 0]^{\mathrm{T}}$. Here, the symbol T denotes the transposed matrix $(a_{ij} = a_{ji}^{\mathrm{T}})$. In Eq. (4.14), $\boldsymbol{L}_{\mathrm{in}}$ shows the normalized Jones vector corresponding to incident light $(\boldsymbol{L}_{\mathrm{in}} = [1, 0]^{\mathrm{T}})$ and A of the rotation matrix $\boldsymbol{R}(A)$ represents a rotation angle of an analyzer, while P is a rotation angle of a polarizer. In order to express the light transmission though the polarizer, we first rotate the $(E_{\mathrm{ip}}, E_{\mathrm{is}})$ coordinates in Fig. 4.1 so that the transmission axis of the polarizer becomes parallel to the E_{ip} axis. After the light passes through the polarizer, the coordinates are rotated again toward the reverse direction to restore the coordinates back into the original position. If we use the Jones matrix, this is expressed by $\boldsymbol{R}(-P)\boldsymbol{P}\boldsymbol{R}(P)$. Nevertheless, the Jones vector cannot describe unpolarized (natural) light (Section 3.4). Thus, with respect to the light emitted from a light source, only the light that transmits the polarizer $\boldsymbol{P}$ is taken into account (see Fig. 3.14). In Eq. (4.14), therefore, the rotation matrix $\boldsymbol{R}(P)$ is eliminated from $\boldsymbol{R}(-P)\boldsymbol{P}\boldsymbol{R}(P)$. Similarly, the rotation matrix $\boldsymbol{R}(-A)$ is eliminated from $\boldsymbol{R}(-A)\boldsymbol{A}\boldsymbol{R}(A)$ since light transmitted through the analyzer is detected independently of the coordinate rotation. In matrix representation, Eq. (4.14) is described as

$$\begin{aligned}\begin{bmatrix} E_{\mathrm{A}} \\ 0 \end{bmatrix} &= \begin{bmatrix} 1 & 0 \\ 0 & 0 \end{bmatrix} \begin{bmatrix} \cos A & \sin A \\ -\sin A & \cos A \end{bmatrix} \begin{bmatrix} \sin\psi\exp(\mathrm{i}\Delta) & 0 \\ 0 & \cos\psi \end{bmatrix} \\ &\quad \times \begin{bmatrix} \cos P & -\sin P \\ \sin P & \cos P \end{bmatrix} \begin{bmatrix} 1 & 0 \\ 0 & 0 \end{bmatrix} \begin{bmatrix} 1 \\ 0 \end{bmatrix}\end{aligned} \tag{4.15}$$

It can be seen from Eq. (4.15) that the matrices of $\boldsymbol{R}(-P)\boldsymbol{P}\boldsymbol{L}_{\mathrm{in}}$ are almost identical to those shown in Fig. 3.14. In Eqs. (4.14) and (4.15), $\boldsymbol{S}$ and $\boldsymbol{R}(A)$ are simply added to Eq. (3.30). As we have seen in Fig. 4.1, when $P = 45°$, Eq. (4.15) is simplified to

$$\begin{bmatrix} E_A \\ 0 \end{bmatrix} = \begin{bmatrix} 1 & 0 \\ 0 & 0 \end{bmatrix} \begin{bmatrix} \cos A & \sin A \\ -\sin A & \cos A \end{bmatrix} \begin{bmatrix} \sin\psi \exp(i\Delta) \\ \cos\psi \end{bmatrix} \quad (4.16)$$

In the above calculation, a proportional constant of $1/\sqrt{2}$ was neglected. By expanding Eq. (4.16), we obtain

$$E_A = \cos A \sin\psi \exp(i\Delta) + \sin A \cos\psi \quad (4.17)$$

From Eq. (3.10), we obtain the light intensity measured by a detector:

$$\begin{aligned} I &= |E_A|^2 \\ &= I_0 (1 - \cos 2\psi \cos 2A + \sin 2\psi \cos\Delta \sin 2A) \\ &= I_0 (1 + S_1 \cos 2A + S_2 \sin 2A) \end{aligned} \quad (4.18)$$

Here, I_0 represents the proportional constant of the reflected light whose intensity is proportional to incident light intensity. It can be seen from Eq. (4.18) that the light intensity varies as a function of the analyzer angle $2A$. This implies that there is no distinction between the upper and lower sides of the transmission axis and the 180° rotation of the analyzer corresponds to one optical rotation (see Fig. 3.15). Thus, the 180° rotation of optical elements is generally referred to as the one optical cycle. In RAE, the Stokes parameters S_1 and S_2 are measured as the Fourier coefficients of $\cos 2A$ and $\sin 2A$ [8], as shown in Eq. (4.18). Now consider that the analyzer rotates continuously with time at a speed of $A = \omega t$, where ω is the angular frequency of the analyzer. If we assume $\alpha = S_1$ and $\beta = S_2$, we can express the light intensity variation in RAE as follows [14,21]:

$$I(t) = I_0(1 + \alpha \cos 2\omega t + \beta \sin 2\omega t) \quad (4.19)$$

Figure 4.5 shows the normalized light intensity ($I_0 = 1$) calculated from Eq. (4.19), plotted as a function of the angle of the rotating analyzer $A = \omega t (0° \leq A \leq 180°)$. In this figure, Δ is varied from 90° to 180° with a constant value of $\psi = 45°$. As shown in Fig. 4.5(a), when reflected light is linear polarization ($\psi = 45°$), the light intensity is zero at $A = 45°$, while the light intensity becomes a maximum at $A = 135°$. This result can be understood easily from an optical configuration shown in Fig. 4.6. Suppose that incident light is the linear polarization of +45° and, upon light reflection on a sample, the light changes into the linear polarization of $-45° (\psi = 45°, \Delta = 180°)$. As shown in Fig. 4.6, only the p-polarized component transmits the analyzer when $A = 0°$. At $A = 45°$, however, the oscillatory direction of the reflected light is perpendicular to the transmission axis of the analyzer and consequently the light intensity measured by the detector becomes

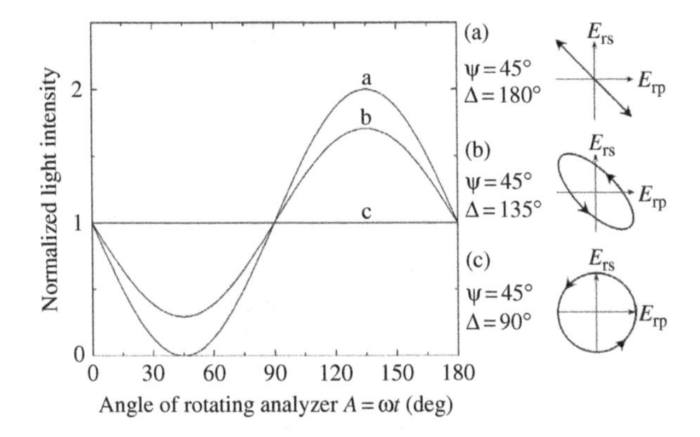

Figure 4.5 Normalized light intensity in rotating-analyzer ellipsometry (RAE), plotted as a function of the angle of rotating analyzer $A = \omega t$. This figure summarizes the calculation results when the polarization states of reflected light are (a) $\psi = 45°$, $\Delta = 180°$, (b) $\psi = 45°$, $\Delta = 135°$, and (c) $\psi = 45°$, $\Delta = 90°$.

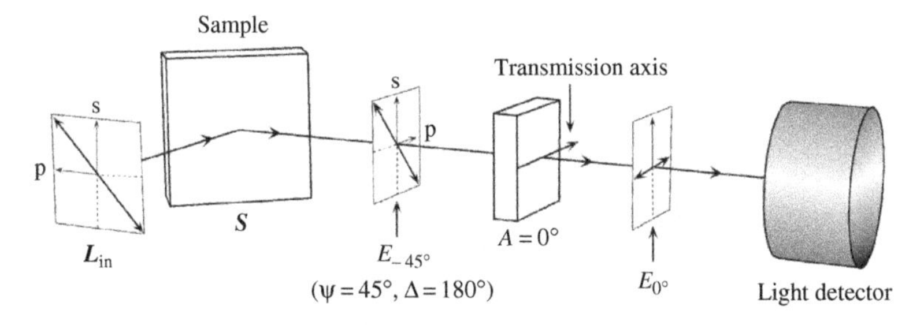

Figure 4.6 Schematic diagram of the measurement in the rotating-analyzer ellipsometry (RAE). In this figure, the reflected light from a sample is linear polarization of $-45°$($\psi = 45°$, $\Delta = 180°$).

zero, as shown in Fig. 4.5(a). This is the same phenomenon that we have seen in Fig. 3.15. When the analyzer rotates further to $A = 135°$, the transmission axis of the analyzer becomes parallel to the polarization direction of the reflected light and the light intensity shows a maximum value. On the other hand, when reflected light is circular polarization, the light intensity is independent of the analyzer angle, as shown in Fig. 4.5(c). This result can be understood from the propagation of circular polarization shown in Fig. 3.2(b). When reflected light is elliptical polarization, normalized light intensities are intermediate between linear and circular polarizations, as confirmed from Fig. 3.2(c).

Figure 4.7 shows the normalized light intensity when ψ is varied with a constant value of $\Delta = 180°$. The result shown in Fig. 4.7(a) is identical to that shown in Fig. 4.5(a). As shown in Fig. 4.7(a), when $(\psi, \Delta) = (45°, 180°)$, we observe $\alpha = 0$ and $\beta = -1$, since $\beta(= S_2)$ originally represents linear polarization of $\pm 45°$. When $\psi = 25°$, on the other hand, the light intensity becomes zero at $A = 25°$. It is clear

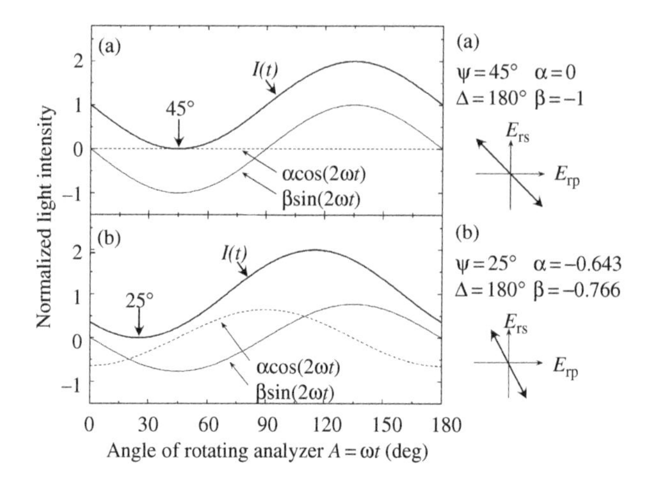

Figure 4.7 Normalized light intensity in rotating-analyzer ellipsometry (RAE), plotted as a function of the angle of rotating analyzer $A = \omega t$. This figure represents the calculation results when the polarization states of reflected light are (a) $\psi = 45°$, $\Delta = 180°$ and (b) $\psi = 25°$, $\Delta = 180°$.

from this result that the analyzer angle of $I(t) = 0$ is given by $A = \psi$ when $\Delta = 180°$. As shown in Fig. 4.7(b), $I(t)$ of $\psi = 25°$ is expressed by superimposing cosine and sine waves. This implies that the linear polarization of $\psi = 25°$ is described by synthesizing S_1 and S_2. From the above results, it can be seen that the shape of $I(t)$ slides in the horizontal direction (analyzer angle) depending on the value of ψ, and the amplitude of $I(t)$ reduces as the state of polarization changes from linear to elliptical polarization. In RAE, therefore, the polarization state of reflected light is determined from a variation of light intensity with the analyzer angle. In this method, however, left-circular polarization cannot be distinguished from right-circular polarization since these polarizations show the same light intensity variation versus the analyzer angle. This is the reason why S_3 cannot be measured and the measurement range for Δ becomes half ($0° \leq \Delta \leq 180°$) in RAE.

So far, we have assumed the polarizer angle to be $P = 45°$. The electric field of detected light for an arbitrary angle of the polarizer can be calculated by expanding Eq. (4.15):

$$E_A = \cos P \cos A \sin \psi \exp(i\Delta) + \sin P \sin A \cos \psi \tag{4.20}$$

Thus, we obtain the light intensity measured by the detector as follows [54,55]:

$$\begin{aligned} I &= |E_A|^2 \\ &= I_0 [(1 - \cos 2P \cos 2\psi) + (\cos 2P - \cos 2\psi) \cos 2A + (\sin 2P \sin 2\psi \cos \Delta) \sin 2A] \end{aligned} \tag{4.21}$$

It can be confirmed that substituting $P = 45°$ into Eq. (4.21) gives Eq. (4.18). If we normalize Eq. (4.21) by the term $(1 - \cos 2P \cos 2\psi)$ and rewrite the equation in the form of Eq. (4.19), we get the following equations [54]:

$$\alpha = \frac{\cos 2P - \cos 2\psi}{1 - \cos 2P \cos 2\psi} \quad \beta = \frac{\sin 2\psi \cos \Delta \sin 2P}{1 - \cos 2P \cos 2\psi} \tag{4.22}$$

where (α, β) are referred to as normalized Fourier coefficients. By transforming $\cos 2\psi$ in Eq. (4.22) using double-angle formulas (see Appendix 1), we get well-known forms for the normalized Fourier coefficients [56]:

$$\alpha = \frac{\tan^2 \psi - \tan^2 P}{\tan^2 \psi + \tan^2 P} \quad \beta = \frac{2 \tan \psi \cos \Delta \tan P}{\tan^2 \psi + \tan^2 P} \tag{4.23}$$

Solving Eq. (4.23) for (ψ, Δ) finally gives the following equations [14,22,57]:

$$\tan \psi = \sqrt{\frac{1+\alpha}{1-\alpha}} |\tan P| \quad \cos \Delta = \frac{\beta}{\sqrt{1-\alpha^2}} \tag{4.24}$$

In ellipsometry measurement using RAE, we first determine (α, β) from the Fourier analysis of measured light intensities (see Section 4.3.2) and then extract (ψ, Δ) values by substituting the measured (α, β) into Eq. (4.24). It should be noted that the P_RSA instrument is mathematically equivalent to the PSA_R instrument since the optical configuration of these instruments is exactly the same. Theoretical equations for P_RSA can be expressed by simply replacing $\tan P$ in Eqs. (4.23) and (4.24) with $\tan A$ [58].

4.2.3 ROTATING-ANALYZER ELLIPSOMETRY WITH COMPENSATOR

In RAE, the Stokes parameter S_3 is not measured and consequently the measurement error increases at $\Delta \cong 0°$ and $180°$ (see Section 4.4.1). If we introduce a compensator into RAE, we can overcome these problems [32,33]. Therefore, RAE with compensator has become increasing popular in recent years. Using the Jones vectors and matrices shown in Eq. (4.14), we can describe RAE with compensator ($PSCA_R$ configuration) as follows:

$$\boldsymbol{L}_{\text{out}} = \boldsymbol{A}\boldsymbol{R}(A)\boldsymbol{C}\boldsymbol{S}\boldsymbol{R}(-P)\boldsymbol{P}\boldsymbol{L}_{\text{in}} \tag{4.25}$$

where $\boldsymbol{C}$ is expressed by

$$\boldsymbol{C} = \begin{bmatrix} \exp(-i\delta) & 0 \\ 0 & 1 \end{bmatrix} \tag{4.26}$$

In Eq. (4.25), the compensator $\boldsymbol{C}$ is simply introduced into Eq. (4.14). In Eq. (4.26), the fast axis of the compensator is chosen in the direction of s-polarization, rather than p-polarization, as confirmed from the Jones matrix shown in Table 3.2. When the fast axis of the compensator is not parallel to the coordinates, the compensator should be expressed as $\boldsymbol{R}(-C)\boldsymbol{C}\boldsymbol{R}(C)$, where C is the rotation angle of the compensator. In Eq. (4.25), however, the compensator is described simply as $\boldsymbol{C}$, in order to simplify equations. In this case, the phase shift of the compensator δ in Eq. (4.26) is expressed as functions of C and the wavelength λ [i.e., $\delta(C, \lambda)$]. If we assume the polarizer angle to be $P = 45°$, Eq. (4.25) is simplified to

$$\begin{aligned}\begin{bmatrix} E_A \\ 0 \end{bmatrix} &= \begin{bmatrix} 1 & 0 \\ 0 & 0 \end{bmatrix} \begin{bmatrix} \cos A & \sin A \\ -\sin A & \cos A \end{bmatrix} \begin{bmatrix} \exp(-i\delta) & 0 \\ 0 & 1 \end{bmatrix} \begin{bmatrix} \sin\psi \exp(i\Delta) \\ \cos\psi \end{bmatrix} \\ &= \begin{bmatrix} 1 & 0 \\ 0 & 0 \end{bmatrix} \begin{bmatrix} \cos A & \sin A \\ -\sin A & \cos A \end{bmatrix} \begin{bmatrix} \sin\psi \exp[i(\Delta - \delta)] \\ \cos\psi \end{bmatrix}\end{aligned} \tag{4.27}$$

It can be seen from the above equation that the compensator introduced into RAE only shifts the Δ value without any effects on ψ. This result shows the fact that the compensator does not change the amplitudes of p- and s-polarizations but changes the relative phase difference between p- and s-polarizations. Accordingly, RAE with compensator can be described by simply replacing Δ with $\Delta' = \Delta - \delta$ in the equations derived for RAE. If we replace Δ with $\Delta' = \Delta - \delta$ in Eq. (4.18), we obtain

$$\begin{aligned} I &= I_0[1 - \cos 2\psi \cos 2A + \sin 2\psi \cos(\Delta - \delta) \sin 2A] \\ &= I_0[1 - \cos 2\psi \cos 2A + (\sin 2\psi \cos\Delta \cos\delta + \sin 2\psi \sin\Delta \sin\delta) \sin 2A] \\ &= I_0[1 + S_1 \cos 2A + (S_2 \cos\delta - S_3 \sin\delta) \sin 2A] \end{aligned} \tag{4.28}$$

For the transformation of Eq. (4.28), the addition theorem was used (see Appendix 1). As shown in Eq. (4.28), by inserting a compensator into RAE, the measurement of the Stokes parameters S_{1-3} becomes possible. It is clear from Eq. (4.28) that the two values, S_1 and $(S_2 \cos\delta - S_3 \sin\delta)$, are measured as the Fourier coefficients in RAE with compensator. In order to obtain S_2 and S_3 separately, however, at least two measurements should be performed with different δ [32]. The phase difference in the compensator can be changed by the rotation of the compensator. For example, it can be seen from Fig. 3.8 that $\delta = 90°$ when the polarization is oriented at 45° and $\delta = 0°$ when the fast or slow axis is parallel to the direction of polarized light. In the PCSA_R configuration, therefore, $\delta = 0°$ at $P = C = 45°$ and $\delta = 90°$ at $P = 45°$ and $C = 0°$. As confirmed from Eq. (4.28), we obtain S_2 when $\delta = 0°$ and S_3 when $\delta = 90°$. Notice from Eq. (4.25) that the equations for the PCSA_R configuration are identical to those for the PSCA_R configuration. The changes of the polarization states in the PSCA configuration will be explained in Section 4.2.4 in greater detail. However, keep in mind that the *slow* axis is parallel to p-polarization when $C = 0°$ in this section, while the *fast* axis is parallel to p-polarization when $C = 0°$ in Section 4.2.4. Thus, $C = 0°$ in this section corresponds to $C = 90°$ in Section 4.2.4.

As we will discuss in Section 4.4.1, the measurement accuracy of RAE is quite high at $\Delta \cong 90°$ and low at $\Delta \cong 0°$ and 180°. In RAE with compensator, however, the value of Δ' obtained from the measurement can be shifted by the compensator. Accordingly, if we measure several Δ' spectra with different δ and synthesize these spectra using the regions of $\Delta' \sim \pm 90°$ only, we can eliminate the error observed at $\Delta \cong 0°$ and 180° in RAE. Since all the Stokes parameters can be measured in RAE with compensator, the degree of polarization defined by Eq. (3.52) can also be determined from this technique [33,59–61].

4.2.4 ROTATING-COMPENSATOR ELLIPSOMETRY (RCE)

If we use the symbols of the Jones matrices shown in Table 3.2, the RCE instrument (PSC_RA configuration) is expressed by

$$\boldsymbol{L}_{\text{out}} = \boldsymbol{AR}(A)\boldsymbol{R}(-C)\boldsymbol{CR}(C)\boldsymbol{SR}(-P)\boldsymbol{PL}_{\text{in}} \tag{4.29}$$

The above equation can be obtained easily by inserting $\boldsymbol{R}(-C)\boldsymbol{CR}(C)$ into Eq. (4.14). Here, we assume polarizer and analyzer angles to be $P = 45°$ and $A = 0°$, since the direct expansion of Eq. (4.29) leads to a complicated equation. If we use these conditions and $\boldsymbol{L}_{\text{in}} = [1, 0]^{\text{T}}$, Eq. (4.29) is described by the following equation:

$$\begin{bmatrix} E_{\text{A}} \\ 0 \end{bmatrix} = \begin{bmatrix} 1 & 0 \\ 0 & 0 \end{bmatrix} \begin{bmatrix} \cos C & -\sin C \\ \sin C & \cos C \end{bmatrix} \begin{bmatrix} 1 & 0 \\ 0 & \exp(-\mathrm{i}\delta) \end{bmatrix} \times \begin{bmatrix} \cos C & \sin C \\ -\sin C & \cos C \end{bmatrix} \begin{bmatrix} \sin\psi \exp(\mathrm{i}\Delta) \\ \cos\psi \end{bmatrix} \tag{4.30}$$

In Eq. (4.30), the fast axis of the compensator is in the direction of p-polarization and the proportional constant is eliminated. By substituting $\delta = 90°$ into Eq. (4.30), we get

$$E_{\text{A}} = \left(\cos^2 C - \mathrm{i}\sin^2 C\right)\sin\psi \exp(\mathrm{i}\Delta) + (1 + \mathrm{i})\cos C \sin C \cos\psi \tag{4.31}$$

From Eq. (4.31), we obtain a light intensity detected by a light detector as follows [18,21]:

$$\begin{aligned} I &= |E_{\text{A}}|^2 \\ &= I_0\,(2 - \cos 2\psi + 2\sin 2\psi \sin\Delta \sin 2C - \cos 2\psi \cos 4C + \sin 2\psi \cos\Delta \sin 4C) \\ &= I_0\,(2 + S_1 - 2S_3 \sin 2C + S_1 \cos 4C + S_2 \sin 4C) \end{aligned} \tag{4.32}$$

It is evident from Eq. (4.32) that, if we rotate the compensator with an angle of C, S_{1-3} can be determined as the Fourier coefficients. As shown in Eq. (4.32), when

$A = 0°$, the term cos2C vanishes [18]. In general, Eq. (4.32) is described by a more general formula [17–21]:

$$I(t) = I_0(1 + \alpha_2 \cos 2\omega t + \beta_2 \sin 2\omega t + \alpha_4 \cos 4\omega t + \beta_4 \sin 4\omega t) \quad (4.33)$$

In Eq. (4.33), the angle of the rotating compensator is described as $C = \omega t$ and $(\alpha_{2,4}, \beta_{2,4})$ show the normalized Fourier coefficients. When $P = 45°$, these Fourier coefficients for an arbitrary analyzer angle A and compensator phase difference δ are given by

$$\alpha_2 = \frac{p \sin \delta \sin 2\varepsilon \sin 2A}{\alpha_0} \quad (4.34a)$$

$$\beta_2 = -\frac{p \sin \delta \sin 2\varepsilon \cos 2A}{\alpha_0} \quad (4.34b)$$

$$\alpha_4 = \frac{p \sin^2 (\delta/2) \cos 2\varepsilon \cos 2(A + \theta)}{\alpha_0} \quad (4.34c)$$

$$\beta_4 = \frac{p \sin^2 (\delta/2) \cos 2\varepsilon \sin 2(A + \theta)}{\alpha_0} \quad (4.34d)$$

$$\alpha_0 = 1 + p \cos^2 (\delta/2) \cos 2\varepsilon \cos 2(A - \theta) \quad (4.34e)$$

The above equations have been derived from the Mueller matrices using the (ε, θ) coordinates [19]. It can be confirmed from Table 3.3 that the same Stokes parameters are described in Eqs. (4.32) and (4.34) when $A = 0°$. In Eq. (4.34), p represents the degree of polarization and, when we express partially polarized light, the Stokes vector (S_{1-3}) includes the coefficient p, as shown in Eq. (3.54). In RCE, the degree of polarization is estimated by obtaining this p value [19]. Although (ψ, Δ) can be derived directly from the Mueller matrix calculation [62], these equations will not be shown here due to their complexity.

Figure 4.8 shows the normalized light intensity calculated from Eq. (4.33), plotted as a function of the angle of the rotating compensator. In this figure, the results when $P = A = 45°$ are shown. In this calculation, the values of (ε, θ) shown in Table 4.1 were used and the degree of polarization was assumed to be $p = 1$. From Eq. (4.34b), it follows that $\beta_2 = 0$ when $A = 45°$. The polarizations indicated by arrows in Fig. 4.8 represent the polarization states of light emerging from the compensator. As shown in Fig. 4.8(a), the light intensity of the linear polarization $[(\psi, \Delta) = (45°, 180°)]$ varies with a period of 90°(cos4C). This behavior can be understood from the schematic diagram illustrated in Fig. 4.9(a). Now suppose that reflected light is linear polarization oriented at −45°, as shown in Fig. 4.9(a). In this case, when the reflected light transmits the compensator with an angle of $C = 0°$, the light is transformed into right-circular polarization (E_R). This phenomenon can be explained easily from Fig. 3.3. Specifically, when $\delta_x - \delta_y = 180°(\Delta = 180°)$, if the phase of the y direction (δ_y) lags further by 90°, the total phase difference becomes

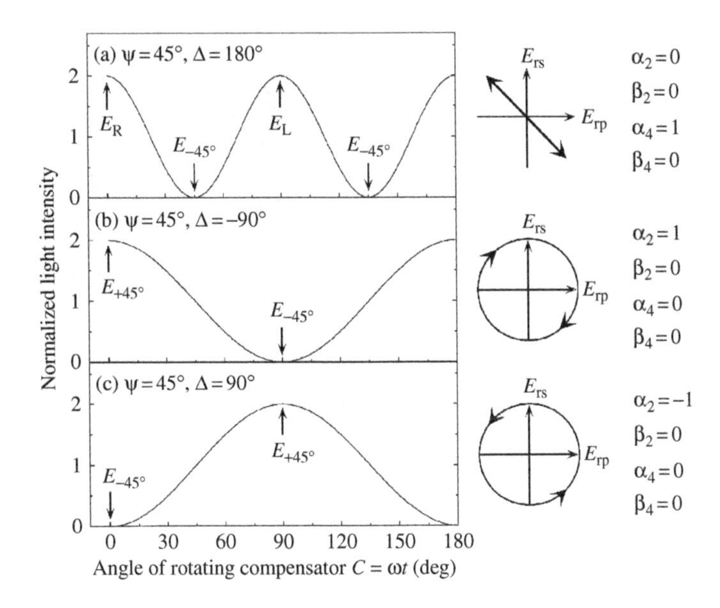

Figure 4.8 Normalized light intensity in the rotating-compensator ellipsometry (RCE), plotted as a function of the angle of rotating compensator $C = \omega t$. This figure summarizes the calculation results when the polarization states of reflected light are (a) $\psi = 45°$, $\Delta = 180°$, (b) $\psi = 45°$, $\Delta = -90°$, and (c) $\psi = 45°$, $\Delta = 90°$. The polarizations denoted by arrows represent the polarization states of reflected light emerging from the rotating compensator.

$\delta_x - \delta_y = 270°$ (right-circular polarization). Recall that the phase of the y component (s-polarization) lags when $C = 0°$ (i.e., the fast axis is parallel to the x direction). As shown in Fig. 4.9(a), this right-circular polarization further transmits the analyzer oriented at 45°. In this case, we can resolve the right-circular polarization into the directions that are parallel and perpendicular to the transmission axis and only the parallel component transmits the analyzer. Finally, a light detector measures its light intensity.

When the compensator rotates to $C = 45°$ in Fig. 4.9(a), the slow axis of the compensator coincides with the linear polarization of the reflected light ($E_{-45°}$). Thus, the reflected light transmits the compensator without any change. However, since the linear polarization (−45°) transmitted through the compensator is perpendicular to the transmission axis of the analyzer ($A = 45°$), the detected light intensity becomes zero at $C = 45°$, as shown in Fig. 4.8(a). When the compensator rotates further to $C = 90°$, the reflected light is transformed into left-circular polarization (E_L) by the compensator. In this case, as the compensator is oriented at $C = 90°$, $\delta_x - \delta_y = 180°$ is converted to $\delta_x - \delta_y = 90°$ (left-circular polarization in Fig. 3.3). As a result, the detected light intensity is restored to a maximum value at $C = 90°$. At $C = 135°$, since the fast axis of the compensator matches with the linear polarization (−45°), the reflected light remains the same after passing through the compensator and the light intensity becomes zero again. In PSC_RA measurement,

(a)

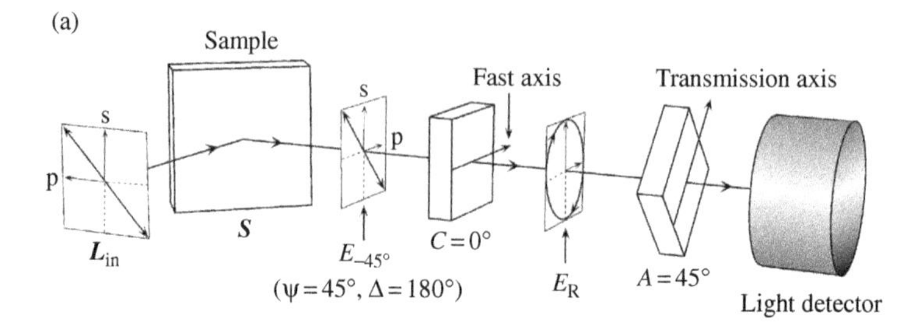

(b)

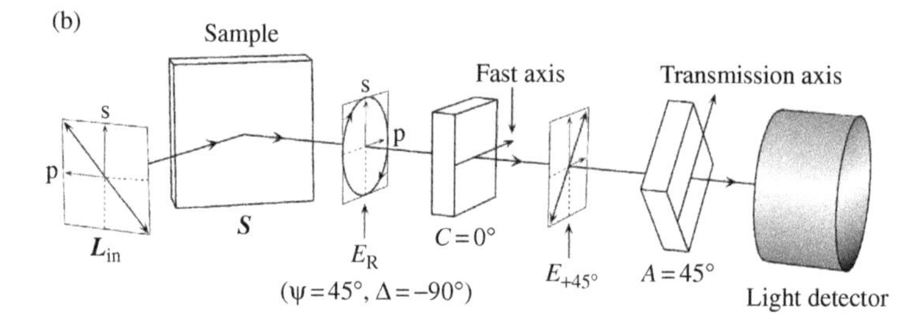

Figure 4.9 Schematic diagrams of the measurements in the rotating-compensator ellipsometry (RCE) when the reflected light waves from a sample are (a) linear polarization of $-45°(\psi = 45°, \Delta = 180°)$ and (b) right-circular polarization ($\psi = 45°, \Delta = -90°$).

therefore, the light intensity of linear polarization is modulated by the rotation of the compensator with a period of 90°. This is the reason why the Stokes parameters S_1 and S_2 are measured as the Fourier coefficients of $\cos 4C$ and $\sin 4C$.

As shown in Fig. 4.8(b), when the reflected light is right-circular polarization, light intensity varies with a period of 180° ($\cos 2C$). If we express the Stokes vector of the reflected light by $\boldsymbol{L}_R = [1, 0, 0, 1]^T$ (right-circular polarization), the polarization state of the light emerging from the compensator is calculated from

$$\boldsymbol{L}_C = \boldsymbol{R}(-C)\boldsymbol{C}\boldsymbol{R}(C)\boldsymbol{L}_R \tag{4.35}$$

From the calculation using the Mueller matrices (see Table 3.2), we obtain

$$\boldsymbol{L}_C = [1, -\sin 2C, \cos 2C, 0]^T \tag{4.36}$$

Eq. (4.36) shows that the compensator ($\delta = 90°$) transforms the right-circular polarization into linear polarization described by $-\sin 2C\ (S_1)$ and $\cos 2C\ (S_2)$, and the direction of linear polarization varies with the rotation angle of the compensator. In other words, when the reflected light is circular polarization, the variation of light intensity versus the compensator angle C becomes similar to the case when an analyzer angle is changed in RAE. As shown in Fig. 4.9(b), right-circular polarization is changed into linear polarization of $+45°(E_{+45°})$ when $C = 0°$. In fact, if we set $C = 0°$ in Eq. (4.36), we obtain $\boldsymbol{L}_C = [1, 0, 1, 0]^T$ which represents

$E_{+45°}$. Furthermore, this transformation corresponds to the one when the propagation direction of light is reversed in Fig. 3.8. In particular, when the propagation direction is reversed, left-circular polarization becomes right-circular polarization, as the phase difference is also reversed. As confirmed from Fig. 4.9(b), the light transformed into +45° linear polarization by the compensator passes through the analyzer ($A = 45°$) without any disturbance. Thus, when the reflected light is right-circular polarization, the detected light intensity is maximized at $C = 0°$. If the compensator rotates further to $C = 90°$, the polarization of the light emerging from the compensator becomes linear polarization of $-45°(E_{-45°})$, as confirmed from Eq. (4.36). Consequently, the detected light intensity becomes zero at $C = 90°$.

When reflected light is left-circular polarization, on the other hand, we obtain the following equation by substituting $\boldsymbol{L}_R = [1, 0, 0, -1]^T$ into Eq. (4.35):

$$\boldsymbol{L}_C = [1, \sin 2C, -\cos 2C, 0]^T \tag{4.37}$$

Thus, when $C = 0°$, the left-circular polarization is changed into linear polarization oriented at $-45°$. In this case, the light intensity becomes zero, as shown in Fig. 4.8(c). From this result, it can be seen that the light intensity variation of left-circular polarization versus C becomes opposite to that of right-circular polarization. In other words, the signs of the normalized Fourier coefficient α_2 are reversed, as shown in Fig. 4.8. This is the reason why left- and right-circular polarizations are distinguished in RCE measurement.

Figure 4.10 shows the calculation results when the polarization states of reflected light waves are elliptical polarization ($\psi = 45°$, $\Delta = 135°$) and linear polarization

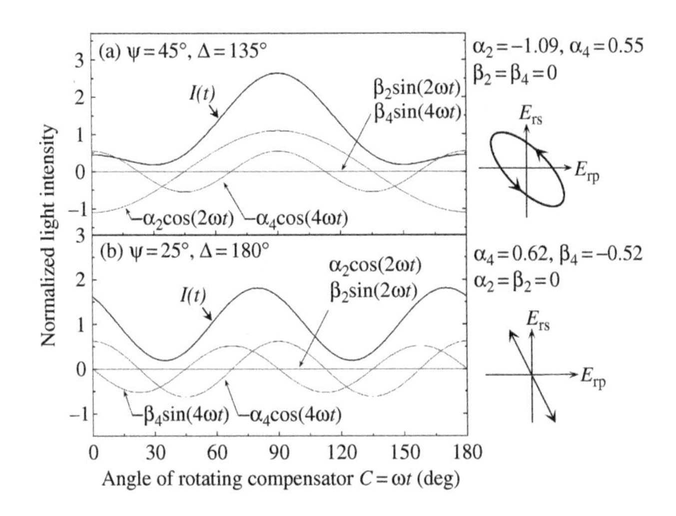

Figure 4.10 Normalized light intensity in rotating-compensator ellipsometry (RCE), plotted as a function of the angle of rotating compensator $C = \omega t$. This figure summarizes the calculation results when the polarization states of reflected light are (a) $\psi = 45°$, $\Delta = 135°$ and (b) $\psi = 25°$, $\Delta = 180°$.

($\psi = 25°$, $\Delta = 180°$). As shown in Fig. 4.10(a), elliptical polarization is described by the superposition of circular polarization (α_2) and linear polarization (α_4). Since this elliptical polarization is rotating counterclockwise, the sign of α_2 is negative, similar to left-circular polarization shown in Fig. 4.8(c). In contrast, linear polarization ($\psi = 25°$) shown in Fig. 4.10(b) is described by adding $\alpha_4(S_2)$ to $\beta_4(S_1)$. Recall that $\psi = 25°$ is also expressed by S_1 and S_2 in RAE (see Fig. 4.7). From the results described above, it can be understood that, in RCE, right- and left-circular polarizations are determined from the signs of α_2, ψ is expressed by the superposition of α_4 and β_4, and elliptical polarization is described by the addition of α_2 to α_4 and β_4.

4.2.5 PHASE-MODULATION ELLIPSOMETRY (PME)

If we use the Jones matrices shown in Table 3.2, a PME instrument with a PSMA configuration is described by

$$\boldsymbol{L}_{\text{out}} = \boldsymbol{A}\boldsymbol{R}(A)\boldsymbol{R}(-M)\boldsymbol{M}\boldsymbol{R}(M)\boldsymbol{S}\boldsymbol{R}(-P)\boldsymbol{P}\boldsymbol{L}_{\text{in}} \tag{4.38}$$

By assuming a polarizer angle of $P = 45°$ and $\boldsymbol{L}_{\text{in}} = [1, 0]^{\text{T}}$, we obtain

$$\begin{aligned}\begin{bmatrix} E_{\text{A}} \\ 0 \end{bmatrix} &= \begin{bmatrix} 1 & 0 \\ 0 & 0 \end{bmatrix} \begin{bmatrix} \cos(A-M) & \sin(A-M) \\ -\sin(A-M) & \cos(A-M) \end{bmatrix} \begin{bmatrix} 1 & 0 \\ 0 & \exp(\text{i}\delta) \end{bmatrix} \\ &\times \begin{bmatrix} \cos M & \sin M \\ -\sin M & \cos M \end{bmatrix} \begin{bmatrix} \sin\psi\exp(\text{i}\Delta) \\ \cos\psi \end{bmatrix}\end{aligned} \tag{4.39}$$

In Eq. (4.39), the proportional constant was neglected and $\boldsymbol{R}(A)\boldsymbol{R}(-M) = \boldsymbol{R}(A-M)$ was used [Eq. (3.18)]. The expansion of Eq. (4.39) leads to a complicated equation, similar to the case of RCE. In PME, however, we set the angle between the analyzer and the photoelastic modulator to $A - M = 45°$ [23–27]. By expanding Eq. (4.39) using $A - M = 45°$, we get

$$E_{\text{A}} = [\cos M - \sin M \exp(\text{i}\delta)]\sin\psi\exp(\text{i}\Delta) + [\cos M \exp(\text{i}\delta) + \sin M]\cos\psi \tag{4.40}$$

In the above equation, the proportional constant is eliminated. From Eq. (4.40), the light intensity in PSMA is expressed as

$$\begin{aligned} I &= |E_{\text{A}}|^2 \\ &= I_0[1 + \sin 2\psi \sin\Delta \sin\delta + (\cos 2\psi \sin 2M + \sin 2\psi \cos\Delta \cos 2M)\cos\delta] \\ &= I_0[1 - S_3 \sin\delta + (-S_1 \sin 2M + S_2 \cos 2M)\cos\delta] \end{aligned} \tag{4.41}$$

It can be seen from Eq. (4.41) that the Stokes parameters S_2 and S_3 are measured when $M = 0°$, while S_1 and S_3 are obtained when $M = 45°$. This is similar to RAE with compensator in which the measured Stokes parameters change depending on the rotational angle of the compensator. Recall from Eq. (3.4) that δ represents the phase shift of the photoelastic modulator. In PME, therefore, the Stokes parameters S_{1-3} are measured as the Fourier coefficient of δ. Notice that Fig. 4.3(d) shows the optical configuration when $M = 45°$.

So far, we have assumed the PSMA configuration shown in Fig. 4.3(d). In actual measurements, however, the PMSA configuration is commonly employed since the optical alignment of the photoelastic modulator becomes easier [23–30]. In the PMSA configuration, the light intensity is given by the following equation, similar to Eq. (4.41) [26,27]:

$$I(t) = I_0\{\alpha_0 + \alpha_1 \sin\delta + \alpha_2 \cos\delta\} \tag{4.42}$$

The α_{0-2} for the arbitrary angles of M, A, and $P-M$ are expressed by the following equations [27]:

$$\begin{aligned}\alpha_0 =& [1-\cos 2\psi \cos 2A + \cos 2(P-M)\cos 2M(\cos 2A - \cos 2\psi) \\ &+ \sin 2A \cos\Delta \cos 2(P-M)\sin 2\psi \sin 2M]\end{aligned} \tag{4.43a}$$

$$\alpha_1 = \sin 2(P-M)\sin 2A \sin 2\psi \sin\Delta \tag{4.43b}$$

$$\begin{aligned}\alpha_2 =& \sin 2(P-M)[(\cos 2\psi - \cos 2A)\sin 2M \\ &+ \sin 2A \cos 2M \sin 2\psi \cos\Delta]\end{aligned} \tag{4.43c}$$

If we set $P-M = 45°$, $M = 0°$, and $A = 45°$ in Eq. (4.43), we obtain $\alpha_0 = 1$, $\alpha_1 = \sin 2\psi \sin\Delta$ and $\alpha_2 = \sin 2\psi \cos\Delta$. In this case, therefore, Eq. (4.43) is reduced to Eq. (4.41). By substituting $\delta = F \sin\omega t$ [Eq. (3.4)] into $\sin\delta$ and $\cos\delta$, we obtain the following equations [24]:

$$\sin\delta = \sin(F\sin\omega t) = 2\sum_{m=0}^{\infty} J_{2m+1}(F)\sin[(2m+1)\omega t] \tag{4.44a}$$

$$\cos\delta = \cos(F\sin\omega t) = J_0(F) + 2\sum_{m=1}^{\infty} J_{2m}(F)\cos(2m\omega t) \tag{4.44b}$$

Here, the terms J_k are Bessel functions with respect to F. If we use Eq. (4.44), the Fourier analysis becomes complicated. Thus, only the low frequency components in Eq. (4.44) are generally taken into account. In this case, we get

$$\sin\delta = 2J_1(F)\sin\omega t \ \ (\mathrm{m} = 0) \tag{4.45a}$$

$$\cos\delta = J_0(F) + 2J_2(F)\cos 2\omega t \ \ (\mathrm{m} = 1) \tag{4.45b}$$

If we adjust the voltage applied to the photoelastic modulator so that $F = 138°$, the analysis is simplified further [24]. In this case, we obtain $J_0(F) = 0$ [24],

$2J_1(F) = 1.04$ and $2J_2(F) = 0.86$ [27]. By substituting Eq. (4.45) into Eq. (4.42) and using $J_0(F) = 0$, we get

$$I(t) = I_0\{1 + \sin 2\psi \sin \Delta[2J_1(F) \sin \omega t] + \sin 2\psi \cos \Delta[2J_2(F) \cos 2\omega t]\} \quad (4.46)$$

In Eq. (4.46), the configuration of $P - M = 45°$, $M = 0°$, and $A = 45°$ is assumed. It is evident from Eq. (4.46) that (ψ, Δ) values can be estimated from the Fourier coefficients of $\sin \omega t$ and $\cos 2\omega t$. Figure 4.11 shows a measurement example of the PME instrument [28]. In Fig. 4.11(a), relative light intensities versus measurement time are represented by voltages. Figure 4.11(b) shows the spectrum obtained from the fast Fourier transformation (FFT) of the waveform shown in Fig. 4.11(a). In Fig. 4.11(b), we can observe the two sharp peaks at the positions corresponding to the modulation frequency of the photoelastic modulator (50 kHz) and two times its frequency (100 kHz). From the magnitude of these Fourier coefficients, the polarization state of reflected light is determined in PME. In this respect, the PME instrument is quite similar to the RAE and RCE instruments.

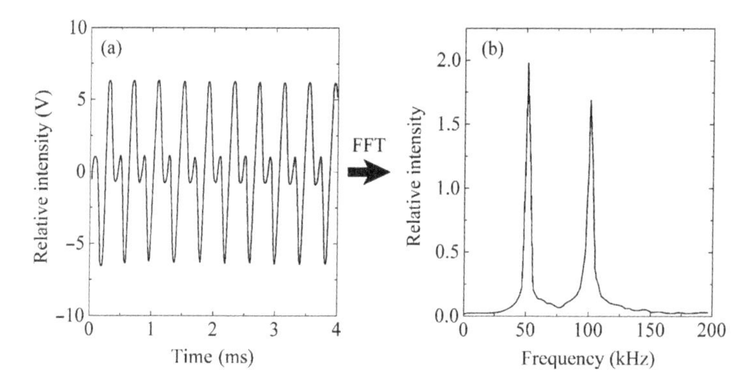

Figure 4.11 (a) Relative light intensity versus measurement time and (b) relative intensity of Fourier coefficients versus frequency in phase-modulation ellipsometry (PME). The result shown in (b) can be obtained from the fast Fourier transformation (FFT) of (a). Reprinted from *Applied Surface Science*, **63**, W. M. Duncan and S. A. Henck, Insitu spectral ellipsometry for real-time measurement and control, 9–16, Copyright (1993), with permission from Elsevier.

4.2.6 INFRARED SPECTROSCOPIC ELLIPSOMETRY

From infrared spectroscopic ellipsometry, free-carrier absorption and various infrared vibrations including photon modes (LO and TO phonons) and local vibration modes can be characterized (see Fig. 1.1). Until the early 1980s, measurements of infrared spectroscopic ellipsometry had been performed using a monochromator combined with an infrared light source [63,64]. In 1981, Röseler reported the first ellipsometry instrument that incorporated the Fourier-transform infrared spectrometer (FTIR) as a light source [31]. This instrument was based on the rotating-polarizer system [P_RSA configuration shown in Fig. 4.12(a)] and was modified later to RAE with compensator [32,33]. This breakthrough led to the development of FTIR-PSA_R [34] and FTIR-PSC_RA shown in Fig. 4.12(b) [35]. In 1986, on the other hand, Graf *et al.* reported the first infrared instrument that combined FTIR with PME [37]. The optical configuration of this instrument is illustrated in Fig. 4.12(c). The calibration procedure of this instrument was completed by Carillas *et al.* in 1993 [38]. In particular, this PME instrument has been applied to real-time spectroscopic measurements [39,40]. As shown in Fig. 4.12, the basic designs of these instruments are the same as those shown in Fig. 4.3, except for FTIR placed on the incident side. In these instruments, however, a wire-grid polarizer (analyzer) shown in Fig. 3.7 is employed. For infrared ellipsometry measurement, compensators of the reflection type [32,33] and the transmission type [35,36] have been employed.

Figure 4.13 summarizes the procedure of ellipsometry measurement in the FTIR-P_RSA configuration [33]. As shown in Fig. 4.13(a), FTIR is basically composed of a light source and the Michelson interferometer, which further consists of a beamsplitter, a fixed mirror, and a moving mirror. The light source of FTIR is generally a heated silicon carbide rod (glowbar light source). In the Michelson interferometer, the light emitted from the light source is divided into two directions and the light beams reflected by each mirror are combined again after passing through the beamsplitter. In infrared spectroscopic ellipsometry, this light is employed as the probe light. As shown in Fig. 4.13(a), let a and b be the distances from the beamsplitter to the fixed and moving mirrors, respectively. In the Michelson interferometer, the light waves reflected with distance a and b overlap and the optical interference occurs between these waves. In this case, the phase difference between the two light waves is given by $\delta = 2\pi x/\lambda$, where x is the difference in optical pass length [see Eq. (2.75)]. If we use the position of the moving mirror d, x is given by $x = 2d$. Consequently, the intensity of light emerging from the interferometer changes with $I \propto \cos(4\pi d/\lambda)$. In general, light intensity measured as a function of d is referred to as the interferogram.

Now suppose that light having a broad spectral range enters the interferometer. In particular, when $d = 0$, light waves reflected by the two mirrors interfere constructively for all the wavelengths since $a = b$, and the intensity of light emerging from the interferometer becomes high, as shown in Fig. 4.13(b). As d increases, however, destructive interference occurs since the light intensity at the position d varies according to the wavelength of light. As a result, the light intensity reduces gradually with increasing d, as shown in Fig. 4.13(b). Thus, it is evident that the interferogram includes all the contributions of different wavelengths. In FTIR, a light intensity spectrum is determined from the Fourier transformation of the interferogram, as shown in Fig. 4.13(c). In FTIR-P_RSA, several FTIR measurements are performed with different polarizer angles. In Fig. 4.13(c), four intensity spectra are measured at $P = 0°, 45°, 90°$, and $135°$. If we know the light intensities at

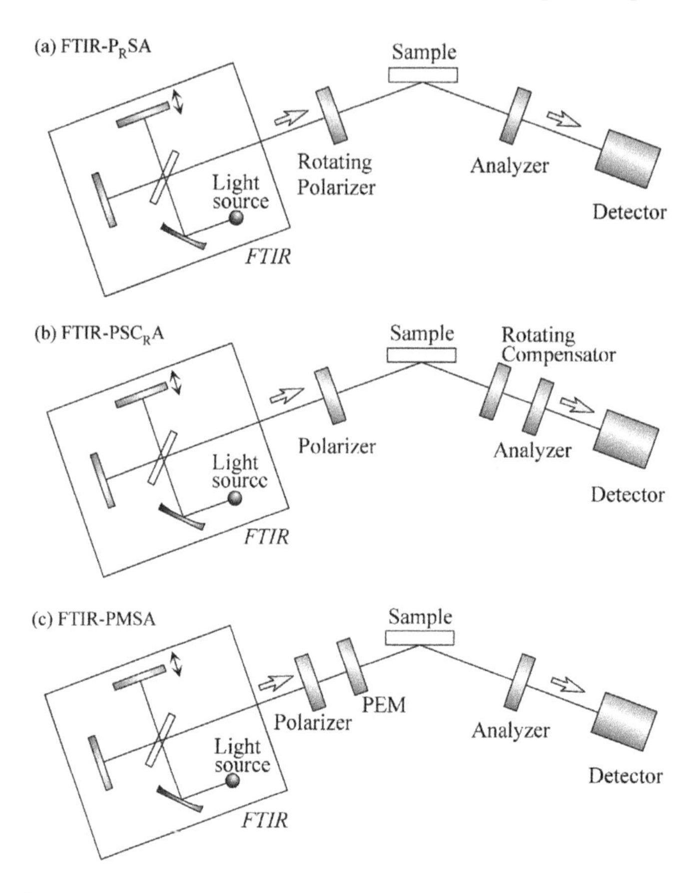

Figure 4.12 Optical configurations of infrared ellipsometry instruments: (a) the rotating-polarizer ellipsometry (FTIR-P$_R$SA), (b) the rotating-compensator ellipsometry (FTIR-PSC$_R$A), and (c) the phase-modulation ellipsometry (FTIR-PMSA). These instruments employ the Fourier-transform infrared spectrometer (FTIR) as a light source.

these angles, we can calculate the Stokes parameters (S_1 and S_2) directly from Table 3.3 [33]:

$$\frac{S_1}{S_0} = \frac{I_{0°} - I_{90°}}{I_{0°} + I_{90°}} = -\cos 2\psi \tag{4.47a}$$

$$\frac{S_2}{S_0} = \frac{I_{45°} - I_{135°}}{I_{0°} + I_{90°}} = \sin 2\psi \cos \Delta \tag{4.47b}$$

Recall from Fig. 4.2 that I_x and I_y in Table 3.3 correspond to $I_{0°}$ and $I_{90°}$, respectively. It should be noted that $I_{-45°} = I_{135°}$ since $P = -45°$ and $135°$ are the same setting. It can be seen from Eq. (4.47) that (ψ, Δ) values can be obtained from the measured values of S_{0-2}. As mentioned earlier, the Stokes parameter S_3 cannot be determined

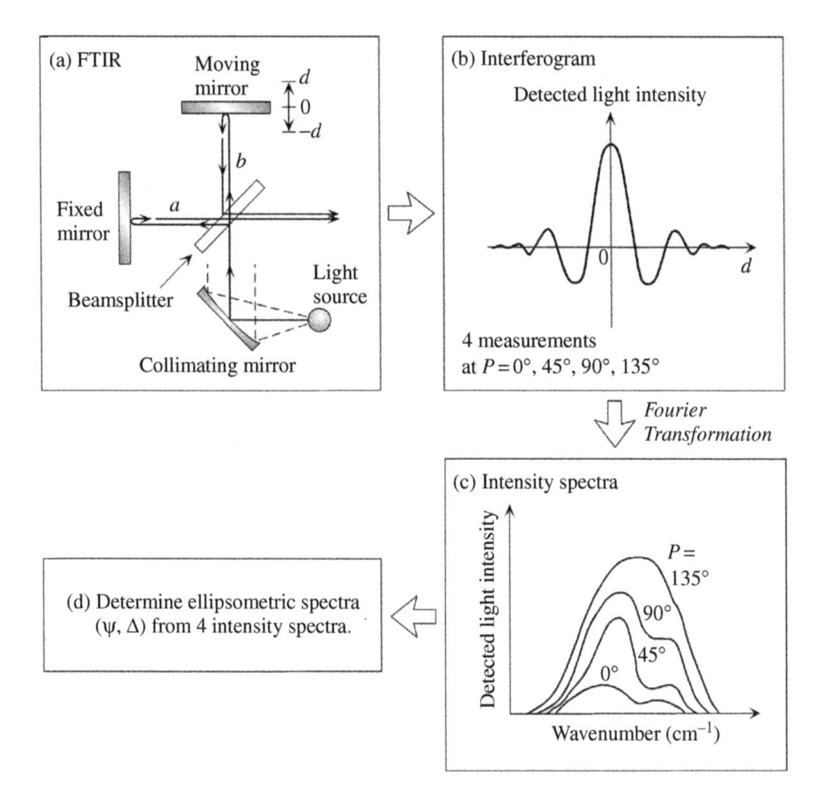

Figure 4.13 Measurement procedure of infrared ellipsometry using the FTIR-P_RSA configuration. Reprinted from Thin Solid Films, **234**, A. Röseler, IR spectroscopic ellipsometry: instrumentation and results, 307–313, Copyright (1993), with permission from Elsevier.

with the P_RSA configuration and the compensator is required to estimate S_3 (see Section 4.2.3). The measurement procedure for the FTIR-PSC_RA configuration is essentially the same. In this instrument, several intensity spectra are measured at different compensator and polarizer angles [35].

As discussed above, the measurement methods in the infrared region are slightly different from those in the visible/UV region, although the optical configurations are the same. In particular, since the modulation frequency of the interferometer is of the same order as that of the rotating elements, FTIR measurement becomes difficult if we rotate the optical elements continuously. Accordingly, in these infrared instruments that employ rotating elements, several measurements are necessary to obtain (ψ, Δ) spectra. It should be noted that the distinction between RCE and RAE with compensator is rather ambiguous in the infrared instruments, as the optical elements are not rotating continuously.

In contrast to the rotating-element ellipsometers, an infrared phase-modulation ellipsometer can be applied to real-time measurement, since the modulation frequency of a photoelastic modulator is quite different from that of an interferometer. In this technique, the interferometer and photoelastic modulator are operated simultaneously during the measurement and consequently the photoelastic modulator modifies the interferogram of FTIR [38]. The photoelastic modulator used in the infrared region is composed of ZnSe crystal cemented together with quartz crystal and the structure is similar to the one shown in Fig. 3.9(a) [37,38]. The resonant frequency of this photoelastic modulator is $\omega = 37\,\text{kHz}$. In PME, light intensity signals appearing at ω and 2ω are obtained from the Fourier analysis, as we have seen in Fig. 4.11. In the case of FTIR-PME, we obtain the interferograms corresponding to ω and 2ω from the Fourier analysis [38]. Thus, if we convert these interferograms into intensity spectra, (ψ, Δ) spectra can be determined from a similar analysis to that described in Section 4.2.5 [37,38,40].

Unfortunately, the infrared instruments described above generally suffer from relatively low signal-to-noise ratio, mainly due to the low light intensity of the glowbar light source. Although the sensitivity can be improved by increasing measurement time, this cannot be applied to real-time measurement. In real-time measurement using FTIR-PME, therefore, a cascade arc lamp has been applied [39]. This arc lamp emits light corresponding to a blackbody radiation of 12000 K [39], which is much higher than the temperature of the glowbar source ($\sim$1200 K) [36,38]. Moreover, in order to increase light intensity, an infrared beam is generally extracted from a light source with a large diameter ($\sim$5 cm) [33,38]. In this case, the light beam has to be focused so that the light enters into a detector window. Nevertheless, if the light beam is focused too much, an incident angle changes and measurement error may increase by a depolarization effect (see Section 4.4.4).

As we have seen in Table 1.2, the characterization of low absorption coefficients ($\alpha < 100\,\text{cm}^{-1}$) is rather difficult in the ellipsometry technique and measurement noise generally appears in the region of $\alpha < 10^3\,\text{cm}^{-1}$ (see Section 4.4.3). In hydrogenated amorphous silicon (a-Si:H), for example, the absorption coefficient of a Si–H stretching mode in the infrared region ($\sim 2000\,\text{cm}^{-1}$) is $\alpha \sim 1000\,\text{cm}^{-1}$ ($k \sim 0.05$) when hydrogen content is $\sim$15 at.% [65]. Thus, the characterization of this mode is rather difficult using ellipsometry. For the characterization of this infrared mode, therefore, infrared attenuated total reflection (ATR) spectroscopy, which provides monolayer sensitivity, has been more popular [66,67]. Recently, a real-time monitoring system that incorporates spectroscopic ellipsometry (visible/UV region) and ATR has been applied to characterize a-Si:H film growth [68,69] (see Fig. 8.10). In contrast to this particular mode, other modes including Si–O [70] and C=C [71] show much higher light absorption ($k \sim 1$) in the infrared region. Moreover, optical light absorption by free carriers [72] and optical phonon modes (LO and TO modes) [73] is also quite high. Thus, these characterizations can be performed using infrared spectroscopic ellipsometry. It should be emphasized that optical light transitions in the infrared region vary significantly according to individual infrared modes. Accordingly, when infrared spectroscopic ellipsometry is applied to material characterization, one needs to confirm that ellipsometry instruments have enough sensitivity for measurements.

4.2.7 MUELLER MATRIX ELLIPSOMETRY

Recently, Mueller matrix ellipsometry has been applied widely for the characterization of optically anisotropic samples [74]. The important feature of Mueller matrix ellipsometry is its ability to allow direct measurement of the Mueller matrix corresponding to a sample. As confirmed from Table 3.2, if a sample is optically isotropic, the Mueller matrix of the sample has only three independent parameters (i.e., the Stokes parameters S_{1-3}). Thus, an RCE measurement, for example, is sufficient to characterize the whole Mueller matrix. In contrast, the Mueller matrix for an anisotropic sample is expressed from six independent parameters [4]. As we will see in Chapter 6, all the ellipsometry instruments can still be applied for this characterization. Nevertheless, when an anisotropic sample has a depolarization effect, we need at least seven parameters to define the Mueller matrix [4]. In this case, the Jones matrix cannot be used to describe a sample and thus the sample is characterized from the Mueller matrix determined by Mueller matrix ellipsometry [74].

Since the late 1990s, extensive efforts have been made for anisotropic sample characterization, and Mueller matrix ellipsometers of various types have been developed [47–51]. However, instrument design for Mueller matrix ellipsometry began as early as the late 1970s [41–43]. Figure 4.14 illustrates the basic designs for Mueller matrix ellipsometry, which have been reviewed by Hauge [21]. In this figure, the Mueller matrix corresponding to a sample is denoted by dots, and large dots represent the elements of the Mueller matrix obtained from the measurement. As shown in this figure, all the instruments have a light source and a polarizer on the incident side and an analyzer and a detector on the reflection side. In the instrument shown in Fig. 4.14(a), the polarizer and the analyzer are rotated simultaneously during the measurement. However, this instrument does not allow the measurement of the 4th row and column of the Mueller matrix. If we introduce a rotating compensator after the sample [Fig. 4.14(b)], measurement of the first three columns becomes possible. In contrast, when a rotating compensator is placed on the incident side [Fig. 4.14(c)], the first three rows can be measured. Notice that the optical configurations of P_RSC_RA and PC_RSA_R shown in Figs. 4.14(b) and (c) are identical to those of RAE with compensator and RCE. Accordingly, these instruments can also be applied to determine the first three rows or columns of the Mueller matrix. In RCE (PC_RSA configuration), for example, the first three rows can be obtained from several measurements performed at different rotation angles of the analyzer [75]. As shown in Fig. 4.14, in the Mueller matrix ellipsometer that uses the rotating optical elements, only the dual rotating-compensator instrument provides complete measurement of the Mueller matrix [Fig. 4.14(d)]. If the compensators are rotated simultaneously using different rotation frequencies, all the elements of the Mueller matrix can be extracted from a single measurement. The design of this instrument was originally proposed in the late 1970s [41–43], and the constructions of this instrument were reported in the early 1990s [45,46]. The first Mueller matrix ellipsometer that allows real-time spectroscopic measurement (PC_RSC_RA) was reported in 2000 [49,50].

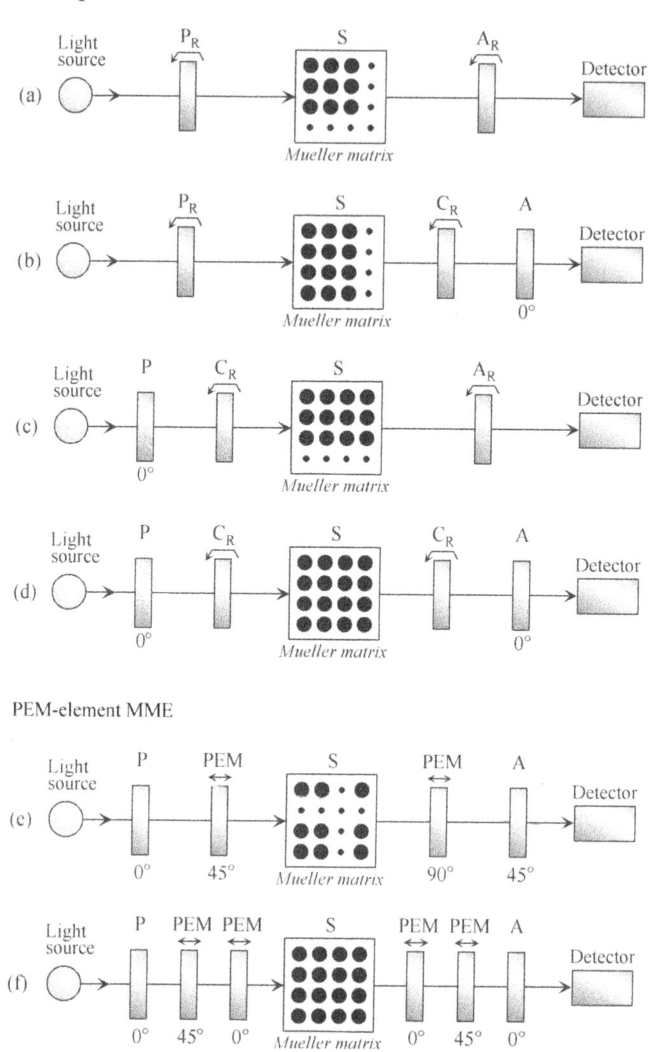

Figure 4.14 Optical configurations for Mueller matrix ellipsometry (MME): rotating-element MME [(a)–(d)] and PEM-element MME [(e)–(f)]. In this figure, the Mueller matrix corresponding to a sample is denoted by dots, and large dots represent the elements of the Mueller matrix obtained from the measurement. Reprinted from *Surface Science*, **96**, P. S. Hauge, Recent developments in instrumentation in ellipsometry, 108–140, Copyright (1980), with permission from Elsevier.

In Mueller matrix ellipsometry utilizing the photoelastic modulator, several photoelastic modulators are necessary, as shown in Fig. 4.14. The development of the Mueller matrix instrument shown in Fig. 4.14(e) was reported in 1997 [47]. In this instrument, two photoelastic modulators operating at different resonant frequencies are employed. Although only nine elements of the Mueller matrix are determined in this method, all the elements of the Mueller matrix can be obtained from a total of four measurements performed at different angles of the optical elements [47]. Since the instrument shown in Fig. 4.14(f) is complicated, the development of this instrument has not been reported. On the other hand, the conventional PME described in Section 4.2.5 can also be applied for Mueller matrix measurement. In this case, the first three columns can be measured, similar to the one in Fig. 4.14(b). In this instrument, these three columns are characterized from a total of eight measurements performed at different angle settings of the optical elements [76]. Recently, a different Mueller matrix ellipsometer that utilizes a photoelastic modulator has also been reported [51].

4.2.8 NULL ELLIPSOMETRY AND IMAGING ELLIPSOMETRY

Historically, null ellipsometry was the basis of the first ellipsometry instrument, developed originally by Drude. The measurement principle of this instrument is different from the ellipsometers described so far. Figure 4.15 illustrates a measurement example of null ellipsometry. As shown in this figure, the optical configuration of the null ellipsometer is exactly the same as RCE (i.e., PCSA configuration). In this method, however, the polarizer and the analyzer are rotated so that the detected light intensity becomes zero, and (ψ, Δ) values are estimated from the rotation angles of the polarizer and analyzer. Thus, if we judge light intensity by human eye, measurement can be performed without using electrical equipment. This is the reason why Drude was able to construct the ellipsometer more than 100 years ago.

If we use Jones matrices, the null ellipsometer shown in Fig. 4.15 (PCSA) is expressed by

$$\boldsymbol{L}_{\text{out}} = \boldsymbol{AR}(A)\boldsymbol{SR}(-C)\boldsymbol{CR}(C)\boldsymbol{R}(-P)\boldsymbol{PL}_{\text{in}} \tag{4.48}$$

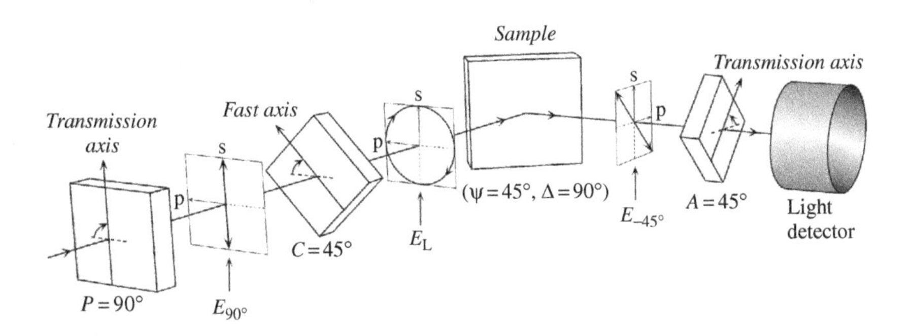

Figure 4.15 Schematic diagram of measurement in null ellipsometry. In this figure, the (ψ, Δ) values of a sample are assumed to be $\psi = 45°$ and $\Delta = 90°$. In this measurement, the detected light intensity is zero.

Therefore, the equation itself is identical to that of RCE, although the compensator is installed after the sample in Eq. (4.29) (PSC_RA configuration). Here, we will use Eq. (4.7) to express the Jones matrix of the sample. In this case, the expansion of Eq. (4.48) yields

$$E_A = r_p \cos A\,[\cos C \cos(P-C) - \rho_c \sin C \sin(P-C)] + r_s \sin A\,[\sin C \cos(P-C) + \rho_c \cos C \sin(P-C)] \quad (4.49)$$

where ρ_c represents the phase shift of the compensator $[\rho_c = \exp(-i\delta)]$. In Eq. (4.49), $\boldsymbol{R}(C-P) = \boldsymbol{R}(C)\boldsymbol{R}(-P)$ was converted to $\boldsymbol{R}(P-C)$ using $\sin(C-P) = -\sin(P-C)$. As mentioned above, the detected light intensity is zero ($E_A = 0$) in null ellipsometry. In addition, the rotation angle and phase shift of a compensator are generally 45° and 90°, respectively. Thus, by substituting $E_A = 0$, $C = 45°$, and $\delta = 90°(\rho_c = -i)$ into Eq. (4.49), we get

$$\rho \equiv \frac{r_p}{r_s} = -\tan A \frac{1 - i\tan(P - 45°)}{1 + i\tan(P - 45°)} \quad (4.50)$$

Since

$$\exp(-i2\theta) = \frac{1 - i\tan\theta}{1 + i\tan\theta} \quad (4.51)$$

and $\rho = \tan\psi \exp(i\Delta)$, we obtain the following equation from Eq. (4.50) [1]:

$$\tan\psi \exp(i\Delta) = \tan(-A) \exp[i(-2P + 90°)] \quad (4.52)$$

Accordingly, the (ψ, Δ) values are expressed as

$$\psi = -A \ (-A > 0) \quad \Delta = -2P + 90° \quad (4.53)$$

In null ellipsometry, therefore, (ψ, Δ) values are determined from the rotation angles of (P, A) when $E_A = 0$. If $A' = 180° - A$ and $P' = P + 90°$, we have another solution. By substituting these A' and P' into Eq. (4.52), we get

$$\psi = A \ (A > 0) \quad \Delta = -2P - 90° \quad (4.54)$$

There are two other solutions when $C = -45°$. Thus, there are a total of four settings that satisfy $E_A = 0$. Accurate measurement can be performed by averaging these values (four-zone averaging) [1].

Figure 4.15 shows a measurement example when a solution is given by Eq. (4.54). Here, the (ψ, Δ) values of a sample are assumed to be $(45°, 90°)$. Using these values, we obtain $A = 45°$ and $P = 90°(-270°)$ from Eq. (4.54). In this case, linear polarization in the direction of s-polarization ($E_{90°}$) is generated by the polarizer and is transformed into left-circular polarization (E_L) by a compensator, as illustrated in Fig. 4.15. The generation of E_L by the compensator is shown in Fig. 3.8, although the direction of the x axis is opposite in Fig. 3.8. In Fig. 4.15, E_L is changed into linear polarization of $-45°$ upon light reflection on the sample. Recall that $\Delta = 90°$ represents $\delta_{rp} - \delta_{rs} = 90°$ and therefore the phase of s-polarization lags by 90° upon light reflection. Since the phase difference in E_L is 90° ($\delta_x - \delta_y = 90°$ in Fig. 3.3), δ becomes 180° after light reflection. The direction of linear polarization generated by the light reflection is perpendicular to the transmission axis of the analyzer and consequently the detected light intensity becomes zero, as we expected. In null ellipsometry, therefore, the polarizer is rotated so that light becomes linear polarization upon light reflection. It should be emphasized that the null ellipsometer is still the most accurate instrument since this instrument is free from various detector errors (see Table 4.4). In this method, however, quite a long time is required for a spectral measurement since a pair of the angles (P, A) has to be found for each wavelength.

This null method has been applied to imaging ellipsometry, which allows the measurement of a two-dimensional plane [5–7]. Fig. 4.16(a) shows a schematic diagram of an imaging ellipsometry instrument (PCSA configuration) [5]. In this instrument, a parallel beam generated by a collimating system illuminates a sample and the reflected light intensity is measured by a CCD camera. The wavelength of the probe light is chosen by a filter. As shown in Fig. 4.16(a), the CCD camera is tilted slightly, in order to obtain a clear image of the sample surface. Figure 4.16(b) shows the measurement principle of imaging ellipsometry. Imagine that a probe light is illuminated to the whole surface of a substrate partially covered with a thin film. Here, let us assume that the polarization states of light waves reflected on the thin film and substrate are described by $\tan\psi_f \exp(i\Delta_f)$ and $\tan\psi_s \exp(i\Delta_s)$, respectively. If we adjust the polarizer and analyzer angles to (ψ_s, Δ_s) using the condition given by Eq. (4.53) or (4.54), the intensity of the light reflected from the substrate becomes zero ($I_s = 0$). However, the light intensity does not become zero with respect to the light reflected from the thin film ($I_f \neq 0$) since the polarization state of the light is different ($\psi_f \neq \psi_s$ and $\Delta_f \neq \Delta_s$). If $\psi_s \sim \psi_f$ and $\Delta_s \sim \Delta_f$, I_f is quite low since $I_s \sim I_f$, while I_f becomes high when $\psi_s \neq \psi_f$ and $\Delta_s \neq \Delta_f$. In this measurement, therefore, detected light intensity varies according to the (ψ_f, Δ_f) values. As we will see in Section 5.1.2, (ψ, Δ) values change significantly with thin film thickness. Accordingly, with the help of numerical simulation and film-thickness calibration, the intensity obtained from this measurement can be converted to thickness [5,77]. In other words, thickness variation is measured as intensity variation in imaging ellipsometry. When a film thickness is thin (< 5 nm), film thickness can be estimated directly from light intensity using $I_f = I_0 d^2$, where I_0 and d show the proportional constant and film thickness, respectively [5,77]. If conventional null measurement is performed on the thin film surface, the complex refractive index of the thin film can also be deduced. Figure 4.16(c) shows an image of a step structure obtained from the imaging ellipsometer shown in Fig. 4.16(a) [5]. The steps in this image are SiO_2 formed on a Si substrate. In Fig. 4.16(c),

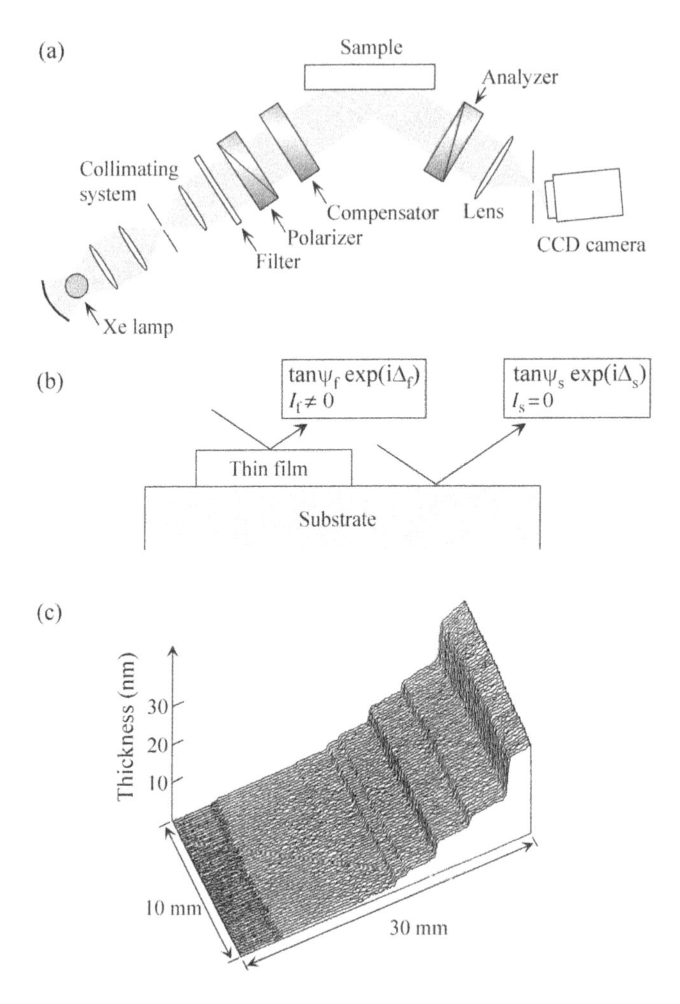

Figure 4.16 (a) Schematic diagram of an imaging ellipsometry instrument (PCSA configuration), (b) light reflection on a substrate partially covered with a thin film, and (c) image of a step structure obtained from the imaging ellipsometer shown in (a). Drawing (a) and (c): Reprinted with permission from *Review of Scientific Instruments*, **67**, G. Jin, R. Jansson, and H. Arwin, Imaging ellipsometry revisited: Developments for visualization of thin transparent layers on silicon substrates, 2930–2936 (1996). Copyright 1996, American Institute of Physics.

a thickness variation ranging from 1 to 25 nm is visualized quite clearly. In this instrument, a lateral resolution better than 5 μm has been reported [5]. Recently, a higher lateral resolution of 0.5 μm has also been reported [7].

Naturally, imaging ellipsometry has all the advantages of the ellipsometry technique shown in Table 1.2. In particular, the very high thickness sensitivity (~0.1Å) of this technique is quite advantageous in various characterizations. In addition, since an ellipsometric image is obtained in this technique, the measurement is faster than a scanning-type measurement. Recently, the imaging ellipsometer has been applied to the inspection of DNA chips. So far, imaging ellipsometers of the rotating-analyzer (polarizer) type have been developed [78–80]. In this case, the two ellipsometric images for (ψ, Δ) are measured. Furthermore, a color-imaging ellipsometer, in which three different wavelengths are measured simultaneously using three CCD cameras, has also been reported [6].

4.3 INSTRUMENTATION FOR ELLIPSOMETRY

In general, the installation of spectroscopic ellipsometry instruments requires extra care. In addition, in order to perform proper ellipsometry measurements, various calibrations and error corrections are necessary. This section will address the installation of ellipsometry systems and the Fourier analysis performed in ellipsometry measurements. In this section, we will also cover various calibration and correction procedures used in ellipsometry instruments.

4.3.1 INSTALLATION OF ELLIPSOMETRY SYSTEM

Figure 4.17 shows a spectroscopic ellipsometry instrument of the RCE type (PC_RSA configuration) installed in a plasma deposition system. In an actual ellipsometry instrument, an iris is placed in front of the light source, in order to create a point source. A collimator lens (or achromatic lens) placed behind the iris is used to generate a parallel beam. The basic optical configuration of the light detection

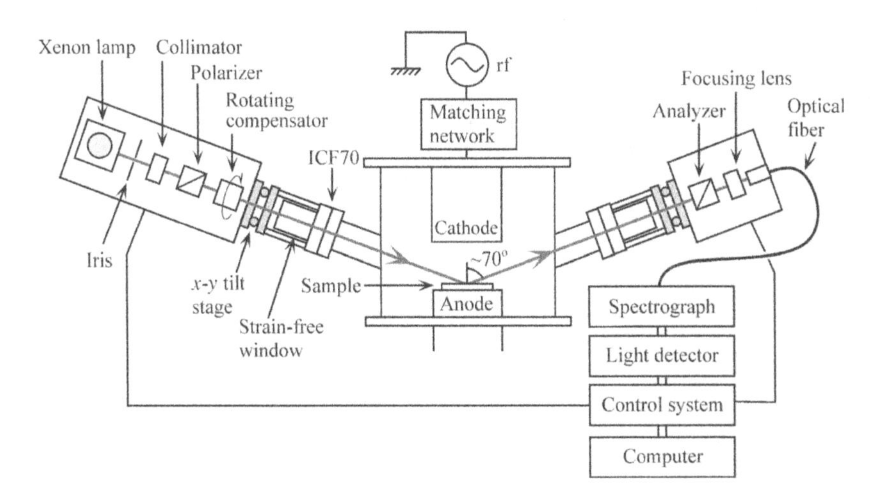

Figure 4.17 Schematic diagram of a spectroscopic ellipsometry instrument (PC_RSA configuration) installed in a plasma deposition system.

system is shown in Fig. 4.4, although an optical fiber is employed in the instrument shown in Fig. 4.17. Optical elements employed for an ellipsometry instrument are generally controlled by a computer and the position of each optical element is monitored by signal generators (encoders) attached to these optical elements [14]. In particular, when we use rotating-optical elements, light intensities at specific rotation angles are measured by synchronizing pulse signals generated from the optical elements with the light detector (see Fig. 4.4) [19,81].

In general, the incident angle of ellipsometry measurement is set at the Brewster angle (see Section 2.3.4). Ideally, an ellipsometry instrument should be installed to an optical bench that enables the incident angle to be changed. When an instrument is installed to a processing system, however, it is often difficult to use an optical bench since the space around the processing system is limited. Therefore, in a commercialized ellipsometry instrument, optical elements are enclosed in optical units, which can be fixed to a processing system using ICF (ConFrat) flanges. When there is not enough space around a processing system, a straight ICF tube may be attached to the processing system so that these ellipsometry units can be installed. As shown in Fig. 4.17, if there are x–y tilt stages for incident and detector units, the optical stage for a sample holder (anode in Fig. 4.17) is not always required. Nevertheless, in the case of real-time monitoring, slight variation in the sample position often leads to a large beam deviation, since the optical pass length between the incident and detector units generally becomes quite long (1–2 m). Accordingly, if there is no optical stage for the sample, care is required so that the sample position does not change even when the sample temperature is varied to a processing temperature. Because the diameter of the light beam is typically $\sim 5\,\mathrm{mm}$ in spectroscopic ellipsometry, a sample larger than the spot size is necessary for the measurement. However, keep in mind that an illuminated surface area increases with increasing incidence angle.

The most difficult part in installing an ellipsometry system is the installation of the optical windows. Figure 4.18 illustrates the schematic diagram of a strain-free window commonly used in spectroscopic ellipsometry measurement in the visible/UV region. In this optical window, fused quartz is attached to an ICF flange [82]. If there is residual strain in an optical window, a phase difference is generated by the photoelastic effect (see Section 3.2.3) and this affects

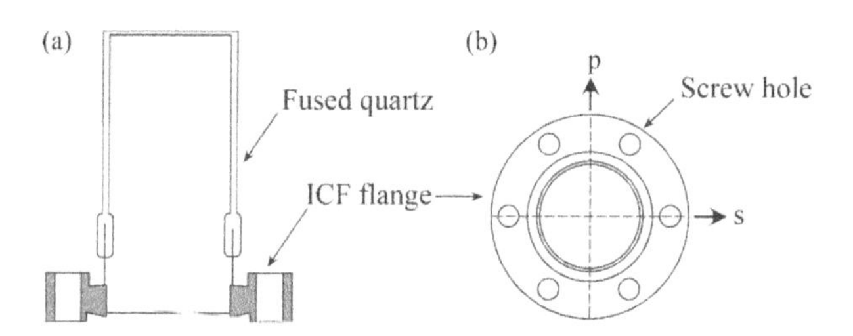

Figure 4.18 Structure of a strain-free window used for a processing system: (a) side view and (b) top view.

ellipsometry measurement seriously. For ellipsometry measurement, therefore, strain-free windows should be employed. Nevertheless, when we attach the optical windows to the ICF flanges, uniaxial stress is generated quite easily [82]. Now imagine that the arrows shown in Fig. 4.18(b) are the directions of p- and s-polarizations. If we tighten only the screws in the direction of s-polarization, for example, a phase difference is generated between p- and s-polarizations, and the measured Δ spectrum shifts to one direction. In other words, a phase difference induced by the strained window is added directly to the measured value of Δ [54,62]. The phase difference induced by the strained window is quite large, compared with the precision of ellipsometry systems, and a phase shift equivalent to a SiO_2 layer thickness of ± 5 Å can be generated rather easily.

In order to avoid a measurement error induced by the optical window, measurement of a standard sample (for example, a crystalline Si substrate covered with a thermal oxide) is performed without using the optical window, to confirm true measurement values. Second, after the installation of the optical window, any variation in the measured spectra is checked. Finally, the window is fixed to a flange ensuring that the value of Δ does not shift. If we observe a shift in Δ, this should be corrected by tightening the screws in the direction of p- or s-polarization. Since there are two optical windows, this procedure is performed twice. When residual strain in the optical window cannot be eliminated, an external clamp can be employed to apply strain in the reverse direction [82]. On the other hand, when we grow thin films in processing systems, the degradation of measurement sensitivity often occurs by a reduction in the transmitted light intensity due to film deposition inside optical windows. In addition, if there is optical anisotropy in films deposited inside windows, (ψ, Δ) spectra themselves may change. These unfavorable effects can be avoided by separating optical windows from the processing system or by introducing an inert gas that does not affect the processing.

The optical alignment of an ellipsometry instrument can be performed using the straight-through configuration ($\theta = 90°$). In this optical alignment, all the optical elements in instruments are aligned to a central position using a reference light, such as a He–Ne laser. Ideally, if we measure (ψ, Δ) spectra in the straight-through configuration without introducing a sample, all wavelengths should show $\psi = 45°$ and $\Delta = 0°$; i.e., the amplitude ratio between p- and s-polarizations is one ($\psi = 45°$) and there is no phase difference between p- and s-polarizations ($\Delta = 0°$). From the (ψ, Δ) values obtained in the measurement, we can confirm whether the instrument is operating properly. When we install a commercialized instrument into a processing system, however, it is often difficult to perform the optical alignment using a laser light. In this case, the optical alignment is carried out so that the signal-to-noise ratio in the measurement is maximized. It should be emphasized that the spectral noise in ellipsometry measurement is quite sensitive to optical alignment and a slight beam deviation sometimes generates large spectral noise. Thus, fine adjustment is generally necessary for optical alignment.

In optical alignment, the position adjustment of the light source is quite important. In a spectroscopic ellipsometry instrument that performs measurement in the visible/UV region, a high-pressure xenon lamp is commonly employed as the light source. This lamp, shown in Fig. 4.19, is composed of two electrodes (cathode and anode). In this lamp, however, the part close to the cathode (A in Fig. 4.19) and the part close to anode (B in Fig. 4.19) show different light intensities in the infrared and UV regions. Accordingly, when we perform optical alignment of the xenon lamp, the relative noises in the infrared and UV regions should be taken into account. In particular, the measurement noise in the UV region often changes drastically depending on the position of the xenon lamp, since the light emission intensity of this lamp reduces rapidly in the UV region (> 4 eV) [58]. In spectroscopic ellipsometry instruments that allow measurement up to the deep UV region (~6 eV), xenon and deuterium lamps have been utilized as light sources [58,83].

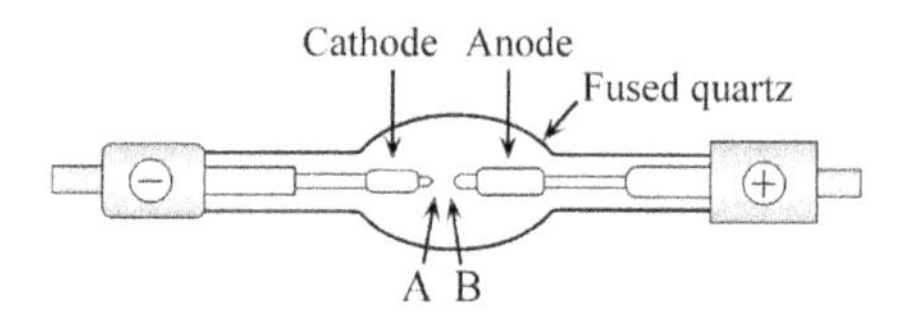

Figure 4.19 Structure of a high-pressure xenon lamp used for spectroscopic ellipsometry measurements.

4.3.2 FOURIER ANALYSIS

In ellipsometry measurements, the Stokes parameters are measured as normalized Fourier coefficients, as discussed in Section 4.2. Here, we will see the actual Fourier analysis performed in rotating-element ellipsometers including RAE and RCE.

Figure 4.20 shows a schematic diagram of the Fourier analysis used in spectroscopic ellipsometry instruments. In this figure, the light intensity (I_e) measured by an RAE instrument is shown versus measurement time (or the rotation angle of an analyzer). As shown in Fig. 4.20(a), if we use Eqs. (2.5) and (2.6), the measurement time of one optical cycle ($A = 0°–180°$) is expressed as π/ω, where

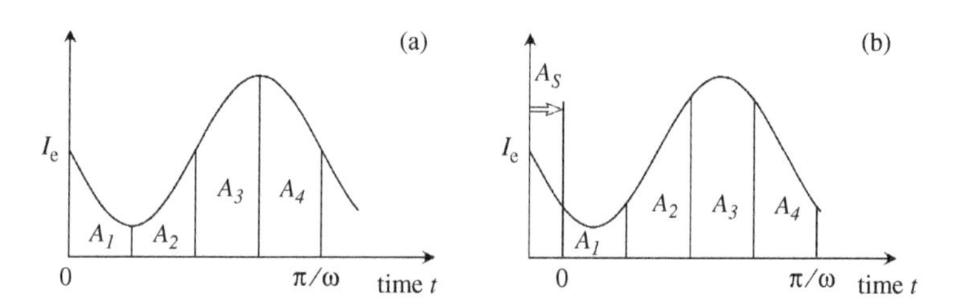

Figure 4.20 Light intensity measurement by a photodiode array used in spectroscopic ellipsometry instruments: (a) $A_s = 0°$ and (b) $A_s \neq 0°$.

ω is the angular frequency of the rotating analyzer. In ellipsometry instruments that employ a photomultiplier tube, the normalized Fourier coefficients are determined from measured light intensities of ~ 40 points per one optical cycle [11]. In real-time instruments that apply a photodiode array, on the other hand, the normalized Fourier coefficients are calculated from the integrated values of I_e [19,20,84], as shown in Fig. 4.20. If we use this method, we can reduce the amount of signal processing and increase signal intensities. As shown in Eq. (4.19), in the case of the RAE instrument, the three values (I_0, α, β) are obtained from the measurement. Accordingly, if we divide the total integrated light intensity in one optical cycle into a minimum of three sections, we can calculate these values. As confirmed from Eq. (4.33), a minimum of five integrals is required in RCE [19]. In reported instruments, either four integrals (P_RSA) [84,85] or five integrals (PSC_RA) [19] are employed.

As shown in Fig. 4.20(a), when an RAE measurement is performed using the four integrals of light intensity, each integrated light intensity $A_j (j = 1\text{–}4)$ is expressed by the following equation [84,85]:

$$A_j = \int_{(j-1)\pi/4\omega}^{j\pi/4\omega} I_e(t)\, \mathrm{d}t \tag{4.55}$$

If we assume that the normalized Fourier coefficients obtained from an actual measurement are α' and β', $I_e(t)$ is given by rewriting Eq. (4.19):

$$I_e(t) = I_0'(1 + \alpha' \cos 2\omega t + \beta' \sin 2\omega t) \tag{4.56}$$

By calculating Eq. (4.55), we obtain the above α' and β' as follows [84,85]:

$$\alpha' = \left(\frac{\pi}{2}\right) \frac{A_1 - A_2 - A_3 + A_4}{A_1 + A_2 + A_3 + A_4} \qquad \beta' = \left(\frac{\pi}{2}\right) \frac{A_1 + A_2 - A_3 - A_4}{A_1 + A_2 + A_3 + A_4} \tag{4.57}$$

Such a Fourier analysis is called a Hadamard analysis. As confirmed from the above equation, the normalized Fourier coefficients (α', β') are obtained by normalizing the measured light intensity using the total light intensity of one optical cycle. If $I_e(t)$ is given by a cosine function, it follows that $A_1 = A_4$ and $A_2 = A_3$. In this case, β' in Eq. (4.57) becomes zero since β' represents the coefficient for a sine function, as confirmed from Eq. (4.56).

Figure 4.20(a) shows a case where the rotation angle of an analyzer starts from $A = 0°$. In contrast, Fig. 4.20(b) shows a case where the readout of the integrated light intensity starts from $A = A_s$. In the case of Fig. 4.20(b), a correction for A_s is required. In ellipsometry instruments that employ a photodiode array, the light intensities of multiwavelengths are measured simultaneously, as mentioned earlier. Nevertheless, the readout of the integrated light intensities at each pixel is performed sequentially, and the analyzer rotates continuously during this period. In these instruments, therefore, the correction for A_s is performed for each pixel [84,85], as will be shown in Section 4.3.3.

4.3.3 CALIBRATION OF OPTICAL ELEMENTS

The 0° positions of optical elements used in ellipsometry instruments have to be adjusted accurately to the coordinates of p- and s-polarizations. If the 0° positions of the optical elements are not correct, the resulting (ψ, Δ) values deviate from true values. Accordingly, calibration of the rotation angles of the optical elements has to be carried out. Figure 4.21 shows the measurement configuration of an RAE instrument. In ellipsometry measurement, the plane of incidence (see Section 2.3.2) is chosen as the basis for the 0° position. Now consider that a polarizer angle given by an instrument (P) deviates from the plane of incidence by P_s, as shown in Fig. 4.21. In this case, $P - P_s$ provides the accurate position (angle) of the polarizer ($P_s < 0°$ in Fig. 4.21). Similarly, if the angle of a rotating analyzer ($A = \omega t$) deviates by A_s, the accurate angle of the rotating analyzer is described by $\omega t - A_s (A_s < 0°$ in Fig. 4.21).

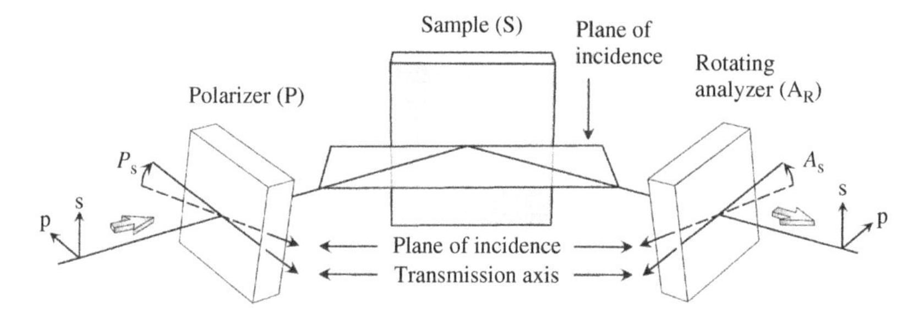

Figure 4.21 Optical configuration of a rotating-analyzer ellipsometry (RAE) instrument. In this figure, the 0° positions of a polarizer and rotating analyzer deviate from the plane of incidence by P_s and A_s, respectively.

As mentioned earlier, since A_s of each pixel in a photodiode array differs, the correction for A_s is performed mathematically [84,85]. When the angle of the rotating analyzer deviates by A_s, the theoretical equation for the light intensity $I_t(t)$ is given by rewriting Eq. (4.19) as follows [11,14]:

$$I_t(t) = I_0[1 + \eta\alpha \cos 2(\omega t - A_s) + \eta\beta \sin 2(\omega t - A_s)] \tag{4.58}$$

where η is a correction coefficient that represents nonideal behavior of a light detector. In particular, η shows the attenuation of the ac components (α, β) relative to the dc component (I_0). By expanding Eq. (4.58) using the addition theorem (see Appendix 1) and comparing with Eq. (4.56), we get the following relation between (α', β') obtained from an actual measurement and (α, β) in the theoretical equation:

$$\begin{bmatrix} \alpha' \\ \beta' \end{bmatrix} = \eta \begin{bmatrix} \cos 2A_s & -\sin 2A_s \\ \sin 2A_s & \cos 2A_s \end{bmatrix} \begin{bmatrix} \alpha \\ \beta \end{bmatrix} \tag{4.59}$$

It can be seen from Eq. (4.59) that the relation is expressed by the matrix of the coordinate rotation with respect to A_s (see Section 3.3.2). Conversely, when we estimate (α, β) from the experimental values of (α', β'), the following equation is used [11]:

$$\begin{bmatrix} \alpha \\ \beta \end{bmatrix} = \frac{1}{\eta} \begin{bmatrix} \cos 2A_s & \sin 2A_s \\ -\sin 2A_s & \cos 2A_s \end{bmatrix} \begin{bmatrix} \alpha' \\ \beta' \end{bmatrix} \tag{4.60}$$

Notice that the matrix corresponding to the coordinate rotation in Eq. (4.60) is the inverse matrix of Eq. (4.59). As confirmed from the above equation, if A_s and η are known, we can obtain (α, β) from experimental (α', β'). In actual measurements, (ψ, Δ) are obtained from the values of (α, β) [11,14].

When there is light absorption in a sample ($k > 0$), the values of (P_s, A_s, η) can be obtained from the residual calibration method [11,14,86]. As we will see in Section 5.1.1, when $k > 0$, the reflected light from a sample becomes elliptical or circular polarization. In the residual calibration method, the value of P_s is determined by utilizing this characteristic. In other words, when the polarizer angle is set at 0° accurately, the s-polarized component is nonexistent and only the p-polarized light illuminates the sample. In this case, the reflected light becomes linear polarization of the p-polarization only. In contrast, when the polarizer angle deviates from 0°, the reflected light becomes elliptical polarization due to the presence of the s-polarized component.

In the residual calibration method, the 0° position of the polarizer is estimated from the residual function defined by the following equation [11,14,86]:

$$R(P) = 1 - (\alpha'^2 + \beta'^2) \tag{4.61a}$$

$$= 1 - \eta^2(\alpha^2 + \beta^2) \tag{4.61b}$$

$$= (1 - \eta^2) + \eta^2 \left[\frac{\sin 2\psi \sin \Delta \sin 2(P - P_s)}{1 - \cos 2\psi \cos 2(P - P_s)} \right]^2 \tag{4.61c}$$

In Eq. (4.59), when $A_s = 0°$, it follows that $\alpha' = \eta\alpha$ and $\beta' = \eta\beta$. Thus, $(\alpha^2 + \beta^2)$ in Eq. (4.61b) has a coefficient of η^2. We can obtain Eq. (4.61c) by substituting Eq. (4.22) into Eq. (4.61b) and rewriting the polarizer angle using $P \rightarrow P - P_s$ [55]. When the position of the polarizer coincides with the plane of incidence, we get $\alpha(S_1) = 1$ and $\beta(S_2) = 0$. In particular, in the case of an ideal light detector ($\eta = 1$), we obtain $R(P) = 0$ from Eq. (4.61b). If the reflected light from a sample is circular polarization shown in Fig. 4.5(c), it follows that $R(P) = 1 (\alpha = \beta = 0)$. Accordingly, we can determine P_s from the position where $R(P)$ is minimized. Moreover, since Eq. (4.61c) is described as a function of $(P - P_s)$, the value of P_s can be estimated directly from $R(P)$, independent of A_s.

Figure 4.22(a) shows a measured example of the residual function $R(P)$, plotted as a function of the angle of the polarizer P [86]. This result can be obtained by measuring (α', β') at each position of P shown in the figure and then inserting (α', β') into Eq. (4.61a). As shown in Fig. 4.22(a), $R(P)$ is minimized at $P_{min} = 0.26°$. If the position P deviates from P_{min}, the reflected light becomes elliptical polarization

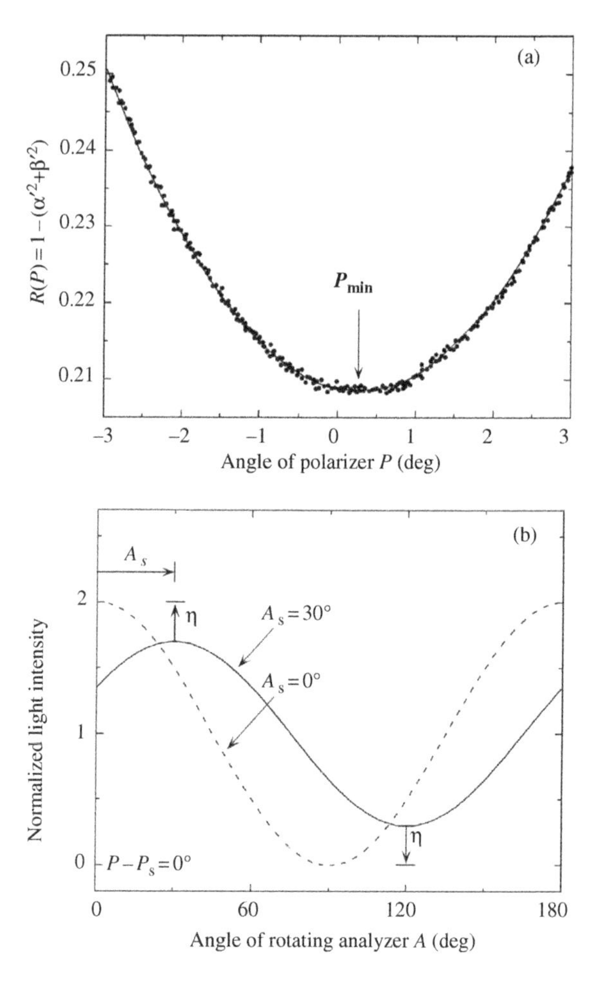

Figure 4.22 (a) Measured value of residual function $R(P)$, plotted as a function of the angle of polarizer P, and (b) normalized light intensity plotted as a function of the angle of rotating analyzer A. In (b), the results for $A_s = 0°$ (dotted line) and 30° (solid line) are shown. The η in (b) represents the correction coefficient for the ac attenuation. Drawing (a): from *Journal of the Optical Society of America*, **64**, D. E. Aspnes, Effects of component optical activity in data reduction and calibration of rotating-analyzer ellipsometers, 812–819 (1974). Reproduced by permission of the Optical Society of America.

and consequently the $R(P)$ value increases. The solid line in Fig. 4.22(a) represents a fitting result obtained using $R(P) = c_0 + c_1 P + c_2 P^2$. Here, c_{0-2} indicate the fitting coefficients. Using these coefficients, P_{min} is calculated from $P_{min} = -c_1/(2c_2)$. In the residual calibration method, P_s is estimated from this result, provided that $P_s = P_{min}$. When the polarizer (analyzer) is a Rochon prism made of quartz (see Section 3.2.1), correction for the optical activity is required [84,86]. It is evident from Eq. (4.61b) that $R(P) = 1 - \eta^2$ at P_{min} since $\alpha = 1$ and $\beta = 0$ at P_{min}. Thus, from the value of $R(P)$ at P_{min}, the value of η can be estimated [11]. In other words, we determine the correction coefficient of a light detector so that $R(P)$ becomes zero at P_{min}.

Figure 4.22(b) shows the normalized light intensity when $P - P_s = 0°$, plotted as a function of the angle of the rotating analyzer $A = \omega t$. The solid line in this figure shows the case where the analyzer offset is $A_s = 30°$, while the dotted line shows the ideal case ($A_s = 0°$). As shown in Fig. 4.22(b), the waveform shifts toward the right-hand side when $A_s > 0°$ and, if there is attenuation of the ac components, the amplitude of the waveform becomes less than one. Attenuation of the amplitude is corrected from the η value. As mentioned earlier, $\alpha = 1$ and $\beta = 0$ hold at $P - P_s = 0°$. By substituting these values into Eq. (4.59), we get

$$\begin{bmatrix} \alpha' \\ \beta' \end{bmatrix} = \eta \begin{bmatrix} \cos 2A_s \\ \sin 2A_s \end{bmatrix} \tag{4.62}$$

If we use the above equation, the value of A_s can be estimated from the following equation [11]:

$$A_s = \frac{1}{2} \tan^{-1} \left(\frac{\beta'}{\alpha'} \right) \tag{4.63}$$

When samples show low light absorption ($k \sim 0$) and thus $\Delta \cong 0°$ and $\pm 180°$, the residual calibration method cannot be applied. This arises from the fact that $\sin \Delta$ in Eq. (4.61c) becomes zero in this condition and consequently the value of $R(P)$ becomes independent of P [55]. In this case, (P_s, A_s, η) are determined from other calibration methods including zone-difference calibration [14,55] and regression calibration [56]. In zone-difference calibration method, P_s is obtained from a calibration function that is different from the residual function. However, the procedure of zone-difference calibration is essentially similar to that of residual calibration. In particular, zone-difference calibration provides superior calibration, compared with residual calibration, in the regions of $|\Delta| < 30°$ and $|\Delta| > 150°$ [55]. In the regression calibration method, on the other hand, P is varied over a wide range and, from the regression analysis of (α', β') obtained from this range, (P_s, A_s, η) are determined [56]. This method employs Eq. (4.59) and the equation obtained by rewriting Eq. (4.23) using $P \rightarrow P - P_s$. In this case, (α', β') are expressed as functions of $(P, \psi, \Delta, P_s, A_s, \eta)$. In this calibration method, each value of $(P, \psi, \Delta, P_s, A_s, \eta)$ is obtained from fitting to (α', β') using $(P, \psi, \Delta, P_s, A_s, \eta)$ as free parameters.

Figure 4.23 shows the results of the regression calibration, obtained from the samples of (a) a $SiO_2(250\,Å)/Si$ substrate and (b) a SiO_2 substrate (Corning 7059) [56]. It has been reported that the fitting results are almost identical to the measured values. In Fig. 4.23, the values of the residual function calculated from Eq. (4.61a) are also shown as $R(P)$. Since the sample in Fig. 4.23(a) shows $\Delta \cong 90°(\lambda = 5000\,Å)$, the calibration can also be performed using the residual calibration method.

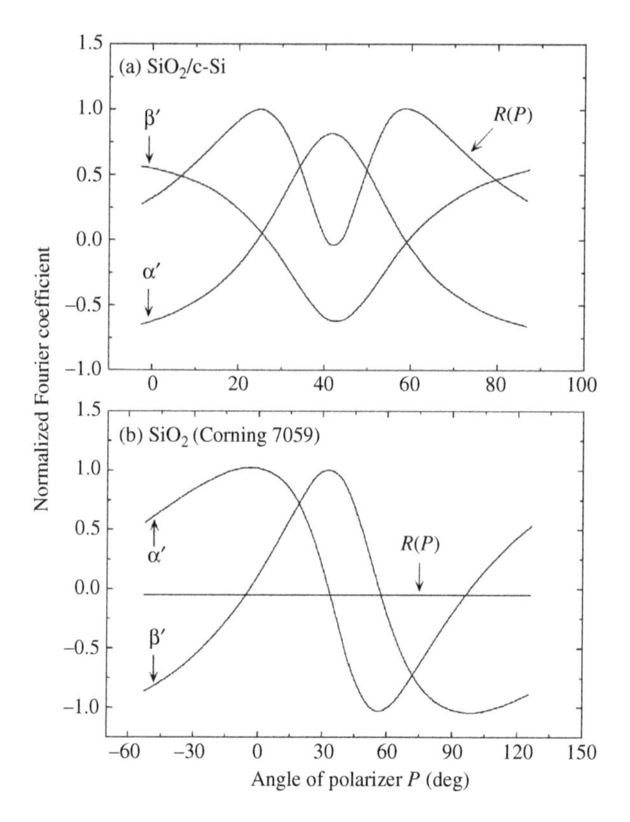

Figure 4.23 Measurement and analysis in the regression calibration method: (a) a SiO_2(250 Å)/Si substrate and (b) a SiO_2 substrate (Corning 7059). Reprinted from *Thin Solid Films*, **234**, B. Johs, Regression calibration method for rotating element ellipsometers, 395–398, Copyright (1993), with permission from Elsevier.

In this method, P_s is obtained from the measurement around $P = 42°$ where $R(P)$ is minimized, as shown in Fig. 4.23(a). In the regression calibration method, on the other hand, (P_s, A_s, η) are obtained simultaneously from the regression analysis of (α', β') measured at $P = 0°$–$90°$. It is clear from Fig. 4.23(b) that the residual calibration method cannot be performed when $\Delta = 0°$ or $180°$, since the residual function shows a constant value of $R(P) = 0$. If we use the regression calibration method, the calibration coefficients can be obtained from the fitting, even in this case.

The calibration of the RCE instrument has also been performed using a method based on the residual and zone-difference calibration methods [19] or a regression calibration method that incorporates a compensator [87]. In the case of PME, since there are no mechanical moving parts, the calibration of the optical elements can be carried out relatively easily [26,27]. The calibration described above can be performed without changing the optical configuration. In addition, the above calibration should be performed for each sample after finishing optical alignment. Furthermore, when a sample temperature is changed to a processing temperature, the calibration should be carried out again, since the plane of incidence often deviates by temperature variation. In general, the above calibration is automated completely by a computer and can be performed within a few minutes.

4.3.4 CORRECTION OF MEASUREMENT ERRORS

In general, spectroscopic ellipsometry instruments require various error corrections. Table 4.4 summarizes the error corrections and controls performed in spectroscopic ellipsometry instruments. In the RCE instrument, the phase shift of the compensator versus wavelength has to be determined. The phase shift of the compensator can be measured using the straight-through configuration (PC_RA) [88]. Figure 4.24 shows the phase difference δ of a MgF_2 compensator determined by this method [88]. As shown in Fig. 4.24, δ increases linearly versus photon energy since δ is inversely proportional to wavelength [Eq. (3.3)]. In the RCE instrument, (ψ, Δ) spectra are measured using the phase difference determined for each wavelength. If we parameterize the phase difference of a compensator versus wavelength, the phase difference can be determined in a conventional measurement configuration [87]. In the PME instrument, on the other hand, the accurate control of the phase difference generated by the photoelastic modulator is quite important. With respect to the photoelastic modulator, the control of the phase shift in the photoelastic modulator, control of the phase shift for different wavelengths, and temperature control have been performed [27]. In addition, the higher harmonic correction shown in Table 4.4 (PMSA) represents the correction of the phase difference generated by high-order frequency components in the photoelastic modulator [27].

It can be seen from Table 4.4 that a light detector requires many corrections in all the instruments. The dark signal correction in Table 4.4 is performed by installing a mechanical shutter in the light pass and by subtracting a dark signal from a measured signal [26,84]. The nonlinear response of detectors represents the phenomenon that a detected light signal is not proportional to an input light intensity, and the correction is performed by obtaining the correction coefficients for the detected light intensities [14,84,85]. When a light detector is a photomultiplier tube, this nonlinearity has been suppressed by optimizing the control circuit [89] or by controlling the applied voltage [27]. Moreover, if residual stress is present in the window of the photomultiplier tube, a detected light intensity may show polarization dependence due to the photoelasticity of the window [14]. The correction coefficient for ac attenuation (η in Section 4.3.3) represents the one for the nonlinear response and the polarization dependence of the light detector [14]. Image persistence in a photodiode array detector indicates the observation that certain signals remain after the readout of light intensity signals. The effect of image persistence can be

Table 4.4 Calibration, correction and control in spectroscopic ellipsometry instruments

	Instrument type[a]			
	PSA_R	P_RSA	$PSC_RA(PC_RSA)$	PMSA
Optical element	• Calibration for rotation angle [14] • Optical activity correction [86] (only for quartz prism)	• Calibration for rotation angle [14] • Optical activity correction [86] (only for quartz prism)	• Calibration for rotation angle [19] and phase shift of a compensator [88]	• Calibration for rotation angle [26] • Phase control of PEM [27] • Higher harmonic correction for PEM [27]
Light detector[b]	• Correction for dark signal and attenuation of ac components [14] • Control of nonlinear response (PMT) [89]	• Correction for dark signal [84], nonlinear response (PDA) [85], image persistence [85] and signal integration error [90]	• Same as PSA_R and P_RSA	• Correction for dark signal [26] and signal response [27] • Control of applied voltage for PMT [27]
Spectrograph	• Correction for diffraction efficiency [56] (only PSA_R)	• Correction for stray light [85]		
Light source		• Correction for source polarization [84] (only P_RSA)		
Optical window	• Correction for photoelasticity [54]		• Correction for photoelasticity [62]	• Correction for photoelasticity [25]

[a] Polarizer (P), sample (S), analyzer (A), compensator (C), photoelastic modulator PEM (M). The subscript R shows that its optical element rotates [b] Photomultiplier tube (PMT) and photodiode array (PDA).

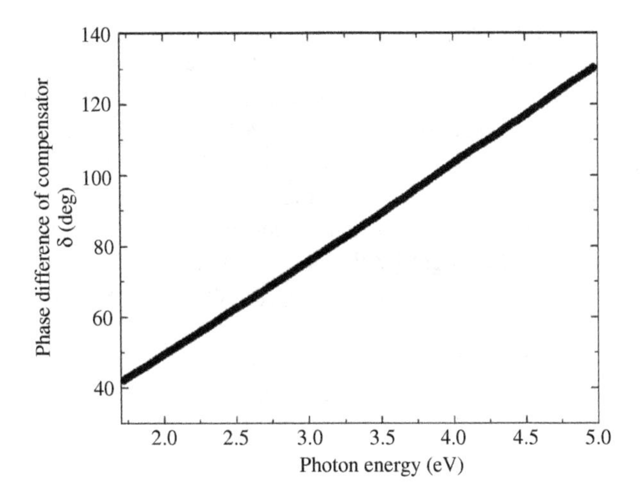

Figure 4.24 Phase difference δ of a MgF_2 compensator versus photon energy. From *Journal of the Optical Society of America A*, **18**, J. Lee, P. I. Rovira, I. An, and R. W. Collins, Alignment and calibration of the MgF_2 biplate compensator for applications in rotating-compensator multichannel ellipsometry, 1980–1985 (2001). Reproduced by permission of the Optical Society of America.

suppressed by correcting the remaining light intensity signals [85,90]. In addition, light intensities are not measured during the readout of light intensity signals. This error (signal integration error) can also be corrected [90]. The above corrections for a photodiode array depend strongly on the type of detectors, and some of these corrections may not be necessary. In a PME instrument, on the other hand, due to the high modulation frequency of the photoelastic modulator, correction for signal response has been carried out [27].

With respect to the spectrograph and light source, the PSA_R instrument requires different system corrections, compared with the P_RSA instrument. In the case of the PSA_R instrument, for example, different polarizations enter into a spectrograph due to the continuous rotation of the analyzer. Nevertheless, since the diffraction efficiency of a spectrograph has polarization dependence (see Section 3.2.4), correction for this dependence is necessary [56]. In the P_RSA instrument, on the other hand, there is no need to perform this correction since the analyzer position is fixed. In the case of the P_RSA instrument, however, due to the rotation of the polarizer placed on the incident side, correction for source polarization (see Section 3.2.4) is required [84,90]. With respect to the RCE and PME instruments, corrections for the spectrograph and source polarization are not performed, since the positions of the polarizer and analyzer are fixed. Thus, there are basically no differences in system corrections between PC_RSA and PSC_RA. In the PC_RSA configuration, however, since the compensator is installed on the incident side, the optical alignment and calibration of the compensator are easier, compared with PSC_RA [62]. In PC_RSA, however, the interpretation of a depolarization spectrum may become difficult, since the polarization state of the incident light changes continuously. As shown in

Table 4.4, the phase difference generated by the photoelasticity of optical windows can be corrected analytically [14,25,54,62]. However, it is preferable to attach the optical windows by the procedure explained in Section 4.3.1. The stray light correction shown in the spectrograph of Table 4.4 (P_RSA) represents correction for light scattered inside the spectrograph and light detector [85,90]. Although stray light correction is generally required in the UV region (> 4 eV) where light intensity is relatively small [85], this correction is rather difficult to perform. Thus, it is essential to minimize stray light by the optimization of the optical system [90]. In RAE [54,91] and RCE [62], we can employ a measurement method in which the measurement is performed twice using different polarizer angles and final results are obtained from these averaged values. In this method, we perform the measurement twice at $P = +45°$ and $-45°$, for example, and obtain the final result as a simple average of these spectra. If we use this method, the system errors induced by the misalignment and imperfection of optical elements and the strain of optical windows can be suppressed [54,62,91].

Unfortunately, only a few studies have been reported for calibration and error correction procedures in infrared spectroscopic ellipsometry [33,37,38,40]. Fourier-transform infrared spectrometer (FTIR) employed as a light source in infrared spectroscopic ellipsometry has been reported to show source polarization owing to the optical properties of the beamsplitter incorporated in FTIR (see Fig. 4.13) [31,33]. As confirmed from Table 4.4, the FTIR-P_RSA configuration [see Fig. 4.12(a)] requires error correction for source polarization [33]. In FTIR-PME, on the other hand, calibration for the wavelength dependence of the photoelastic modulator is quite important [37,38,40]. In addition, a mercury cadmium telluride (HgCdTe) detector, which has been used widely in infrared spectroscopy because of its high sensitivity, has been reported to show a nonlinear response [33]. Moreover, correction for the wire-grid polarizer (analyzer) used in infrared ellipsometers may also be required, since the degree of polarization reduces with decreasing wavelength due to the imperfection of the polarizer (analyzer) [33].

4.4 PRECISION AND ERROR OF MEASUREMENT

In this section, we will address the precision and error of each ellipsometry instrument, actual sensitivity for film thickness and optical constants, and the depolarization effect of samples. The understanding of these is of significant importance for the appropriate data analysis of spectroscopic ellipsometry. Furthermore, the precision and error of spectroscopic ellipsometry measurement depend not only on the types of instruments, but also sample structures. In particular, when samples depolarize incident light, extra care is needed in the interpretation of measured spectra.

4.4.1 VARIATION OF PRECISION AND ERROR WITH MEASUREMENT METHOD

In some ellipsometry instruments, measurement errors increase drastically in specific measurement regions. In RAE, for example, the measurement error increases at $\Delta \cong 0°$ and 180° (see Table 4.3). The increase in the measurement error in RAE can be explained from the relation between $\cos\Delta$ and Δ [Fig. 4.25 (a)]. Recall from Section 4.2.2 that, in RAE, the Stokes parameter S_3 cannot be measured and only $\cos\Delta$ is obtained with respect to Δ (see Table 3.3). As we have seen in Eq. (4.24), RAE provides the value of $\cos\Delta$, instead of Δ, from the measurement of the normalized Fourier coefficients (α, β). In this case, the value of Δ is obtained from the conversion of $\cos\Delta$ to Δ. However, $\cos\Delta$ is not a linear function of Δ and consequently the measurement error varies according to the absolute value of $\cos\Delta$. As shown in Fig. 4.25(a), a small measurement error in $\cos\Delta$ leads to a large error in Δ when $|\cos\Delta| = 1$ (i.e., $\Delta \cong 0°$ and $\pm 180°$), while measurement sensitivity is rather linear when $\cos\Delta = 0\,(\Delta = 90°)$. This is the reason why the measurement error increases at $\Delta \cong 0°$ and $\pm 180°$ and the measurement sensitivity becomes maximum at $\Delta = 90°$ in RAE.

Similarly, the measurement error in PME increases in specific regions. Specifically, if we use the measurement configuration of $M = 0°$, for example, the Stokes parameters S_2 and S_3 are measured (see Section 4.2.5). In this case, only $\sin 2\psi$ is measured with respect to ψ (see Table 3.3). Consequently, the measurement error increases at $\psi = 45°(2\psi = 90°)$ in this configuration [Fig. 4.25(b)]. In PME, however, if we perform the measurement twice with different measurement

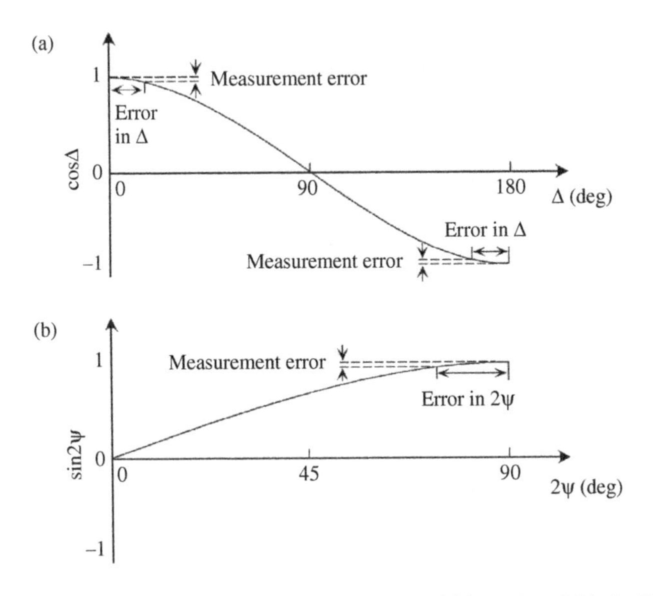

Figure 4.25 Measurement errors in the specific regions of (a) $\cos\Delta$ and (b) $\sin 2\psi$.

configurations, all the Stokes parameters can be measured. Thus, the increase in the measurement error can be suppressed by using this method. In real-time monitoring, however, this method is unfavorable due to the constraint of measurement time. In this respect, RAE with compensator has a similar problem since the measurement has to be performed more than twice to obtain all the Stokes parameters (see Section 4.2.3).

If we use the Poincaré sphere shown in Fig. 4.26, the above results can be interpreted more systematically [13]. As mentioned above, $S_1(\alpha)$ and $S_2(\beta)$ are measured in RAE. By transforming Eq. (4.24) using $S_1^2+S_2^2+S_3^2=1$ [Eq. (3.50)], we get the following equation [14]:

$$\cos\Delta = \frac{S_2}{\sqrt{1-S_1^2}} = \frac{S_2}{\sqrt{S_2^2+S_3^2}} \tag{4.64}$$

As discussed in Section 3.4.1, S_1 and S_2 originally represent linear polarizations with different orientation angles, and thus the component of elliptical polarization cannot be determined without measuring S_3. In RAE, therefore, the value of Δ is estimated by assuming that reflected light is totally polarized, as shown in Eq. (4.64). In other words, the Δ value in RAE is obtained by projecting the measured values S_1 and S_2 onto the surface of the Poincaré sphere [13].

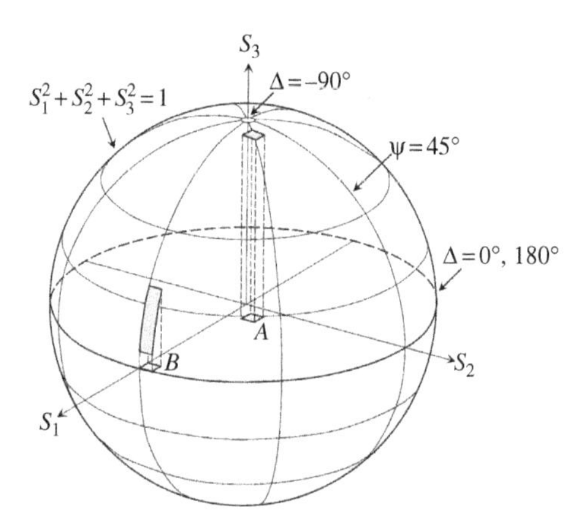

Figure 4.26 Representation of measurement errors in the rotating-analyzer (polarizer) ellipsometer using the Poincaré sphere. From *OYO BUTURI*, **57**, S. Kawabata, Ellipsometry and some difficulties in the measurements of extremely thin films *(in Japanese)*, 1868–1876 (1988). Reproduced by permission of Japan Society of Applied Physics.

Here, let us assume that the values S_1 and S_2 measured by RAE have errors of δS_1 and δS_2, respectively. In this case, the measurement error is expressed by an area, as shown in Fig. 4.26. When we project the area A ($\Delta \sim \pm 90°$) on the surface of the Poincaré sphere, the area does not change much. However, if we project the area B, the projected area becomes quite large. In other words, in RAE, the measurement error increases around the equator of the Poincaré sphere (i.e., $\Delta \cong 0°$ and $\pm 180°$), even if the actual measurement errors δS_1 and δS_2 are the same. Although S_1 and S_2 are projected in the negative direction of Δ in Fig. 4.26 (toward the north pole of the Poincaré sphere), S_1 and S_2 are projected in the opposite direction in actual measurements (toward the south pole of the Poincaré sphere). As we will discuss in Section 5.1, when the light absorption of samples is small ($k \sim 0$) or optical interference is large, Δ becomes $\Delta \cong 0°$ and $\pm 180°$. When we characterize these samples using RAE, the measurement error increases.

The increase in the measurement error in PME can also be interpreted from the Poincaré sphere. As mentioned earlier, if we use the configuration of $M = 0°$, S_2 and S_3 are measured. In this configuration, therefore, a point on the plane of S_2 and S_3 is projected on the surface of the Poincaré sphere in the direction of S_1. In this case, the measurement error increases at around $\psi = 45°$ (circle parallel to the S_2 and S_3 axes). In RCE, on the other hand, since all the Stokes parameters (S_{0-3}) are measured simultaneously, the measured value is expressed as a point on the surface of the Poincaré sphere. Therefore, the measurement error of RCE is determined by the errors of measured Fourier coefficients themselves, and the measurement sensitivity of RCE is independent of the values of (ψ, Δ). In the regions where RAE and PME show an increase in measurement errors, extra care is required in data analysis (see Section 5.5.2).

Figure 4.27 shows the depolarization and (ψ, Δ) spectra obtained from PSC_RA and P_RSA instruments [19]. The sample for this figure is a nanocrystalline diamond film (2000 Å) formed on a Si substrate. In Fig. 4.27, the P_RSA measurement is performed using $A = 30°$, while $P = 45°$ and $A = 30°$ are employed for the PSC_RA measurement. Figure 4.27(a) shows the degree of polarization p obtained from PSC_RA. The reduction in p at the high energy side is caused by the depolarization effect of the sample (see Section 4.4.4). The result from P_RSA in Fig. 4.27(b) is represented by the region $-180° \leq \Delta \leq 180°$ using the result from PSC_RA. As shown in Fig. 4.27(b), the Δ values obtained from P_RSA are not continuous at $\Delta \cong 0°$ and $\pm 180°$. This result confirms the increase in the measurement error in $P_RSA(PSA_R)$ at $\Delta \cong 0°$ and $\pm 180°$. The stepwise variation of Δ shown in Fig. 4.27(b) represents the optical interference of the diamond thin film (see Section 5.1.2). On the other hand, the difference in ψ observed below 3.0 eV originates from the difference in the incident angle and inhomogeneity of the sample.

In the case of RAE, the errors $\delta\psi$ and $\delta\Delta$ of (ψ, Δ) can be expressed from the measurement errors of the Fourier coefficients ($\delta\alpha$ and $\delta\beta$) [54]:

$$\delta\psi = -\frac{(1 - \cos 2P \cos 2\psi)^2}{2 \sin^2 2P \sin 2\psi}\delta\alpha \tag{4.65a}$$

$$\begin{aligned}\delta\Delta = &-\frac{\cos\Delta(\cos 2P - \cos 2\psi)(1 - \cos 2P \cos 2\psi)}{\sin^2 2P \sin\Delta \sin^2 2\psi}\delta\alpha \\ &-\frac{(1 - \cos 2P \cos 2\psi)}{\sin 2P \sin\Delta \sin 2\psi}\delta\beta\end{aligned} \tag{4.65b}$$

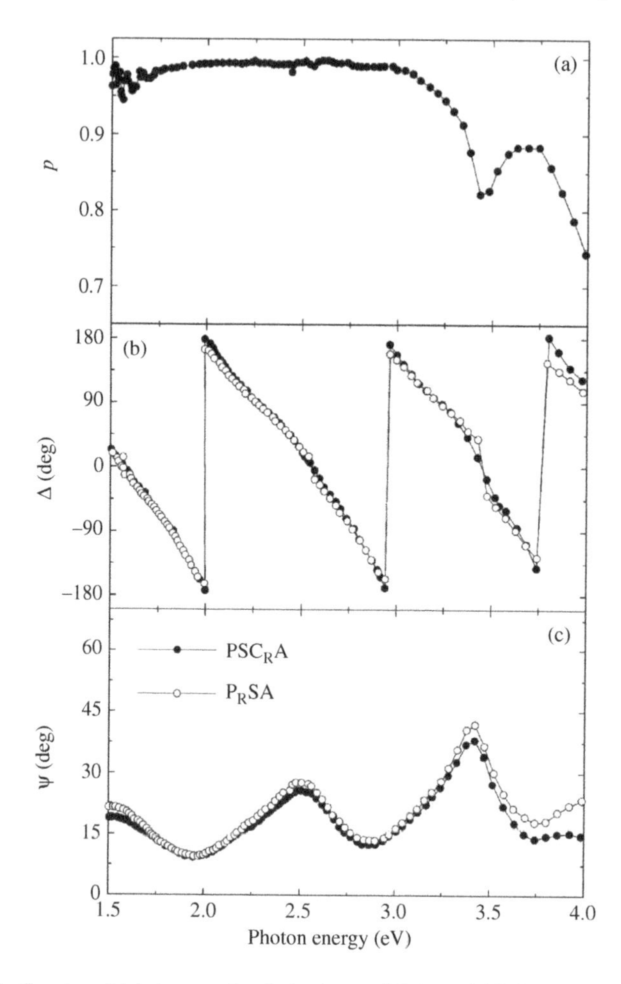

Figure 4.27 Spectra of (a) degree of polarization p, (b) Δ, and (c) ψ, measured from PSC_RA and P_RSA. The measurement sample is a nanocrystalline diamond film (2000 Å) formed on a Si substrate. Reprinted with permission from *Review of Scientific Instruments*, **69**, J. Lee, P. I. Rovira, I. An, and R. W. Collins, Rotating-compensator multichannel ellipsometry: Applications for real time Stokes vector spectroscopy of thin film growth, 1800–1810 (1998). Copyright 1998, American Institute of Physics.

The above equations can be derived as follows: we first determine the Jacobian matrix $\partial(\alpha, \beta)/\partial(\psi, \Delta)$, corresponding to the slopes of each coefficient, from Eq. (4.22), and then obtain $\partial(\psi, \Delta)/\partial(\alpha, \beta)$ from the inversion matrix of the Jacobian matrix, and finally expand the matrix to obtain Eq. (4.65). From the denominator of Eq. (4.65b), it can be seen that $\delta\Delta$ increases drastically at $\Delta \cong 0°$ and $\pm 180°$ by the reduction of the $\sin\Delta$ value. When the reflected light is circular polarization, on the other hand, the measurement errors $\delta\alpha$ and $\delta\beta$ become relatively small since $\alpha = \beta = 0$. Thus, if we make the values α and β smaller, the measurement precision increases. In particular, when the angle of the polarizer P coincides with ψ of a sample, we obtain $\alpha = 0$, as confirmed from Eq. (4.23). Therefore, the measurement precision improves in this condition [14,57]. When we measure crystalline Si by RAE, the measurement error is minimized at around $P = 30°$ [57].

Recall that the RAE instrument shows a minimum error when $\Delta = 90°$. If we substitute $\Delta = 90°$ and $P = \psi = 45°$ into Eq. (4.65) and assume $\delta\alpha = \delta\beta$, we obtain

$$\delta\psi = \delta\Delta/2 \tag{4.66}$$

As confirmed from the above equation, $\delta\Delta$ is two times higher than $\delta\psi$ in RAE (see Table 1.3) [92]. This originates from the fact that RAE measures $\cos 2\psi$ and $\sin 2\psi$ with respect to ψ, while $\cos\Delta$ is measured for Δ. In other words, the measurement error for ψ becomes smaller by the coefficient 2 of 2ψ. Eq. (4.66) also holds for the measurement error of PME [92]. Although the measurement error of RCE has been discussed theoretically [62], the interpretation is more complicated, since four Fourier coefficients have to be taken into account in this case. In general, the above measurement errors decrease with increasing integration time (t) of light intensity in the measurement. In particular, when the noise of measured spectra is determined by the fluctuation of light intensity in a light source (shot-noise limit), the measurement error has been reported to decrease with $t^{-1/2}$ [57,92].

4.4.2 PRECISION OF (ψ, Δ)

Figure 4.28 shows measurement values obtained from RAE when the measurement for the same sample is repeated [11]. In this measurement, the sample is an optically polished Ni crystal and the measurement wavelength is 4000 Å. Each data point in Fig. 4.28 shows the results obtained from a measurement time of 7 s (the integration number is 1000). In this instrument, the thermal fluctuation of air around a light source governs the measurement noise. The sample in Fig. 4.28 shows $\Delta \sim \pm 90°$ and thus provides an ideal measurement condition for RAE. The (α, β) in Fig. 4.28 represent the normalized Fourier coefficients. From this measurement, $(\delta\psi, \delta\Delta) = (\pm 0.0005°, \pm 0.001°)$ have been reported [11]. This RAE instrument is an *ex situ* instrument in which a monochromator is placed behind a light source and a photomultiplier tube is used as the light detector. In spectroscopic ellipsometry instruments that allow real-time monitoring, measurement errors of $(\delta\psi, \delta\Delta) = (0.01°, 0.02°)$ [58,81] and $\delta\varepsilon = 0.02$ [81] have been reported with a measurement time of ~ 2 s ($P_R SA$ configuration). When a Cr substrate is employed as an ideal sample ($\Delta = 111°$), a higher precision of $(\delta\psi, \delta\Delta) = (0.003°, 0.007°)$ has been reported with an integration time of 3.2 s [16]. It should be emphasized that the precision and error of ellipsometry

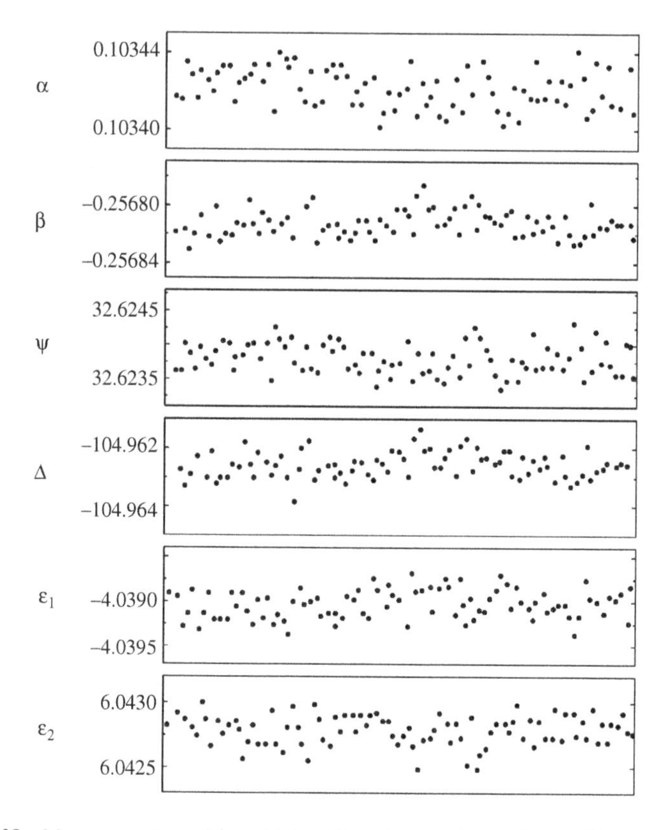

Figure 4.28 Measurement precision obtained from the rotating-analyzer ellipsometer when the same measurement is repeated. The measurement sample is a Ni crystal. From *Applied Optics*, **14**, D. E. Aspnes and A. A. Studna, High precision scanning ellipsometer, 220–228 (1975). Reproduced by permission of the Optical Society of America.

instruments depend strongly on measurement wavelength particularly in real-time instruments [81]. In general, since the light intensity of a xenon lamp decreases rapidly in the UV region ($>4\,\text{eV}$) [58], measurement precision degrades at high energies.

From the above results, it is clear that spectroscopic ellipsometry allows high-precision measurements for (ψ, Δ) and optical constants. As mentioned earlier, spectroscopic ellipsometry measures relative light intensities modulated by optical elements, instead of the absolute light intensities of reflected p- and s-polarizations. Accordingly, measurement errors induced by various imperfections in the instruments become very small in ellipsometry measurement, if we compare them with absolute reflectance measurements. This is the reason why optical constants and film thickness can be estimated with high precision using the ellipsometry technique.

4.4.3 PRECISION OF FILM THICKNESS AND ABSORPTION COEFFICIENT

In this section, we will estimate the measurement precision for film thickness and absorption coefficient from the measurement errors of (ψ, Δ) described in Section 4.4.2. The measurement precision for the thickness of a thin film formed on a substrate can be deduced from the optical interference effect of the thin film. As we have seen in Eq. (2.77), in a thin film/substrate structure, a phase difference is generated between two light beams that are reflected on the film surface and at the film/substrate interface. If we assume that the phase difference induced by this optical interference corresponds to the phase difference Δ observed in ellipsometry, we obtain the following equation from Eq. (2.77):

$$\delta d = \frac{\lambda}{4\pi N_1 \cos\theta_1}\delta\Delta \tag{4.67}$$

Here, we further assume that the thin film is $SiO_2[N_1 = 1.5\text{–}i0, \theta_1 = 40°(\theta_0 \sim 70°)]$. Thus, by substituting $\delta\Delta = 0.02°$ and $\lambda = 4000\,Å$ into Eq. (4.67), we get the following [93]:

$$\delta d \sim 0.1\,Å. \tag{4.68}$$

This result implies that the thickness precision in ellipsometry is better than one monolayer, even when the measurement error is rather conventional $\delta\Delta = 0.02°$. If we assume the diameter of an atom to be 2 Å, this thickness precision corresponds to 5 atoms on a surface composed of 100 atoms. Thus, ellipsometry provides very high precision for film thickness measurement. In actual ellipsometry analysis that employs an optical model, however, the film thickness is estimated from fitting to measurement spectra, and the confidence limit obtained from the fitting analysis is generally larger than the above measurement precision (see Section 7.1.1). Accordingly, the result shown in Eq. (4.68) should be considered as the sensitivity in ellipsometry measurement. As confirmed from Eq. (4.67), δd increases linearly with λ. In other words, the measurement sensitivity for film thickness degrades at longer wavelengths.

On the other hand, the measurement sensitivity (precision) for the absorption coefficient α of a sample can be estimated from the following equation [22]:

$$\delta\alpha = \frac{2\pi}{n\lambda}\delta\varepsilon_2 \tag{4.69}$$

The above equation can be derived easily by transforming $\alpha = 4\pi k/\lambda$ [Eq. (2.36)] using $\varepsilon_2 = 2nk$ [Eq. (2.47b)]. If we perform the calculation by substituting $\delta\varepsilon_2 = 0.02\,(\lambda = 4000\,Å)$ [81] and $n = 5$ into Eq. (4.69), we obtain

$$\delta\alpha \sim 600\,\text{cm}^{-1} \tag{4.70}$$

This precision for the absorption coefficient is worse than the one obtained from conventional transmittance measurements. In particular, in transmittance measurements, the sensitivity of the measurement is improved by increasing the film thickness of a sample. However, this method cannot be applied to spectroscopic ellipsometry since ellipsometry basically characterizes light reflection [93]. Thus, the characterization of low absorption coefficients is rather difficult using ellipsometry, although spectroscopic ellipsometry allows the measurement of high absorption coefficients. This is the common disadvantage of the ellipsometry technique (see Table 1.2). In general, when we evaluate absorption coefficients from spectroscopic ellipsometry, noise appears in the region of $\alpha < 10^3\,\mathrm{cm}^{-1}$ [94], as confirmed from Eq. (4.70). Figure 4.29 shows the spectrum of the absorption coefficient obtained from a hydrogenated amorphous silicon (a-Si:H) thin film [95]. In this characterization, an absorption coefficient of $\alpha > 10^5\,\mathrm{cm}^{-1}$ is estimated from spectroscopic ellipsometry (SE), and the intermediate region ($10^3 < \alpha < 10^5\,\mathrm{cm}^{-1}$) is measured using transmittance/reflectance measurement (T&R). In the region of low absorption coefficients ($\alpha < 10^3\,\mathrm{cm}^{-1}$), the measurement is performed using a technique that characterizes photocurrents ($\Delta\sigma_p$). As shown in Fig. 4.29, when the characterization of low absorption coefficients is important, other measurement methods including transmittance/reflectance measurements should also be employed.

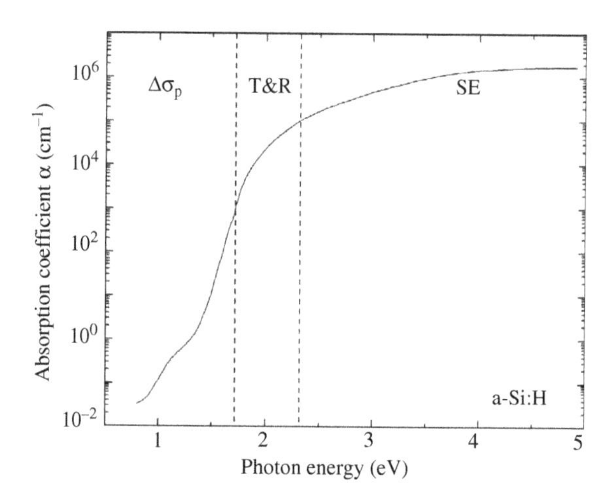

Figure 4.29 Absorption coefficient α of a hydrogenated amorphous silicon (a-Si:H) film determined by spectroscopic ellipsometry (SE), transmittance/reflectance measurement (T&R) and a photocurrent characterization method ($\Delta\sigma_p$). From Optical properties of solids, R. W. Collins and K. Vedam, in *Encyclopedia of Applied Physics, vol. 12*, 285–336, Wiley-VCH (1995). Reproduced by permission of Wiley-VCH.

4.4.4 DEPOLARIZATION EFFECT OF SAMPLES

When samples have a depolarization effect, totally polarized light used as a probe in ellipsometry is transformed into partially polarized light (see Section 3.4.3). In this case, the measurement error of the RAE instrument becomes quite large [19,60], as the RAE instrument assumes that reflected light is totally polarized (see Section 4.4.1). Similarly, when a PME measurement is carried out using only one measurement configuration (i.e., when S_1 or S_2 is not measured), the depolarization effect of samples increases measurement errors. Thus, sample structures also change the precision and error in these ellipsometry measurements.

We can summarize physical phenomena that generate partially polarized light upon light reflection as follows:

(a) surface light scattering caused by a large surface roughness of a sample [19],
(b) incident angle variation originating from the weak collimation of probe light [59,61,96],
(c) wavelength variation caused by the finite bandwidth of the monochromator [61,97],
(d) thickness inhomogeneity in a thin film formed on a substrate [19,60,61,96,98],
(e) backside reflection that occurs when the light absorption of a substrate is quite weak ($k \sim 0$) [53,59,96,99,100].

The above depolarization phenomena (a)–(e) are illustrated schematically in Fig. 4.30(a)–(e). It is evident from Fig. 4.30 that depolarization occurs by the generation of different polarizations upon light reflection. Such a depolarization phenomenon is generally referred to as quasi-depolarization (see Section 3.4.3).

As shown in Fig. 4.30(a), when surface roughness is very large, reflected light becomes partially polarized light due to multiple light scattering on a sample surface. Now imagine that the reflected light of linear polarization is overlapped with circular polarization generated by multiple light scattering, as shown in Fig. 4.30(a). In the RAE instrument, the polarization state of this reflected light is misinterpreted as elliptical polarization, since this instrument assumes totally polarized light for reflected light. In other words, the curve of Fig. 4.5(a) is mixed with (c) and is measured as (b). Furthermore, when reflected light becomes unpolarized light by light scattering on the surface, the measured light intensity in RAE becomes constant, as shown in Fig. 4.5 (c). In this particular case, the sample is misinterpreted as $\psi = 45°$ and $\Delta = 90°$. In the RCE instrument, on the other hand, accurate sample characterization can still be performed, even when reflected light is partially polarized, since (ψ, Δ) are evaluated together with the measurement of the degree of polarization (see Section 4.2.4). By applying the PSC_RA instrument, the characterization of textured structures has been performed [101]. In the depolarization spectrum shown in Fig. 4.27(a), the degree of polarization decreases significantly at around 4.0 eV. This reduction has been explained by multiple light scattering on the sample surface [19]. In particular, when the photon energy is

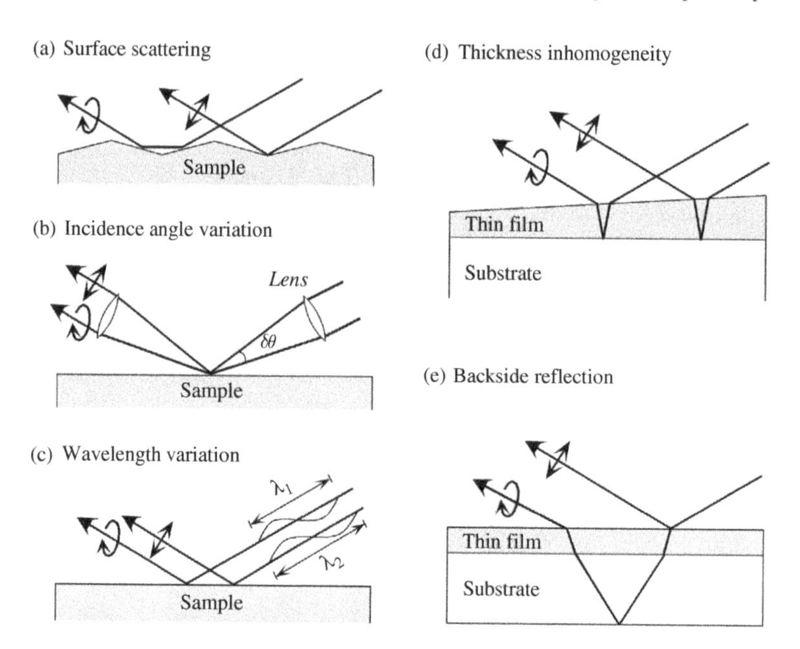

Figure 4.30 Depolarization of incident light by (a) surface scattering, (b) incidence angle variation, (c) wavelength variation, (d) thickness inhomogeneity, and (e) backside reflection.

large (i.e., when the wavelength of light is shorter), light scattering on a surface occurs more easily. As confirmed from Figs. 4.27(b) and (c), the (ψ, Δ) values obtained from PSC_RA in this region differ significantly from those obtained from P_RSA. Specifically, the Δ values estimated from P_RSA shift from the value of linear polarization (180°) to the value of circular polarization (90°) due to the depolarization effect of the sample. This result confirms that the measurement error of RAE increases when reflected light is partially polarized. In general, when samples have inhomogeneous structures including porous and patterned structures, light scattering occurs. If structural inhomogeneity is larger than $\sim 30\%$ of a measurement wavelength, reflected light is expected to become partially polarized light [19].

As shown in Fig. 4.30(b), when the probe light is focused, depolarization occurs by variation of incidence angle. In infrared spectroscopic ellipsometry, this depolarization effect should be taken into account [59], since an infrared light beam is generally focused (see Section 4.2.6). The depolarization shown in Fig. 4.30(c) originates from the characteristics of a monochromator. In particular, the light diffracted by a grating monochromator has a finite bandwidth and thus different wavelengths are measured simultaneously by the light detector. If the bandwidth of the monochromator is too broad, depolarization occurs due to the wavelength dependence of the optical properties of a sample. As shown in Fig. 4.30(d), when the thickness of a thin film formed on a substrate is inhomogeneous, the probe light is depolarized. As confirmed from Eq. (2.77), if the film thickness varies, the phase difference generated by optical interference also changes. Thus, thickness inhomogeneity induces the depolarization of reflected light by the principle similar to Fig. 3.10. The reduction in the degree of polarization at around 3.4 eV in Fig. 4.27(a) is caused by the thickness inhomogeneity of the sample [19]. When the thickness inhomogeneity increases, a depolarization spectrum shows large variations over a wide energy range [19,60,98]. From this spectrum, the quantitative analysis of thickness inhomogeneity can also be performed [19,60,98] (see Section 5.4.4).

As shown in Fig. 4.30(e), backside reflection of a substrate depolarizes reflected light. This depolarization phenomenon can be explained as follows. In general, light employed for spectroscopic measurements is not perfect monochromatic light, but quasi-monochromatic light that shows continuous variation of the phase with time [1]. In a thin film formed on a substrate, optical interference induced by the quasi-monochromatic light is quite similar to that generated by perfect monochromatic light, since the optical pass length of light is quite short. This is generally referred to as the coherent condition. With respect to backside reflection, on the other hand, the optical pass length becomes quite long since the thickness of the substrate is relatively thick (a few millimeters). In this case, the phase of the quasi-monochromatic light cannot be defined anymore due to the variation of the phase with time (incoherent condition). Consequently, the reflected light is partially polarized when the reflected light from the backside is included. This backside reflection, which occurs in a transparent substrate ($k \sim 0$), can be removed by mechanical polishing of the rear surface (see Section 5.4.3). If we use this method, we can simplify the data analysis procedure [101]. The data analysis procedures for depolarizing samples will be explained in Section 5.4.4 in detail.

REFERENCES

[1] R. M. A. Azzam and N. M. Bashara, *Ellipsometry and Polarized Light*, North-Holland, Amsterdam (1977).

[2] S. Teitler and B. H. Henvis, Refraction in stratified, anisotropic media, *J. Opt. Soc. Am.*, **60** (1970) 830–834.

[3] R. M. A. Azzam and N. M. Bashara, Generalized ellipsometry for surfaces with directional preference: application to diffraction gratings, *J. Opt. Soc. Am.*, **62** (1972) 1521–1523; R. M. A. Azzam and N. M. Bashara, Application of generalized ellipsometry to anisotropic crystals, *J. Opt. Soc. Am.*, **64** (1974) 128–133.

[4] G. E. Jellison, Jr, Data analysis for spectroscopic ellipsometry, in *Handbook of Ellipsometry*, edited by H. G. Tompkins and E. A. Irene, Chapter 3, 237–296, William Andrew, New York (2005).

[5] G. Jin, R. Jansson, and H. Arwin, Imaging ellipsometry revisited: Developments for visualization of thin transparent layers on silicon substrates, *Rev. Sci. Instrum.*, **67** (1996) 2930–2936.

[6] D. Tanooka, E. Adachi, and K. Nagayama, Color-imaging ellipsometer: high-speed characterization of in-plane distribution of film thickness at nano-scale, *Jpn. J. Appl. Phys.*, **40** (2001) 877–880.

[7] Q. Zhan and J. R. Leger, High-resolution imaging ellipsometer, *Appl. Opt.*, **41** (2002) 4443–4450.

[8] W. Budde, Photoelectric analysis of polarized light, *Appl. Opt.*, **1** (1962) 201–205.
[9] B. D. Cahan and R. F. Spanier, A high speed precision automatic ellipsometer, *Surf. Sci.*, **16** (1969) 166–176.
[10] P. S. Hauge and F. H. Dill, Design and operation of ETA, an automated ellipsometer, *IBM J. Res. Develop.*, **17** (1973) 472–489.
[11] D. E. Aspnes and A. A. Studna, High precision scanning ellipsometer, *Appl. Opt.*, **14** (1975) 220–228.
[12] For a review, see R. H. Muller, Present status of automatic ellipsometers, *Surf. Sci.*, **56** (1976) 19–36.
[13] S. Kawabata, Ellipsometry and some difficulties in the measurements of extremely thin films, *in Japanese, OYO BUTURI*, **57** (1988) 1868–1876.
[14] For a review, see R. W. Collins, Automatic rotating element ellipsometers: calibration, operation, and real-time applications, *Rev. Sci. Instrum.*, **61** (1990) 2029.
[15] Y.-T. Kim, R. W. Collins, and K. Vedam, Fast scanning spectroelectrochemical ellipsometry: *in situ* characterization of gold oxide, *Surf. Sci.*, **223** (1990) 341–350.
[16] I. An, Y. M. Li, H. V. Nguyen, and R. W. Collins, Spectroscopic ellipsometry on the millisecond time scale for real-time investigations of thin-film and surface phenomena, *Rev. Sci. Instrum.*, **63** (1992) 3842–3848.
[17] P. S. Hauge and F. H. Dill, A rotating-compensator Fourier ellipsometer, *Opt. Commun.*, **14** (1975) 431–437.
[18] P. S. Hauge, Generalized rotating-compensator ellipsometry, *Surf. Sci.*, **56** (1976) 148–160.
[19] J. Lee, P. I. Rovira, I. An, and R. W. Collins, Rotating-compensator multichannel ellipsometry: Applications for real time Stokes vector spectroscopy of thin film growth, *Rev. Sci. Instrum.*, **69** (1998) 1800–1810.
[20] For a review, see R. W. Collins, J. Koh, H. Fujiwara, P. I. Rovira, A. S. Ferlauto, J. A. Zapien, C. R. Wronski, and R. Messier, Recent progress in thin film growth analysis by multichannel spectroscopic ellipsometry, *Appl. Surf. Sci.*, **154–155** (2000) 217–228.
[21] For a review, see P. S. Hauge, Recent developments in instrumentation in ellipsometry, *Surf. Sci.*, **96** (1980) 108–140.
[22] D. E. Aspnes, Spectroscopic ellipsometry of solids, in *Optical Properties of Solids: New Developments*, edited by B. O Seraphin, Chapter 15, 801–846, North-Holland, Amsterdam (1976).
[23] S. N. Jasperson and S. E. Schnatterly, An improved method for high reflectivity ellipsometry based on a new polarization modulation technique, *Rev. Sci. Instrum.*, **40** (1969) 761–767.
[24] S. N. Jasperson, D. K. Burge, and R. C. O'Handley, A modulated ellipsometer for studying thin film optical properties and surface dynamics, *Surf. Sci.*, **37** (1973) 548–558.
[25] V. M. Bermudez and V. H. Ritz, Wavelength-scanning polarization-modulation ellipsometry: some practical considerations, *Appl. Opt.*, **17** (1978) 542–552.
[26] B. Drévillon, J. Perrin, R. Marbot, A. Violet, and J. L. Dalby, Fast polarization modulated ellipsometer using a microprocessor system for digital Fourier analysis, *Rev. Sci. Instrum.*, **53** (1982) 969–977.
[27] O. Acher, E. Bigan, and B. Drévillon, Improvements of phase-modulated ellipsometry, *Rev. Sci. Instrum.*, **60** (1989) 65–77.
[28] W. M. Duncan and S. A. Henck, Insitu spectral ellipsometry for real-time measurement and control, *Appl. Surf. Sci.*, **63** (1993) 9–16.
[29] S. A. Henck, W. M. Duncan, L. M. Lowenstein and S. W. Butler, *In situ* spectral ellipsometry for real-time thickness measurement: etching multilayer stacks, *J. Vac. Sci. Technol. A*, **11** (1993) 1179–1185.

[30] M. Kildemo, S. Deniau, P. Bulkin, and B. Drévillon, Real time control of the growth of silicon alloy multilayers by multiwavelength ellipsometry, *Thin Solid Films*, **290–291** (1996) 46–50.
[31] A. Röseler, Spectroscopic ellipsometry in the infrared, *Infrared Physics*, **21** (1981) 349–355.
[32] A. Röseler and W. Molgedey, Improvement in accuracy of spectroscopic IR ellipsometry by the use of IR retarders, *Infrared Physics*, **24** (1984) 1–5.
[33] A. Röseler, IR spectroscopic ellipsometry: instrumentation and results, *Thin Solid Films*, **234** (1993) 307–313.
[34] F. Ferrieu, Infrared spectroscopic ellipsometry using a Fourier transform infrared spectrometer: some applications in thin-film characterization, *Rev. Sci. Instrum.*, **60** (1989) 3212–3216.
[35] T. E. Tiwald, D. W. Thompson, J. A. Woollam, S. V. Pepper, Determination of the mid-IR optical constants of water and lubricants using IR ellipsometry combined with an ATR cell, *Thin Solid Films*, **313–314** (1998) 718–721.
[36] C. Defranoux, T. Emeraud, S. Bourtault, J. Venturini, P. Boher, M. Hernandez, C. Laviron, T. Noguchi, Infrared spectroscopic ellipsometry applied to the characterization of ultra shallow junction on silicon and SOI, *Thin Solid Films*, **455–456** (2004) 150–156.
[37] R. T. Graf, F. Eng, J. L. Koenig and H. Ishida, Polarization modulation Fourier transform infrared ellipsometry of thin polymer films, *Appl. Spectrosc.*, **40** (1986) 498–503.
[38] A. Canillas, E. Pascual and B. Drévillon, Phase-modulated ellipsometer using a Fourier transform infrared spectrometer for real time applications, *Rev. Sci. Instrum.*, **64** (1993) 2153–2159.
[39] For a review, see B. Drévillon, *In situ* spectroscopic ellipsometry studies of interfaces of thin films deposited by PECVD, *Thin Solid Films*, **241** (1994) 234–239.
[40] K. Tachibana, T. Shirafuji and S. Muraishi, Construction and performance of a Fourier-transform infrared phase-modulated ellipsometer for in-process surface diagnostics, *Jpn. J. Appl. Phys.*, **35** (1996) 3652–3657.
[41] P. S. Hauge, Automated Mueller matrix ellipsometry, *Opt. Commun.*, **17** (1976) 74–76.
[42] P. S. Hauge, Mueller matrix ellipsometry with imperfect compensators, *J. Opt. Soc. Am.*, **68** (1978) 1519–1528.
[43] R. M. A. Azzam, Photopolarimetric measurement of the Mueller matrix by Fourier analysis of a single detected signal, *Opt. Lett.*, **2** (1978) 148–150.
[44] R. M. A. Azzam, K. A. Giardina, and A. G. Lopez, Conventional and generalized Mueller-matrix ellipsometry using the four-detector photopolarimeter, *Opt. Eng.*, **30** (1991) 1583–1589.
[45] D. H. Goldstein, Mueller matrix dual-rotating retarder polarimeter, *Appl. Opt.*, **31** (1992) 6676–6683.
[46] D. A. Ramsey and K. C. Ludema, The influences of roughness on film thickness measurements by Mueller matrix ellipsometry, *Rev. Sci. Instrum.*, **65** (1994) 2874–2881.
[47] G. E. Jellison, Jr, and F. A. Modine, Two-modulator generalized ellipsometry: experiment and calibration, *Appl. Opt.*, **36** (1997) 8184–8189; G. E. Jellison, Jr, and F. A. Modine, Two-modulator generalized ellipsometry: theory, *Appl. Opt.*, **36** (1997) 8190–8198.
[48] E. Compain, B. Drévillon, J. Huc, J. Y. Parey, J. E. Bouree, Complete Mueller matrix measurement with a single high frequency modulation, *Thin Solid Films*, **313–314** (1998) 47–52.
[49] J. Lee, J. Koh, and R. W. Collins, Multichannel Mueller matrix ellipsometer for real-time spectroscopy of anisotropic surfaces and films, *Opt. Lett.*, **25** (2000) 1573–1575.

[50] For a review, see C. Chen, I. An, G. M. Ferreira, N. J. Podraza, J. A. Zapien, R. W. Collins, Multichannel Mueller matrix ellipsometer based on the dual rotating compensator principle, *Thin Solid Films*, **455–456** (2004) 14–23.
[51] A. De Martino, E. Garcia-Caurel, B. Laude, B. Drévillon, General methods for optimized design and calibration of Mueller polarimeters, *Thin Solid Films*, **455–456** (2004) 112–119.
[52] T.Tadokoro, K. Akao, T. Yoshihara, S. Okutani, M. Kimura, T. Akahane, and H. Toriumi, Dynamics of surface-stabilized ferroelectric liquid crystals at the alignment layer surface studied by total-reflection ellipsometry, *Jpn. J. Appl. Phys.*, **40** (2001) L453–455.
[53] M. Kildemo, R. Ossikovski, and M. Stchakovsky, Measurement of the absorption edge of thick transparent substrates using the incoherent reflection model and spectroscopic UV-visible-near IR ellipsometry, *Thin Solid Films*, **313–314** (1998) 108–113.
[54] J. M. M. de Nijs and A. van Silfhout, Systematic and random errors in rotating-analyzer ellipsometry, *J. Opt. Soc. Am. A*, **5** (1988) 773–781.
[55] J. M. M. de Nijs, A. H. M. Holtslag, A. Hoeksta, and A. van Silfhout, Calibration method for rotating-analyzer ellipsometers, *J. Opt. Soc. Am. A*, **5** (1988) 1466–1471.
[56] B. Johs, Regression calibration method for rotating element ellipsometers, *Thin Solid Films*, **234** (1993) 395–398.
[57] D. E. Aspnes, Optimizing precision of rotating-analyzer ellipsometers, *J. Opt. Soc. Am.*, **64** (1974) 639–646.
[58] J. A. Zapien, R. W. Collins, and R. Messier, Multichannel ellipsometer for real time spectroscopy of thin film deposition from 1.5 to 6.5 eV, *Rev. Sci. Instrum.*, **71** (2000) 3451–3460.
[59] A. Röseler, Problem of polarization degree in spectroscopic photometric ellipsometry (polarimetry), *J. Opt. Soc. Am. A*, **9** (1992) 1124–1131.
[60] U. Richter, Application of the degree of polarization of film thickness gradients, *Thin Solid Films*, **313–314** (1998) 102–107.
[61] S. Zollner, T.-C. Lee, K. Noehring, A. Konkar, N. D. Theodore, W. M. Huang, D. Monk, T. Wetteroth, S. R. Wilson and J. N. Hilfiker, Thin-film metrology of silicon-on-insulator materials, *Appl. Phys. Lett.*, **76** (2000) 46–48.
[62] R. Kleim, L. Kuntzler, and A. E. Ghemmaz, Systematic errors in rotating-compensator ellipsometry, *J. Opt. Soc. Am. A*, **11** (1994) 2550–2559.
[63] R. W. Stobie, B. Rao, and M. J. Dignam, Automatic ellipsometer with high sensitivity and special advantages for infrared spectroscopy of adsorbed species, *Appl. Opt.*, **14** (1975) 999–1003.
[64] A. S. Siddiqui and D. M. Treherne, Optical properties of some transition metals at infrared frequencies, *Infrared Physics*, **17** (1977) 33–42.
[65] A. A. Langford, M. L. Fleet, B. P. Nelson, W. A. Lanford, and N. Maley, Infrared absorption strength and hydrogen content of hydrogenated amorphous silicon, *Phys. Rev. B*, **45** (1992) 13367–13377.
[66] Y. J. Chabal, High resolution infrared spectroscopy of adsorbates on semiconductor surfaces: hydrogen on Si(100) and Ge(100), *Surf. Sci.*, **168** (1986) 594–608.
[67] Y. J. Chabal, G. S. Higashi, K. Raghavachari, and V. A. Burrows, Infrared spectroscopy of Si(111) and Si(100) surfaces after HF treatment: hydrogen termination and surface morphology, *J. Vac. Sci. Technol. A*, **7** (1989) 2104–2109.
[68] H. Fujiwara, M. Kondo, and A. Matsuda, Depth profiling of silicon-hydrogen bonding modes in amorphous and microcrystalline Si:H thin films by real-time infrared spectroscopy and spectroscopic ellipsometry, *J. Appl. Phys.*, **91** (2002) 4181–4190.
[69] H. Fujiwara and M. Kondo, Real-time monitoring and process control in amorphous/crystalline silicon heterojunction solar cells by spectroscopic ellipsometry and infrared spectroscopy, *Appl. Phys. Lett.*, **86** (2005) 32112.

[70] E. D. Palik (editor), *Handbook of Optical Constants of Solids*, Academic Press, San Diego (1985).
[71] K. Hinrichs, A. Röseler, M. Gensch, E. H. Korte, Structure analysis of organic films by mid-infrared ellipsometry, *Thin Solid Films*, **455–456** (2004) 266–271.
[72] S. Zangooie, M. Schubert, D. W. Thompson, and J. A. Woollam, Infrared response of multiple-component free-carrier plasma in heavily doped p-type GaAs, *Appl. Phys. Lett.*, **78** (2001) 937–939.
[73] M. Schubert, *Infrared Ellipsometry on Semiconductor Layer Structures: Phonons, Plasmons, and Polaritons*, Springer, Heidelberg (2004).
[74] For example, see M. Fried, K. Hingerl, and J. Humlíček, Eds, *The Proceedings of the 3rd International Conference on Spectroscopic Ellipsometry; Thin Solid Films*, **455–456** (2004) 1–836.
[75] J. N. Hilfiker, B. Johs, C. M. Herzinger, J. F. Elman, E. Montback, D. Bryant, and P. J. Bos, Generalized spectroscopic ellipsometry and Mueller-matrix study of twisted nematic and super twisted nematic liquid crystals, *Thin Solid Films*, **455–456** (2004) 596–600.
[76] A. Laskarakis, S. Logothetidis, E. Pavlopoulou, and M. Gioti, Mueller matrix spectroscopic ellipsometry: formulation and application, *Thin Solid Films*, **455–456** (2004) 43–49.
[77] H. Arwin, S. W-Klintström, and R. Jansson, Off-null ellipsometry revisited: basic considerations for measuring surface concentrations at solid/liquid interfaces, *J. Coll. Int. Sci.*, **156** (1993) 377–382.
[78] R. F. Cohn, J. W. Wagner, and J. Kruger, Dynamic imaging microellipsometry: theory, system design, and feasibility demonstration, *Appl. Opt.*, **27** (1988) 4664–4671.
[79] A. Albersdörfer, G. Elender, G. Mathe, K. R. Neumaier, P. Paduschek, and E. Sackmann, High resolution imaging microellipsometry of soft surfaces at 3 μm lateral and 5 Å normal resolution, *Appl. Phys. Lett.*, **72** (1998) 2930–2932.
[80] P. Boher, O. Thomas, J. P. Piel, and J. L. Stehle, A new multiple wavelength ellipsometric imager: design, limitations and applications, *Thin Solid Films*, **455–456** (2004) 809–818.
[81] For a review, see R. W. Collins, I. An, H. Fujiwara, J. Lee, Y. Lu, J. Koh, and P. I. Rovira, Advances in multichannel spectroscopic ellipsometry, *Thin Solid Films*, **313–314** (1998) 18–32.
[82] A. A. Studna, D. E. Aspnes, L. T. Florez, B. J. Wilkens, J. P. Harbison, and R. E. Ryan, Low-retardance fused-quartz window for real-time optical applications in ultrahigh vacuum, *J. Vac. Sci. Technol. A*, **7** (1989) 3291–3294.
[83] J. N. Hilfiker, C. L. Bungay, R. A. Synowicki, T. E. Tiwald, C. M. Herzinger, B. Johs, G. K. Pribil and J. A. Woollam, Progress in spectroscopic ellipsometry: Applications from vacuum ultraviolet to infrared, *J. Vac. Sci. Technol. A*, **21** (2003) 1103–1108.
[84] N. V. Nguyen, B. S. Pudliner, I. An, and R. W. Collins, Error correction for calibration and data reduction in rotating-polarizer ellipsometry: applications to a novel multichannel ellipsometer, *J. Opt. Soc. Am. A*, **8** (1991) 919–931.
[85] I. An and R. W. Collins, Waveform analysis with optical multichannel detectors: Applications for rapid-scan spectroscopic ellipsometry, *Rev. Sci. Instrum.*, **62** (1991) 1904–1911.
[86] D. E. Aspnes, Effects of component optical activity in data reduction and calibration of rotating-analyzer ellipsometers, *J. Opt. Soc. Am.*, **64** (1974) 812–819.
[87] B. D. Johs and D. W. Thompson, United States Patent 5872630 (1999).
[88] J. Lee, P. I. Rovira, I. An, and R. W. Collins, Alignment and calibration of the MgF_2 biplate compensator for applications in rotating-compensator multichannel ellipsometry, *J. Opt. Soc. Am. A*, **18** (2001) 1980–1985.

[89] D. E. Aspnes and A. A. Studna, Methods for drift stablization and photomultiplier linearization for photometric ellipsometers and polarimeters, *Rev. Sci. Instrum.*, **49** (1978) 291–297.

[90] I. An, Y. Cong, N. V. Nguyen, B. S. Pudliner and R. W. Collins, Instrumentation considerations in multichannel ellipsometry for real-time spectroscopy, *Thin Solid Films*, **206** (1991) 300–305.

[91] R. M. A. Azzam and N. M. Bashara, Analysis of systematic errors in rotating-analyzer ellipsometers, *J. Opt. Soc. Am.*, **64** (1974) 1459–1469.

[92] D. E. Aspnes, Precision bounds to ellipsometer systems, *Appl. Opt.*, **14** (1975) 1131–1136.

[93] R. W. Collins and Y.-T. Kim, Ellipsometry for thin-film and surface analysis, *Anal. Chem.*, **62** (1990) 431–442.

[94] G. E. Jellison, Jr and F. A. Modine, Optical functions of silicon at elevated temperatures, *J. Appl. Phys.*, **76** (1994) 3758–3761.

[95] R. W. Collins and K. Vedam, Optical properties of solids, in *Encyclopedia of Applied Physics*, vol. 12, 285–336, Wiley-VCH (1995).

[96] J. -Th. Zettler, Th. Trepk, L. Spanos, Y. -Z. Hu, and W. Richter, High precision UV-visible-near-IR Stokes vector spectroscopy, *Thin Solid Films*, **234** (1993) 402–407.

[97] S. F. Nee, Depolarization and retardation of a birefringent slab, *J. Opt. Soc. Am. A*, **17** (2000) 2067–2073.

[98] G. E. Jellison, Jr and J. W. McCamy, Sample depolarization effects from thin films of ZnS on GaAs as measured by spectroscopic ellipsometry, *Appl. Phys. Lett.*, **61** (1992) 512–514.

[99] Y. H. Yang, and J. R. Abelson, Spectroscopic ellipsometry of thin films on transparent substrates: a formalism for data interpretation, *J. Vac. Sci. Technol. A*, **13** (1995) 1145–1149.

[100] R. Joerger, K. Forcht, A. Gombert, M. Köhl, and W. Graf, Influence of incoherent superposition of light on ellipsometric coefficients, *Appl. Opt.*, **36** (1997) 319–327.

[101] P. I. Rovira and R. W. Collins, Analysis of specular and textured SnO_2 : F films by high speed four-parameter Stokes vector spectroscopy, *J. Appl. Phys.*, **85** (1999) 2015–2025.

5 Data Analysis

In order to evaluate the optical constants and thickness of samples from spectroscopic ellipsometry, it is necessary to perform ellipsometry data analysis that consists of three major parts; i.e., dielectric function modeling, the construction of an optical model, and fitting to measured (ψ, Δ) spectra. In this chapter, we will review various dielectric function models as well as optical models used in ellipsometry analysis. This chapter will also address the effective medium approximation, which is commonly employed for surface roughness analysis. Here, we first discuss variations of (ψ, Δ) in transparent and absorbing films and, in the last part of the chapter, we discuss the fitting procedure in ellipsometry data analysis.

5.1 INTERPRETATION OF (ψ, Δ)

As we have seen in Chapter 4, ψ and Δ represent the amplitude ratio and phase difference between p- and s-polarizations, respectively. However, (ψ, Δ) generally show complicated variations with changes in optical constants and film thicknesses. In this section, we will discuss variations of (ψ, Δ) in transparent and absorbing films in detail. The understanding of these behaviors is quite helpful, not only for the interpretation of measured spectra, but also for the construction of optical models.

5.1.1 VARIATIONS OF (ψ, Δ) WITH OPTICAL CONSTANTS

As the simplest example, we first consider a case where (ψ, Δ) are determined by light reflection at an air/sample interface. In other words, we assume that the sample is composed of a substrate with infinite thickness only. In this case, using Eqs. (2.79) and (4.1), we can express $\rho = \tan\psi \exp(i\Delta)$ as follows:

$$\rho = \frac{r_p}{r_s} = \left(\frac{N_1\cos\theta_0 - N_0\cos\theta_1}{N_1\cos\theta_0 + N_0\cos\theta_1}\right) \Big/ \left(\frac{N_0\cos\theta_0 - N_1\cos\theta_1}{N_0\cos\theta_0 + N_1\cos\theta_1}\right) \tag{5.1}$$

Spectroscopic Ellipsometry: Principles and Applications H. Fujiwara

Here, N_0 and N_1 are the complex refractive indices of air ($N_0 = 1 - i0$) and the sample, respectively. The transmission angle θ_1 can be calculated from the incidence angle θ_0 using Snell's law (see Section 2.3.1). If we use Eqs. (4.5) and (4.6), we can calculate (ψ, Δ) from ρ. It should be noted that Δ is given by $\Delta = -\arg(\rho)$ for the definition of $N \equiv n + ik$ since $\rho = \tan\psi \exp(-i\Delta)$ (see Section 4.1.1).

Figure 5.1 shows the reflectance for p- and s-polarizations (R_p, R_s), phase of reflected p- and s-polarizations (δ_{rp}, δ_{rs}), and (ψ, Δ) at air/c-Si (crystalline Si) and air/Cu interfaces, obtained from Eq. (5.1). In this calculation, the complex refractive

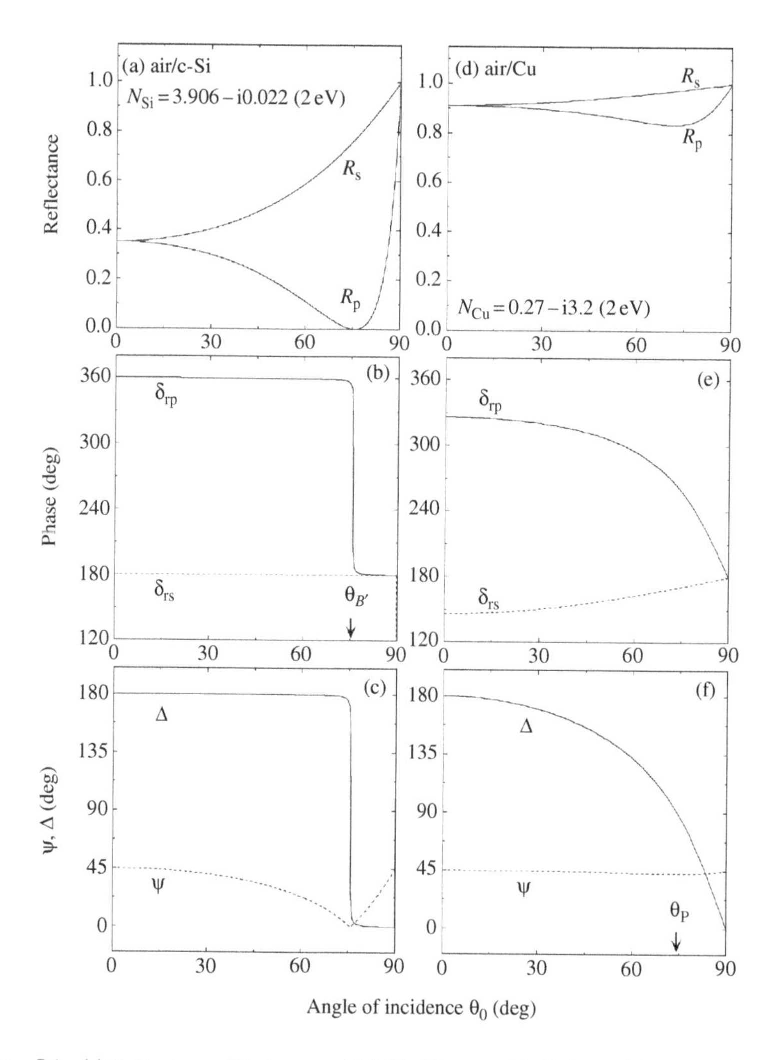

Figure 5.1 (a) Reflectance, (b) phase, and (c) (ψ, Δ) at an air/c-Si (crystalline silicon) interface and (d) reflectance, (e) phase, and (f) (ψ, Δ) at an air/Cu interface, plotted as a function of the angle of incidence θ_0.

indices of c-Si and Cu are assumed to be $N_1 = 3.906 - i0.022$ (2 eV) [1] and $N_1 = 0.27 - i3.2$ (2 eV) [2], and the angle of incidence θ_0 is varied. The trends of the reflectance and phase at the air/c-Si interface are similar to those at the air/glass interface shown in Figs. 2.16(c) and 2.18(a), since the extinction coefficient k of c-Si is almost zero at 2 eV. In the case of c-Si, however, the variation of δ_{rp} is smoother at around the pseudo-Brewster angle $\theta_{B'}$ (see Section 2.3.4). As shown in Fig. 5.1(c), since Δ is given by $\delta_{rp} - \delta_{rs}$ [Eq. (4.4)], we find that $\Delta \sim 180°$ at $\theta_0 < \theta_{B'}$ and $\Delta \sim 0°$ at $\theta_0 > \theta_{B'}$. In the case of glass ($k = 0$), we observe $\Delta = 180°$ at $\theta_0 < \theta_B$ and $\Delta = 0°$ at $\theta_0 > \theta_B$ [Fig. 2.16(c)]. The change in ψ shown in Fig. 5.1(c) can be understood from the results of R_p and R_s [3]. In particular, at normal incidence ($\theta_0 = 0°$), there is no distinction between p- and s-polarizations ($R_p = R_s$) and thus we obtain $\psi = 45°$ from $\psi = \tan^{-1}[(R_p/R_s)^{1/2}]$ [Eq. (4.5)]. At $\theta_0 = \theta_{B'}$, on the other hand, we obtain $\psi \sim 0°$ since $R_p \sim 0$.

In general, metals have a small refractive index n and large extinction coefficient k. Thus, metals show very high reflectance, as confirmed from Eq. (2.72). In particular, ψ of metals generally becomes $\sim 45°$, independent of θ_0, since R_p is the almost same as R_s [see Figs. 5.1(d) and (f)]. However, Δ varies drastically versus θ_0 due to the continuous changes of δ_{rp} and δ_{rs} with θ_0, as shown in Figs. 5.1(e) and (f). The angle when $\Delta = 90°$ is called the principal angle θ_P, but the difference between θ_P and $\theta_{B'}$ is less than 1° [3]. From the above calculations, it can be seen that, when $n \sim 0$ and $k \gg 0$, Δ varies without a large change in ψ since $R_p \sim R_s$. When $n \gg 0$ and $k \sim 0$, on the other hand, ψ varies due to the large change in reflectance.

Figure 5.2 shows (ψ, Δ) when the values of (n, k) are varied independently. This calculation was performed from Eq. (5.1) using $\theta_0 = 70°$. In particular, n_1 is varied with a constant value of $k_1 = 0.01$ ($N_1 = n_1 - i0.01$) in Fig. 5.2(a), while k_1 is varied with a constant value of $n_1 = 4.0$ ($N_1 = 4.0 - ik_1$) in Fig. 5.2(b). It should be emphasized that the calculation in Fig. 5.2 is highly hypothetical since the values of n and k are interrelated and never change independently, as will be shown in Section 5.2.6. In Fig. 5.2(a), when n_1 increases from 3.5 to 4.5, ψ increases by ~7°, whereas Δ shows a change of ~0.5° only. This implies that the position (angle) of $\theta_{B'}$ varies according to n_1 and consequently ψ changes. In contrast, if k_1 increases up to 0.5 with constant n_1, Δ reduces greatly with a small change in ψ. These results confirm that ψ and Δ vary depending on n_1 and k_1, respectively.

Figure 5.3 shows (a) (ψ, Δ) at $\theta_0 = 70°$, (b) (n, k), and (c) absorption coefficient of c-Si. The (n, k) in Fig. 5.3(b) show reported optical constants [4], whereas the absorption coefficient in Fig. 5.3(c) was calculated from k using Eq. (2.36). From a comparison between Figs. 5.3(a) and (b), it can be seen that the trends of variations observed in Fig. 5.3(a) are quite similar to those shown in Fig. 5.2 in the region of En < 4.0 eV. Accordingly, when a sample structure is simple, variations in ψ and Δ can be attributed to changes in n and k, respectively.

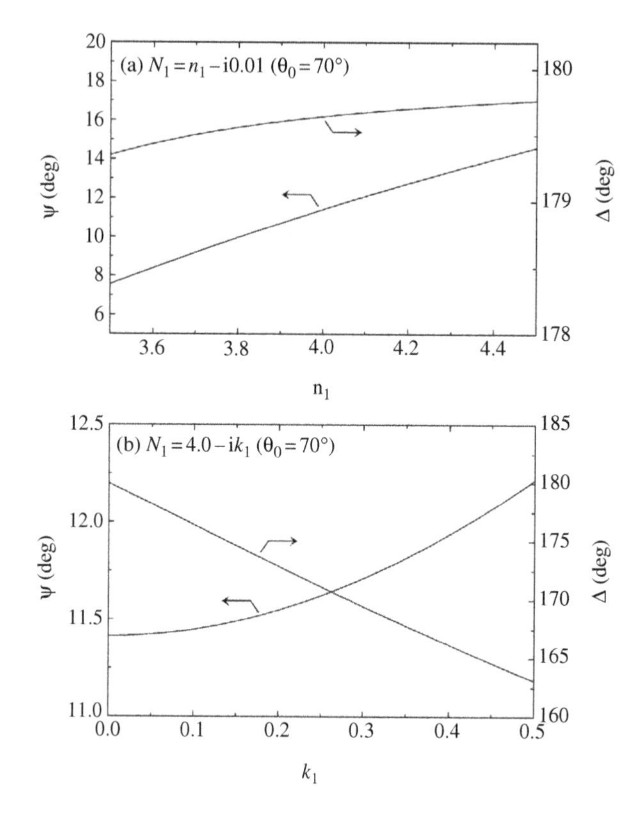

Figure 5.2 Variations of (ψ, Δ) with the complex refractive index of a sample at an air/sample interface: (a) variation with n_1 when $N_1 = n_1 - \mathrm{i}0.01$ and (b) variation with k_1 when $N_1 = 4.0 - \mathrm{i}k_1$.

5.1.2 VARIATIONS OF (ψ, Δ) IN TRANSPARENT FILMS

If we apply Eq. (2.86), $\rho = \tan\psi \exp(\mathrm{i}\Delta)$ of an ambient/thin film/substrate structure (Fig. 2.23) can be expressed as

$$\rho = \frac{r_\mathrm{p}}{r_\mathrm{s}} = \left[\frac{r_{01,\mathrm{p}} + r_{12,\mathrm{p}}\exp(-i2\beta)}{1 + r_{01,\mathrm{p}}r_{12,\mathrm{p}}\exp(-i2\beta)}\right] \Bigg/ \left[\frac{r_{01,\mathrm{s}} + r_{12,\mathrm{s}}\exp(-i2\beta)}{1 + r_{01,\mathrm{s}}r_{12,\mathrm{s}}\exp(-i2\beta)}\right] \tag{5.2}$$

where β represents the film phase thickness given by $\beta = 2\pi d N_1 \cos\theta_1/\lambda$ [Eq. (2.78)]. In the case of transparent films, we obtain $N_1 = n_1$ since $k_1 = 0$ and thus $\beta = 2\pi d n_1 \cos\theta_1/\lambda$. Figure 5.4 shows (a) ψ, (b) Δ, (c) reflectance and (d) $\psi - \Delta$ trajectory in an air/SiO_2 thin film/c-Si structure. In this calculation, $N_1 = n_1 = 1.46$ (SiO_2 thin film) and $N_2 = 3.87 - \mathrm{i}0.0146$ (c-Si substrate) [4] were used as the values at $\lambda = 6328$ Å corresponding to the wavelength of a He–Ne laser. Here, the angle of incidence is 70° and we will use $\theta_0 = 70°$ in the following parts. If we calculate

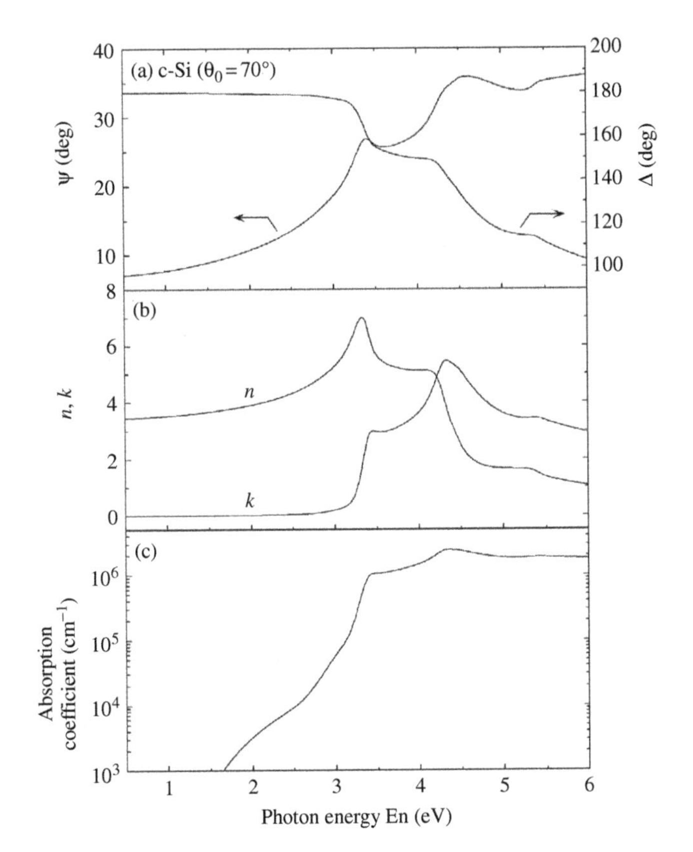

Figure 5.3 (a) (ψ, Δ) at $\theta_0 = 70°$, (b) complex refractive index (n, k), and (c) absorption coefficient of crystalline silicon. Data (b) from Herzinger *et al.* (Ref. [4].)

Δ from Eq. (5.2), the range of Δ becomes $-180° \leq \Delta \leq 180°$. In Fig. 5.4, however, the range of Δ is converted to $0° \leq \Delta \leq 360°$ by adding 360° to $-180° \leq \Delta < 0°$ to simplify the figure. In Figs. 5.4(a)–(c), each value oscillates with a period of 2832 Å, as the SiO_2 thickness increases. Thus, the ψ–Δ trajectory rotates in the direction indicated by arrows in Fig. 5.4(d). This behavior originates from the fact that the complex value of the term $\exp(-i2\beta)$ in Eq. (5.2) rotates on the complex plane with increasing film thickness [3].

As we have seen in Section 2.4.1, 2β shows the phase shift of light traveling inside a thin film. When $2dn_1 \cos\theta_1/\lambda = 1$, it follows from $\beta = 2\pi dn_1 \cos\theta_1/\lambda$ that $2\beta = 2\pi$. In this condition, we obtain $\exp(-i2\beta) = 1$. In other words, the phase shift of a secondary beam becomes exactly one wavelength. Accordingly, the thickness period d_i in optical interference is expressed by

$$d_i = \lambda/(2n_1 \cos\theta_1) \tag{5.3}$$

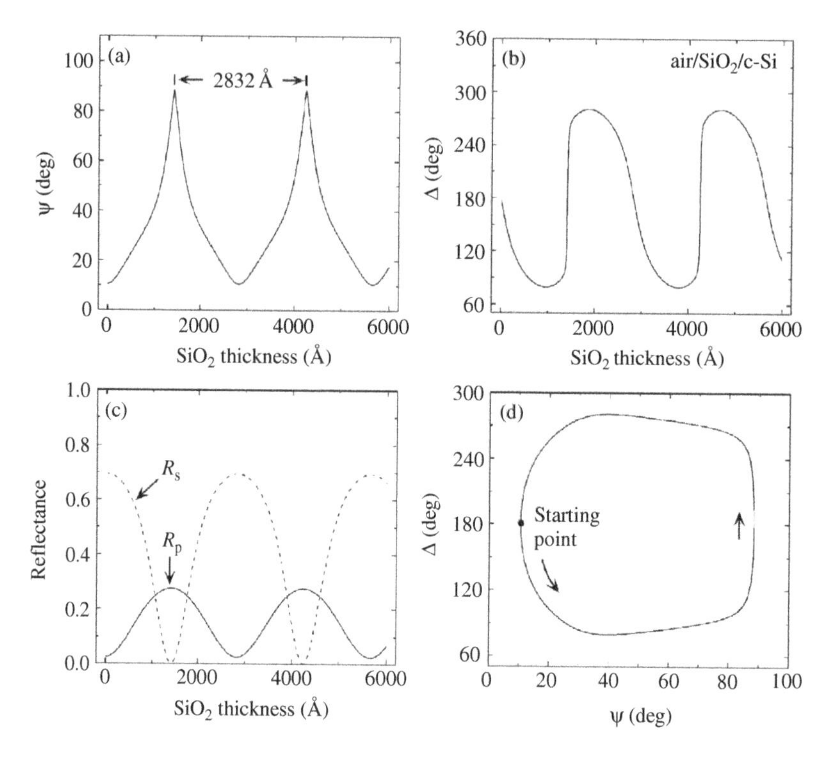

Figure 5.4 (a) ψ, (b) Δ, (c) reflectance, and (d) ψ − Δ trajectory in an air/SiO_2/c-Si structure. In the calculation, $N_1 = 1.46$ (SiO_2) and $N_2 = 3.87 - i0.0146$ (c-Si) [4] were used as the values at $\lambda = 6328$ Å. The (ψ, Δ) show the values when $\theta_0 = 70°$.

From Eq. (5.3), it is clear that d_i becomes smaller with increasing n_1. In particular, Eq. (5.3) shows that optical interference occurs with the condition $n_1 \cos\theta_1 d_i =$ const. when λ is fixed. In reflectance measurements performed at normal incidence, this condition becomes $n_1 d_i =$ const. and the *nd* product is generally called the optical thickness.

If we apply Eq. (5.3) to the calculation for Fig. 5.4, we obtain $d_i = 2832$ Å. When the film thickness is $d_i/2$, on the other hand, the phase shift becomes half the wavelength ($2\beta = 180°$). At this thickness, the optical interference of s-polarization occurs destructively, since the phase of the s-polarization reflected at the film surface (primary beam) is exactly opposite to that of the secondary beam. Consequently, the reflectance of s-polarization becomes $R_s \sim 0$ at $d_i/2$ [Fig. 5.4(c)]. In the case of p-polarization, the phase shift at the air/SiO_2 interface becomes $\delta_{rp} = 180°$ since $\theta_0 > \theta_B$ [Fig. 2.16(c)], while the phase change at the SiO_2/c-Si interface is $\delta_{rp} = 360°$ since $\theta_1 < \theta_{B'}$. Thus, if the phase of the secondary beam lags by 180° (half a wavelength) due to traveling inside the thin film, the primary and secondary waves overlap constructively. Strictly speaking, when the number of light reflections at the SiO_2/c-Si interface is odd, the phase shift between the primary and multiply reflected waves is zero (constructive), while the phase shift is 180° when the number of reflections is even (destructive). As a result, R_p is maximized at $d = d_i/2$, as shown in Fig. 5.4(c). At $d = d_i/2$, therefore, R_p/R_s shows a quite large value, while R_p/R_s is minimized at $d = d_i$. Consequently, ψ values become ~90° and ~0° at $d = d_i/2$ and $d = d_i$, respectively, since $\psi = \tan^{-1}[(R_p/R_s)^{1/2}]$ [Eq. (4.5)]. It can be seen from Fig. 5.4(b) that Δ changes significantly at $d = d_i/2$ since the phase is reversed at this thickness. As we have seen in Fig. 5.1, when light is reflected at an ambient/sample interface, (ψ, Δ) are always confined to the ranges $0° \leq \psi \leq 45°$ and $0° \leq \Delta \leq 180°$. When there is optical interference of a transparent film, however, (ψ, Δ) change in the wider ranges of $0° \leq \psi \leq 90°$ and $0° \leq \Delta \leq 360°$ (or $-180° \leq \Delta \leq 180°$).

The above changes in (ψ, Δ) are also observed in spectra in which the wavelength λ is varied with a constant film thickness d. Figure 5.5 shows (ψ, Δ) spectra obtained from an air/thin film/substrate structure, plotted as functions of

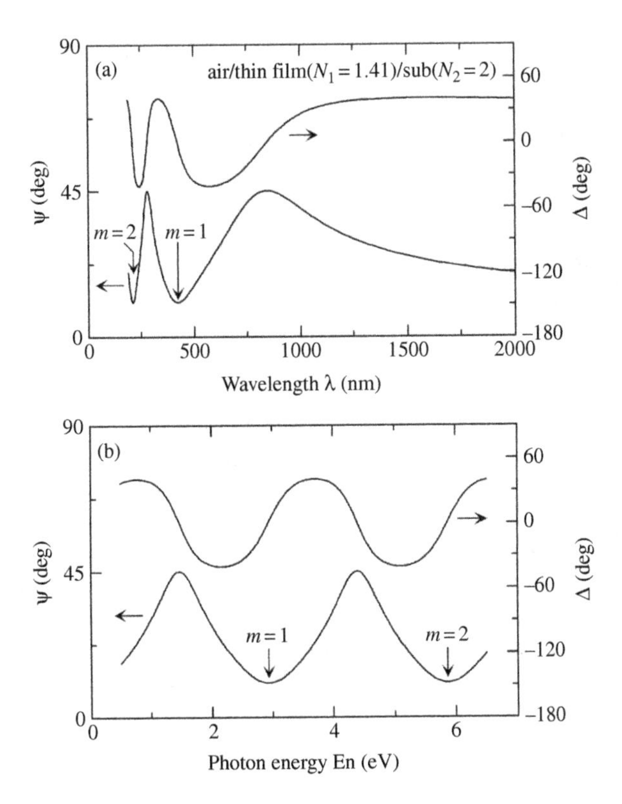

Figure 5.5 (ψ, Δ) spectra obtained from an air/thin film/substrate structure, plotted as functions of (a) wavelength and (b) photon energy. In this calculation, $N_1 = n_1 = 1.41$(thin film), $N_2 = n_2 = 2$ (substrate), $d = 2000$ Å, and $\theta_0 = 70°$ are assumed. At the positions of $m = 1$ and 2, the phase shift between primary and secondary beams is zero.

(a) wavelength and (b) photon energy. In this calculation, $N_1 = n_1 = 1.41(\varepsilon = 2)$, $N_2 = n_2 = 2(\varepsilon = 4)$, $d = 2000\,\text{Å}$, and $\theta_0 = 70°(\theta_1 = 42°)$ were used. If we apply Eq. (5.3), we can express the condition in which the phase shift between the primary and secondary beams becomes zero:

$$2dn_1 \cos\theta_1/\lambda = m \; (m = 1, 2, 3 \ldots) \tag{5.4}$$

Here, m is an integer, and the wavelengths and photon energies corresponding to $m = 1$ and 2 are indicated by arrows in Fig. 5.5. In Fig. 5.5(a), the period of interference fringes becomes smaller with decreasing wavelength, as confirmed from Eq. (5.4). In contrast, when the interference effect is represented by photon energy, its period is constant since En is inversely proportional to λ [Eq. (2.27)]. The variations of (ψ, Δ) shown in Fig. 4.27 also represent the optical interference effect in a transparent film (diamond thin film). In Figs. 4.27 and 5.5, the value of ψ decreases when $2\beta = 2\pi$ and increases when $2\beta = \pi$ by the same principle discussed above. Notice that $\Delta = 0°$ at the peak position of ψ in these results.

Figure 5.6(a) shows the variations of (ψ, Δ) with the SiO_2 thickness in the region of 0–50 Å in Fig. 5.4. In Fig. 5.6(a), when the SiO_2 thickness is very thin ($< 10\,\text{Å}$), ψ is almost constant and Δ decreases linearly. This implies that the phase of the waves lags linearly with increasing thickness in the optical interference of very thin films. If we use this region, the thickness of a thin film can be estimated directly from the linear approximation of Δ. Figure 5.6(b) shows an example of sample characterization by this method [5]. In particular, the oxygen adsorption processes on Si(111) surfaces are characterized in Fig. 5.6(b). In this experiment, the Si(111) surface is formed by cleaving c-Si in vacuum and oxygen is supplied to this surface. In Fig. 5.6(b), the shift of Δ value ($|\Delta - \Delta_0|$) is shown versus time of oxygen supply. Here, Δ_0 represents a measurement value obtained before the oxygen supply. Since the oxygen adsorption stops automatically after uniform coverage of the Si surface, $|\Delta - \Delta_0|$ is proportional to the oxygen coverage on the silicon surface. It has been reported that the surface coverage ratio of oxygen can be characterized with a precision of ± 0.02 when the measurement error is $\delta\Delta = \pm 0.01$ [5].

When the refractive index of a thin film (n_1) is known, the thin-film thickness can be estimated from the above method. Nevertheless, the evaluation of a refractive index from a known film thickness is rather difficult. This reason can be explained from calculation results shown in Fig. 5.7. Figure 5.7(a) shows the ψ–Δ trajectory when the refractive index of a thin film ($N_1 = n_1$) is varied from 1.46 to 3.0 in the calculation for Fig. 5.4. The open circles in the figure represent the values every 100 Å, and the (ψ, Δ) values move toward the direction of arrows with increasing film thickness. In Fig. 5.7(a), although Δ shows large changes for the variation in the film thickness, ψ is almost constant in the region of $d < 100\,\text{Å}$. This arises from the fact that R_p and R_s show almost no changes at $d < 100\,\text{Å}$, even when n_1 is varied over a wide range [Fig. 5.7(b)]. Recall from Eq. (4.5) that $\psi = \tan^{-1}[(R_p/R_s)^{1/2}]$. Consequently, the evaluation of n_1 becomes difficult when the thickness of a transparent film is very thin ($d < 100\,\text{Å}$). In other words, a slight measurement error in ψ affects the estimated value of n_1 greatly [6].

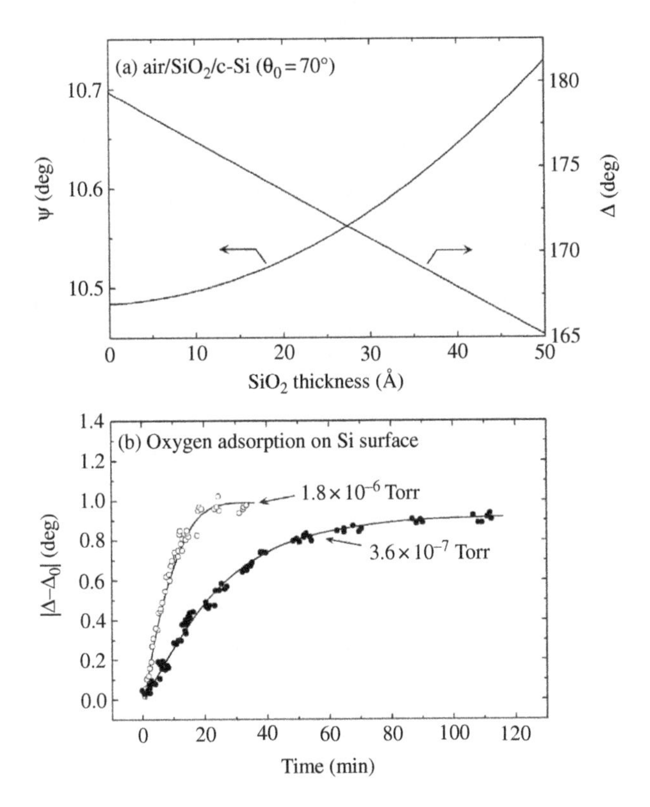

Figure 5.6 (a) Variations of (ψ, Δ) in an air/SiO_2/c-Si structure, plotted as a function of SiO_2 thickness and (b) oxygen adsorption process on a Si(111) surface, evaluated from the measured value of $|\Delta - \Delta_0|$. In (b), Δ_0 represents a measurement value obtained before the oxygen absorption. Drawing (b): Reprinted from *Journal of Physics and Chemistry of Solids*, **26**, R. J. Archer and G. W. Gobeli, Measurement of oxygen adsorption on silicon by ellipsometry, 343–351, Copyright (1965), with permission from Elsevier.

5.1.3 VARIATIONS OF (ψ, Δ) IN ABSORBING FILMS

The (ψ, Δ) of absorbing films $(k_1 > 0)$ can also be obtained from Eq. (5.2). Figure 5.8 shows (ψ, Δ) in an air/absorbing film/c-Si structure, plotted as a function of the thickness of the absorbing film. Here, the absorbing film is assumed to be a hydrogenated amorphous silicon (a-Si:H) film with a complex refractive index of $N_1 = 5 - i0.85$ ($\lambda = 5000$ Å). For c-Si, $N_2 = 4.3 - i0.04$ ($\lambda = 5000$ Å) is assumed. In the case of absorbing films, (ψ, Δ) oscillate with a period of d_i, but the amplitude of the oscillation gradually reduces. This shows that the interference effect becomes weaker in thicker films due to increasing light absorption. If we apply Eq. (2.37), the penetration depth of light in this absorbing film ($k_1 = 0.85$) is estimated to be $d_p = 468$ Å. If measurement errors are $(\delta\psi, \delta\Delta) = (0.01°, 0.02°)$ (see Section 4.4.2),

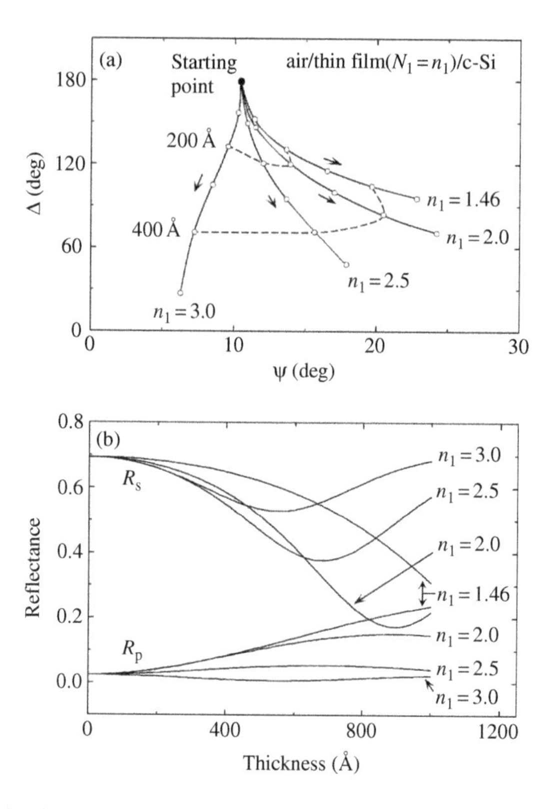

Figure 5.7 (a) $\psi - \Delta$ trajectory and (b) reflectance in an air/thin film/c-Si structure. In this calculation, $\theta_0 = 70°$ and $N_2 = 3.87 - i0.0146$ (c-Si) [4] were used, and the refractive index of the thin film ($N_1 = n_1$) was changed from 1.46 to 3.0. In (a), open circles show the values every 100 Å.

we can perform the measurement up to a thickness of $d = 2500$–3000 Å ($d \sim 5d_p$), as confirmed from Fig. 5.8. In other words, the influence of an underlying layer (substrate) should be taken into account in data analysis up to $d \sim 5d_p$ [7]. When the thickness of an absorbing film is $d > 5d_p$, (ψ, Δ) values are determined by the light reflection at an air/film interface only, since light propagating inside the film is absorbed almost completely.

Figure 5.9(a) shows the $\psi - \Delta$ trajectory obtained from Fig. 5.8. As shown in this figure, (ψ, Δ) change in a circular manner and the values converge at (ψ, Δ) = (18.24°, 166.23°) when $d > 5d_p$. These values correspond to N_1 and can be obtained directly by applying Eq. (5.1). On the other hand, Fig. 5.9(b) shows the result when light absorption in a thin film is small ($k_1 = 0.25$). In this calculation, the value of a-Si:H ($N_1 = 4.51 - i0.25$) at $\lambda = 6328$ Å was used. In Fig. 5.9, the calculation results up to $d = 5000$ Å are shown. Since $5d_p \sim 1$ μm when $k_1 = 0.25$, (ψ, Δ) are not converged in Fig. 5.9(b). It is clear from Fig. 5.9 that the number of rotations

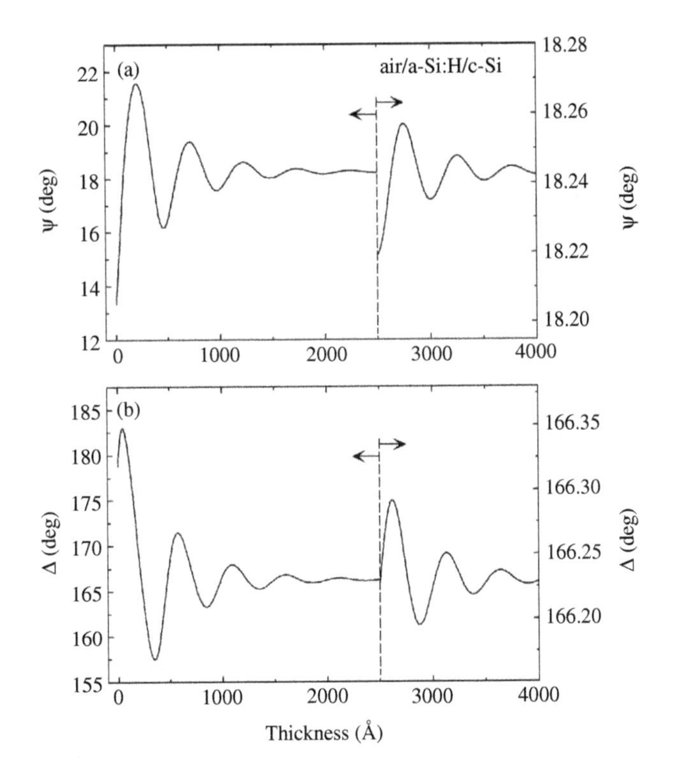

Figure 5.8 (a) ψ and (b) Δ in an air/absorbing film/c-Si structure, plotted as a function of film thickness. The absorbing film is assumed to be a hydrogenated amorphous silicon (a-Si:H) film. In this calculation, $N_1 = 5 - i0.85$ (a-Si:H) and $N_2 = 4.3 - i0.04$ (c-Si) were used as the values of $\lambda = 5000$ Å. The (ψ, Δ) show the values when $\theta_0 = 70°$.

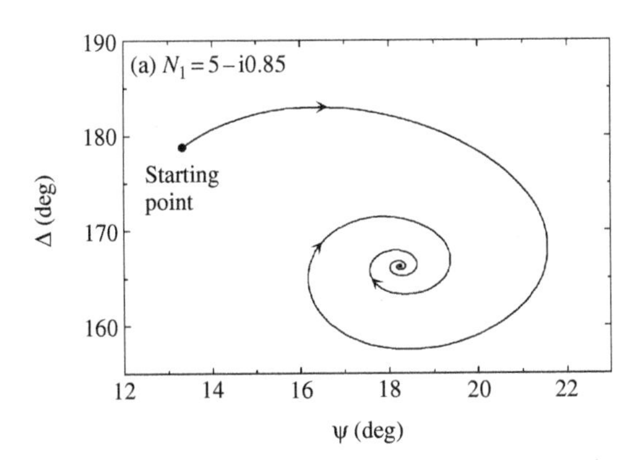

Figure 5.9 $\psi - \Delta$ trajectories in an air/a-Si:H/c-Si structure: (a) $N_1 = 5 - i0.85$ ($\lambda = 5000$ Å) and (b) $N_1 = 4.51 - i0.25$ ($\lambda = 6328$ Å). The (ψ, Δ) show the values when $\theta_0 = 70°$.

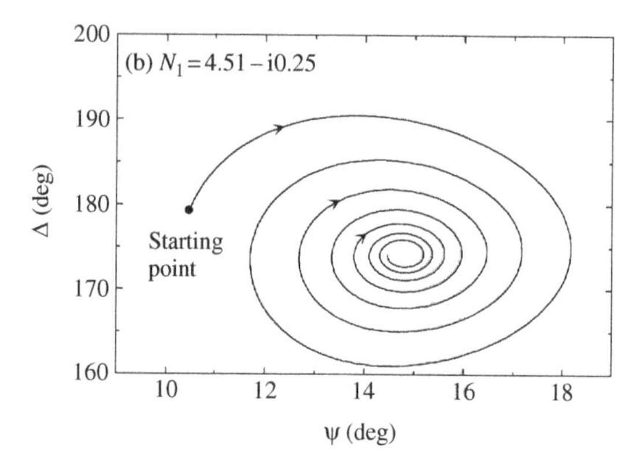

Figure 5.9 (Continued).

on the $\psi - \Delta$ plane increases with decreasing k_1 owing to weaker light absorption in the thin film. As mentioned above, the condition $d > 5d_p$ varies depending on the value of k. Since α and k generally increase at higher energies (see Fig. 5.3), a film thickness that satisfies $d > 5d_p$ reduces at higher energies [see Fig. 1.3(b)]. It should be noted that a $\psi - \Delta$ trajectory converges on a point when the complex refractive index of an absorbing film is constant versus film thickness. In contrast, when the complex refractive index varies with film thickness, converging values also change [8–10]. Accordingly, a $\psi - \Delta$ trajectory plot is quite useful when we confirm whether the complex refractive index of a thin film is constant versus film thickness.

5.2 DIELECTRIC FUNCTION MODELS

In the data analysis of spectroscopic ellipsometry, the dielectric function of a sample is required. When the dielectric function of a sample is not known, modeling of the dielectric function is necessary. There are many dielectric function models, and we need to select an appropriate model according to the optical properties of the sample. Figure 5.10 illustrates dielectric function models used frequently in ellipsometry data analysis. Notice that the effect of atomic polarization is neglected in this figure (see Fig. 2.11). For dielectric function modeling in a transparent region ($\varepsilon_2 \sim 0$), the Sellmeier or Cauchy model is used. When there is free-carrier absorption, the data analysis is generally performed using the Drude model. To express the electric polarization in the visible/UV region, various models including the Lorentz model, Tauc–Lorentz model [11], harmonic oscillator approximation (HOA), and model dielectric function (MDF) have been used. The dielectric function shown in Fig. 5.10 represents the one calculated from the Lorentz model.

Figure 5.11 shows the dielectric functions of group-IV semiconductor crystals (Ge, Si, and diamond) [1]. As confirmed from this figure, actual dielectric functions

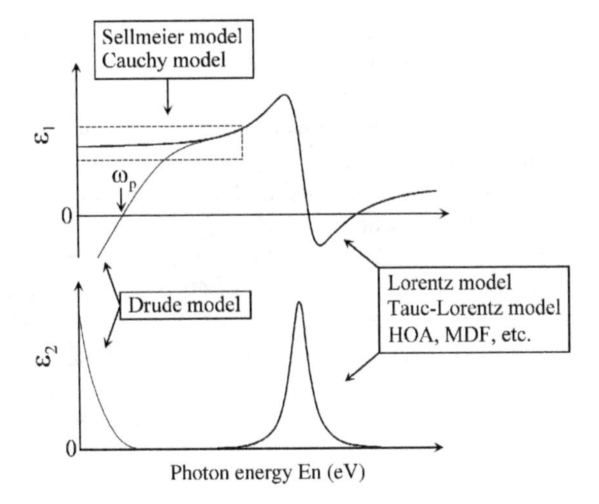

Figure 5.10 Dielectric function models used in ellipsometry data analysis.

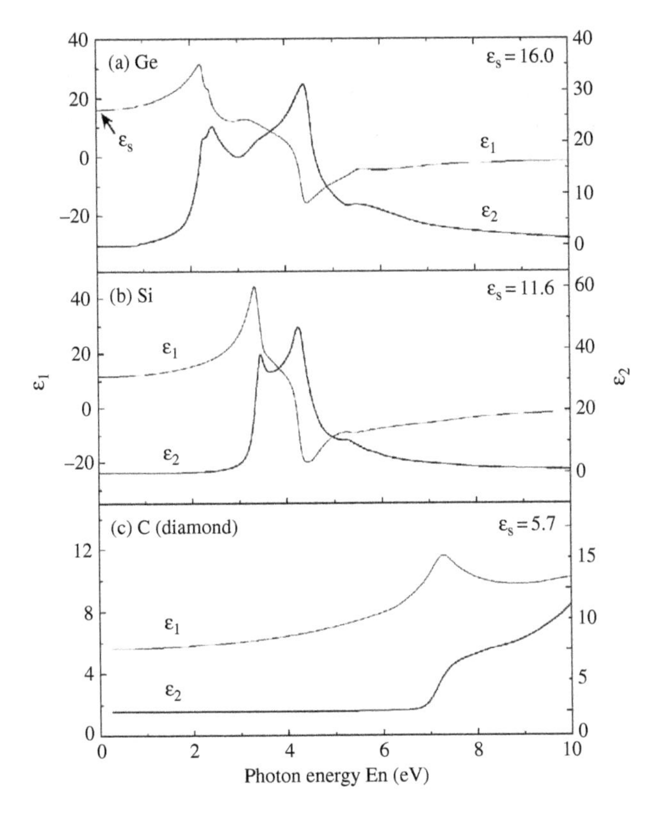

Figure 5.11 Dielectric functions of group-IV semiconductor crystals: (a) Ge, (b) Si, and (c) C (diamond). The ε_s in the figure shows the static dielectric constant. Data from Adachi (Ref. [1].)

show rather complicated structures. Since the bandgap of these crystals (indirect transitions) increases with the order of Ge < Si < C (diamond), the dielectric function also shifts toward higher energies with this order. As we have discussed in Section 2.2.3, the static dielectric constant ε_s reduces with increasing bandgap. Since the measurement range of spectroscopic ellipsometry is generally 1–5 eV, the Sellmeier or Cauchy model can be applied to the analysis of diamond. We can also model the dielectric function of silicon rather easily if we perform the analysis using the transparent region of En < 2.5 eV [12]. In the case of Ge, however, the analysis becomes difficult since the dielectric function of Ge has a complicated structure.

This section will explain various dielectric function models including the Lorentz, Sellmeier, Cauchy, Tauc-Lorentz, and Drude models. Basically, all these models are derived from the Lorentz model. The HOA and MDF that are employed for modeling of crystalline semiconductors will be described in Section 7.2.2.

5.2.1 LORENTZ MODEL

The Lorentz model is a classical model and, in the electric polarization shown in Fig. 5.12(a), a negatively charged electron is bound to a positively charged atomic nucleus with a spring (Section 2.2.1). If light is shone, the ac electric field of the light $[E = E_0 \exp(i\omega t)]$ will induce dielectric polarization in the x direction of Fig. 5.12(a). The Lorentz model assumes a physical model in which the electron

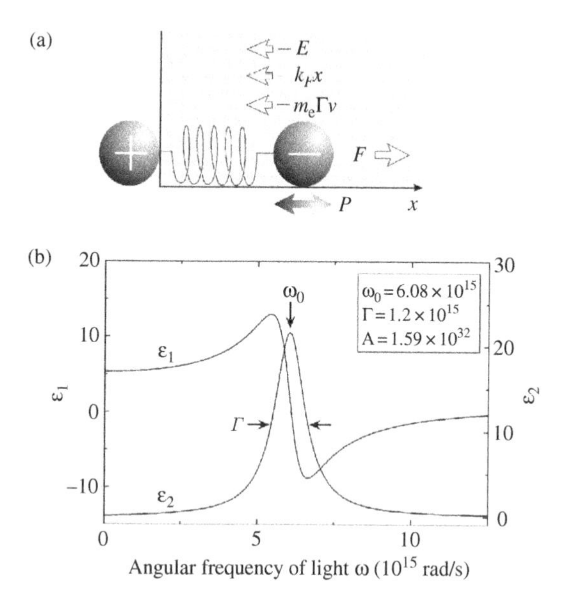

Figure 5.12 (a) Physical model of the Lorentz model and (b) dielectric function calculated from the Lorentz model.

oscillates in viscous fluid. In this case, the position of the atomic nucleus is fixed, since the mass of the atomic nucleus is far larger than that of the electron. If we use Newton's second law, the physical model shown in Fig. 5.12(a) is expressed as

$$m_{\mathrm{e}}\frac{\mathrm{d}^2x}{\mathrm{d}t^2} = -m_{\mathrm{e}}\Gamma\frac{\mathrm{d}x}{\mathrm{d}t} - m_{\mathrm{e}}\omega_0^2x - eE_0\exp(\mathrm{i}\omega t) \tag{5.5}$$

where m_{e} and e show the mass and charge of the electron, respectively. In Eq. (5.5), the first term on the right represents the viscous force of the viscous fluid. In general, the viscous force is proportional to the speed of an object when the speed is slow. The Γ in Eq. (5.5) represents a proportional constant of the viscous force, known as the damping coefficient. The second term on the right expresses that the electron moved by the electric field of light is restored according to Hook's law ($F = -k_F x$), and ω_0 shows the resonant frequency of the spring ($\omega_0 = \sqrt{k_F/m_{\mathrm{e}}}$). The last term on the right shows the electrostatic force ($F = qE$). As shown in Fig. 5.12(a), the direction of the force F applied to the electron is opposite to that of the electric field [see Fig. 2.7(a)], and the restoration force ($-k_F x$) and viscous force ($-m_{\mathrm{e}}\Gamma v$) act in the reverse direction to F. Eq. (5.5) represents the forced oscillation of the electron by the external ac electric field. By this forced oscillation, the electron oscillates at the same frequency as the ac electric field [i.e., $\exp(\mathrm{i}\omega t)$]. Thus, if we assume that the solution of Eq. (5.5) is described by the form $x(t) = a\exp(\mathrm{i}\omega t)$, the first and second derivatives of $x(t)$ are given by $\mathrm{d}x/\mathrm{d}t = \mathrm{i}a\omega\exp(\mathrm{i}\omega t)$ and $\mathrm{d}^2x/\mathrm{d}t^2 = -a\omega^2\exp(\mathrm{i}\omega t)$, respectively. By substituting these into Eq. (5.5) and rearranging the terms, we get

$$a = -\frac{eE_0}{m_{\mathrm{e}}}\frac{1}{(\omega_0^2 - \omega^2) + \mathrm{i}\Gamma\omega} \tag{5.6}$$

On the other hand, if the number of electrons per unit volume is given by N_{e}, the dielectric polarization is expressed as $P = -eN_{\mathrm{e}}x(t)$ [see Eq. (2.39)]. From $x(t) = a\exp(\mathrm{i}\omega t)$, we obtain $P = -eN_{\mathrm{e}}a\exp(\mathrm{i}\omega t)$. By substituting $P = -eN_{\mathrm{e}}a\exp(\mathrm{i}\omega t)$ and $E = E_0\exp(\mathrm{i}\omega t)$ into Eq. (2.44), we obtain the dielectric constant ε as follows:

$$\varepsilon = 1 + \frac{e^2N_{\mathrm{e}}}{\varepsilon_0 m_e}\frac{1}{(\omega_0^2 - \omega^2) + \mathrm{i}\Gamma\omega} \tag{5.7}$$

The above equation represents the Lorentz model. If we multiply by $(\omega_0^2 - \omega^2 - \mathrm{i}\Gamma\omega)$ both the numerator and the denominator of Eq. (5.7), we get

$$\varepsilon_1 = 1 + \frac{e^2N_{\mathrm{e}}}{\varepsilon_0 m_e}\frac{(\omega_0^2 - \omega^2)}{(\omega_0^2 - \omega^2)^2 + \Gamma^2\omega^2} \tag{5.8a}$$

$$\varepsilon_2 = \frac{e^2N_{\mathrm{e}}}{\varepsilon_0 m_e}\frac{\Gamma\omega}{(\omega_0^2 - \omega^2)^2 + \Gamma^2\omega^2} \tag{5.8b}$$

Here, let us simply assume that only one electron of a Si atom induces electric polarization. In this case, we obtain $A = e^2 N_e/(\varepsilon_0 m_e) = 1.59 \times 10^{32}$ using $N_e = 5 \times 10^{28}\,\mathrm{m}^{-3}$ and the values shown in Table 2.1. If we further assume $\omega_0 = 6.08 \times 10^{15}$ and $\Gamma = 1.2 \times 10^{15}$ and substitute these values into Eq. (5.7) or (5.8), we obtain the dielectric function shown in Fig. 5.12(b). From this figure, it can be seen that ε_2 shows a maximum value at ω_0 corresponding to the resonant frequency of the spring and the half width of the ε_2 peak becomes Γ. As confirmed from Table 2.2, ω is proportional to the photon energy En, and if we express ω_0 in Fig. 5.12(b) using En, we obtain En = 4 eV from $\omega = 1.519 \times 10^{15}$ En.

In actual data analysis, we commonly express the Lorentz model using the photon energy En:

$$\varepsilon = 1 + \sum_j \frac{A_j}{\mathrm{En}_{0j}^2 - \mathrm{En}^2 + \mathrm{i}\Gamma_j \mathrm{En}} \tag{5.9}$$

In Eq. (5.9), the dielectric function is described as the sum of different oscillators and the subscript j denotes the jth oscillator. In general, A in Eq. (5.9) is called the oscillator strength. In a quantum-mechanical expression, on the other hand, the Lorentz model is described by the following equation [13]:

$$\varepsilon = 1 + \sum_j \frac{A_j \mathrm{En}_{0j}}{\mathrm{En}_{0j}^2 - \mathrm{En}^2 + \mathrm{i}\Gamma_j \mathrm{En}} \tag{5.10}$$

Thus, even in this case, Eq. (5.9) derived from the classical model holds, although the expression for the oscillator strength is slightly different between Eqs. (5.9) and (5.10).

The above equations correspond to the ones when the complex refractive index is defined by $N \equiv n - \mathrm{i}k$. When we use the convention of $N \equiv n + \mathrm{i}k$, the electric field of light in Eq. (5.5) is expressed as $\exp(-\mathrm{i}\omega t)$ since the phase of light becomes $(Kx - \omega t + \delta)$ in the case of $N \equiv n + \mathrm{i}k$. Accordingly, for $N \equiv n + \mathrm{i}k$, we need to replace +i in the above equations with −i (see Appendix 2).

5.2.2 INTERPRETATION OF THE LORENTZ MODEL

Here, we will discuss the propagation of light in media in greater detail using the results derived from the Lorentz model. Although the Lorentz model is a classical model, it is sufficient to understand the interaction between atoms and propagating light. The Lorentz model also provides insight into the reflection and transmission of light.

In order to understand the propagation of light inside a medium, we first consider the transmission of light through an atomic plate (Fig. 5.13). Here, the atomic plate is a hypothetical 2-dimensional plate consisting of many ordered atoms and, when light enters this atomic plate, electric dipole radiation occurs (see

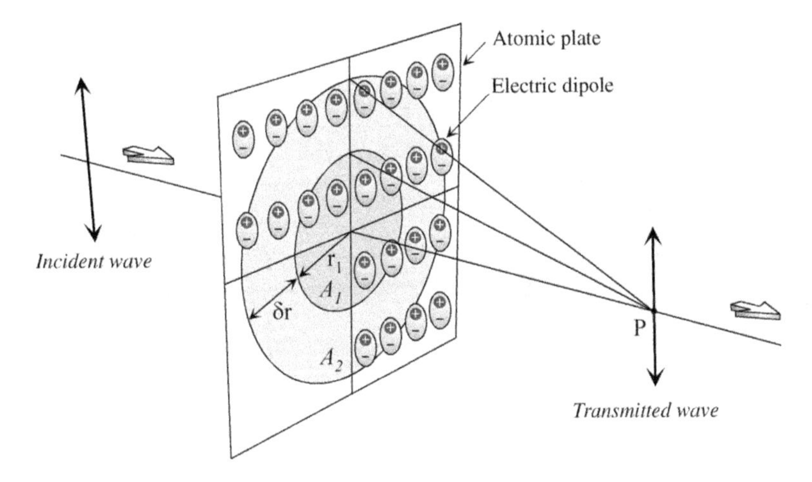

Figure 5.13 Transmission of a light wave through an atomic plate.

Section 2.2.2). It can be seen from Fig. 5.13 that the electric field formed at a point P is described by the sum of all the waves radiated from the atomic plate. If we apply a graphical method, referred to as phasor addition, we can estimate the total electric field formed at P relatively easily [14,15]. In this method, the electric field of light is expressed by the phasor E shown in Fig. 5.14. The phasor is described by the following equation using the polar coordinates on the complex plane [3,16]:

$$E = |E| \exp[\mathrm{i}(\omega t + \delta)] \tag{5.11}$$

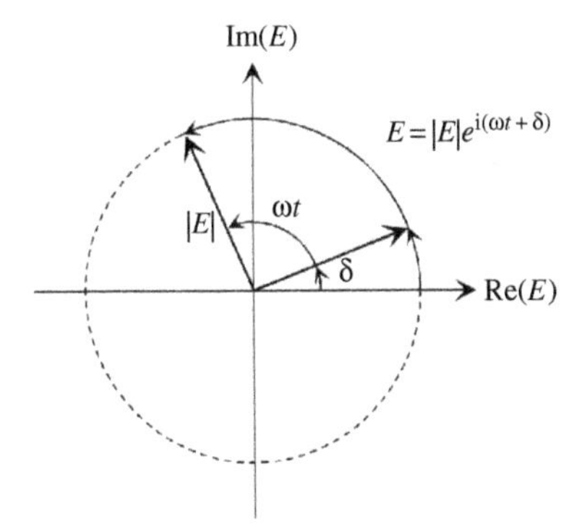

Figure 5.14 Representation of light by the phasor on the complex plane. The x and y axes show real and imaginary components of an electric field. In the figure, $|E|$ and δ denote the amplitude and initial phase, respectively. The phasor rotates counterclockwise with time as ωt.

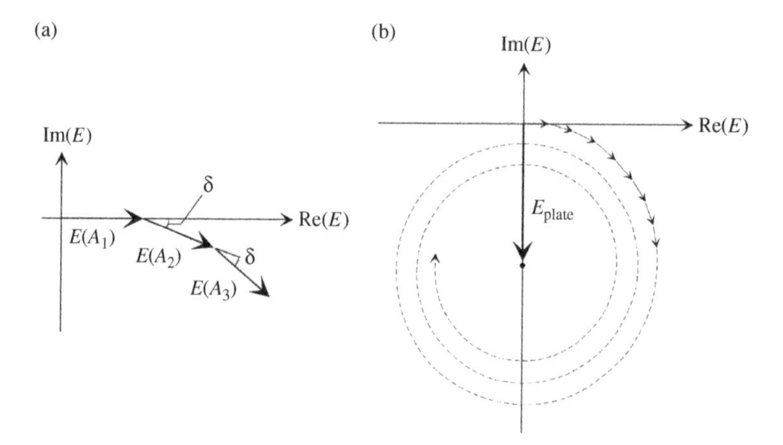

Figure 5.15 Representation of light waves radiated from an atomic plate: (a) superposition of the three waves radiated from the areas A_1–A_3 and (b) integration of $E(A_n)$ by phasor addition. The $E(A_n)$ shows the phasor of light radiated from the area A_n in Fig. 5.13. In (b), E_{plate} represents the total electric field of light waves. Drawing (b): from *Physics of Light: Why Light Refracts, Reflects and Transmits* (in Japanese), K. Kobayashi, (2002). Reproduced by permission of University of Tokyo Press.

where $|E|$ and δ represent the amplitude (length) and initial phase of the phasor. As shown in Fig. 5.14, the phasor rotates counterclockwise with time. In Fig. 2.3, this corresponds to the increase in θ. If we use the phasor, the superposition of waves is simply described by the addition of phasors [Fig. 5.15(a)].

Now consider the electric dipole radiation from the area A_n shown in Fig. 5.13. Here, the area A_1 represents the circular area with a radius of r_1, while $A_n (n > 1)$ shows the area surrounded by the two circles with radii of r_{n-1} and $r_{n-1} + \delta r$. Let us assume that the light emitted from A_n creates an electric field $E(A_n)$ at the point P. From Eq. (5.11), it follows that $E(A_n) = |E(A_n)| \exp\{i[\omega t + \delta(A_n)]\}$. If we set $t = 0$ ($\omega t = 0°$) and $\delta = 0°$ for $E(A_1)$, $E(A_1)$ is parallel to Re(E), as shown in Fig. 5.15(a). However, the light emitted from A_2 has to travel a longer distance to reach the point P and thus $\delta(A_2)$ lags slightly, compared with $\delta(A_1)$. As confirmed from Fig. 5.13, when δr is constant, the phase difference between $E(A_n)$ and $E(A_{n-1})$ also becomes constant. In Fig. 5.15(a), this constant phase lag is denoted as δ. Notice from Fig. 5.14 that the phase lag is described by a clockwise rotation. On the other hand, there is no electric dipole radiation in the oscillatory direction of electric dipoles, as we have seen in Fig. 2.9. Thus, the length of $E(A_n)$ [i.e., $|E(A_n)|$] becomes shorter with increasing n of A_n. Figure 5.15(b) illustrates the integration of $E(A_n)$ by phasor addition [15]. Since $|E(A_n)|$ gradually decreases with a constant δ, the sum of $E(A_n)$ converges on a point indicated by the phasor E_{plate} in Fig. 5.15(b). This E_{plate} represents the total electric field of the light waves radiated from the atomic plate. It is obvious that the direction of E_{plate} is $-90°$ ($\delta_{\text{plate}} = -90°$). Accordingly, the integration of the dipole radiation from the atomic plate leads to a phase change of $-90°$ [15,16].

We are now in a position to refer to the Lorentz model in more detail. From Eq. (5.6), it can be seen that the amplitude a of the electron oscillation is described by a complex number. This implies that there is a phase difference between the oscillating electron and the electric field of the light that induces the forced oscillation. Thus, we consider $a = |a_{dip}|\exp(i\delta_{dip})$, where $|a_{dip}|$ and δ_{dip} represent the amplitude and phase of the electric dipole. By using Eq. (2.17), we obtain $\mathrm{Re}(a) = |a_{dip}|\cos\delta_{dip}$ and $\mathrm{Im}(a) = |a_{dip}|\sin\delta_{dip}$. If we apply these to Eq. (5.6), δ_{dip} is expressed by

$$\delta_{dip} = \tan^{-1}\left[\frac{\mathrm{Im}(a)}{\mathrm{Re}(a)}\right] = \tan^{-1}\left(-\frac{\Gamma\omega}{\omega_0^2 - \omega^2}\right) \tag{5.12}$$

Notice from Eq. (5.12) that δ_{dip} is given by the ratio of the numerators in Eq. (5.8b). On the other hand, if we use Eq. (2.10), $|a_{dip}|$ is expressed by

$$|a_{dip}| = \sqrt{[\mathrm{Re}(a)]^2 + [\mathrm{Im}(a)]^2} = \left(\frac{E_0 e}{m_e}\right)\frac{1}{\sqrt{(\omega_0^2 - \omega^2)^2 + \Gamma^2\omega^2}} \tag{5.13}$$

In general, the term excluding $(E_0 e/m_e)$ of Eq. (5.13) is called the susceptibility and represents how well the oscillator responds to forced oscillation.

Figure 5.16 shows (a) phase of the electric dipole δ_{dip} and (b) normalized amplitude of the electric dipole $|a_{dip}|$ calculated from Eqs. (5.12) and (5.13) using the parameters shown in Fig. 5.12. As shown in Fig. 5.16(a), δ_{dip} indicates negative values that represent phase lag. Thus, the result in Fig. 5.16(a) implies that the electric dipole cannot follow the oscillation of light at high ω and the phase lag increases with increasing ω. The δ_{dip} in Fig. 5.16(a) shows the phase of one electric dipole and, when we consider the dipole radiation from an atomic plate, the phase lags further by 90°, as we have seen in Fig. 5.15(b). Accordingly, the phase of the atomic plate (δ_{plate}) is expressed by $\delta_{plate} = \delta_{dip} - 90°$. In Fig. 5.16(a), δ_{plate} is indicated on the right side. On the other hand, $|a_{dip}|$ is maximized at $\omega = \omega_0$, as shown in Fig. 5.16(b). Thus, when the angular frequency of light coincides with the resonant frequency of an oscillator, the oscillation amplitude becomes a maximum. In particular, δ_{dip} at $\omega = \omega_0$ is $-90°$ and, in this condition, resonant oscillation occurs. In the oscillation of a pendulum, for example, if we keep accelerating an object downward when the position of the object becomes the highest, the amplitude of the oscillation increases gradually. As shown in Fig. 5.16(b), $|a_{dip}|$ at $\omega \gg \omega_0$ is lower than that at $\omega \ll \omega_0$ since the oscillator cannot follow the forced oscillation at $\omega \gg \omega_0$. As a result, ε_1 reduces by a value corresponding to the electric polarization at $\omega \gg \omega_0$ (see Fig. 2.11).

Although we have considered only dipole radiation so far, the actual propagation of light in a medium is determined by the interaction between incident light and light radiated from an atomic plate [14–16]. If we take this interaction into account, we can understand the dielectric function derived from the Lorentz model more clearly. Figure 5.17 shows ε_1 spectrum calculated from the Lorentz model. The insets of

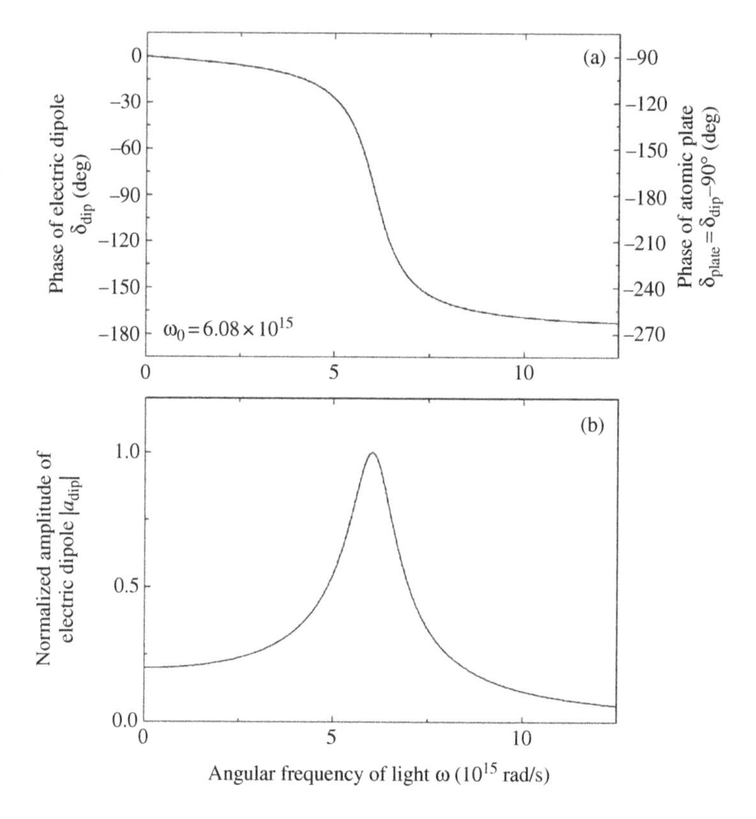

Figure 5.16 (a) Phase of the electric dipole δ_{dip} and atomic plate δ_{plate} and (b) normalized amplitude of the electric dipole $|a_{dip}|$, plotted as a function of the angular frequency of light ω.

this figure show the phasor representations of incident light (E_i), light radiated from an atomic plate (E_{plate}), and transmitted light (E_t). From Eq. (5.11), E_i, E_{plate}, and E_t are given by $E_i = |E_i|\exp[i(\omega t + \delta_i)]$, $E_{plate} = |E_{plate}|\exp[i(\omega t + \delta_{plate})]$, and $E_t = |E_t|\exp[i(\omega t + \delta_t)]$, respectively. If we use phasor addition, the transmitted light is expressed by $E_t = E_i + E_{plate}$ [14,15], as shown in the insets. Here, we assume that δ_{plate} is given by Fig. 5.16(a) and $|a_{dip}|$ shown in Fig. 5.16(b) is proportional to $|E_{plate}|$ (i.e., $|E_{plate}| \propto |a_{dip}|$). In this condition, E_t is described by

$$\begin{aligned} E_t &= E_i + E_{plate} \\ &= E_i + |a_{dip}|\exp[i(\omega t + \delta_{plate})] \end{aligned} \tag{5.14}$$

In the insets of Fig. 5.17, however, the length of $|a_{dip}|$ has been modified for clarity, although δ_{plate} is accurate.

If we set $t = 0$ and $\delta_i = 0°$, the phasor of E_i lies parallel to Re(E), as shown in Fig. 5.17. When $\omega \ll \omega_0$, the phase of the atomic plate (δ_{plate}) lags by 90°, compared

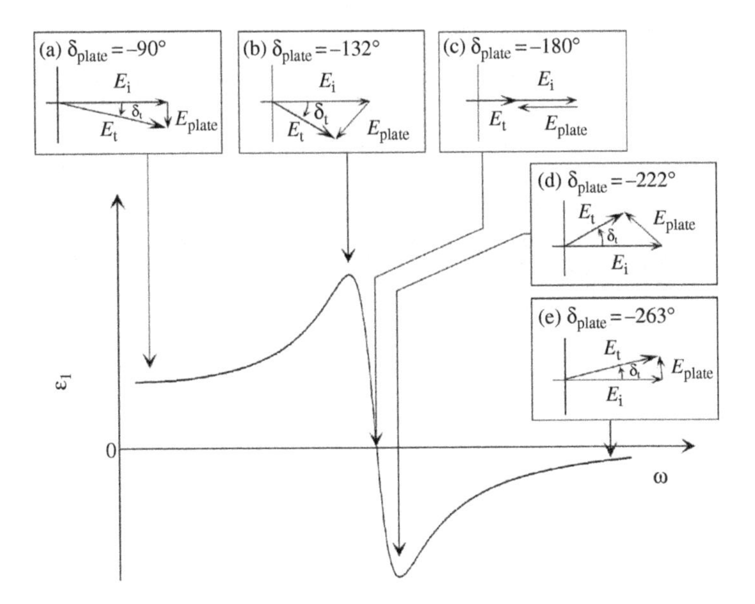

Figure 5.17 ε_1 spectrum calculated from the Lorentz model. The insets show the phasor representations of incident light (E_i), light radiated from an atomic plate (E_{plate}), and transmitted light (E_t). In the insets (a)–(e), δ_{plate} indicates the phase of the atomic plate shown in Fig. 5.16(a), and δ_t represents the phase of the transmitted light.

with that of the incident light (δ_i), as shown in Fig. 5.16(a). Thus, the phasor of E_{plate} is rotated by $-90°$ at the tip of E_i in Fig. 5.17(a). In this case, δ_t becomes negative by the superposition of E_i and E_{plate}. In other words, the phase of the transmitted wave lags as the result of the interaction between E_i and E_{plate} [14–16]. Recall from Fig. 5.14 that clockwise rotation indicates phase lag since the rotation direction is opposite to ωt. As confirmed from Fig. 5.16, with increasing ω, the phase lag and amplitude of E_{plate} increase. Consequently, δ_t becomes a maximum at $\delta_{plate} = -132°$ [Fig. 5.17(b)]. If ω increases further, δ_t reduces rapidly as the angle between E_{plate} and E_i becomes smaller. At $\omega = \omega_0$, E_{plate} is completely parallel to E_i [Fig. 5.17(c)]. When $\omega > \omega_0$, δ_t shows positive values (i.e., phase lead) and the phase lead is maximized at $\delta_{plate} = -222°$ [Fig. 5.17(d)]. However, this phase lead gradually reduces, as ω increases [Fig. 5.17(e)].

As shown in Fig. 5.17, the phasor E_{plate} rotates clockwise at the tip of E_i, and δ_t varies by the superposition of E_i and E_{plate}. From the above results, it is evident that the variation of ε_1 with ω can be interpreted from the phase of the transmitted light. The variation of ε_2 can be understood from the amplitude of the transmitted light ($|E_t|$). In particular, when $\delta_{plate} = -180°$, $|E_t|$ is minimized, as confirmed from Fig. 5.17(c). This corresponds to light absorption by the medium. Figure 5.18 shows the phase of transmitted light δ_t and amplitude of transmitted light $|E_t|$ obtained from Eq. (5.14). In this calculation, $|E_i| \gg |a_{dip}|$ is assumed. In Fig. 5.18, we can see clearly that the shapes of δ_t and $|E_t|$ are exactly opposite to ε_1 and ε_2 shown in

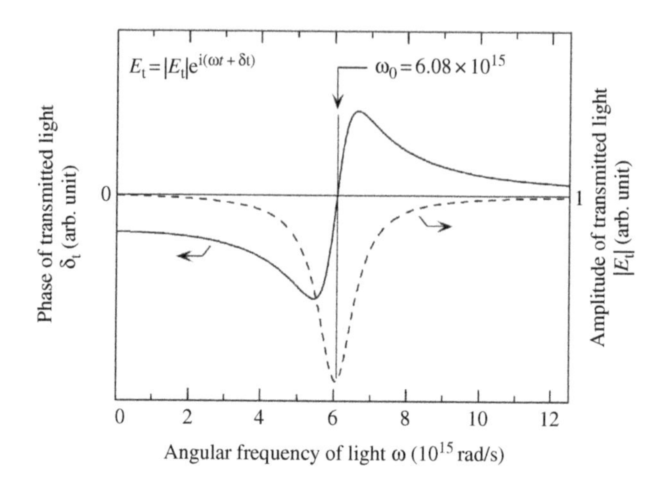

Figure 5.18 (a) Phase of transmitted light δ_t and (b) amplitude of transmitted light $|E_t|$, plotted as a function of the angular frequency of light ω. This figure can be obtained from the result shown in Fig. 5.16.

Fig. 5.12, although the vertical scales are rather qualitative in Fig. 5.18. This result confirms that ε_1 and ε_2 represent the phase and amplitude of a transmitted wave, respectively.

The above results can also be applied to interpret the propagation of light in media. Figure 5.19 illustrates the propagation of light through atomic plates at (a) $\omega \ll \omega_0$ and (b) $\omega \gg \omega_0$ [15]. As we have seen in Fig. 3.1, a wave slides to the left when the wave lags (i.e., $\delta < 0$), while a wave moves toward the right when the wave leads ($\delta > 0$). As shown in Fig. 5.18, the value of δ_t is negative when $\omega \ll \omega_0$. In this case, the wave slides to the left whenever light transmits the atomic plate. Consequently, the wavelength in the medium becomes shorter than that in vacuum [Fig. 5.19(a)]. As confirmed from Fig. 5.18, the large value of ε_1 implies larger δ_t. Thus, with increasing ε_1, the wavelength in media becomes shorter. Since $n = \sqrt{\varepsilon_1}$ when $k = 0$ [Eq. (2.47a)], the wavelength also becomes shorter with increasing n. This phenomenon has been illustrated in Fig. 2.5. In contrast, when $\omega \gg \omega_0$, the wave slides to the right when light transmits the atomic plate. Consequently, the wavelength becomes longer in this case. From the result shown in Fig. 5.12, we find $n < 1$ at $\omega > 7 \times 10^{15}$ rad/s. In this region, therefore, the propagation speed is faster than the speed of light. However, this propagation speed represents the phase speed determined by the interaction between E_i and E_{plate}. Accordingly, although the speed of light in media appears to change depending on n, the speed of light itself does not change [14–16].

The above discussion can be extended to explain light reflection. In the phasor representation, the reflection of light is described by superimposing all the radiated waves from atomic plates [15]. As shown in Fig. 5.16(a), δ_{plate} is $-90°$ when $\omega \ll \omega_0$. Thus, the direction of E_{plate} is downward at $\omega \ll \omega_0$ [Fig. 5.20(a)]. This phasor represents light reflection from the top layer. For light reflected from the

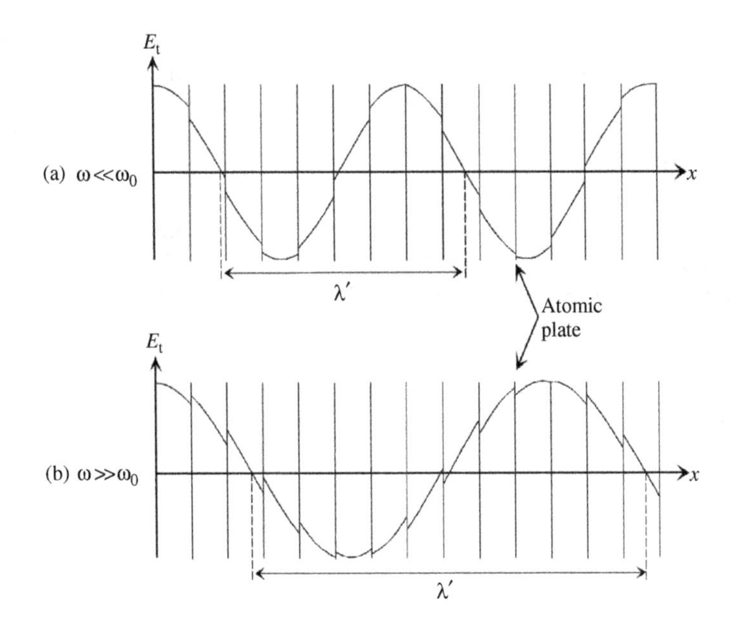

Figure 5.19 Propagation of light through atomic plates at (a) $\omega \ll \omega_0$ and (b) $\omega \gg \omega_0$. λ' represents the wavelength in a medium. In the figure, the phase of light changes by $-10°$ in (a) and $+10°$ in (b) whenever the light transmits the atomic plate. From *Physics of Light: Why Light Refracts, Reflects and Transmits* (in Japanese), K. Kobayashi, (2002). Reproduced by permission of University of Tokyo Press.

second layer, a phase lag should be taken into account, since the light has to travel a longer distance of $2d$, where d denotes the distance from the first layer. Thus, the phasor of the second layer is slightly tilted, compared with the one from the top layer. Here, we assume that the extinction coefficient of the medium is $k \sim 0$.

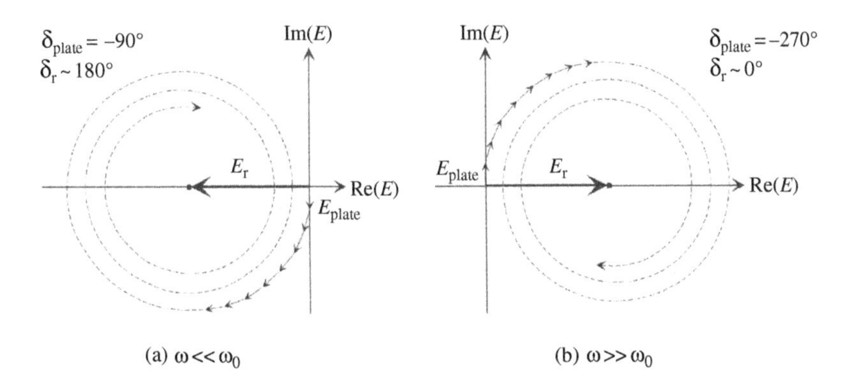

Figure 5.20 Representations of light waves reflected from atomic plates: (a) $\omega \ll \omega_0$ and (b) $\omega \gg \omega_0$. E_r represents the total electric field of reflected waves. Drawing (a): from *Physics of Light: Why Light Refracts, Reflects and Transmits* (in Japanese), K. Kobayashi, (2002). Reproduced by permission of University of Tokyo Press.

In this case, $|E_{\text{plate}}|$ reduces very slightly with increasing d. If we integrate all the phasors up to the nth layer, we obtain the total electric field of the reflected light (E_r). As shown in Fig. 5.20(a), the phase of E_r (δ_r) becomes $\sim 180°$ [15]. Clearly, the length of E_r represents the amplitude of the reflected light. Accordingly, light reflection basically increases as $|E_{\text{plate}}|$ increases. On the other hand, δ_{plate} is $-270°$ at $\omega \gg \omega_0$. In this case, the phasor of the first layer is in the direction of 90°. Similarly, if we integrate all the phasors, we obtain $\delta_r \sim 0°$ [15], as shown in Fig. 5.20(b). At $\omega \gg \omega_0$, therefore, the light shows the reflection by a free end shown in Fig. 2.17(b). From the above results, we can understand why phase change upon light reflection is either 0° or 180° in transparent films.

5.2.3 SELLMEIER AND CAUCHY MODELS

The Sellmeier model corresponds to a region where $\varepsilon_2 \sim 0$ in the Lorentz model and this model can be derived by assuming $\Gamma \to 0$ at $\omega \ll \omega_0$. In this condition, if we transform Eq. (5.7) using $\omega/c = 2\pi/\lambda$ [Eq. (2.29)], we obtain

$$\varepsilon = \varepsilon_1 = 1 + \frac{e^2 N_e}{\varepsilon_0 m_e (2\pi c)^2} \frac{\lambda_0^2 \lambda^2}{\lambda^2 - \lambda_0^2} \tag{5.15}$$

Notice that $\varepsilon_2 = k = 0$ when $\Gamma = 0$. The Sellmeier model is expressed by rewriting the above equation:

$$\varepsilon_1 = n^2 = A + \sum_j \frac{B_j \lambda^2}{\lambda^2 - \lambda_{0j}^2} \qquad \varepsilon_2 = 0 \tag{5.16}$$

where A and B_j represent analytical parameters used in data analysis and λ_0 corresponds to ω_0. An equation that assumes $B_j \lambda_{0j}^2 \lambda^2$ for the numerator of Eq. (5.16) has also been used.

On the other hand, the Cauchy model is given by

$$n = A + \frac{B}{\lambda^2} + \frac{C}{\lambda^4} + \ldots \qquad k = 0 \tag{5.17}$$

The above equation can be obtained from the series expansion of Eq. (5.15). Although the Cauchy model is an equation relative to the refractive index n, this is an approximate function of the Sellmeier model.

5.2.4 TAUC–LORENTZ MODEL

The Tauc–Lorentz model has been employed to model the dielectric function of amorphous materials [11,17]. Recently, this model has also been applied to dielectric function modeling of transparent conductive oxides [18,19]. As we have seen Fig. 5.12(b), the shape of ε_2 peaks calculated from the Lorentz model is completely symmetric. However, the ε_2 peaks of amorphous materials generally show asymmetric shapes. In the Tauc–Lorentz model [11], therefore, ε_2 is modeled from the product of a unique bandgap of amorphous materials (Tauc gap [20]) and the Lorentz model.

The Tauc gap E_g of amorphous materials is given by the following equation [20]:

$$\varepsilon_2 = A_{\mathrm{Tauc}}(\mathrm{En} - E_g)^2/\mathrm{En}^2 \tag{5.18}$$

The ε_2 of the Tauc-Lorentz model is expressed by multiplying ε_2 of Eq. (5.10) by Eq. (5.18) [11]:

$$\varepsilon_2 = \frac{A\,\mathrm{En}_0 C(\mathrm{En} - E_g)^2}{(\mathrm{En}^2 - \mathrm{En}_0^2)^2 + C^2\mathrm{En}^2}\frac{1}{\mathrm{En}} \quad (\mathrm{En} > E_g) \tag{5.19a}$$

$$= 0 \quad (\mathrm{En} \le E_g) \tag{5.19b}$$

The ε_1 of the Tauc–Lorentz model is given by the following equation [11]:

$$\begin{aligned}\varepsilon_1 = {} & \varepsilon_1(\infty) + \frac{AC}{\pi\xi^4}\frac{a_{\ln}}{2\alpha\mathrm{En}_0}\ln\left(\frac{\mathrm{En}_0^2 + E_g^2 + \alpha E_g}{\mathrm{En}_0^2 + E_g^2 - \alpha E_g}\right) \\ & - \frac{A}{\pi\xi^4}\frac{a_{\tan}}{\mathrm{En}_0}\left[\pi - \tan^{-1}\left(\frac{2E_g + \alpha}{C}\right) + \tan^{-1}\left(\frac{-2E_g + \alpha}{C}\right)\right] \\ & + 2\frac{A\,\mathrm{En}_0}{\pi\xi^4\alpha}E_g(\mathrm{En}^2 - \gamma^2)\left[\pi + 2\tan^{-1}\left(2\frac{\gamma^2 - E_g^2}{\alpha C}\right)\right] \\ & - \frac{A\,\mathrm{En}_0 C}{\pi\xi^4}\frac{\mathrm{En}^2 + E_g^2}{\mathrm{En}}\ln\left(\frac{|\mathrm{En} - E_g|}{\mathrm{En} + E_g}\right) + \frac{2A\,\mathrm{En}_0 C}{\pi\xi^4}E_g\ln\left[\frac{|\mathrm{En} - E_g|(\mathrm{En} + E_g)}{\sqrt{(\mathrm{En}_0^2 - E_g^2)^2 + E_g^2 C^2}}\right]\end{aligned} \tag{5.20}$$

where

$$a_{\ln} = (E_g^2 - \mathrm{En}_0^2)\mathrm{En}^2 + E_g^2 C^2 - \mathrm{En}_0^2(\mathrm{En}_0^2 + 3E_g^2) \tag{5.21a}$$

$$a_{\tan} = (\mathrm{En}^2 - \mathrm{En}_0^2)(\mathrm{En}_0^2 + E_g^2) + E_g^2 C^2 \tag{5.21b}$$

$$\xi^4 = (\mathrm{En}^2 - \gamma^2)^2 + \alpha^2 C^2/4 \tag{5.21c}$$

$$\alpha = \sqrt{4\mathrm{En}_0^2 - C^2} \tag{5.21d}$$

$$\gamma = \sqrt{\mathrm{En}_0^2 - C^2/2} \tag{5.21e}$$

The above equation has been derived by using the Kramers–Kronig relations (see Section 5.2.6). Although the equation for ε_1 is rather complicated, the dielectric function of this model is expressed from a total of five parameters $[\varepsilon_1(\infty), A, C, \mathrm{En}_0, E_g]$. Figure 5.21 shows (a) dielectric function and (b) (n, k) spectra of an amorphous silicon (a-Si) calculated from the Tauc–Lorentz model [11]. The values of the analytical parameters in this calculation are $A = 122\,\mathrm{eV}$, $C = 2.54\,\mathrm{eV}$, $\mathrm{En}_0 = 3.45\,\mathrm{eV}$, $E_g = 1.2\,\mathrm{eV}$ and $\varepsilon_1(\infty) = 1.15$. It can be seen from Fig. 5.21(a) that $\varepsilon_2 = 0$ at $\mathrm{En} \leq E_g$ and the ε_2 peak position is given by En_0. The A and C of the Tauc–Lorentz model represent the amplitude and half width of the ε_2 peak, respectively, similar to the Lorentz model.

So far, the dielectric function of amorphous materials has also been described using other models including the Cody–Lorentz model [21], Forouhi–Bloomer model [22], MDF theory [23], and band model [24]. Furthermore, the tetrahedral model [25] and a model that extends the tetrahedral model using the effective medium approximation [26,27] have also been proposed.

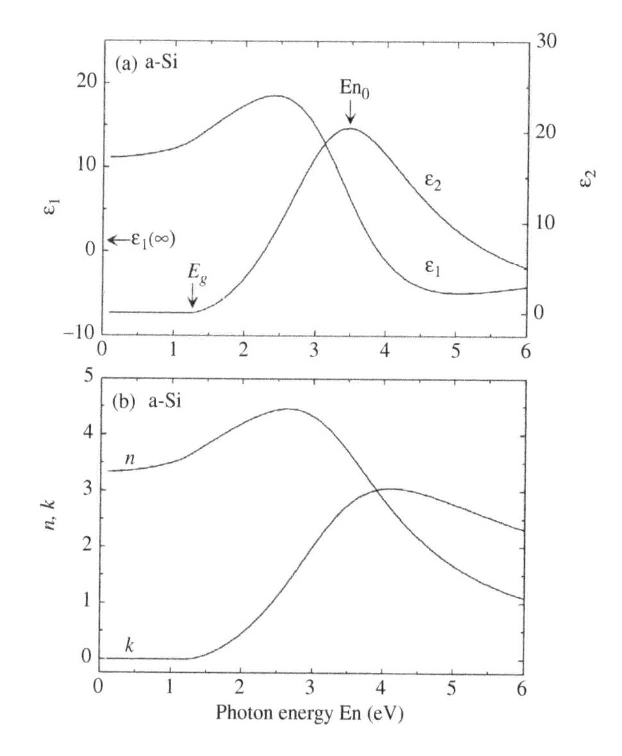

Figure 5.21 (a) Dielectric function and (b) (n, k) spectra of an amorphous silicon (a-Si) calculated from the Tauc–Lorentz model. Drawing (b): Reprinted with permission from *Applied Physics Letters*, **69**, G. E. Jellison, Jr and F. A. Modine, Parameterization of the optical functions of amorphous materials in the interband region, 371–373 (1996). Copyright 1996, American Institute of Physics.

5.2.5 DRUDE MODEL

Free electrons in metals and free carriers in semiconductors absorb light and alter dielectric functions. The Drude model has been applied widely to describe such light absorption. Here, we will examine free-carrier absorption in semiconductors. Figure 5.22(a) shows a schematic diagram of the band structure in semiconductors (see Section 7.2.1). In this figure, $\boldsymbol{K}$ represents the wave vector (momentum vector). As shown in Fig. 5.22(a), when the free-carrier concentration in a semiconductor is quite high (typically $> 10^{18}\,\mathrm{cm}^{-3}$), excess free electrons (holes) fill the conduction (valence) band and the semiconductor shows metallic character. If an electron propagating in the semiconductor with the wave vector $\boldsymbol{K}$ is scattered by a point defect, the wave vector changes into $\boldsymbol{K}'$ [Fig. 5.22(b)] and consequently free-carrier absorption occurs [Fig. 5.22(a)]. The absorption coefficient of free-carrier absorption (α_{FCA}) increases with the free-carrier concentration N_{f}, and the optical transition takes place more easily at low En.

If an electrostatic force F_{e} is applied to a semiconductor, a free electron in the semiconductor drifts along the direction of F_{e} with a speed of $\langle v \rangle$ and is scattered repeatedly by defects with a time interval of $\langle \tau \rangle$. Here, $\langle v \rangle$ and $\langle \tau \rangle$ represent the mean values of v and τ, respectively, and $\langle \tau \rangle$ is often called the mean scattering time. During the time $\langle \tau \rangle$, the free electron is accelerated by F_{e}. Thus, this free electron is expressed by

$$F_{\mathrm{f}} = m^* \frac{\mathrm{d}v}{\mathrm{d}t} = m^* \frac{\langle v \rangle}{\langle \tau \rangle} \tag{5.22}$$

where m^* shows the effective mass of semiconductors. On the other hand, $\langle v \rangle$ is proportional to the applied electric field E:

$$\langle v \rangle = -\mu E \tag{5.23}$$

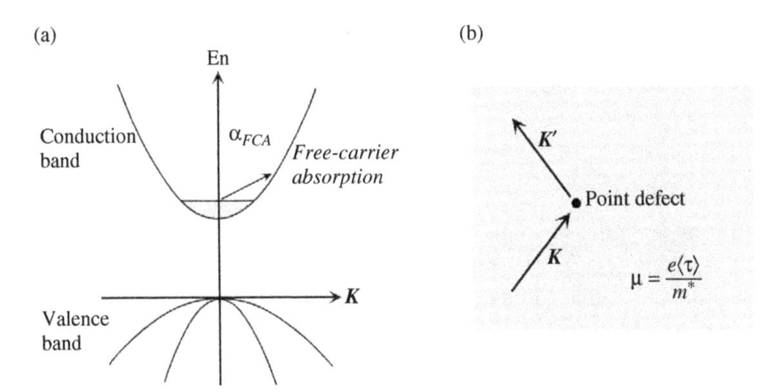

Figure 5.22 (a) Representation of free-carrier absorption in a semiconductor and (b) scattering of a free electron by a point defect.

where μ represents a proportional constant referred to as the drift mobility. The negative sign of Eq. (5.23) indicates that the direction of movement of the electrons is opposite to the direction of E. By combining Eqs. (5.22) and (5.23) and using $F = -eE$, we get

$$\mu = e\langle\tau\rangle/m^* \tag{5.24}$$

Thus, μ is proportional to $\langle\tau\rangle$ and reduces with increasing m^*. In semiconductors with high mobilities, α_{FCA} becomes smaller since the frequency of carrier scattering reduces. From the analysis of free-carrier absorption, N_f, μ, and σ can be estimated simultaneously, where σ shows the conductivity given by $\sigma = eN_f\mu$.

If we use Eq. (5.22), the equation of motion for free carriers is given by

$$m^*\frac{d^2x}{dt^2} = -F_f + F_e = -\frac{m^*}{\langle\tau\rangle}\frac{dx}{dt} - eE_0\exp(i\omega t) \tag{5.25}$$

Notice from Eq. (5.25) that the term of F_f has a minus sign. This represents that the accelerated carrier completely loses its acceleration (or force) by scattering. By setting $\Gamma \equiv \langle\tau\rangle^{-1}$, we obtain ε from Eq. (5.25):

$$\varepsilon = \varepsilon_\infty\left(1 - \frac{\omega_p^2}{\omega^2 - i\omega\Gamma}\right) \tag{5.26}$$

Here, ε_∞ shows the high-frequency dielectric constant (see Fig. 2.11) and ω_p represents the plasma angular frequency expressed by

$$\omega_p = \left(\frac{e^2N_f}{\varepsilon_0\varepsilon_\infty m^*}\right)^{1/2} \tag{5.27}$$

We can obtain Eq. (5.26) quite easily by substituting $\omega_0 = 0$ into Eq. (5.7). In the case of semiconductors, ω_p is located in the infrared region, while ω_p is in the visible/UV region in metals. If we use Eq. (5.24), we get

$$\Gamma = \frac{e}{m^*\mu} \tag{5.28}$$

In metals, we generally presume $\Gamma = 0$.

Figure 5.23 shows (a) dielectric function and (b) (n, k) spectra obtained from the Drude model. In this calculation, a transparent conductive oxide was assumed, and the values of $N_f = 1 \times 10^{20}\,\mathrm{cm}^{-3}$, $\mu = 30\,\mathrm{cm}^2/\mathrm{Vs}$, $m^* = 0.2m_0$ and $\varepsilon_\infty = 4$ were used. The actual values used in the calculation are shown in Fig. 5.23(b). From this calculation, we obtain $\omega_p = 6.34 \times 10^{14}\,\mathrm{rad/s}$ (En $= 0.42\,\mathrm{eV}$), and ε_1 becomes zero around ω_p. On the other hand, we observe $\varepsilon_1 = \varepsilon_\infty$ at $\omega \gg \omega_p$. At $\omega < \omega_p$, ε_2 increases rapidly due to the increase in free-carrier absorption. In

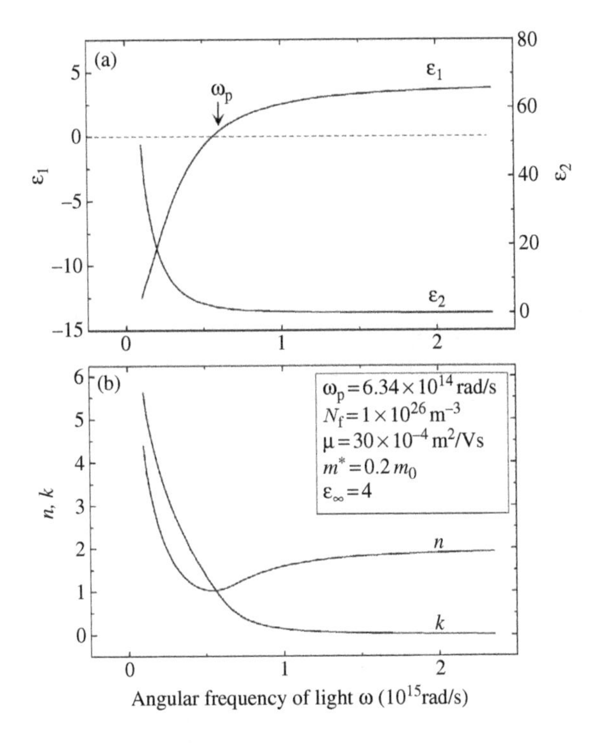

Figure 5.23 (a) Dielectric function and (b) (n, k) spectra of a transparent conductive oxide calculated from the Drude model. In (a), ω_p represents the plasma angular frequency.

Fig. 5.23(b), n reduces around ω_p and (n, k) increase drastically at $\omega < \omega_p$. Thus, reflectance becomes quite high at $\omega < \omega_p$. In metals, if we neglect the effect of bound electrons in d bands (see Section 7.3.1), the refractive index becomes zero at ω_p. Accordingly, the propagation speed of light is infinite at ω_p. This implies that the phase of the free electrons completely matches with that of the incident light and the oscillation of all the free electrons synchronizes with the electric field of the incident light [13]. Thus, the electric field of the incident light is completely screened by the free electrons present at the metal surface. Consequently, the electric field of the light cannot penetrate into the metal and the light is totally reflected. The reduction in n shown in Fig. 5.23(b) can also be explained from similar effects.

In ellipsometry data analysis, we employ an equation that describes Eq. (5.26) using En [19]:

$$\varepsilon = \varepsilon_\infty - \frac{A}{\mathrm{En}^2 - \mathrm{i}\Gamma\mathrm{En}} \tag{5.29}$$

where

$$A = \varepsilon_\infty E_\mathrm{p}^2 \tag{5.30}$$

$$E_\mathrm{p} = \hbar\omega_p = \left(\frac{\hbar^2 e^2 N_\mathrm{f}}{\varepsilon_0 \varepsilon_\infty m^*}\right)^{1/2} \tag{5.31}$$

$$\Gamma = \hbar\gamma = \frac{\hbar e}{m^* \mu} \tag{5.32}$$

Here, E_p and γ represent the plasma energy and the broadening parameter in angular frequency shown in Eq. (5.28). When we define the complex refractive index by $N \equiv n + \mathrm{i}k$, $-\mathrm{i}$ in Eqs. (5.26) and (5.29) should be replaced with $+\mathrm{i}$ (see Appendix 2).

5.2.6 KRAMERS–KRONIG RELATIONS

As we have seen in Section 5.2.2, ε_1 and ε_2 are not independent of each other and, if ε_1 varies, ε_2 also changes. The relation between ε_1 and ε_2 is described by the well-known Kramers–Kronig relations:

$$\varepsilon_1(\omega) = 1 + \frac{2}{\pi} P \int_0^\infty \frac{\omega' \varepsilon_2(\omega')}{\omega'^2 - \omega^2} \mathrm{d}\omega' \tag{5.33a}$$

$$\varepsilon_2(\omega) = -\frac{2\omega}{\pi} P \int_0^\infty \frac{\varepsilon_1(\omega') - 1}{\omega'^2 - \omega^2} \mathrm{d}\omega' \tag{5.33b}$$

P in the above equation shows the principal value of the integral:

$$P \int_0^\infty \mathrm{d}\omega' \equiv \lim_{\delta \to 0} \left(\int_0^{\omega - \delta} \mathrm{d}\omega' + \int_{\omega + \delta}^\infty \mathrm{d}\omega' \right) \tag{5.34}$$

A similar equation also holds between n and k [23]. Eq. (5.33) can be derived from the procedure described in Appendix 5.

When $\varepsilon_2(\omega)$ is known from $\omega = 0$ to ∞, for example, $\varepsilon_1(\omega)$ can be estimated directly by applying Eq. (5.33). In this case, it is required that $\varepsilon_2(\omega) = 0$ at $\omega \to \infty$. The Tauc–Lorentz model (see Section 5.2.4) satisfies this condition and thus $\varepsilon_1(\omega)$ is calculated from $\varepsilon_2(\omega)$ using the Kramers–Kronig relations [11]. As discussed previously [13], the Kramers–Kronig relations follow causality, which means that light absorption occurs *after* light enters media. Accordingly, physically correct models satisfy the Kramers–Kronig relations. In the Lorentz, Tauc–Lorentz, and Drude models, the Kramers–Kronig relations hold. Nevertheless, the Sellmeier and Cauchy models do not satisfy the Kramers–Kronig relations, since ε_1 is obtained by assuming $\varepsilon_2 = 0$.

5.3 EFFECTIVE MEDIUM APPROXIMATION

Spectroscopic ellipsometry is quite sensitive to surface and interface structures. Thus, it is necessary to incorporate these structures into an optical model in data analysis. If we apply the effective medium approximation (EMA), the complex refractive indices of surface roughness and interface layers can be calculated relatively easily. Furthermore, from ellipsometry analysis using EMA, we can characterize volume fractions in composite materials. This section will review various effective medium theories and explain modeling of surface roughness layers.

5.3.1 EFFECTIVE MEDIUM THEORIES

As we have seen in Section 2.2.2, the dielectric constant represents the magnitude of dielectric polarization formed in a dielectric by an external ac electric field. Now consider a spherical dielectric inserted into a capacitor [Fig. 5.24(a)]. If an external electric field is applied to this capacitor, polarization charges will be created on the outer surface of the dielectric by dielectric polarization. Consequently, the atoms inside the dielectric will receive an electric field induced by the polarization charges (E') in addition to the external electric field (E). In other words, by dielectric polarization, the electric field inside the dielectric (cavity) becomes stronger than the electric field applied to the capacitor. In particular, the electric field generated by the polarization charges is referred to as the Lorentz cavity field [28]. If we use the notation of Fig. 5.24(b), the Lorentz cavity field E' is given by the following equation [28]:

$$\begin{aligned} E' &= \int_0^{\pi} (L^{-2})(2\pi L \sin\theta)(L d\theta)(P\cos\theta)(\cos\theta) \\ &= 4\pi P/3 \end{aligned} \tag{5.35}$$

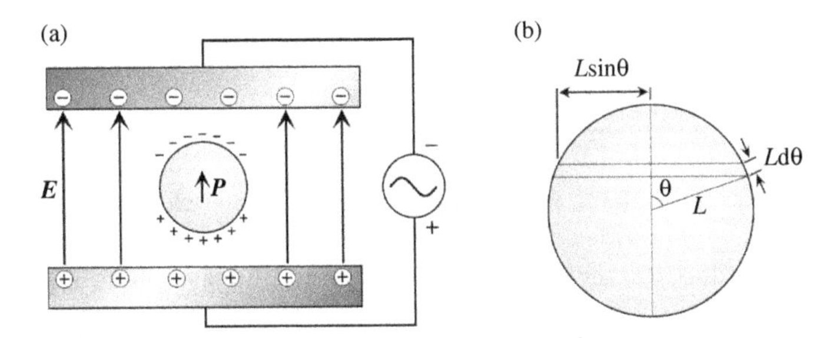

Figure 5.24 (a) Spherical dielectric inserted into a capacitor and (b) calculation method of polarization charges generated on the outer surface of a spherical dielectric. In (b), L and θ represent the radius of the sphere and the angle from the center of the sphere, respectively.

Here, P is the dielectric polarization given by Eq. (2.39). In the above equation, $(P\cos\theta)$ represents the polarization charge density on the surface of the circle shown in Fig. 5.24(b), and $(\cos\theta)$ describes the electric field formed in the center of the sphere. Eq. (5.35) assumes that E' is independent of the size of L. The result shown in Eq. (5.35) indicates the one in CGS units and, in SI units, we obtain $E' = P/(3\varepsilon_0)$ from the conversion of $4\pi \rightarrow 1/\varepsilon_0$ [28]. Thus, the local field $E_{\mathrm{loc}} = E + E'$ in SI units is expressed by

$$E_{\mathrm{loc}} = E + \frac{P}{3\varepsilon_0} \tag{5.36}$$

From Eq. (5.36), it is clear that E_{loc} increases with increasing dielectric polarization. Here, we assume electric polarization, and let N_{e} be the number of electrons in the dielectric. In this case, the dielectric polarization is given by $P = N_{\mathrm{e}}\alpha E_{\mathrm{loc}}$, where α represents the polarizability, which expresses the proportion of the dielectric polarization. Substituting Eq. (5.36) into $P = N_{\mathrm{e}}\alpha E_{\mathrm{loc}}$ yields

$$P = N_{\mathrm{e}}\alpha E \Big/ \left(1 - \frac{N_{\mathrm{e}}\alpha}{3\varepsilon_0}\right) \tag{5.37}$$

If we substitute Eq. (5.37) further into Eq. (2.44), we get a well-known formula, known as the Clausius–Mossotti relation:

$$\frac{\varepsilon - 1}{\varepsilon + 2} = \frac{N_{\mathrm{e}}\alpha}{3\varepsilon_0} \tag{5.38}$$

When the above dielectric is composed of two phases (components) a and b, we obtain

$$\frac{\varepsilon - 1}{\varepsilon + 2} = \frac{1}{3\varepsilon_0}(N_a\alpha_a + N_b\alpha_b) \tag{5.39}$$

The Lorentz–Lorenz (LL) relation is expressed from Eqs. (5.38) and (5.39) as follows [29]:

$$\frac{\varepsilon - 1}{\varepsilon + 2} = f_a\frac{\varepsilon_a - 1}{\varepsilon_a + 2} + (1 - f_a)\frac{\varepsilon_b - 1}{\varepsilon_b + 2} \tag{5.40}$$

where ε_a and ε_b represent the dielectric constants of the phases a and b, respectively, and f_{a} and $(1 - f_{\mathrm{a}})$ show each volume fraction. In this effective medium theory, ambient surrounding the dielectric is vacuum or air, similar to Fig. 5.24(a). When this spherical dielectric is present in a host material with a dielectric constant ε_{h}, Eq. (5.40) is rewritten as

$$\frac{\varepsilon - \varepsilon_{\mathrm{h}}}{\varepsilon + 2\varepsilon_{\mathrm{h}}} = f_a\frac{\varepsilon_a - \varepsilon_{\mathrm{h}}}{\varepsilon_a + 2\varepsilon_{\mathrm{h}}} + (1 - f_a)\frac{\varepsilon_b - \varepsilon_{\mathrm{h}}}{\varepsilon_b + 2\varepsilon_{\mathrm{h}}} \tag{5.41}$$

In an effective medium theory, known as the Maxwell Garnett (MG) model, the dielectric constant of mixed phase materials is described by assuming $\varepsilon_a = \varepsilon_h$ in Eq. (5.41) [29]:

$$\frac{\varepsilon - \varepsilon_a}{\varepsilon + 2\varepsilon_a} = (1 - f_a)\frac{\varepsilon_b - \varepsilon_a}{\varepsilon_b + 2\varepsilon_a} \tag{5.42}$$

As shown in Fig. 5.25(a), the MG model assumes a structure in which the phase of ε_b is surrounded by the phase of ε_a, and their volume ratio determines f_a [30]. In the case of the MG model, however, if we exchange ε_a with ε_b, the resulting ε varies. On the other hand, Bruggeman assumed $\varepsilon = \varepsilon_h$ in Eq. (5.41) and proposed the effective medium approximation (EMA) expressed by the following equation [29]:

$$f_a \frac{\varepsilon_a - \varepsilon}{\varepsilon_a + 2\varepsilon} + (1 - f_a)\frac{\varepsilon_b - \varepsilon}{\varepsilon_b + 2\varepsilon} = 0 \tag{5.43}$$

In the EMA shown in Fig. 5.25(b), f_a and $(1 - f_a)$ represent the probabilities of finding ε_a and ε_b in a spherical space [30]. This model can be extended easily to describe a material consisting of many phases:

$$\sum_{i=1}^{n} f_i \frac{\varepsilon_i - \varepsilon}{\varepsilon_i + 2\varepsilon} = 0 \tag{5.44}$$

There are other models in which two dielectrics are placed in parallel [Fig. 5.25(c)] and in series [Fig. 5.25(d)]. The dielectric constant ε of Fig. 5.25(c) is given by

$$\varepsilon = f_a\varepsilon_a + f_b\varepsilon_b \tag{5.45}$$

The above equation is quite similar to the formula used for capacitance calculation. In the parallel configuration, there is no interaction between ε_a and ε_b, and the

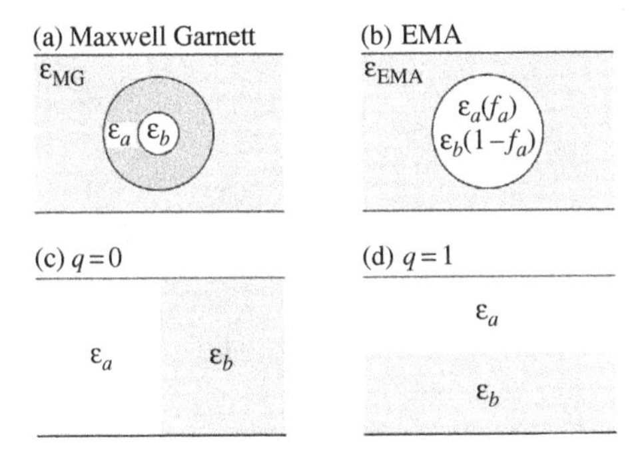

Figure 5.25 Physical models for effective medium theories: (a) Maxwell Garnett, (b) effective medium approximation (EMA), (c) $q = 0$, and (d) $q = 1$.

screening factor becomes $q = 0$ ($0 \leq q \leq 1$). On the other hand, ε in Fig. 5.25(d) is calculated from

$$\varepsilon^{-1} = f_a \varepsilon_a^{-1} + f_b \varepsilon_b^{-1} \tag{5.46}$$

In this configuration, the screening effect is maximized ($q = 1$).

In the case of a two-phase material, all the effective medium models shown in Fig. 5.25 can be expressed by a single formula [29]:

$$\varepsilon = \frac{\varepsilon_a \varepsilon_b + \kappa \varepsilon_h (f_a \varepsilon_a + f_b \varepsilon_b)}{\kappa \varepsilon_h + (f_a \varepsilon_b + f_b \varepsilon_a)} \tag{5.47}$$

Here, κ is defined by $\kappa = (1 - q)/q$ using the screening factor q. In models that assume spherical dielectrics (i.e., the MG and EMA models), the screening factor is given by $q = 1/3$. In Eq. (5.47), the MG and EMA models are described by setting $\varepsilon_h = \varepsilon_a$ and $\varepsilon_h = \varepsilon$, respectively. Eqs. (5.45) and (5.46) can also be expressed by inserting $q = 0$ and $q = 1$ into Eq. (5.47), respectively.

Figure 5.26 shows the dielectric constant of a two-phase composite (a and b), plotted as a function of the volume fraction of the component b ($f_b = 1 - f_a$). In this example, the dielectric constants of the phases a and b were assumed to be $\varepsilon_a = 2$ and $\varepsilon_b = 4$, respectively, and ε for EMA, $q = 0$, and $q = 1$ were calculated. Naturally, when $f_b = 0$ and 1, we observe $\varepsilon = \varepsilon_a$ and $\varepsilon = \varepsilon_b$, respectively. In the case of $q = 0$, ε increases linearly with increasing f_b, since there is no interaction between the two components. In the model of $q = 1$, however, the values obtained

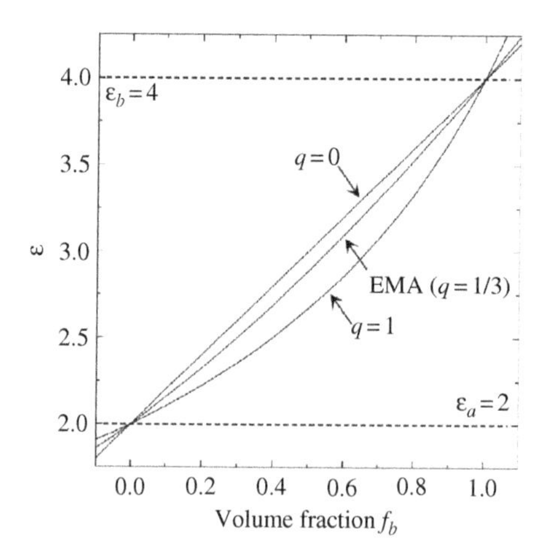

Figure 5.26 Dielectric constant ε of a two-phase composite (a and b) calculated from EMA, $q = 0$, and $q = 1$, plotted as a function of the volume fraction of the component b. In this calculation, the dielectric constants of the phases a and b were assumed to be $\varepsilon_a = 2$ and $\varepsilon_b = 4$ with their volume fractions of f_a and $f_b = 1 - f_a$, respectively.

at $0 < f_b < 1$ are always lower than those of $q = 0$ due to the screening effect. In other words, when there is a screening effect, ε is more influenced by the phase having a lower dielectric constant. As confirmed from Fig. 5.26, the values of EMA are intermediate between $q = 0$ and $q = 1$, but are closer to those of $q = 0$ since the screening factor of EMA is $q = 1/3$. Quite interestingly, even when $f_b < 0$ and $f_b > 1$, ε can be calculated from the above models, and we sometimes obtain these volume fractions in actual data analyses. Although the volume fractions $f_b < 0$ and $f_b > 1$ are unphysical, their meanings are straightforward; i.e., we obtain $f_b < 0$ when dielectric constants used in an analysis are too high and, conversely, we obtain $f_b > 0$ when dielectric constants are too low.

When ε in Fig. 5.26 is determined from an ellipsometry data analysis, f_b can be estimated by applying an effective medium model, if ε_a and ε_b are known. We can also employ f_b as an analytical parameter in data analysis (see Section 7.2.3). Among various effective medium theories, the effective medium approximation has been reported to provide the best fit to (ψ, Δ) spectra, with respect to the analysis of surface roughness layers [31,32]. Nevertheless, although physical values estimated from analyses vary, similar results (or trends) can also be obtained using other models [32]. Accordingly, ellipsometry results are basically independent of effective medium theories used in data analyses [32]. At present, the effective medium approximation has mainly been applied to ellipsometry analysis. In some materials, however, the best result has been obtained from the model of $q = 0$ [33]. Thus, an appropriate effective medium model may vary according to the optical properties of composite materials.

5.3.2 MODELING OF SURFACE ROUGHNESS

Figure 5.27 shows an optical model corresponding to a sample with surface roughness. In this optical model, a flat surface roughness layer with a thickness of d_s is placed on a bulk layer, and the complex refractive indices of the surface roughness and bulk layers are given by N_1 and N_2, respectively. In general, it is rather difficult to estimate the complex refractive index of the surface roughness layer particularly

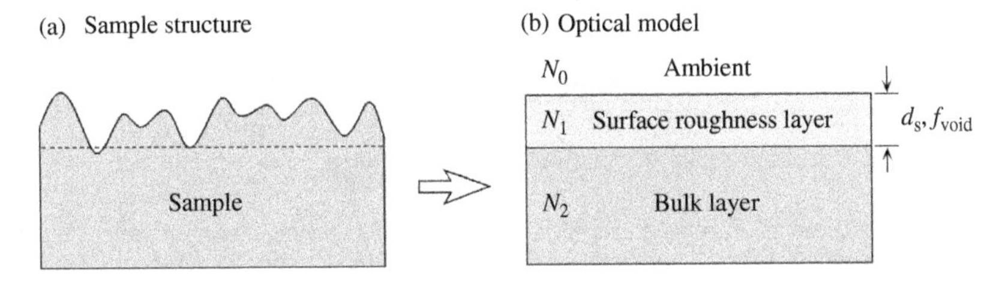

Figure 5.27 (a) Sample with surface roughness, and (b) optical model composed of surface roughness and bulk layers. In (b), d_s and f_{void} represent the thickness of the surface roughness layer and the volume fraction of the ambient (voids) present within the surface roughness layer, respectively.

when its thickness is several atomic layers. However, the surface roughness layer is composed of the bulk material (N_2) and ambient (N_0), as confirmed from Fig. 5.27. Thus, if we apply effective medium theories for these two phases, N_1 can be estimated relatively easily. For the calculation of N_1 using an effective medium theory, the volume fraction of the ambient (or voids) within a surface roughness layer (f_{void}) is required. Although f_{void} can be used as an analytical parameter [17,31,32], the analysis can also be performed assuming $f_{\text{void}} = 0.5$ (void volume fraction: 50 vol.%) [34]. If we substitute $\varepsilon_a = N_0^2$, $\varepsilon_b = N_2^2$ and $f_a = f_{\text{void}} = 0.5$ into Eq. (5.43), ε of the EMA model is obtained. From this ε, N_1 is determined using $N_1 = \sqrt{\varepsilon}$. When $N_0 = 1$, $N_2 = n_2 = 5$ and $f_{\text{void}} = 0.5$, for example, we get $N_1 = n_1 = 2.83$. Accordingly, n_1 roughly becomes half of n_2 due to the presence of

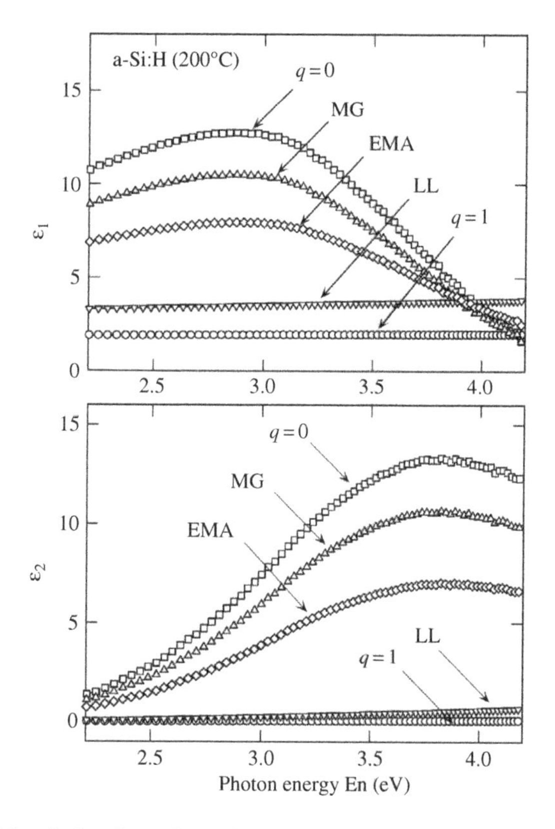

Figure 5.28 Dielectric functions of a surface roughness layer calculated from various effective medium theories. In this figure, the sample is a-Si:H held at 200 °C, and MG, EMA, and LL denote the Maxwell Garnett, effective medium approximation and Lorentz–Lorenz models, respectively. Adapted with permission from H. Fujiwara, J. Koh, P. I. Rovira, and R. W. Collins, Assessment of effective-medium theories in the analysis of nucleation and microscopic surface roughness evolution for semiconductor thin-films, *Phys. Rev. B*, **61** (2000) 10832–10844. Copyright 2000, the American Physical Society.

voids (50 vol.%) within the surface roughness layer. Notice that only N_2 is required for the calculation of N_1 if $N_0 = 1$ and $f_{\text{void}} = 0.5$ are assumed. Thus, data analysis is simplified considerably if we apply effective medium theories. Furthermore, when the ambient in Fig. 5.27(b) is a bulk layer with a different complex refractive index ($N_0 \neq N_2$), the complex refractive index of an interface layer can be determined from effective medium theories [17,34].

Figure 5.28 shows the dielectric functions of a surface roughness layer calculated from different effective medium theories [32]. The sample for this figure is a-Si:H (200 °C), and $f_{\text{void}} = 0.5$ is assumed for the calculation. In the dielectric functions obtained from the $q = 0$, MG, and EMA models, the values of ε are simply smaller, compared with ε of the bulk layer. The thickness of a surface roughness layer evaluated from these dielectric functions varies by a maximum of 10 Å. Although the dielectric functions of LL and $q = 1$ are completely different from that of the bulk layer, there is no significant influence on the characterization of d_s. As mentioned earlier, the EMA model provides the best fit in the analysis of surface roughness layers [31,32].

Figure 5.29 shows (ψ, Δ) spectra calculated from the optical model in Fig. 5.27(b). Here, a bulk layer is a c-Si substrate [4], and EMA ($f_{\text{void}} = 0.5$) was used for the calculation of the surface roughness layer. As shown in this

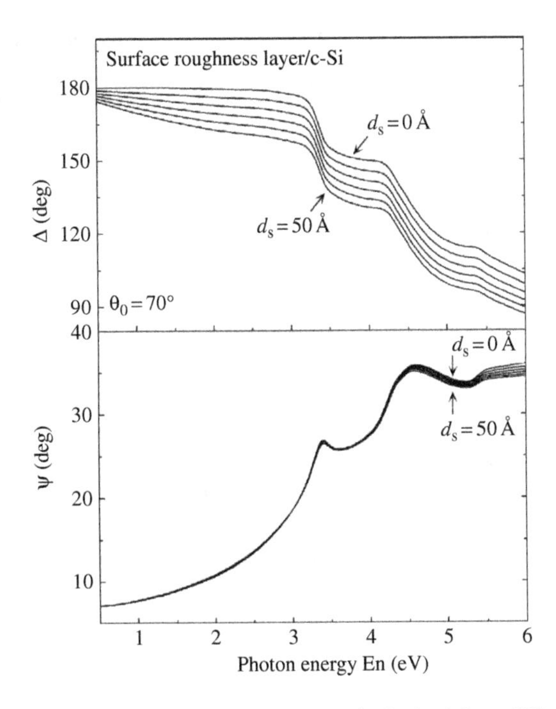

Figure 5.29 (ψ, Δ) spectra of c-Si (crystalline silicon) obtained from different thicknesses of the surface roughness layer (d_s). In this calculation, d_s was varied from 0 Å to 50 Å with a step of 10 Å. The incidence angle used for the calculation is $\theta_0 = 70°$.

figure,with increasing d_s from $0\,Å$ to $50\,Å$, the Δ values reduce rapidly, while ψ shows almost no changes. This result represents the same phenomenon as the one observed in Fig. 5.7(a). In particular, when N_1 is calculated from the effective medium approximation, N_1 becomes roughly half of N_2. In this case, the optical response is quite similar to the case when a layer with a small refractive index is formed on c-Si. Consequently, only Δ shows a large change in the region of $d_s < 100\,Å$. As confirmed from Fig. 5.29, a small surface roughness of $d_s = 10\,Å$ induces a large variation in Δ. Accordingly, when we characterize the dielectric functions of samples, the effect of surface roughness should be taken into account. It is also possible to evaluate d_s from changes in Δ.

5.3.3 LIMITATIONS OF EFFECTIVE MEDIUM THEORIES

Effective medium theories are quite effective in reducing the number of analytical parameters in an optical model and in calculating complex refractive indices of surface roughness and interface layers. When we apply these effective medium theories, however, the following conditions should be satisfied:

(a) the sizes of the phases (dielectrics) in a composite material are sufficiently greater than atomic sizes, but smaller than $\lambda/10$ of the wavelength,
(b) the dielectric functions of phases are independent of size and shape.

The condition of $\lambda/10$ in (a) above has been estimated from theoretical calculation [29,35]. When surface roughness is analyzed by EMA, for example, it is required that the morphology of surface roughness is less than $310\,Å$ at En $= 4\,\text{eV}$ ($\lambda \sim 3100\,Å$). (b) above is a precondition that was used when we derived Eq. (5.35). If we employ an effective medium theory that incorporates size effect, the restriction of the (b) above can be avoided in some cases [36].

When the above conditions are satisfied, d_s estimated from spectroscopic ellipsometry (SE) has been reported to show quite good agreement with the root mean square roughness d_{rms} characterized by atomic force microscopy (AFM) [37] [Fig. 5.30]. Similar results have also been reported [38,39]. However, the values estimated from these measurements are not exactly the same. In particular, d_s shows an intercept of $\sim 4\,Å$ when $d_{rms} = 0\,Å$ [34,37,39]. This implies that ellipsometry results include the effect of microscopic roughness on the atomic scale, which cannot be detected by AFM. On the other hand, slopes given by d_s/d_{rms} vary according to the scan size of AFM measurement [38]. This effect can be explained by a statistical factor of AFM measurement. In particular, when the scan size of AFM measurement is small, d_{rms} cannot represent the whole surface roughness and d_{rms} becomes smaller, which in turn increases the slope of d_s/d_{rms}. If we fix the scan area of AFM, d_s and d_{rms} show a clear relationship over a wide region. Conversely, we can perform ellipsometry data analysis by referring to d_{rms} values.

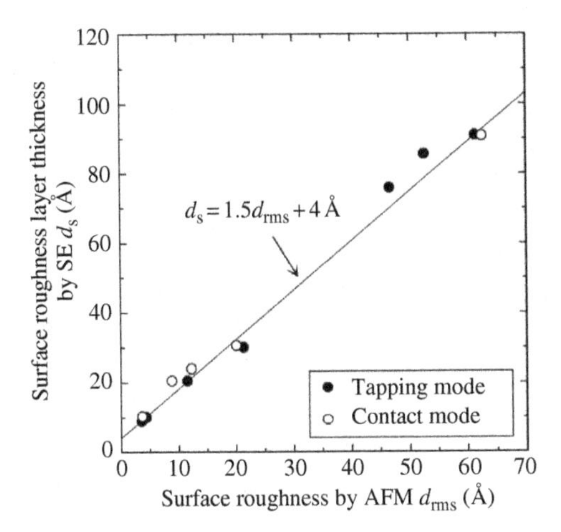

Figure 5.30 Surface roughness layer thickness (d_s) estimated from spectroscopic ellipsometry (SE), plotted as a function of the root mean square roughness d_{rms} characterized from atomic force microscopy (AFM). For the AFM measurements, tapping and contact modes were used. The relation between d_s and d_{rms} is given by $d_s = 1.5d_{rms} + 4$ Å. Reprinted with permission from J. Koh, Y. Lu, C. R. Wronski, Y. Kuang, R. W. Collins, T. T. Tsong, and Y. E. Strausser, Correlation of real time spectroellipsometry and atomic force microscopy measurements of surface roughness on amorphous semiconductor thin films, *Applied Physics Letters*, **69**, 1297–1299 (1996). Copyright 1996, American Institute of Physics.

In contrast, the following shows examples when the application of effective medium theories is difficult:

(a) calculation of the dielectric functions of semiconductor alloys [40,41],
(b) characterization of surface roughness when its morphology is larger than $\lambda/10$ [39],
(c) evaluation of surface roughness in transparent films with small refractive indices,
(d) characterization of two-dimensional island growth on substrates [42,43].

In the case of (a), above effective medium theories cannot be employed since they are originally derived from dielectric polarization induced in microscopically large dielectrics. In particular, since constituent atoms in an alloy are mixed randomly, quantum effects have to be taken into account to express the optical properties of the alloy. Let us consider that the semiconductors A and B show ε_2 peaks at $\text{En} = a$ and b, respectively. In general, the ε_2 peak shifts smoothly from $\text{En} = a$ to b, depending on the composition of the alloy A_xB_{1-x}. Nevertheless, if we calculate the dielectric function of this alloy by applying EMA, we will find two ε_2 peaks at $\text{En} = a$ and b in the calculated dielectric function, which is quite different from the true dielectric function [41]. When we perform compositional analysis of alloys, dielectric function modeling for alloy composition is required (see Section 7.2.3).

When the morphology of surface roughness is larger than $\lambda/10$, a surface roughness layer cannot be treated as a uniform layer shown in Fig. 5.27(b) and, in this condition, d_s is generally underestimated [39,44]. The problem of surface roughness characterization in transparent films with small refractive indices arises from measurement sensitivity, rather than the limitation of effective medium theories. In these transparent materials, it is generally difficult to distinguish the surface roughness layer from the bulk layer, because the refractive indices of these layers are similar. In the case of $SiO_2 (n = 1.49)$, for example, the refractive index of the surface roughness layer becomes $n = 1.24$. In addition, in transparent thin

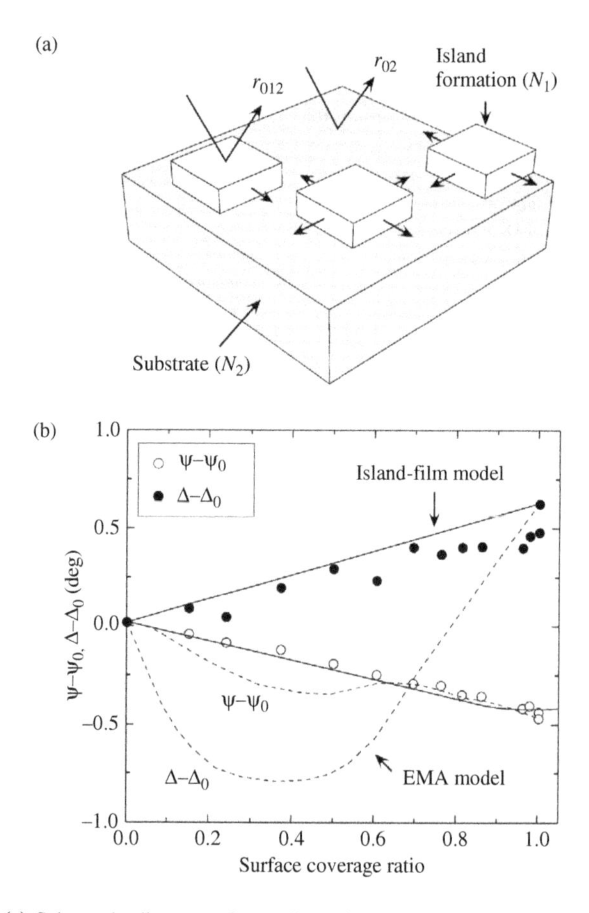

Figure 5.31 (a) Schematic diagram of two-dimensional island growth on a substrate and (b) $(\psi - \psi_0, \Delta - \Delta_0)$ versus surface coverage ratio of Pd when two-dimensional island growth occurs on a Ag(111) substrate. In (b), (ψ_0, Δ_0) show the values observed before the Pd growth. The solid and dotted lines represent calculation results obtained from the island-film and EMA models, respectively. Drawing (b): Reprinted from *Surface Science*, **135**, R. H. Muller and J. C. Farmer, Macroscopic optical model for the ellipsometry of an underpotential deposit: Lead on copper and silver, 521–531, Copyright (1983), with permission from Elsevier.

films, there is the relation of $n_1 \cos \theta_1 d = \text{const.}$ (see Section 5.1.2). Therefore, even when surface roughness is present, this influence is often averaged out by the whole transparent film, when the refractive index difference between surface roughness and bulk layers is small. In this case, it is advisable to perform the analysis using a high energy region where the transparent film shows light absorption. Alternatively, we may perform the analysis using d_s estimated from d_{rms}.

Figure 5.31 shows (a) schematic diagram of two-dimensional island growth mentioned in the above (d), and (b) (ψ–ψ_0, Δ–Δ_0) versus surface coverage ratio of Pd when two-dimensional island growth occurs on a Ag(111) substrate [42]. In Fig. 5.31(b), Pd has been deposited by an electrochemical method using electrolyte, and the measured values represent shifts from (ψ_0, Δ_0) observed before the Pd growth. As confirmed from the dotted lines in this figure, calculated values using EMA differ significantly from the measured values. Thus, EMA cannot be used to express the two-dimensional island growth shown in Fig. 5.31(b).

When a thin film has a two-dimensional structure, (ψ, Δ) can be calculated from the following method [42,43]. Let N_1 and N_2 be the complex refractive indices of an island material and a substrate, respectively. It can be seen from Fig. 5.31(a) that the amplitude reflection coefficients of the island film and substrate are given by r_{012} and r_{02}, respectively. When the surface coverage ratio of the islands on the substrate is given by s, the total amplitude-reflection coefficient can be expressed by

$$r_{\text{total}} = s r_{012} + (1 - s) r_{02} \tag{5.48}$$

By calculating r_{total} for p- and s-polarizations, we obtain (ψ, Δ) from $\rho = r_{\text{total,p}}/r_{\text{total,s}}$. In the following parts, this model will be referred to as the island-film model [42,43]. The solid lines in Fig. 5.31(b) represent (ψ, Δ) calculated from the island-film model, and we can see quite good agreement between the calculated and experimental results.

5.4 OPTICAL MODELS

The construction of an optical model is one of the most important steps in ellipsometry data analysis. If samples depolarize incident light, however, we have to use optical models that allow the calculation of (ψ, Δ) spectra from the Stokes parameters. In order to simplify the optical model, sample structures can also be modified. In this section, we will address the construction of optical models, optimization of sample structures, and optical models for depolarizing samples.

5.4.1 CONSTRUCTION OF OPTICAL MODELS

Figure 5.32 summarizes optical models used in ellipsometry data analysis. In these optical models, the analysis becomes complicated in the order (a) → (h). In general,

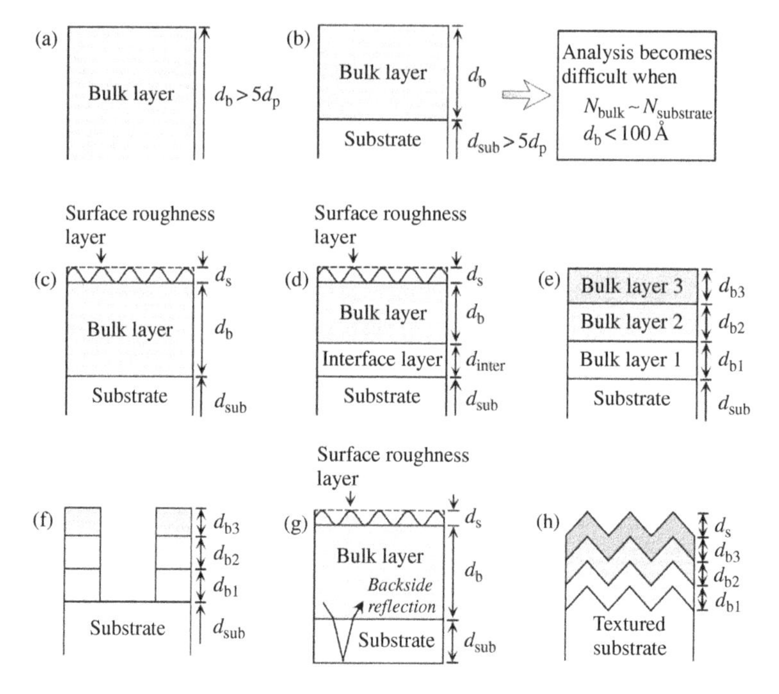

Figure 5.32 Optical models used in data analysis of spectroscopic ellipsometry. The analysis becomes complicated in the order (a) → (h). d_p in the figure represents the penetration depth of light.

when the dielectric functions of each layer are known, we can perform the thickness analysis relatively easily. In contrast, when a dielectric function is not known, the analysis sometimes becomes difficult even in the case of (c).

In the optical model shown in Fig. 5.32(a), only the light reflection at an ambient/bulk layer interface is taken into account, and $\rho = \tan\psi \exp(i\Delta)$ is given by Eq. (5.1). When we apply this model, the bulk layer thickness d_b must satisfy the condition $d_b > 5d_p$, where d_p represents the penetration depth of light (see Section 5.1.3). In the case of this model, ε can be estimated directly from measured (ψ, Δ), as we will see in Section 5.4.2. The optical model in Fig. 5.32(b) illustrates an ambient/bulk layer/substrate structure described in Sections 5.1.2 and 5.1.3. In this optical model, the thickness of a substrate (d_{sub}) is $d_{sub} > 5d_p$. When the optical constants of the bulk layer are almost the same as those of the substrate, however, the analysis of the bulk layer becomes difficult. If a thin film is a homoepitaxial layer, for example, sample characterization is almost impossible, except for the surface roughness layer, since $N_{bulk} = N_{substrate}$. In addition, when $d_b < 100$ Å, the evaluation of N_{bulk} also becomes difficult, although we can still determine the layer thickness if N_{bulk} is known (see Section 5.1.2). In contrast, when N_{bulk} differs significantly from $N_{substrate}$ (i.e., semiconductor/metal structures, etc.), N_{bulk} can be estimated even if $d_b \sim 50$ Å. The dielectric function of the surface roughness layer shown in Fig. 5.32(c) can be calculated from the dielectric function of the bulk layer by applying the effective medium approximation. We can obtain the amplitude reflection coefficient of this multilayer structure (r_{0123}) from the method described in Section 2.4.2. In this case, ρ is given by

$$\rho = r_{0123,\mathrm{p}}/r_{0123,\mathrm{s}} \tag{5.49}$$

Even when there are many layers in an optical model, the amplitude reflection coefficient can be calculated from this method. As shown in Fig. 5.32(d), when an interface layer is formed on a substrate, we introduce the interface layer into an optical model [45]. If a substrate surface is microscopically rough, we may obtain the dielectric function of the interface layer by applying EMA [17,34]. The optical model shown in Fig. 5.32(f) has been employed for the characterization of the etching process in LSI devices [46–49]. The island-film model described in Section 5.3.3 can be applied for the data analysis of this structure (see Section 8.3.2). When the light absorption in a substrate is small ($k \sim 0$) and backside reflection of the transparent substrate is present [Fig. 5.32(g)], we need to employ an analytical model that incorporates the effect of backside reflection (see Section 5.4.4). So far, the analysis of the thin film formed on a textured substrate shown in Fig. 5.32(h) has also been reported using spectroscopic ellipsometry of the RCE type [18]. When we analyze sample structures shown in Figs. 5.32(f)–(h), the depolarization effect of samples should be taken into account (see Sections 4.4.4 and 5.4.4).

5.4.2 PSEUDO-DIELECTRIC FUNCTION

The pseudo-dielectric function $\langle\varepsilon\rangle$ represents a dielectric function obtained directly from the measured values (ψ, Δ) and is calculated from an optical model that assumes a perfectly flat substrate with infinite thickness. If there is surface roughness on a sample, the pseudo-dielectric function varies depending on the size of surface roughness (Fig. 5.33). Accordingly, the pseudo-dielectric function is basically different from the dielectric function of a material itself. When optical

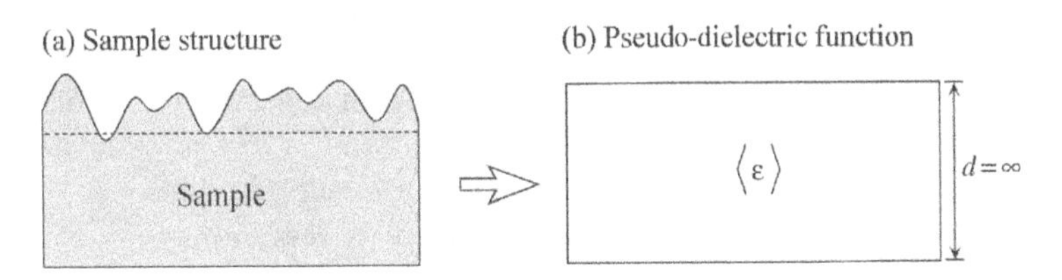

Figure 5.33 (a) Structure of a sample with surface roughness and (b) optical model corresponding to the pseudo-dielectric function.

interference is present in a thin-film structure, this interference effect appears in $\langle\varepsilon\rangle$ spectra.

The pseudo-dielectric function can be calculated using an equation derived from the following procedure [50]. Let us consider light reflection at an air/sample interface only, similar to Eq. (5.1). In this case, by applying the Fresnel equation shown in Eq. (2.65), we get

$$\begin{aligned}\rho &= r_p/r_s \\ &= \frac{\varepsilon_i \sin^2\theta_i - N_{ii}N_{tt}}{\varepsilon_i \sin^2\theta_i + N_{ii}N_{tt}} \\ &= \frac{\sin^2\theta_i - \cos\theta_i\left[\varepsilon_t/\varepsilon_i - \sin^2\theta_i\right]^{1/2}}{\sin^2\theta_i + \cos\theta_i\left[\varepsilon_t/\varepsilon_i - \sin^2\theta_i\right]^{1/2}}\end{aligned} \tag{5.50}$$

If the dielectric constants of an ambient (air) and a sample are given by $\varepsilon_i = 1$ and $\varepsilon_t = \langle\varepsilon\rangle$, respectively, we obtain the pseudo-dielectric function $\langle\varepsilon\rangle$ from Eq. (5.50) as follows:

$$\langle\varepsilon\rangle = \varepsilon_t = \sin^2\theta_i\left[1 + \tan^2\theta_i\left(\frac{1-\rho}{1+\rho}\right)^2\right] \tag{5.51}$$

It is evident from Eq. (5.51) that $\langle\varepsilon\rangle$ is calculated directly from the measured value $\rho = \tan\psi\exp(i\Delta)$ using the incidence angle θ_i. In the case of Fig. 5.32(a), the dielectric function of a sample (ε_t) is determined immediately by applying Eq. (5.51). When the complex refractive index is given by $N \equiv n - ik$, the pseudo-dielectric function is defined by $\langle\varepsilon\rangle \equiv \langle\varepsilon_1\rangle - i\langle\varepsilon_2\rangle$, while $\langle\varepsilon\rangle \equiv \langle\varepsilon_1\rangle + i\langle\varepsilon_2\rangle$ when $N \equiv n + ik$. In order to obtain $\langle\varepsilon\rangle \equiv \langle\varepsilon_1\rangle + i\langle\varepsilon_2\rangle$, the definition of $\rho = \tan\psi\exp(-i\Delta)$ should be used (see Appendix 2).

Figure 5.34 shows $\langle\varepsilon\rangle$ obtained from changing d_s on a c-Si substrate. This result was calculated from Eq. (5.51) using the (ψ, Δ) spectra shown in Fig. 5.29. As shown in the figure, $\langle\varepsilon\rangle$ changes drastically with increasing d_s. In particular, the ε_2 peak at En = 4.25 eV shows a large change, since light absorption is quite strong at En = 4.25 eV and the penetration depth of light becomes very small. The variation of $\langle\varepsilon\rangle$ with d_s originates from the change in Δ shown in Fig. 5.29. It should be emphasized that $\langle\varepsilon\rangle = \varepsilon_t$ when $d_s = 0$ Å. In other words, $\langle\varepsilon\rangle$ of $d_s = 0$ Å represents the dielectric function of c-Si itself. When a sample is contaminated by organic layers, the $\langle\varepsilon_2\rangle$ value at En = 4.25 eV reduces, similar to Fig. 5.34. Thus, the value of $\langle\varepsilon_2\rangle$ becomes larger if there is less surface roughness and contamination [29,51]. In crystalline semiconductors, surface chemical treatments are often performed, in order to determine accurate dielectric functions from their pseudo-dielectric functions [29,51].

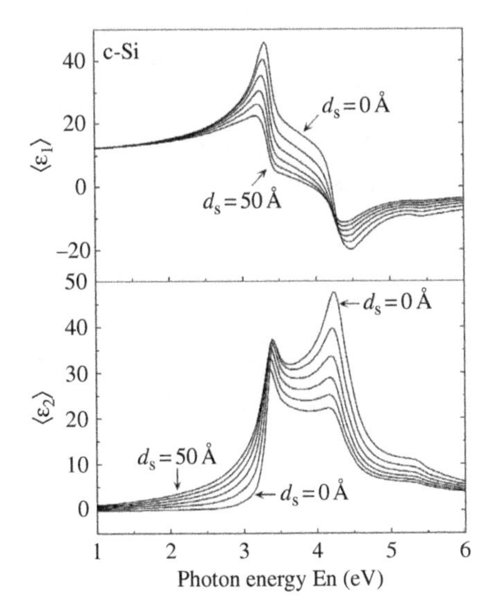

Figure 5.34 Pseudo-dielectric functions of c-Si (crystalline silicon) obtained from different thicknesses of the surface roughness layer (d_s). These pseudo-dielectric functions were calculated from (ψ, Δ) spectra shown in Fig. 5.29.

5.4.3 OPTIMIZATION OF SAMPLE STRUCTURES

If we optimize the sample structure, the optical model can be simplified (Fig. 5.35). Suppose that there is a multilayer structure on a substrate, and a bulk layer is formed on the multilayer [Fig. 5.35(a)]. When we need the dielectric function of this bulk layer, we can increase the thickness of the bulk layer up to $d_b > 5d_p$. In this case, we can employ an optical model composed of surface roughness and bulk layers only. As shown in Fig. 5.35(b), the effect of backside reflection can be eliminated if we roughen the backside and coat the backside with a black paint (graphite paint, etc.) that absorbs light [18,52]. In the case of optical devices, however, device characterization often becomes difficult with this treatment. When we cannot perform this treatment, we analyze data using the optical model described in Section 5.4.4. On the other hand, when there is very large surface roughness on a sample, we may remove the surface roughness by mechanical polishing if the bulk layer is sufficiently thick [Fig. 5.35(c)]. In some samples, however, damaged layers will be created on the surface by mechanical polishing, and analysis for these layers may be required.

5.4.4 OPTICAL MODELS FOR DEPOLARIZING SAMPLES

When samples depolarize incident light, the values of ($\psi\Delta$) will vary according to the degree of polarization of the reflected light. In the analysis of depolarizing samples, therefore, we need to use an optical model that incorporates depolarization

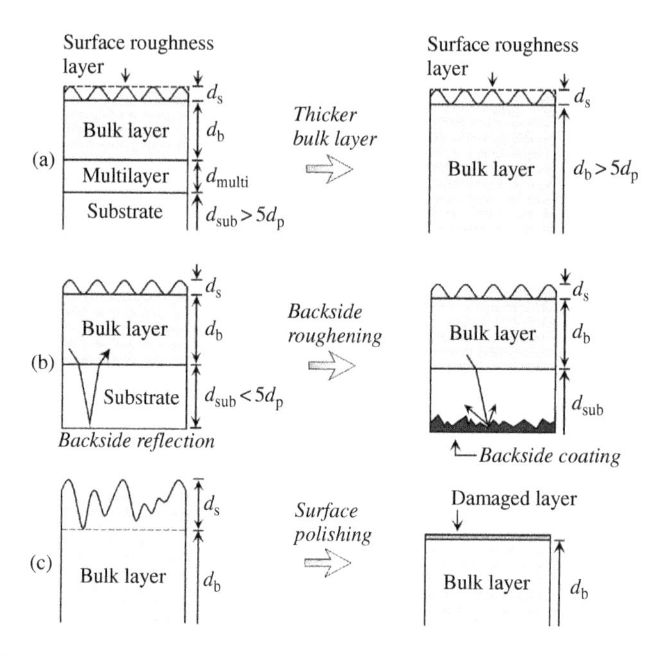

Figure 5.35 Optimization of sample structures: (a) simplification of the optical model by thicker bulk layer formation, (b) elimination of backside reflection by backside roughening, and (c) removal of surface roughness by mechanical polishing. In this figure, d_p denotes the penetration depth of light.

effects. Here, we will treat depolarizations induced by backside reflection and thickness inhomogeneity.

Figure 5.36 represents an optical model that includes the effect of backside reflection on a transparent substrate ($k \sim 0$). In this model, there are two thin films on the front side (thin film 1) and rear side (thin film 2) of a transparent

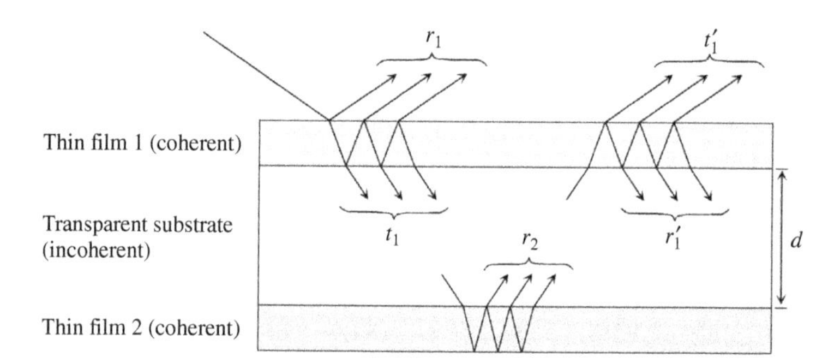

Figure 5.36 Reflection and transmission of light in a (thin film 1)/(transparent substrate)/(thin film 2) structure. The thicknesses of the thin films and substrate are assumed to satisfy the coherent and incoherent conditions, respectively. In the figure, r (t) denotes an amplitude reflection (transmission) coefficient, and d shows the thickness of the transparent substrate.

substrate. Here, we assume that these thin films are optically thin and are well within the coherent condition, while the thickness of the transparent substrate is thick enough to satisfy the incoherent condition (see Section 4.4.4). In Fig. 5.36, $r_1(t_1)$ represents the amplitude reflection (transmission) coefficient that corresponds to r_{012} (t_{012}) in Fig. 2.23. The r_2 shown in Fig. 5.36 expresses the backside reflection on the transparent substrate, and the light reflected at the backside further transmits the thin film 1 with the amplitude transmission coefficient of t_1'. If we use the procedure described in Section 2.4.1, we can describe the total amplitude-reflection coefficient of this optical model (r_t) as follows:

$$r_t = r_1 + t_1 t_1' r_2 e^{-i2\beta_s} + t_1 t_1' r_2 (r_1' r_2) e^{-i4\beta_s} + t_1 t_1' r_2 (r_1'^2 r_2^2) e^{-i6\beta_s} + \cdots \quad (5.52a)$$

$$= r_1 + t_1 t_1' r_2 e^{-i2\beta_s} \sum_{m=0}^{n} (r_1' r_2)^m \exp(-i2m\beta_s) \quad (5.52b)$$

$$= r_1 + \frac{t_1 t_1' r_2 \exp(-i2\beta_s)}{1 - r_1' r_2 \exp(-i2\beta_s)} \quad (5.52c)$$

where

$$\beta_s = \frac{2\pi d}{\lambda} N_s \cos\theta_s = \frac{2\pi d}{\lambda} \left(N_s^2 - \sin^2\theta_i\right)^{1/2} \quad (5.52d)$$

Notice that Eqs. (5.52c) and (5.52d) are quite similar to Eqs. (2.82) and (2.78), respectively. In Eq. (5.52d), β_s, N_s, θ_s, and d show the phase variation, complex refractive index ($N_s = n_s - ik_s$), transmission angle, and thickness for the transparent substrate, respectively, and θ_i denotes the angle of incidence. Now recall from Section 4.4.4 that information on the phase of light is completely lost in the incoherent system. In this case, we have to calculate (ψ, Δ) from light intensities; in other words, the Stokes parameters. It is clear from Eq. (4.11) that the Stokes parameters can be calculated from the light intensities $r_p r_p^*$, $r_s r_s^*$, and $r_p^* r_s$. From Eq. (5.52c), these light intensities are expressed by the following equations [53–57]:

$$r_{t,p} r_{t,p}^* = |r_{t,p}|^2 = (r_{1,p} r_{1,p}^*) + \frac{(t_{1,p} t_{1,p}^*)(t_{1,p}' t_{1,p}'^*)(r_{2,p} r_{2,p}^*) \exp[-4|\mathrm{Im}(\beta_s)|]}{1 - (r_{1,p}' r_{1,p}'^*)(r_{2,p} r_{2,p}^*) \exp[-4|\mathrm{Im}(\beta_s)|]} \quad (5.53a)$$

$$r_{t,s} r_{t,s}^* = |r_{t,s}|^2 = (r_{1,s} r_{1,s}^*) + \frac{(t_{1,s} t_{1,s}^*)(t_{1,s}' t_{1,s}'^*)(r_{2,s} r_{2,s}^*) \exp[-4|\mathrm{Im}(\beta_s)|]}{1 - (r_{1,s}' r_{1,s}'^*)(r_{2,s} r_{2,s}^*) \exp[-4|\mathrm{Im}(\beta_s)|]} \quad (5.53b)$$

$$r_{t,p}^* r_{t,s} = (r_{1,p}^* r_{1,s}) + \frac{(t_{1,p}^* t_{1,s})(t_{1,p}'^* t_{1,s}')(r_{2,p}^* r_{2,s}) \exp[-4|\mathrm{Im}(\beta_s)|]}{1 - (r_{1,p}'^* r_{1,s}')(r_{2,p}^* r_{2,s}) \exp[-4|\mathrm{Im}(\beta_s)|]} \quad (5.53c)$$

In Eq. (5.53), the term $\exp[-4|\mathrm{Im}(\beta_s)|]$ describes the light absorption in the transparent substrate. At normal incidence, it follows from (5.52d) that

$$\exp[-4|\mathrm{Im}(\beta_s)|] = \exp\left(-\frac{4\pi k_s}{\lambda} 2d\right) = \exp(-2\alpha d) \quad (5.54)$$

In this condition, therefore, Eq. (5.54) simply shows Beer's law [Eq. (2.35)] when the light pass length is $2d$. From Eqs. (4.11) and (5.53), we can express the Stokes parameters as $S_0 = r_{t,p}r_{t,p}^* + r_{t,s}r_{t,s}^*$, $S_1 = r_{t,p}r_{t,p}^* - r_{t,s}r_{t,s}^*$, $S_2 = 2\mathrm{Re}(r_{t,p}^* r_{t,s})$, and $S_3 = 2\mathrm{Im}(r_{t,p}^* r_{t,s})$. Using these values, we obtain final (ψ, Δ) from Eq. (4.12). Since all the Stokes parameters are now known, the degree of polarization p can also be calculated from Eq. (3.52) [56,57]. If we apply Eq. (5.52b), (ψ, Δ, p) are expressed as a function of m. m can also be employed as an analytical parameter in data analysis [55]. If thin film 1 is composed of two thin layers, r_1 and t_1 in Fig. 5.36 are obtained from r_{0123} and t_{0123} [Eqs. (2.92) and (2.93)]. Accordingly, the above model can be extended quite easily to describe a multilayer structure.

Figure 5.37 shows the degree of polarization p in an air/glass substrate/Al structure [57]. The glass substrate is a Corning 7059 glass (0.85 mm) and the backside of the substrate is coated with Al film to enhance the light reflection on the backside and thus to increase the depolarization. In this figure, the closed circles show experimental data obtained from PME and the solid line represents a depolarization spectrum calculated from the above procedure. It can be seen from Fig. 5.37 that the experimental spectrum agrees quite well with the calculated spectrum. At En > 4.2 eV, however, the experimental result shows an artifact ($p > 1$) due to the error in calibration [57]. In Fig. 5.37, the increase in p at En > 4.2 eV is caused by increasing light absorption in the glass substrate. The (ψ, Δ) spectra calculated from this optical model have also been reported to show quite good agreement with experimental spectra [57].

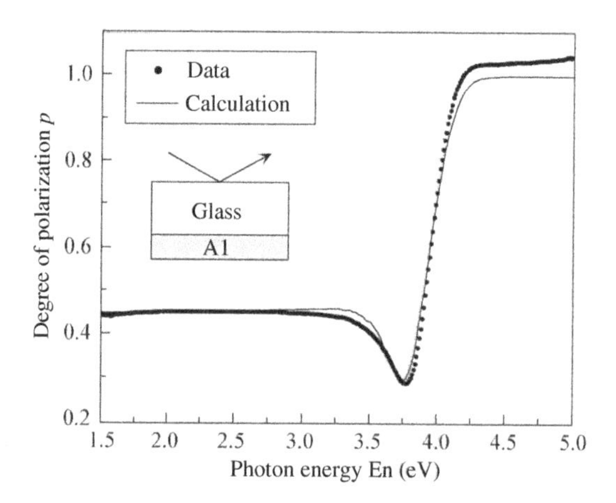

Figure 5.37 Degree of polarization p in an air/glass substrate/Al structure. In the figure, the closed circles show experimental data and the solid line represents a depolarization spectrum calculated from the optical model shown in Fig. 5.36. Reprinted from *Thin Solid Films*, **313–314**, M. Kildemo, R. Ossikovski, and M. Stchakovsky, Measurement of the absorption edge of thick transparent substrates using the incoherent reflection model and spectroscopic UV-visible-near IR ellipsometry, 108–113, Copyright (1998), with permission from Elsevier.

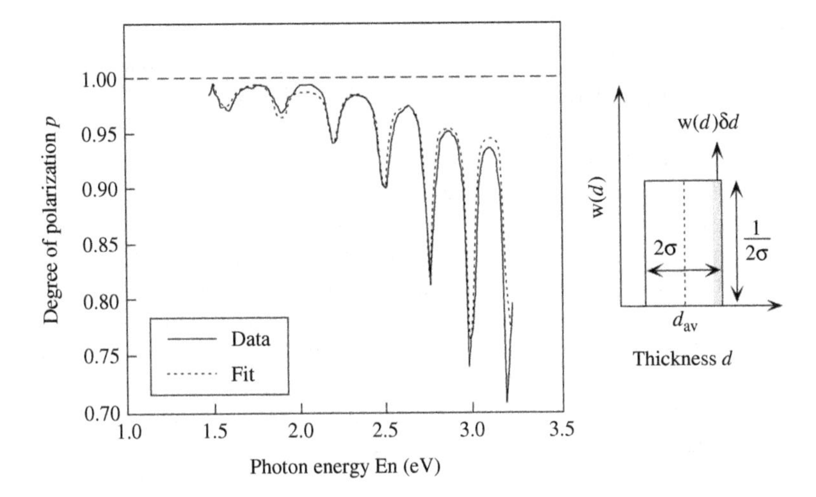

Figure 5.38 Depolarization spectrum obtained from a nonuniform ZnS film (827 nm) deposited on a GaAs substrate. The inset of this figure shows a thickness distribution given by Eq. (5.55). The solid and dotted lines represent the experimental data and fitting result, respectively. Reprinted with permission from *Applied Physics Letters*, **61**, G. E. Jellison, Jr and J. W. McCamy, Sample depolarization effects from thin films of ZnS on GaAs as measured by spectroscopic ellipsometry, 512–514 (1992). Copyright 1992, American Institute of Physics.

Depolarization generated by thickness nonuniformity can be modeled more easily [54,58–60]. Figure 5.38 shows the depolarization spectrum obtained from a nonuniform ZnS film (827 nm) deposited on a GaAs substrate [58]. The inset of this figure shows a thickness distribution given by

$$w(d) = \begin{cases} 1/(2\sigma) & \text{for } |d - d_{av}| \leq \sigma \\ 0 & \text{for } |d - d_{av}| > \sigma \end{cases} \tag{5.55}$$

Here, $w(d)$ represents the probability of finding the thickness d. The d_{av} and σ show the average film thickness and width of the distribution function, respectively, and σ becomes smaller when a thin film is uniform. The depolarization caused by film thickness inhomogeneity can also be calculated from the Stokes parameters. In this case, we first determine the light intensities $r_{\text{p}}r_{\text{p}}^*$, $r_{\text{s}}r_{\text{s}}^*$, and $r_{\text{p}}^*r_{\text{s}}$ at different film thicknesses d_m, and then calculate the Stokes parameters at each thickness $[s_{0-3}(d_m)]$. Using these values, we obtain the final Stokes parameters as follows [54,60]:

$$S_j = \sum_{m=1}^{n} s_j(d_m)w(d_m)\delta d_m \ (j = 0, 1, 2, 3). \tag{5.56}$$

As shown in the inset of Fig. 5.38, $w(d_m)\delta d_m$ shows a small rectangular area. Thus, Eq. (5.56) represents the integration of the rectangular areas from a minimum thickness (d_1) to a maximum thickness (d_n) with the coefficient of $s_j(d_m)$. Notice that $\sum w(d_m)\delta d_m = 1$. Using S_{0-3} obtained from Eq. (5.56), final (ψ, Δ, p) are calculated from Eqs. (4.12) and (3.52). The solid and dotted lines in Fig. 5.38 show experimental and calculated results, respectively. Quite good agreement between these spectra supports the validity of this analysis. It has been reported that p in Fig. 5.38 reduces when the Stokes parameters go through a maximum as a function of wavelength [58]. Quite surprisingly, the thickness distribution estimated from Fig. 5.38 is only $\sigma = 13.6 \pm 0.6$ nm. Thus, a thickness inhomogeneity of ~2% is sufficient to cause the relatively large depolarization shown in Fig. 5.38 when the total film thickness is ~1 μm. The thickness nonuniformity can be modeled more explicitly using a thickness gradient [59,60]. In this case, the orientation of the thickness gradient should be taken into account [59,60]. Eq. (5.56) can also be applied to calculate depolarizations induced by incidence angle and wavelength variations [Figs. 4.30(b) and (c)]. The calculation can be performed by simply replacing d_m in Eq. (5.56) with θ_m and λ_m, which show the variation in the incidence angle and wavelength, respectively.

5.5 DATA ANALYSIS PROCEDURE

Ellipsometry data analysis is generally performed by using linear regression analysis, and optical constants and film structures are determined by minimizing fitting errors calculated from a fitting error function. If we use a method known as mathematical inversion, the dielectric function of a sample can be extracted rather easily. In this section, we will overview data analysis procedures that incorporate linear regression analysis and mathematical inversion.

5.5.1 LINEAR REGRESSION ANALYSIS

Figure 5.39 shows the data analysis procedure in spectroscopic ellipsometry. In this procedure, an optical model corresponding to a sample is constructed first and then the dielectric functions of each layer are selected. When the dielectric functions are not known, the dielectric function models described in Section 5.2 are employed. The dielectric function of a substrate can be obtained from the pseudo-dielectric function if there are no overlayers and surface roughness is negligible (see Section 5.4.2). The incidence angle in ellipsometry measurement can be determined from a measurement using a standard sample. Alternatively, the incidence angle can be used as an analytical parameter. Using analytical parameters such as d_s and d_b, calculated (ψ, Δ) are fitted to experimental spectra. In the next step, the fitting error σ is evaluated. When the fitting error is large, the optical model or dielectric functions are optimized. Finally, from the optical model and dielectric functions that minimize σ, the optical constants and film thicknesses of the sample are determined.

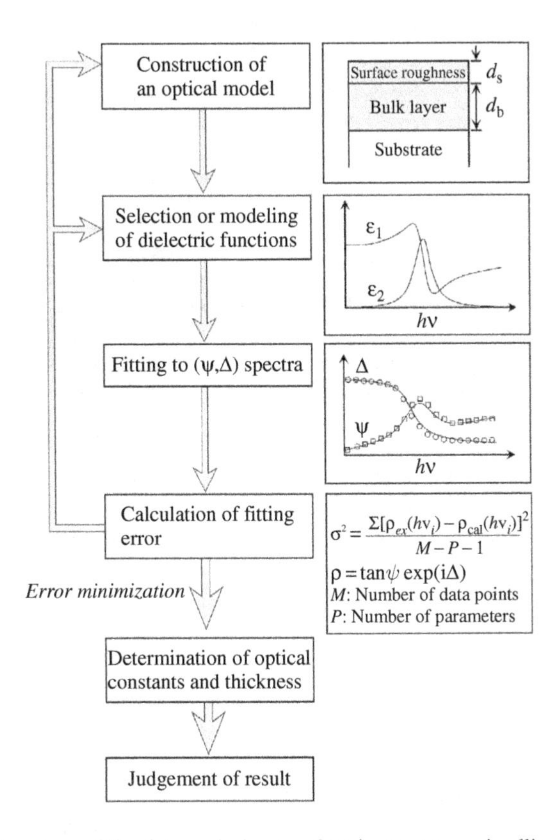

Figure 5.39 Flowchart of the data analysis procedure in spectroscopic ellipsometry.

As confirmed from Fig. 5.39, ellipsometry analysis is performed from fitting using an optical model. Nevertheless, an optical model used in ellipsometry analysis merely represents an approximated sample structure, and obtained results are not necessarily correct even when the fit is sufficiently good. Accordingly, when the optical constants or film structures of a sample are not known well, the ellipsometry results must be justified using other measurement methods. This is the greatest disadvantage of the ellipsometry technique. However, once an analytical method is established, it becomes possible to perform high-precision characterization in a short time using spectroscopic ellipsometry. In order to verify sample structures estimated from spectroscopic ellipsometry, various characterization techniques including scanning electron microscope (SEM), transmission electron microscope (TEM), and AFM have been used. Recently, X-ray reflectivity technique has also been employed to confirm ellipsometry results [61,62]. If we perform ellipsometry analysis using a data set obtained from different incidence angles or thin film thicknesses, more reliable ellipsometry results can be obtained [4].

Figure 5.40 shows a schematic diagram of linear regression analysis. In this analysis, the analytical solution x_r is determined from a value that minimizes the

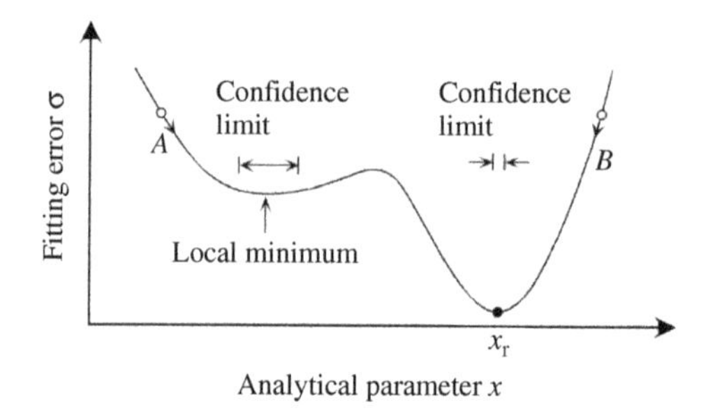

Figure 5.40 Schematic diagram of linear regression analysis. In this figure, x_r denotes the analytical solution obtained from the linear regression analysis.

fitting error σ versus the analytical parameter x. Although $\sigma(x)$ is generally a nonlinear function, this analysis is referred to as linear regression analysis, since the analytical parameter x itself is linear. For linear regression analysis, Newton's method [63] or the Levenberg–Marquardt method [64] are employed, and analytical parameters used in actual data analysis are determined simultaneously by applying these methods. The confidence limit shown in Fig. 5.40 represents the precision of the regression analysis and becomes smaller when the absolute value of σ is small and the variation of σ around x_r is steep. Unfortunately, linear regression analysis depends greatly on initial values used in the analysis. For example, when A and B shown in Fig. 5.40 are used as initial values, the value of the local minimum is obtained for the case of A, while the value of x_r can be obtained from the initial value of B. In general, σ shows gradual change around a local minimum and the resulting confidence limit becomes large. Accordingly, when confidence limits or fitting errors are large, it is advisable to change initial values.

In some cases, reasonable fitting or sufficiently low σ cannot be obtained in data analysis. This can be attributed to the following reasons:

(a) measured (ψ, Δ) spectra are inaccurate,
(b) dielectric functions used in data analysis are inappropriate,
(c) the optical model used in data analysis is inappropriate, and
(d) the sample has a depolarization effect.

(a) above can be confirmed relatively easily from measurement and data analysis of a standard sample such as a c-Si substrate covered with a thermal oxide. We often encounter the situation of (b) above particularly when the dielectric function of a sample is not known well. In this case, the dielectric function of a sample has to be changed to obtain a better fit. We can limit an analyzed wavelength region when the dielectric function of a sample shows complicated structures in a certain wavelength region. The situation of (c) above also occurs frequently. When fitting errors are large, optical models are often too simplified. The fit may be improved drastically by the introduction of a surface roughness or interface layer. Furthermore, when the optical properties of a thin film change in the growth direction, fitting errors increase if we represent the thin film by a single bulk layer, instead of a multilayer. As we have seen in Section 4.4.4, the effect of depolarization on measured spectra is quite large, and the optical models described in Section 5.4.4 should be employed when we analyze depolarizing samples.

5.5.2 FITTING ERROR FUNCTION

For the calculation of fitting errors, various functions have been used. The following functions have been employed widely for the calculation of fitting errors [65–67]:

$$\sigma = \frac{1}{\sqrt{M-P-1}}\left\{\sum_{j=1}^{M}\left[\rho_{\mathrm{ex}}(h\nu_j)-\rho_{\mathrm{cal}}(h\nu_j)\right]^2\right\}^{1/2} \tag{5.57}$$

$$\sigma = \frac{1}{\sqrt{M-P-1}}\left\{\sum_{j=1}^{M}\left(\left[\tan\psi_{\mathrm{ex}}(h\nu_j)-\tan\psi_{\mathrm{cal}}(h\nu_j)\right]^2 + \left[\cos\Delta_{\mathrm{ex}}(h\nu_j)-\cos\Delta_{\mathrm{cal}}(h\nu_j)\right]^2\right)\right\}^{1/2} \tag{5.58}$$

where the subscripts ex and cal represent experimental and calculated values at $\mathrm{En}_j = h\nu_j$, respectively. In Eqs. (5.57) and (5.58), M and P show the number of measurement points and analytical parameters, respectively. σ above is generally called the unbiased estimator.

Recently, the following fitting error function has been proposed [65,66]:

$$\chi = \frac{1}{\sqrt{M-P-1}}\left\{\sum_{j=1}^{M}\left[\frac{\rho_{\mathrm{ex}}(h\nu_j)-\rho_{\mathrm{cal}}(h\nu_j)}{\delta\rho(h\nu_j)}\right]^2\right\}^{1/2} \tag{5.59}$$

where $\delta\rho$ represents measurement errors for ellipsometry instruments. χ in Eq. (5.59) is generally referred to as the biased estimator. A similar function has also been used [4]:

$$\chi = \frac{1}{\sqrt{M-P-1}}\left\{\sum_{j=1}^{M}\left(\left[\frac{\psi_{\mathrm{ex}}(h\nu_j)-\psi_{\mathrm{cal}}(h\nu_j)}{\delta\psi(h\nu_j)}\right]^2 + \left[\frac{\Delta_{\mathrm{ex}}(h\nu_j)-\Delta_{\mathrm{cal}}(h\nu_j)}{\delta\Delta(h\nu_j)}\right]^2\right)\right\}^{1/2} \tag{5.60}$$

where $(\delta\psi, \delta\Delta)$ show measurement errors in (ψ, Δ). In Eqs. (5.59) and (5.60), the fitting errors are divided by the measurement errors. Thus, by using these functions, we can suppress the increase in χ in regions where $\delta\rho$ and $(\delta\psi, \delta\Delta)$ increase.

In conventional spectroscopic ellipsometers, $(\delta\psi, \delta\Delta)$ increase in the UV region (see Section 4.4.2). In this case, if we perform linear regression analysis using σ, the results may be affected by large σ values in the UV region. This influence can be suppressed by using χ. We observe $\chi \sim 1$ when fitting is sufficiently good, while we obtain $\chi \gg 1$ when calculated spectra do not fit well. When measurement errors are overestimated, we get $\chi < 1$. It has been reported that confidence limits reduce if we apply χ, instead of σ [65]. Nevertheless, when fitting to experimental spectra is sufficiently good, ellipsometry results are basically independent of the fitting error functions described above.

As mentioned earlier, measurement errors increase around $\Delta = 0°$ and $180°$ in the rotating-analyzer (polarizer) ellipsometer (see Section 4.4.1). When transparent films are characterized using this instrument, σ and χ values will be governed by the large fitting errors in the regions $\Delta \sim 0°$ and $180°$. This unfavorable situation can be avoided by increasing $\delta\Delta$ intentionally in the regions $\Delta < 20°$ and $\Delta > 160°$, for example. More simply, we can exclude these regions in data analysis. In the phase-modulation ellipsometer, since measurement errors also increase in specific regions, similar procedures may be necessary.

5.5.3 MATHEMATICAL INVERSION

When the dielectric function of a sample is not known, dielectric function modeling is required. Nevertheless, complete modeling of the dielectric function over a wide wavelength range is often difficult. When data analysis is performed in a selected wavelength region to improve fitting, we obtain the dielectric function of this region only. In this case, however, if we use a method known as mathematical inversion, the dielectric function in a whole measured range can be extracted rather easily. Figure 5.41 illustrates the ellipsometry data analysis using the mathematical inversion. In this example, the optical model of a sample is given by an air/thin film/substrate structure (Fig. 2.23). Recall from Eq. (1.1) that the optical model shown in Fig. 5.41(a) is expressed as

$$\tan\psi \exp(i\Delta) = \rho(N_0, N_1, N_2, d_b, \theta_0) \tag{5.61}$$

If known parameters are assumed to be measured (ψ, Δ), N_0, N_2 and θ_0, unknown parameters become N_1 and d_b. Now suppose that the dielectric function of the thin film changes smoothly in the region from En $= a$ to b, and the dielectric function of this region can be expressed from the Cauchy model using the parameters (A, B, C) [Eq. (5.17)]. In this case, fitting from En $= a$ to b can be performed easily using d_b and $(A, B, C)_{\text{Cauchy}}$ as free parameters [Fig. 5.41(b)]. Since d_b can be obtained from the analysis, now the only unknown parameter in Eq. (5.61) is $N_1 = n_1 - ik_1$. Thus, if we solve Eq. (5.61), the measured (ψ, Δ) can be converted directly into (n_1, k_1) [Fig. 5.41(c)]. This procedure is referred to as mathematical inversion. Actual mathematical inversion, however, is performed by using linear regression analysis. Thus, mathematical inversion is also called optical constant fit. As shown in Fig. 5.41(d), from mathematical inversion, the optical constants over the whole

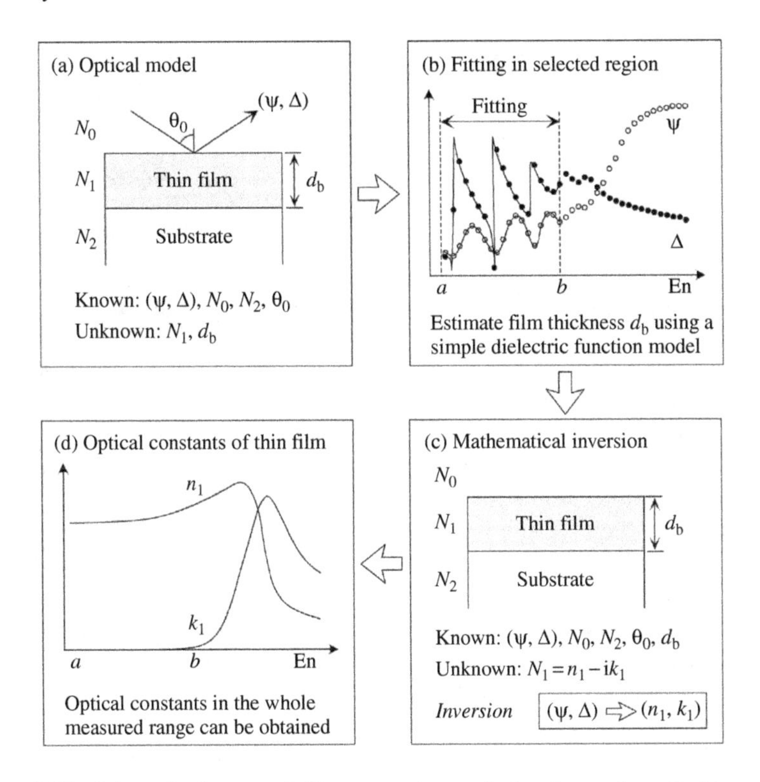

Figure 5.41 Schematic diagram of ellipsometry data analysis using mathematical inversion.

measured range can be determined. This data analysis procedure is quite effective in determining the dielectric function of a sample particularly when dielectric function modeling is difficult in some specific regions. We can further parameterize the extracted dielectric function using various dielectric function models. From the parameterized dielectric function, we may perform the fitting shown in Fig. 5.41(b) again, to determine the dielectric function more accurately.

Figure 5.42 shows an optical model in which two layers are formed on a substrate. When the only unknown parameter in the optical model is N_2, (n_2, k_2) can also be obtained from (ψ, Δ) using mathematical inversion. In other words, mathematical inversion can always be employed to extract the optical constants of one layer if all the other parameters are known. Now suppose that layer 1 shown in Fig. 5.42 represents a surface roughness layer. In this case, if d_1 and d_2 are obtained from the procedure shown in Fig. 5.41, N_2 can be extracted assuming $f_{\text{void}} = 0.5$. Similarly, when a known oxide layer (layer 1) is formed on a thin film (layer 2), we can employ mathematical inversion to extract the optical constants of the thin film. It should be noted that mathematical inversion cannot be performed when an inversion layer is located deeper than $5d_{\text{p}}$, since the optical response from the inversion layer becomes negligible in this case.

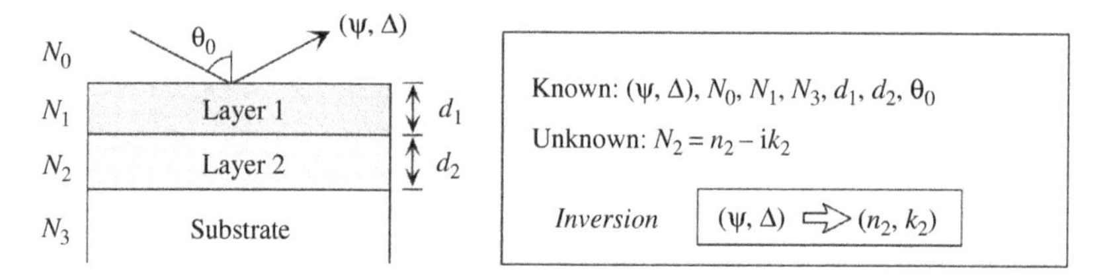

Figure 5.42 Optical model for an ambient/layer 1/layer 2/substrate structure. In this optical model, known parameters are assumed to be (ψ, Δ), N_0, N_1, N_3, d_1, d_2, and θ_0.

By applying mathematical inversion, layer thicknesses can also be determined [68,69]. Figure 5.43(a) shows ρ spectra obtained from an air/surface roughness layer/bulk layer/SiO_2 layer/c-Si substrate structure [69]. Here, the SiO_2 layer is a thermal oxide and the bulk layer is a n-type polycrystalline Si (poly-Si). The optical model of this sample corresponds to the one shown in Fig. 5.32(d). In the ρ spectra shown in Fig. 5.43(a), $\mathrm{Re}(\rho) = \tan\psi\cos\Delta$ and $\mathrm{Im}(\rho) = \tan\psi\sin\Delta$ oscillate versus En due to the optical interference induced by the SiO_2 layer. At higher energies, the light absorption within the poly-Si layer increases and the penetration depth of the light becomes smaller. Thus, the interference effect becomes weaker at higher energies (see Section 5.1.3). Figure 5.43(b) shows the ε_2 spectra of the poly-Si layer obtained from the mathematical inversion of the ρ spectra [69]. The ε_2 spectrum shown at the top represents the final spectrum extracted using the layer thicknesses of $d_s = 6.5$ Å, $d_b = 5370$ Å, $d_{SiO} = 1020$ Å, where d_s, d_b, and d_{SiO}

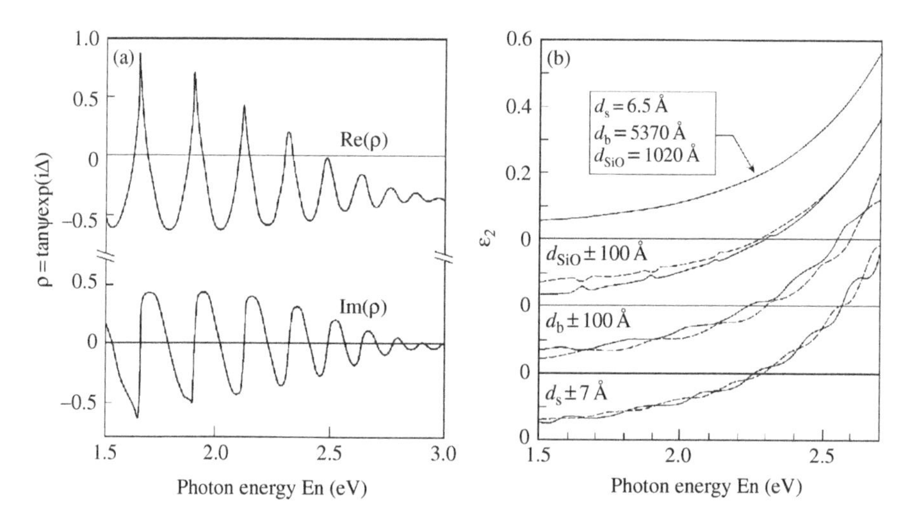

Figure 5.43 (a) ρ spectra of a P-doped polycrystalline Si formed on a SiO_2/c-Si substrate and (b) dielectric function of the P-doped polycrystalline Si extracted using mathematical inversion. Adapted with permission from D. E. Aspnes, A. A. Studna, and E. Kinsbron, Dielectric properties of heavily doped crystalline and amorphous silicon from 1.5 to 6.0 eV, *Phys. Rev. B*, **29** (1984) 768–779. Copyright 1984, the American Physical Society.

denote the thicknesses of the surface roughness layer (poly-Si), bulk layer (poly-Si), and SiO_2 layer, respectively. In this analysis, the void volume fraction in the surface roughness layer is $f_{\text{void}} = 0.6$. In the energy region shown in Fig. 5.43, the dielectric functions of c-Si and poly-Si are quite similar and thus the ε_2 spectrum in Fig. 5.43(b) corresponds to the one shown in Fig. 5.11.

On the other hand, the other three spectra in Fig. 5.43(b) show ε_2 spectra when each layer thickness is changed by the values indicated in the figure. In particular, the dotted lines show the calculation results when the layer thicknesses are increased from the accurate values, while the solid lines represent the results when the thicknesses are decreased. It can be seen from Fig. 5.43(b) that the extracted dielectric functions show anomalous structures when the layer thicknesses deviate from the accurate values. For example, when the SiO_2 layer thickness is different, peaks appear at the positions where the Re(ρ) values are large. Thus, the influence of optical interferences appears in inverted spectra when the thicknesses of layers are not accurate. Similar effects can also be seen when d_b and d_s are varied. In particular, in the case of d_s, the interference effect becomes stronger at higher energies. This shows that the sensitivity of d_s increases at higher energies due to smaller penetration depth of light. As shown above, if we use mathematical inversion, film thickness and the dielectric function can be estimated by eliminating anomalous structures appearing in the dielectric function. It has been reported that anomalous structures can be detected more easily if the second derivative spectra of dielectric functions (i.e., $d^2\varepsilon/d\text{En}^2$) are calculated [68].

In actual analysis, film thicknesses of $d_b > 100\,\text{Å}$ are generally required for mathematical inversion. When the dielectric function of an inversion layer is quite different from that of an underlying layer, however, mathematical inversion can be performed even at $d_b \sim 50\,\text{Å}$. If optical models are too simplified, anomalous structures appear in extracted dielectric functions and correct dielectric functions cannot be determined [68]. As mentioned earlier, even when there is a surface roughness or oxide layer, we can obtain the dielectric function of a sample from mathematical inversion. However, since these layers have a large influence on (ψ, Δ), there is a need to estimate these thicknesses accurately. For this purpose, it is preferable to perform data analysis over a wider wavelength range. The values of d_s can also be confirmed using AFM (see Section 5.3.3). In the case of semiconductors, we observe $\varepsilon_2 = 0$ below the bandgap of semiconductors ($\text{En} < E_g$). From this condition, we can judge the validity of data analysis. It should be emphasized that, when d_s is overestimated, the absolute values of ε_2 peaks obtained from mathematical inversion increase and the region $\varepsilon_2 \sim 0$ becomes $\varepsilon_2 < 0$. In some commercialized programs, the region $\varepsilon_2 < 0$ is corrected automatically to $\varepsilon_2 = 0$ and thus care is required when these programs are used. As we have discussed in Fig. 5.35(a), when a thick film is formed on a substrate, an optical model consisting of only surface roughness and bulk layers can be employed. In this case, the dielectric function of the bulk layer is obtained by simply removing the influence of the surface roughness layer using mathematical inversion. From this method, however, the dielectric function in the absorbing region ($\varepsilon_2 > 0$) is obtained, since the penetration depth of light becomes quite large in the transparent region ($\varepsilon_2 \sim 0$).

REFERENCES

[1] S. Adachi, *Optical Constants of Crystalline and Amorphous Semiconductors: Numerical Data and Graphical Information*, Kluwer Academic Publishers, Norwell (1999).

[2] E. D. Palik (editor), *Handbook of Optical Constants of Solids*, Academic Press, San Diego (1985).

[3] R. M. A. Azzam and N. M. Bashara, *Ellipsometry and Polarized Light*, North-Holland, Amsterdam (1977).

[4] C. M. Herzinger, B. Johs, W. A. McGahan, J. A. Woollam, and W. Paulson, Ellipsometric determination of optical constants for silicon and thermally grown silicon dioxide via a multi-sample, multi-wavelength, multi-angle investigation, *J. Appl. Phys.*, **83** (1998) 3323–3336.

[5] R. J. Archer and G. W. Gobeli, Measurement of oxygen adsorption on silicon by ellipsometry, *J. Phys. Chem. Solids*, **26** (1965) 343–351.

[6] H. G. Tompkins and W. A. McGahan, *Spectroscopic Ellipsometry and Reflectometry: A User's Guide*, John Wiley & Sons, Inc., New York (1999).

[7] M. Wakagi, H. Fujiwara, and R. W. Collins, Real time spectroscopic ellipsometry for characterization of the crystallization of amorphous silicon by thermal annealing, *Thin Solid Films*, **313–314** (1998) 464–468.

[8] J. B. Theeten, F. Hottier and J. Hallais, Ellipsometric assessment of (Ga,Al)As/GaAs epitaxial layers during their growth in an organometallic VPE system, *J. Cryst. Growth*, **46** (1979) 245–252.

[9] D. E. Aspnes, Real-time optical diagnostics for epitaxial growth, *Surf. Sci.*, **307–309** (1994) 1017–1027.

[10] K. Nakamura, K. Yoshino, S. Takeoka and I. Shimizu, Roles of atomic hydrogen in chemical annealing, *Jpn. J. Appl. Phys.*, **34** (1995) 442–449.

[11] G. E. Jellison, Jr and F. A. Modine, Parameterization of the optical functions of amorphous materials in the interband region, *Appl. Phys. Lett.*, **69** (1996) 371–373; Erratum, *Appl. Phys. Lett.*, **69** (1996) 2137.

[12] G. E. Jellison, Jr and F. A. Modine, Optical functions of silicon at elevated temperatures, *J. Appl. Phys.*, **76** (1994) 3758–3761.

[13] F. Wooten, *Optical Properties of Solids*, Academic Press, New York (1972).

[14] R. P. Feynman, R. B. Leighton, and M. Sands, *The Feynman Lectures on Physics*, vol. 1, Addison-Wesley, Massachusetts (1977).

[15] K. Kobayashi, *Physics of Light: Why Light Refracts, Reflects and Transmits*, in *Japanese*, Tokyo University Publisher, Tokyo (2002).

[16] E. Hecht, *Optics*, 4th edition, Addison Wesley, San Francisco (2002).

[17] For a review, see R. W. Collins, J. Koh, H. Fujiwara, P. I. Rovira, A. S. Ferlauto, J. A. Zapien, C. R. Wronski, and R. Messier, Recent progress in thin film growth analysis by multichannel spectroscopic ellipsometry, *Appl. Surf. Sci.*, **154–155** (2000) 217–228.

[18] P. I. Rovira and R. W. Collins, Analysis of specular and textured SnO_2:F films by high speed four-parameter Stokes vector spectroscopy, *J. Appl. Phys.*, **85** (1999) 2015–2025.

[19] H. Fujiwara and M. Kondo, Effects of carrier concentration on the dielectric function of ZnO:Ga and In_2O_3:Sn studied by spectroscopic ellipsometry: analysis of free-carrier and band-edge absorption, *Phys. Rev. B*, **71** (2005) 075109-1–10.

[20] J. Tauc, R. Grigorovici, and A. Vancu, Optical properties and electronic structure of amorphous germanium, *Phys. Stat. Sol.*, **15** (1966) 627–637.

[21] A. S. Ferlauto, G. M. Ferreira, J. M. Pearce, C. R. Wronski, R. W. Collins, X. Deng, and G. Ganguly, Analytical model for the optical functions of amorphous semiconductors from the near-infrared to ultraviolet: applications in thin film photovoltaics, *J. App. Phys.*, **92** (2002) 2424–2436.

[22] A. R. Forouhi and I. Bloomer, Optical dispersion relations for amorphous semiconductors and amorphous dielectrics, *Phys. Rev. B*, **34** (1986) 7018–7026.
[23] S. Adachi, *Optical Properties of Crystalline and Amorphous Semiconductors: Materials and Fundamental Principles*, Kluwer Academic Publishers, Norwell (1999).
[24] J. Leng, J. Opsal, H. Chu, M. Senko, and D. E. Aspnes, Analytic representations of the dielectric functions of materials for device and structural modeling, *Thin Solid Films*, **313–314** (1998) 132–136.
[25] H. R. Philipp, Optical properties of non-crystalline Si, SiO, SiO_x and SiO_2, *J. Phys. Chem. Solids*, **32** (1971) 1935–1945.
[26] D. E. Aspnes and J. B. Theeten, Dielectric function of Si-SiO_2 and Si-Si_3N_4 mixtures, *J. Appl. Phys.*, **50** (1979) 4928–4935.
[27] K. Mui and F. W. Smith, Optical dielectric function of hydrogenated amorphous silicon: Tetrahedron model and experimental results, *Phys. Rev. B*, **38** (1988) 10623–10632.
[28] C. Kittel, *Introduction to Solid State Physics*, 6th edition, John Wiley & Sons, Inc., New York (1986).
[29] For a review, see D. E. Aspnes, Optical properties of thin films, *Thin Solid Films*, **89** (1982) 249–262.
[30] G. A. Niklasson, C. G. Granqvist, and O. Hunderi, Effective medium models for the optical properties of inhomogeneous materials, *Appl. Opt.*, **20** (1981) 26–30.
[31] D. E. Aspnes, J. B. Theeten, and F. Hottier, Investigation of effective-medium models of microscopic surface roughness by spectroscopic ellipsometry, *Phys. Rev. B*, **20** (1979) 3292–3302.
[32] H. Fujiwara, J. Koh, P. I. Rovira, and R. W. Collins, Assessment of effective-medium theories in the analysis of nucleation and microscopic surface roughness evolution for semiconductor thin-films, *Phys. Rev. B*, **61** (2000) 10832–10844.
[33] J. Lee, R. W. Collins, V. S. Veerasamy, J. Robertson, Analysis of amorphous carbon thin films by spectroscopic ellipsometry, *J. Non-Cryst. Solids*, **227–230** (1998) 617–621.
[34] For a review, see R. W. Collins, I. An, H. Fujiwara, J. Lee, Y. Lu, J. Koh, and P. I. Rovira, Advances in multichannel spectroscopic ellipsometry, *Thin Solid Films*, **313–314** (1998) 18–32.
[35] D. E. Aspnes, Bounds on allowed values of the effective dielectric function of two-component composites at finite frequencies, *Phys. Rev. B*, **25** (1982) 1358–1361.
[36] S. Norrman, T. Andersson, C. G. Granqvist, and O. Hunderi, Optical properties of discontinuous gold films, *Phys. Rev. B*, **18** (1978) 674–695.
[37] J. Koh, Y. Lu, C. R. Wronski, Y. Kuang, R. W. Collins, T. T. Tsong, and Y. E. Strausser, Correlation of real time spectroellipsometry and atomic force microscopy measurements of surface roughness on amorphous semiconductor thin films, *Appl. Phys. Lett.*, **69** (1996) 1297–1299.
[38] P. Petrik, L. P. Biró, M. Fried, T. Lohner, R. Berger, C. Schneider, J. Gyulai and H. Ryssel, Comparative study of surface roughness measured on polysilicon using spectroscopic ellipsometry and atomic force microscopy, *Thin Solid Films*, **315** (1998) 186–191.
[39] H. Fujiwara, M. Kondo, and A. Matsuda, Real-time spectroscopic ellipsometry studies of the nucleation and grain growth processes in microcrystalline silicon thin films, *Phys. Rev. B*, **63** (2001) 115306-1–9.
[40] M. Erman, J. B. Theeten, P. Chambon, S. M. Kelso, and D. E. Aspnes, Optical properties and damage analysis of GaAs single crystals partly amorphized by ion implantation, *J. Appl. Phys.*, **56** (1984) 2664–2671.
[41] P. G. Snyder, J. A. Woollam, S. A. Alterovitz, and B. Johs, Modeling Al_xGa_{1-x} As optical constants as functions of composition, *J. Appl. Phys.*, **68** (1990) 5925–5926.
[42] R. H. Muller and J. C. Farmer, Macroscopic optical model for the ellipsometry of an underpotential deposit: Lead on copper and silver, *Surf. Sci.*, **135** (1983) 521–531.

[43] J. C. Farmer and R. H. Muller, Effect of Rhodamine-B on the electrodeposition of lead on copper, *J. Electrochem. Soc.*, **132** (1985) 313–319.
[44] S. J. Fang, W. Chen, T. Yamanaka, and C. R. Helms, Comparison of Si surface roughness measured by atomic force microscopy and ellipsometry, *Appl. Phys. Lett.*, **68** (1996) 2837–2839.
[45] H. Fujiwara, M. Kondo, and A. Matsuda, Interface-layer formation in microcrystalline Si:H growth on ZnO substrates studied by real-time spectroscopic ellipsometry and infrared spectroscopy, *J. Appl. Phys.*, **93** (2003) 2400–2409.
[46] M. Haverlag and G. S. Oehrlein, *In situ* ellipsometry and reflectometry during etching of patterned surfaces: experiments and simulations, *J. Vac. Sci. Technol. B*, **10** (1992) 2412–2418.
[47] N. Blayo, R. A. Cirelli, F. P. Klemens and J. T. C. Lee, Ultraviolet-visible ellipsometry for process control during the etching of submicrometer features, *J. Opt. Soc. Am. A*, **12** (1995) 591–599.
[48] H. L. Maynard, N. Layadi, and J. T. C. Lee, Plasma etching of submicron devices: *in situ* monitoring and control by multi-wavelength ellipsometry, *Thin Solid Films*, **313–314** (1998) 398–405.
[49] S.-J. Cho, P. G. Snyder, C. M. Herzinger and B. Johs, Etch depth control in bulk GaAs using patterning and real time spectroscopic ellipsometry, *J. Vac. Sci. Technol. B*, **20** (2002) 197–202.
[50] D. E. Aspnes, Spectroscopic ellipsometry of solids, in *Optical Properties of Solids: New Developments*, edited by B. O Seraphin, Chapter 15, 801–846, North-Holland, Amsterdam (1976).
[51] T. Yasuda and D. E. Aspnes, Optical-standard surfaces of single-crystal silicon for calibrating ellipsometers and reflectometers, *Appl. Opt.*, **33** (1994) 7435–7438.
[52] J. Lee and R. W. Collins, Real-time characterization of film growth on transparent substrates by rotating-compensator multichannel ellipsometry, *Appl. Opt.*, **37** (1998) 4230–4238.
[53] A. Röseler, Problem of polarization degree in spectroscopic photometric ellipsometry (polarimetry), *J. Opt. Soc. Am. A*, **9** (1992) 1124–1131.
[54] J. -Th. Zettler, Th. Trepk, L. Spanos, Y. -Z. Hu, and W. Richter, High precision UV-visible-near-IR Stokes vector spectroscopy, *Thin Solid Films*, **234** (1993) 402–407.
[55] Y. H. Yang and J. R. Abelson, Spectroscopic ellipsometry of thin films on transparent substrates: a formalism for data interpretation, *J. Vac. Sci. Technol. A*, **13** (1995) 1145–1149.
[56] K. Forcht, A. Gombert, R. Joerger, and M. Köhl, Incoherent superposition in ellipsometric measurements, *Thin Solid Films*, **302** (1997) 43–50.
[57] M. Kildemo, R. Ossikovski, and M. Stchakovsky, Measurement of the absorption edge of thick transparent substrates using the incoherent reflection model and spectroscopic UV-visible-near IR ellipsometry, *Thin Solid Films*, **313–314** (1998) 108–113.
[58] G. E. Jellison, Jr and J. W. McCamy, Sample depolarization effects from thin films of ZnS on GaAs as measured by spectroscopic ellipsometry, *Appl. Phys. Lett.*, **61** (1992) 512–514.
[59] J. Lee, P. I. Rovira, I. An, and R. W. Collins, Rotating-compensator multichannel ellipsometry: applications for real time Stokes vector spectroscopy of thin film growth, *Rev. Sci. Instrum.*, **69** (1998) 1800–1810.
[60] U. Richter, Application of the degree of polarization of film thickness gradients, *Thin Solid Films*, **313–314** (1998) 102–107.
[61] A. C. Zeppenfeld, S. L. Fiddler, W. K. Ham, B. J. Klopfenstein, and C. J. Page, Variation of layer spacing in self-assembled hafnium-1,10-decanediylbis(phosphonate) nultilayers as determined by ellipsometry and grazing angle X-ray diffraction, *J. Am. Chem. Soc.*, **116** (1994) 9158–9165.

[62] L. Sun, J. Fouere, T. Sammet, M. Hatzistergos, and H. Efstathiadis, Spectroscopic ellipsometry (SE) and grazing X-ray reflectometry (GXR) analyses on tungsten carbide films for diffusion barrier in copper metallization schemes, *Thin Solid Films*, **455–456** (2004) 519–524.
[63] W. G. Oldham, Numerical techniques for the analysis of lossy films, *Surf. Sci.*, **16** (1969) 97–103.
[64] W. H. Press, S. A. Teukolsky, W. T. Vetterling, and B. P. Flannery, *Numerical Recipes in C++: The Art of Scientific Computing*, 2nd edition, Cambridge University Press, Cambridge (2002).
[65] G. E. Jellison, Jr, Use of the biased estimator in the interpretation of spectroscopic ellipsometry data, *Appl. Opt.*, **30** (1991) 3354–3360.
[66] G. E. Jellison, Jr, Data analysis for spectroscopic ellipsometry, *Thin Solid Films*, **234** (1993) 416–422.
[67] K. Vedam, P. J. McMarr, and J. Narayan, Nondestructive depth profiling by spectroscopic ellipsometry, *Appl. Phys. Lett.*, **47** (1985) 339–341.
[68] H. Arwin and D. E. Aspnes, Unambiguous determination of thickness and dielectric function of thin films by spectroscopic ellipsometry, *Thin Solid Films*, **113** (1984) 101–113.
[69] D. E. Aspnes, A. A. Studna, and E. Kinsbron, Dielectric properties of heavily doped crystalline and amorphous silicon from 1.5 to 6.0 eV, *Phys. Rev. B*, **29** (1984) 768–779.

6 Ellipsometry of Anisotropic Materials

Recently, spectroscopic ellipsometry has been applied extensively for the study of optically anisotropic materials including organic thin films. In anisotropic materials, the optical constants vary according to the propagation direction of light, and ellipsometry data analysis using conventional Fresnel equations becomes rather difficult. In the analysis of anisotropic materials, therefore, a data analysis procedure referred to as the 4×4 matrix method is generally employed. This method provides an elegant way for describing light reflection and transmission by anisotropic samples. In this chapter, we will overview light reflection and transmission for anisotropic materials, followed by the data analysis procedure using the 4×4 matrix method. This chapter will also address the interpretation of (ψ, Δ) and measurement methods of anisotropic samples.

6.1 REFLECTION AND TRANSMISSION OF LIGHT BY ANISOTROPIC MATERIALS

Reflection and transmission of light by anisotropic materials is quite different from that by isotropic materials. In particular, the optical response of anisotropic materials changes with the orientation of samples and thus the optical constants of anisotropic materials are described by a three-dimensional index ellipsoid or dielectric tensor. Furthermore, the Jones matrix for anisotropic samples also differs from the one for isotropic samples. In this section, we will review light reflection and transmission for anisotropic media and define the index ellipsoid, dielectric tensor, and Jones matrix of anisotropic materials.

6.1.1 LIGHT PROPAGATION IN ANISOTROPIC MEDIA

As we have seen in Section 3.2.1, when electric or atomic polarization is oriented in specific directions, a material shows optical anisotropy and the optical constants

Spectroscopic Ellipsometry: Principles and Applications H. Fujiwara

vary depending on the propagation direction of light. The complex refractive indices of anisotropic materials can be expressed by (N_x, N_y, N_z), where N_x, N_y, and N_z represent the complex refractive indices along the x, y, and z axes, and are given by $N_x = n_x - ik_x$, $N_y = n_y - ik_y$, and $N_z = n_z - ik_z$, respectively. Traditionally, we choose the x, y, and z axes so that the refractive indices become $n_x < n_y < n_z$. According to the complex refractive indices (N_x, N_y, N_z), anisotropic materials can broadly be classified into two types; i.e., uniaxial and biaxial materials. In uniaxial materials, two complex refractive indices are the same and only one direction has a different complex refractive index (i.e., $N_x = N_y \neq N_z$). In biaxial materials, on the other hand, all the complex refractive indices are different (i.e., $N_x \neq N_y \neq N_z$). In order to simplify descriptions, we will assume $k_x = k_y = k_z = 0$ in Section 6.1.

Figure 6.1 shows the propagation of light in (a) uniaxial material $(n_x = n_y < n_z)$ and (b) biaxial material $(n_x < n_y < n_z)$. In anisotropic materials, the propagation speed of light varies with the oscillating direction of the electric field. When the light is traveling along the x axis, for example, the polarized light whose oscillatory

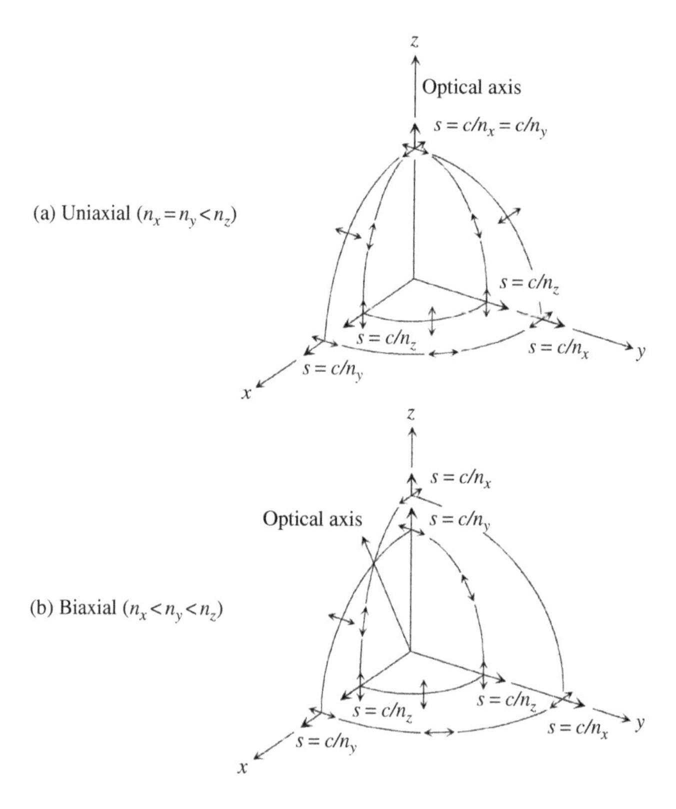

Figure 6.1 Propagation of light in (a) uniaxial material $(n_x = n_y < n_z)$ and (b) biaxial material $(n_x < n_y < n_z)$. In the figures, c represents the speed of light, and s shows the propagation speed of light in anisotropic media.

direction is parallel to the y axis propagates with a speed of $s = c/n_y$ [Eq. (2.28)], while polarized light parallel to the z axis travels at $s = c/n_z$. Notice that $c/n_y > c/n_z$ since $n_y < n_z$ is assumed. In the uniaxial material shown in Fig. 6.1(a), the propagation of light along the z axis is independent of the oscillatory direction and this z axis defines the optical axis. As confirmed from Fig. 6.1(a), the propagation speed is constant when the oscillating direction is perpendicular to the optical axis and such light is particularly called the ordinary ray. On the other hand, the light whose oscillatory direction is parallel to the optical axis is called the extraordinary ray and its propagation speed varies with the direction. When the refractive indices of the ordinary and extraordinary rays are given by n_o and n_e, respectively, it follows that $n_x = n_y = n_o$ and $n_z = n_e$ in Fig. 6.1(a). It can be seen from Fig. 6.1(b) that the propagation of light in a biaxial material is essentially similar. In a biaxial material, however, the propagation of light along the z axis is no longer constant and the optical axis lies at the intersection in the x–z plane.

Uniaxial materials can be divided further into two classes: positive and negative uniaxial materials. A uniaxial material is said to be positive uniaxial when $n_e - n_o > 0$, while it is said to be negative uniaxial when $n_e - n_o < 0$. Figure 6.2 shows the propagation of light in (a) positive uniaxial material and (b) negative uniaxial material. In the figures, the optical axes are in the z direction, and Fig. 6.2(a) corresponds to the cross-section of the x–z plane in Fig. 6.1(a). It is evident that $n_e - n_o > 0$ in the positive uniaxial material and $n_e - n_o < 0$ in the negative uniaxial material, since $n_o = n_x = n_y$ and $n_e = n_z$. Calcite described in Section 3.2.1 is a negative uniaxial material and the light propagation along the x axis in Fig. 6.2(b) is identical to the one illustrated in Fig. 3.4(b).

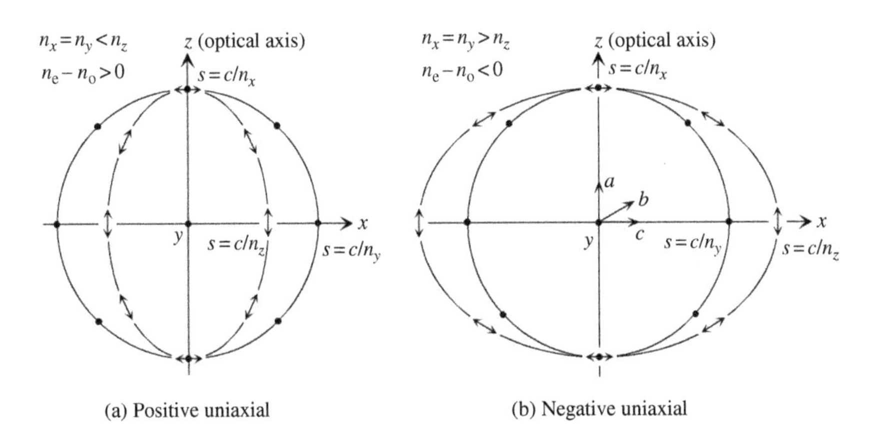

Figure 6.2 Propagation of light in (a) positive uniaxial material ($n_x = n_y < n_z$) and (b) negative uniaxial material ($n_x = n_y > n_z$). In (a), the cross-section of the x–z plane in Fig. 6.1(a) is shown.

Now suppose that polarized light illuminates a uniaxial material and electric dipole radiation occurs at the air/sample interface. In the case of isotropic materials, wavelets radiated from electric dipoles are represented by semi-circles (see Fig. 2.10). In anisotropic samples, however, the shape of the wavelets may not be circlar due to the variation of refractive index with direction [1]. Figure 6.3 illustrates the wavelets radiated in the directions a, b, and c in Fig. 6.2(b). The propagation of the ordinary ray is illustrated in Figs. 6.3(a)–(c), while that of the extraordinary ray is shown in Figs. 6.3(d)–(f). Each line in Fig. 6.3 represents a wavelet that has a constant phase. In Figs. 6.3(d)–(f), however, electric dipole radiation in the oscillatory direction is rather hypothetical, since no dipole radiation occurs along the vibrating direction of dipoles (see Fig. 2.9). It is clear from Figs. 6.3(a)–(c) that the ordinary ray travels at a constant speed, independent of the propagation direction. As we have seen in Figs. 6.1(a) and 6.2, when light is traveling along the optical axis, the ordinary and extraordinary rays show identical behavior [Figs. 6.3(a) and (d)]. However, if light is illuminated at tilted angles relative to the optical axis, the extraordinary ray shows unique behavior [Fig. 6.3(e)]. In particular, the extraordinary ray is not parallel to the incident light and propagates inside the uniaxial material with an inclined angle. On the other hand, when light propagates perpendicular to the optical axis, the propagation speed varies depending on the polarization of light [Figs. 6.3(c) and (f)].

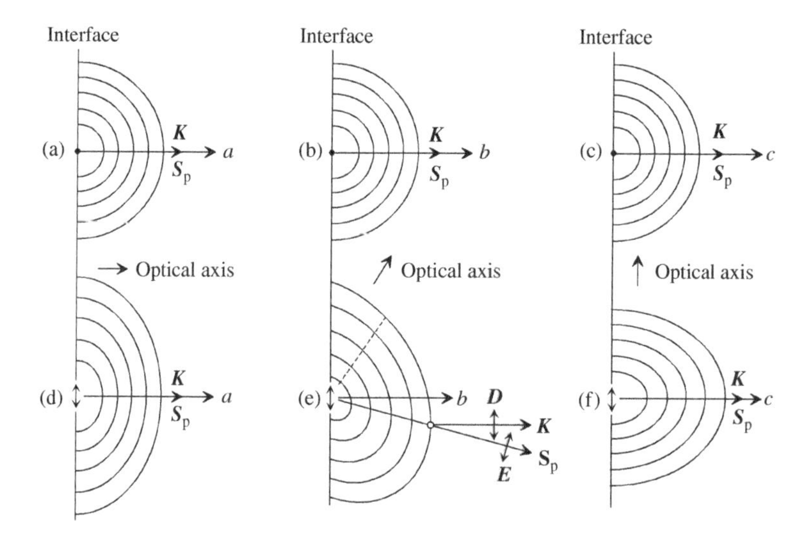

Figure 6.3 Wavelets radiated toward the directions of a, b, and c indicated in Fig. 6.2(b): (a)–(c) ordinary ray and (d)–(f) extraordinary ray. $\boldsymbol{K}$ and $\boldsymbol{S}_\mathrm{p}$ indicate the propagation and pointing vectors, respectively. In (e), $\boldsymbol{D}$ represents the electric displacement.

As shown in Fig. 6.3(e), the propagation direction of the extraordinary ray is described by the pointing vector $\boldsymbol{S}_\mathrm{p}$, and the electric field $\boldsymbol{E}$ is perpendicular to $\boldsymbol{S}_\mathrm{p}$. The pointing vector represents the energy flow in the direction of light propagation and is expressed by $\boldsymbol{S}_\mathrm{p} = \boldsymbol{E} \times \boldsymbol{H}$, where $\boldsymbol{H}$ is the magnetic field [2]. Notice from Fig. 6.3(e) that the direction of $\boldsymbol{S}_\mathrm{p}$ is determined by a point where the propagation speed toward the right-hand side is maximized. In anisotropic materials, the propagation vector $\boldsymbol{K}(|\boldsymbol{K}| = 2\pi n/\lambda)$ is in the direction perpendicular to the tangent plane of the ellipsoid. Here, we introduce a new vector known as the electric displacement $\boldsymbol{D}$. Recall from Eq. (2.41) that $D = \varepsilon_p E$. By using $\varepsilon_p = \varepsilon_0 \varepsilon$ [Eq. (2.43)], we get

$$\boldsymbol{D} = \varepsilon_0 \varepsilon \boldsymbol{E} \tag{6.1}$$

In Eq. (6.1), ε represents a tensor referred to as the dielectric tensor (see Section 6.1.3). As shown in Fig. 6.3(e), $\boldsymbol{D}$ is perpendicular to $\boldsymbol{K}$ and is parallel to the tangent plane of the constant phase surface. In isotropic materials, $\boldsymbol{D}$ and $\boldsymbol{E}$ (or $\boldsymbol{S}_p$ and $\boldsymbol{K}$) are always parallel and ε is expressed by a single value or scalar quantity (i.e., dielectric constant). However, $\boldsymbol{D}$ and $\boldsymbol{E}$ are not necessarily parallel in anisotropic materials, and thus ε is described by a tensor. As confirmed from Fig. 6.3, $\boldsymbol{S}_p$ and $\boldsymbol{K}$ of the ordinary ray are parallel, but those of the extraordinary ray become parallel only when polarization is perpendicular or parallel to the optical axis. Obviously, the extraordinary ray shown in Fig. 6.3(e) does not satisfy Snell's law. In anisotropic materials, therefore, light reflection and transmission are described by $\boldsymbol{E}$ and $\boldsymbol{S}_p$, while the propagation of light is expressed by $\boldsymbol{D}$ and $\boldsymbol{K}$.

6.1.2 INDEX ELLIPSOID

The propagation of light in anisotropic media can be described more systematically from an index ellipsoid. In the ellipsometry of anisotropic materials, we generally characterize this index ellipsoid. Now let us consider the energy density of light traveling inside an anisotropic medium. If we apply electromagnetic theory, the energy density of the electric field $\boldsymbol{E}$ is described as follows [2]:

$$U_E = \frac{1}{2} \boldsymbol{E} \cdot \boldsymbol{D} \tag{6.2}$$

Using $\boldsymbol{E} = \boldsymbol{D}/(\varepsilon_0 \varepsilon)$ [Eq. (6.1)], we get the following equation [3]:

$$\begin{aligned} U_E &= \frac{1}{2\varepsilon_0 \varepsilon} \boldsymbol{D} \cdot \boldsymbol{D} \\ &= \frac{1}{2\varepsilon_0} \left(\frac{D_x^2}{\varepsilon_x} + \frac{D_y^2}{\varepsilon_y} + \frac{D_z^2}{\varepsilon_z} \right) \end{aligned} \tag{6.3}$$

By setting $x = D_x/\sqrt{2\varepsilon_0 U_E}$, $y = D_y/\sqrt{2\varepsilon_0 U_E}$, and $z = D_z/\sqrt{2\varepsilon_0 U_E}$, we obtain a well-known formula [3,4]:

$$\frac{x^2}{\varepsilon_x} + \frac{y^2}{\varepsilon_y} + \frac{z^2}{\varepsilon_z} = 1 \tag{6.4}$$

Clearly, Eq. (6.4) represents the equation for a three-dimensional ellipsoid in (x, y, z) coordinates. Since $\varepsilon = n^2$ $(k = 0)$, the semiaxes of this ellipsoid correspond to the refractive indices along the x, y, and z axes. Thus, this ellipsoid is referred to as the index ellipsoid. In Eq. (6.4), the x, y, and z axes are generally called the principal axes, and $(\varepsilon_x, \varepsilon_y, \varepsilon_z)$ are the principal dielectric constants, while (n_x, n_y, n_z) are known as the principal indices of refraction.

Figure 6.4 shows the index ellipsoids of (a) isotropic material $(n_x = n_y = n_z)$, (b) uniaxial material $(n_x = n_y < n_z)$ and (c) biaxial material $(n_x < n_y < n_z)$. As shown in Fig. 6.4(a), the index ellipsoid of isotropic materials becomes a sphere since the principal indices of refraction are independent of the propagation direction of light. The index ellipsoid of Fig. 6.4(b) represents the one for the positive uniaxial material shown in Fig. 6.2(a). Since $n_x = n_y < n_z$ in Fig. 6.4(b), the semiaxis of n_z is longer than that of n_x. In the case of biaxial materials, all the semiaxes have different lengths. If we use the index ellipsoid, the propagation of the polarized light in anisotropic media can be described more easily (Fig. 6.5). In Fig. 6.5, light

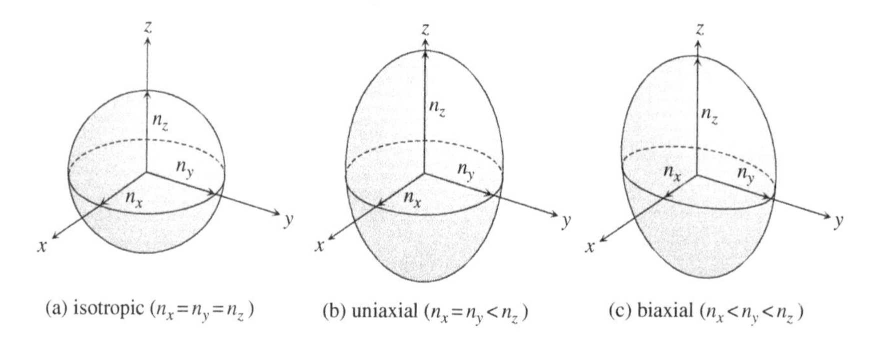

Figure 6.4 Index ellipsoids of (a) isotropic material $(n_x = n_y = n_z)$, (b) uniaxial material $(n_x = n_y < n_z)$, and (c) biaxial material $(n_x < n_y < n_z)$.

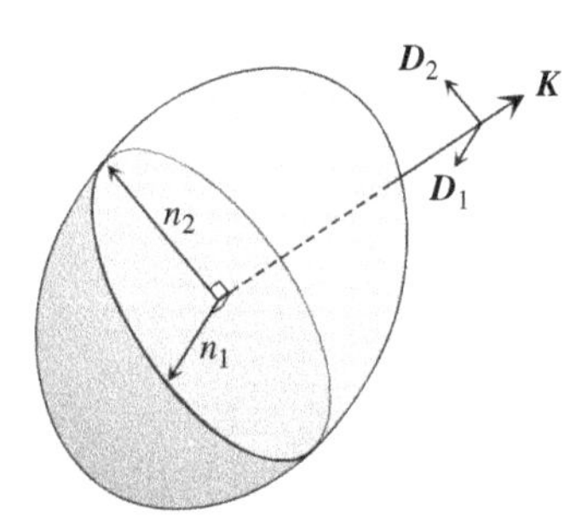

Figure 6.5 Light waves traveling inside an anisotropic medium in the direction of $\boldsymbol{K}$. From the cross-section of the index ellipsoid, the refractive indices for the electric displacements $\boldsymbol{D}_1$ and $\boldsymbol{D}_2$ are given by n_1 and n_2, respectively.

waves whose electric displacements are oriented to $\boldsymbol{D}_1$ and $\boldsymbol{D}_2$ are traveling in the direction of $\boldsymbol{K}$. As we have seen in Fig. 6.3(e), $\boldsymbol{D}$ is perpendicular to $\boldsymbol{K}$ but may not be parallel to $\boldsymbol{E}$. The refractive indices for $\boldsymbol{D}_1$ and $\boldsymbol{D}_2$ can be determined from the cross-section of the index ellipsoid [3,4]. This cross-section is chosen so that the plane includes the origin within the plane and is normal to $\boldsymbol{K}$. Recall from Eq. (6.4) that $D_{x,y,z}$ are parallel to $n_{x,y,z}$ ($\varepsilon_{x,y,z}$). In Fig. 6.5, therefore, the refractive indices for $\boldsymbol{D}_1$ and $\boldsymbol{D}_2$ are given by n_1 and n_2, respectively. For the arbitrary direction of $\boldsymbol{K}$, we can always find two refractive indices in orthogonal directions. It can be seen from Fig. 6.4(a) that the cross-section of isotropic samples is a complete circle. This confirms that the refractive index is independent of the propagation direction in isotropic materials. In the case of the uniaxial material shown in Fig. 6.4(b), the cross-section of the x–y plane is a circle. Thus, when $\boldsymbol{K}$ is parallel to the z axis, the refractive index is constant, as we have seen in Fig. 6.1(a). Notice that one of the semiaxes of the cross-section is always $n_o = n_x = n_y$ in a uniaxial material. From the above discussion, it can be understood that Figs. 6.1 and 6.4 show identical phenomena with respect to light propagation in anisotropic media.

6.1.3 DIELECTRIC TENSOR

The optical constants of anisotropic materials are expressed by the dielectric tensor written as

$$\varepsilon = \begin{bmatrix} \varepsilon_x & 0 & 0 \\ 0 & \varepsilon_y & 0 \\ 0 & 0 & \varepsilon_z \end{bmatrix} = \begin{bmatrix} n_x^2 & 0 & 0 \\ 0 & n_y^2 & 0 \\ 0 & 0 & n_z^2 \end{bmatrix} \tag{6.5}$$

Thus, the dielectric tensor is simply described by the principal dielectric constants (or principal indices of refraction). In the case of isotropic materials ($\varepsilon_x = \varepsilon_y = \varepsilon_z$), the dielectric tensor is given by

$$\varepsilon = \begin{bmatrix} \varepsilon_x & 0 & 0 \\ 0 & \varepsilon_x & 0 \\ 0 & 0 & \varepsilon_x \end{bmatrix} = \varepsilon_x \begin{bmatrix} 1 & 0 & 0 \\ 0 & 1 & 0 \\ 0 & 0 & 1 \end{bmatrix} = \varepsilon_x \boldsymbol{I} \tag{6.6}$$

where $\boldsymbol{I}$ is the identity matrix.

So far, we have treated the cases where the principal axes of index ellipsoids coincide with the (x, y, z) coordinates. In this condition, the off-diagonal elements of a dielectric tensor become zero, as shown in Eq. (6.5). In some anisotropic samples, however, the principal axes may not be parallel to the (x, y, z) coordinates. In this case, it is mathematically convenient to express the dielectric tensor by coordinate transformation. Figure 6.6 shows the transformation of the coordinate system from (x, y, z) to (α, β, γ) using the Euler angles $(\phi_E, \theta_E, \psi_E)$. The three Euler angles represent the rotation angles of the coordinates, and the rotation matrices are

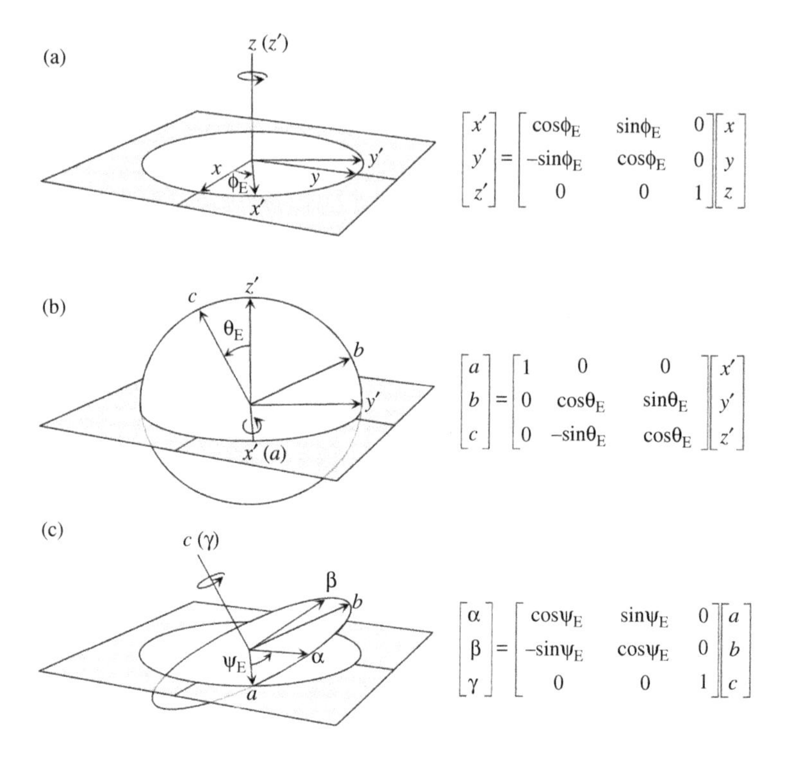

Figure 6.6 Rotation of coordinate systems using the three Euler angles: (a) ϕ_E, (b) θ_E, and (c) ψ_E.

also shown in Fig. 6.6. In this coordinate transformation, we first rotate the x–y plane around the z axis with an angle of ϕ_E to convert the (x, y, z) coordinates to the (x', y', z') coordinates [Fig. 6.6(a)]. Notice that the coordinate rotation is counterclockwise and is exactly the same as the one shown in Fig. 3.11, except for the presence of the z axis. Thus, the rotation matrix of Fig. 6.6(a) is quite similar to Eq. (3.17). In the rotation matrix shown in Fig. 6.6(a), the matrix elements of the x–z and y–z components are zero, since there is no change with respect to the z axis. Secondly, the (x', y', z') coordinates are rotated with an angle of θ_E around the x' axis to form the (a, b, c) coordinates [Fig. 6.6(b)]. In this case, the matrix elements of the x–y and x–z components become zero. Finally, the (a, b, c) coordinates are rotated around the c axis. Since the c axis originally corresponds to the z axis, the rotation matrices of Figs. 6.6(a) and (c) are almost the same. If we combine the rotation matrices shown in Fig. 6.6, the coordinate transformation in Fig. 6.6 is expressed by

$$\boldsymbol{A} = \begin{bmatrix} \cos\psi_E & \sin\psi_E & 0 \\ -\sin\psi_E & \cos\psi_E & 0 \\ 0 & 0 & 1 \end{bmatrix} \begin{bmatrix} 1 & 0 & 0 \\ 0 & \cos\theta_E & \sin\theta_E \\ 0 & -\sin\theta_E & \cos\theta_E \end{bmatrix} \begin{bmatrix} \cos\phi_E & \sin\phi_E & 0 \\ -\sin\phi_E & \cos\phi_E & 0 \\ 0 & 0 & 1 \end{bmatrix} \quad (6.7)$$

The rotation matrix $\boldsymbol{A}$ represents the coordinate transformation from the (x, y, z) coordinates to the (α, β, γ) coordinates (Fig. 6.7). However, the coordinate system of ellipsometry measurement is basically defined by the (x, y, z) coordinates. Accordingly, what we need is the dielectric tensor in the (x, y, z) coordinate system. As shown in Fig. 6.7, the coordinate transformation in the opposite direction is expressed by the inverse transformation matrix $\boldsymbol{A}^{-1}$ (see Section 3.3.2). In general, $\boldsymbol{A}^{-1} = \boldsymbol{A}^{\mathrm{T}}$ holds for coordinate conversion, where the symbol T denotes the transposed matrix $(a_{ij} = a^{\mathrm{T}}_{ji})$. Thus, $\boldsymbol{B} = \boldsymbol{A}^{\mathrm{T}} = \boldsymbol{A}^{-1}$ can be obtained directly from Eq. (6.7) [5,6]:

$$\boldsymbol{B} = \begin{bmatrix} \cos\phi_{\mathrm{E}}\cos\psi_{\mathrm{E}} - \sin\phi_{\mathrm{E}}\cos\theta_{\mathrm{E}}\sin\psi_{\mathrm{E}} & -\cos\phi_{\mathrm{E}}\sin\psi_{\mathrm{E}} - \sin\phi_{\mathrm{E}}\cos\theta_{\mathrm{E}}\cos\psi_{\mathrm{E}} & \sin\phi_{\mathrm{E}}\sin\theta_{\mathrm{E}} \\ \sin\phi_{\mathrm{E}}\cos\psi_{\mathrm{E}} + \cos\phi_{\mathrm{E}}\cos\theta_{\mathrm{E}}\sin\psi_{\mathrm{E}} & -\sin\phi_{\mathrm{E}}\sin\psi_{\mathrm{E}} + \cos\phi_{\mathrm{E}}\cos\theta_{\mathrm{E}}\cos\psi_{\mathrm{E}} & -\cos\phi_{\mathrm{E}}\sin\theta_{\mathrm{E}} \\ \sin\theta_{\mathrm{E}}\sin\psi_{\mathrm{E}} & \sin\theta_{\mathrm{E}}\cos\psi_{\mathrm{E}} & \cos\theta_{\mathrm{E}} \end{bmatrix} \tag{6.8}$$

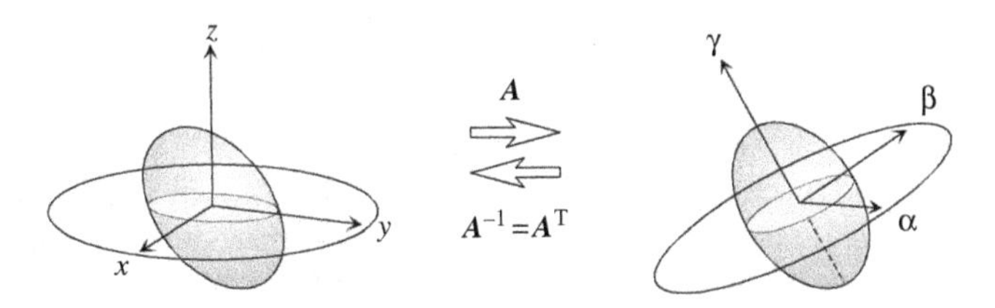

Figure 6.7 Transformation of the (x, y, z) coordinates into the (α, β, γ) coordinates. In the figure, $\boldsymbol{A}$ represents the rotation matrix calculated from Eq. (6.7). The matrix $\boldsymbol{A}^{-1} = \boldsymbol{A}^{\mathrm{T}}$ transforms the coordinates from (α, β, γ) to (x, y, z).

On the other hand, the rotation of a tensor $\boldsymbol{Q}$ using a rotation matrix $\boldsymbol{M}$ is expressed by $\boldsymbol{MQM}^{-1}$ [7]. By applying this, we obtain the dielectric tensor in the (x, y, z) coordinate system as follows [5,6]:

$$\begin{bmatrix} \varepsilon_{xx} & \varepsilon_{xy} & \varepsilon_{xz} \\ \varepsilon_{yx} & \varepsilon_{yy} & \varepsilon_{yz} \\ \varepsilon_{zx} & \varepsilon_{zy} & \varepsilon_{zz} \end{bmatrix} = \boldsymbol{B} \begin{bmatrix} \varepsilon_{\alpha} & 0 & 0 \\ 0 & \varepsilon_{\beta} & 0 \\ 0 & 0 & \varepsilon_{\gamma} \end{bmatrix} \boldsymbol{B}^{\mathrm{T}} \tag{6.9}$$

where $(\varepsilon_{\alpha}, \varepsilon_{\beta}, \varepsilon_{\gamma})$ represent the principal dielectric constants in the (α, β, γ) coordinate system. Notice from Eq. (6.9) that $\boldsymbol{B}^{-1} = \boldsymbol{B}^{\mathrm{T}}$. If $\phi_E = \theta_E = \psi_E = 0°$, we obtain $\varepsilon_{xx} = \varepsilon_{\alpha}$, $\varepsilon_{yy} = \varepsilon_{\beta}$, and $\varepsilon_{zz} = \varepsilon_{\gamma}$, and all the other matrix elements become zero. It should be noted that the dielectric tensor in Eq. (6.9) is symmetric (i.e., $\varepsilon_{ij} = \varepsilon_{ji}$) and has only six independent parameters [4].

6.1.4 JONES MATRIX OF ANISOTROPIC SAMPLES

As we have seen in Eq. (4.13), the off-diagonal elements of the Jones matrix are no longer zero in anisotropic samples. Here, we will see why the off-diagonal elements have finite values in the case of anisotropic samples. First, let us consider light transmission through a transparent anisotropic substrate (Fig. 6.8). In Fig. 6.8, the anisotropic substrate is a retarder ($\lambda/4$ plate) and the optical axis of the retarder is rotated by 45° relative to the E_x axis. It should be emphasized that the light propagation shown in Fig. 6.8 is exactly the same as the one shown in Fig. 3.8, except for the orientations of the E_x and E_y axes. The incident light in Fig. 6.8 is linear polarization parallel to the E_y axis and has no E_x component. In Fig. 6.8, therefore, linear polarization in the E_y direction is transformed into circular polarization. In other words, *the retarder generates the E_x component even though there is no E_x component in the incident light.* From the Jones matrices shown in Table 3.2, the transformation of the polarized light in Fig. 6.8 is expressed as follows:

$$\begin{bmatrix} E_{tx} \\ E_{ty} \end{bmatrix} = \boldsymbol{R}(-C)\boldsymbol{C}\boldsymbol{R}(C) \begin{bmatrix} E_{ix} \\ E_{iy} \end{bmatrix}$$
$$= \begin{bmatrix} \cos^2 C + \sin^2 C \exp(-i\delta) & \cos C \sin C[1-\exp(-i\delta)] \\ \cos C \sin C[1-\exp(-i\delta)] & \cos^2 C \exp(-i\delta) + \sin^2 C \end{bmatrix} \begin{bmatrix} E_{ix} \\ E_{iy} \end{bmatrix} \quad (6.10)$$

where $E_{tx}(E_{ix})$ and $E_{ty}(E_{iy})$ show the electric fields of the transmitted (incident) waves in the x and y directions, respectively (see Fig. 6.8). In Eq. (6.10), C and δ are the rotation angle and phase shift of the retarder, respectively. Notice that the matrix calculated from $\boldsymbol{R}(-C)\boldsymbol{C}\boldsymbol{R}(C)$ represents the Jones matrix of the anisotropic substrate. Thus, the off-diagonal elements of the Jones matrix are not zero. In particular, the values of the off-diagonal elements are maximized at $C = 45°$, since the term $\cos C \sin C$ shows a maximum value at this angle. If we substitute $E_{ix} = 0$, $E_{iy} = 1$, $C = 45°$, and $\delta = 90°$ into Eq. (6.10), we obtain $|E_{tx}| = |E_{ty}| = 1/\sqrt{2}$. This result shows clearly that E_{tx} is generated by E_{iy}.

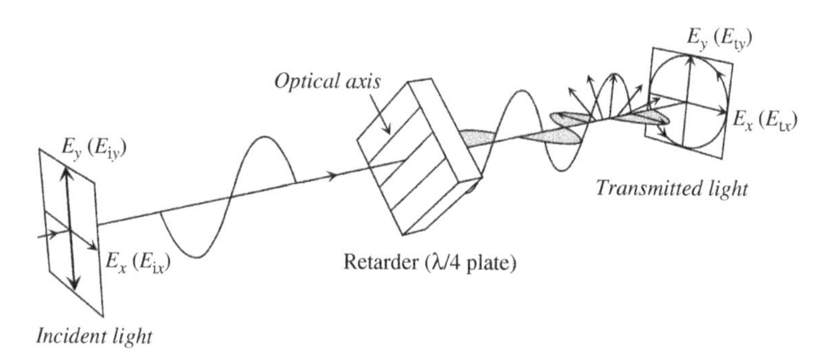

Figure 6.8 Light wave transmitting through a retarder ($\lambda/4$ plate). The optical axis of the retarder is rotated by 45° relative to the E_x axis.

Upon light reflection on anisotropic samples, similar phenomenon occurs. Figure 6.9 illustrates the light reflection on an anisotropic substrate for

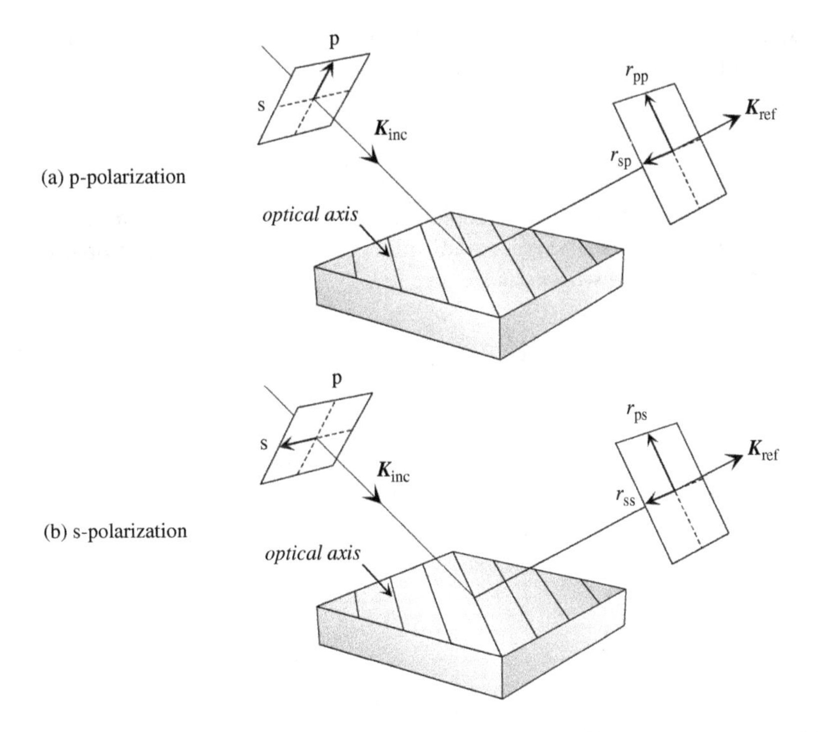

Figure 6.9 Light reflection on an anisotropic substrate for (a) p-polarization and (b) s-polarization. In the figures, the optical axis of the anisotropic substrate lies parallel to the substrate surface and is rotated by 45° from the plane of incidence. $\boldsymbol{K}_{inc}$ and $\boldsymbol{K}_{ref}$ denote the propagation vectors of the incident and reflected waves, respectively.

(a) p-polarization and (b) s-polarization. In the figures, $\boldsymbol{K}_{inc}$ and $\boldsymbol{K}_{ref}$ denote the propagation vectors of the incident and reflected waves, respectively. Here, the optical axis of the anisotropic substrate is parallel to the substrate surface and is rotated, similar to Fig. 6.8. In the case of Fig. 6.9(a), the incident p-polarization generates both p- and s-polarized components. r_{pp} in Fig. 6.9(a) represents the amplitude reflection coefficient of reflected p-polarized light, while r_{sp} shows that of s-polarized light induced by the incident p-polarization. In Fig. 6.9(b), the p-polarized component is generated by incident s-polarization with an amplitude reflection coefficient of r_{ps}. It can be seen from Fig. 6.9 that r_{pp} and r_{ss} correspond to r_p and r_s of isotropic samples. If the p- and s-polarizations shown in Fig. 6.9 illuminate the sample simultaneously, the reflected p- and s-polarizations are expressed by

$$E_{rp} = r_{pp}E_{ip} + r_{ps}E_{is} \tag{6.11a}$$

$$E_{rs} = r_{sp}E_{ip} + r_{ss}E_{is} \tag{6.11b}$$

In Eq. (6.11), the notation of E follows the definitions of Fig. 2.15. In matrix form, Eq. (6.11) is described as follows [8,9]:

$$\begin{bmatrix} E_{rp} \\ E_{rs} \end{bmatrix} = \begin{bmatrix} r_{pp} & r_{ps} \\ r_{sp} & r_{ss} \end{bmatrix} \begin{bmatrix} E_{ip} \\ E_{is} \end{bmatrix} \tag{6.12}$$

Thus, the Jones matrix of anisotropic samples is given by the matrix shown in Eq. (4.13). Eq. (6.12) can be compared with Eq. (4.8) obtained for isotropic samples. Eq. (6.10) can also be written as follows [8,9]:

$$\begin{bmatrix} E_{tp} \\ E_{ts} \end{bmatrix} = \begin{bmatrix} t_{pp} & t_{ps} \\ t_{sp} & t_{ss} \end{bmatrix} \begin{bmatrix} E_{ip} \\ E_{is} \end{bmatrix} \tag{6.13}$$

where E_{tp} and E_{ts} show the electric fields of transmitted p- and s-polarizations, respectively, and t represents the amplitude transmission coefficient. In general, the Jones matrix of anisotropic samples is normalized by r_{ss} [10]:

$$\boldsymbol{S}_{ani} = \begin{bmatrix} r_{pp} & r_{ps} \\ r_{sp} & r_{ss} \end{bmatrix} = r_{ss} \begin{bmatrix} r_{pp}/r_{ss} & r_{ps}/r_{ss} \\ r_{sp}/r_{ss} & 1 \end{bmatrix} = r_{ss} \begin{bmatrix} \rho_{pp} & \rho_{ps} \\ \rho_{sp} & 1 \end{bmatrix} \tag{6.14}$$

where

$$\rho_{pp} = r_{pp}/r_{ss} = \tan\psi_{pp}\exp(i\Delta_{pp}) \tag{6.15a}$$

$$\rho_{ps} = r_{ps}/r_{ss} = \tan\psi_{ps}\exp(i\Delta_{ps}) \tag{6.15b}$$

$$\rho_{sp} = r_{sp}/r_{ss} = \tan\psi_{sp}\exp(i\Delta_{sp}) \tag{6.15c}$$

Accordingly, the Jones matrix of anisotropic samples is defined by six independent parameters (ψ_{pp}, ψ_{ps}, ψ_{sp}, Δ_{pp}, Δ_{ps}, Δ_{sp}), instead of the two parameters for isotropic samples (i.e., ψ, Δ).

Recently, the Jones matrix in which the positions of r_{ps} and r_{sp} are reversed in Eq. (6.12) has also been used [11]. When measurement results are compared, therefore, one needs to confirm which definition is employed. By following the original description [8,9], we will use the notation of Eqs. (4.13) and (6.14) throughout this book. In addition, with respect to ρ_{sp} in Eqs. (6.14) and (6.15c), a different definition has also been proposed [11]:

$$\rho'_{sp} = r_{sp}/r_{pp} = \tan\psi'_{sp}\exp(i\Delta'_{sp}) \tag{6.16}$$

Notice that r_{sp} is normalized by r_{pp}, instead of r_{ss}, in Eq. (6.16). In this case, ρ_{sp} in Eq. (6.14) is described by

$$\rho_{sp} = r_{sp}/r_{ss} = (r_{sp}/r_{pp})(r_{pp}/r_{ss}) = \rho'_{sp}\rho_{pp} \tag{6.17}$$

As confirmed from Eq. (6.10), the off-diagonal elements of the Jones matrix are often equal (i.e., $\rho_{ps} = \rho_{sp}$). For the definition of Eq. (6.16), however, we always observe $\rho_{ps} \neq \rho'_{sp}$. In addition, the representation of the Jones matrix becomes slightly complicated if Eq. (6.17) is applied. Accordingly, the notation of Eq. (6.14) will be used in this book.

In some special orientations, the off-diagonal elements of the Jones matrix become zero even if samples are optically anisotropic. For example, when the optical axis of the retarder shown in Fig. 6.8 is parallel or perpendicular to E_x (i.e., $C = 0°$ or $90°$), the off-diagonal elements of the Jones matrix vanish and the Jones matrix becomes diagonal ($r_{ps} = r_{sp} = 0$), as confirmed from Eq. (6.10). Similarly, the Jones matrix for light reflection becomes diagonal when the optical axis is parallel or perpendicular to the plane of incidence. Figure 6.10 summarizes the orientations of an index ellipsoid when the Jones matrix is diagonal. In these figures, the index ellipsoid of a positive uniaxial material ($n_x = n_y < n_z$) is shown and the index ellipsoid is rotated using the Euler angles (ϕ_E, θ_E, ψ_E). n_o and n_e in Fig. 6.10 represent the refractive indices for the ordinary and extraordinary rays, respectively. In ellipsometry, the x axis is chosen to be parallel to the plane of incidence. In addition, in order to match the direction of the y axis with that of

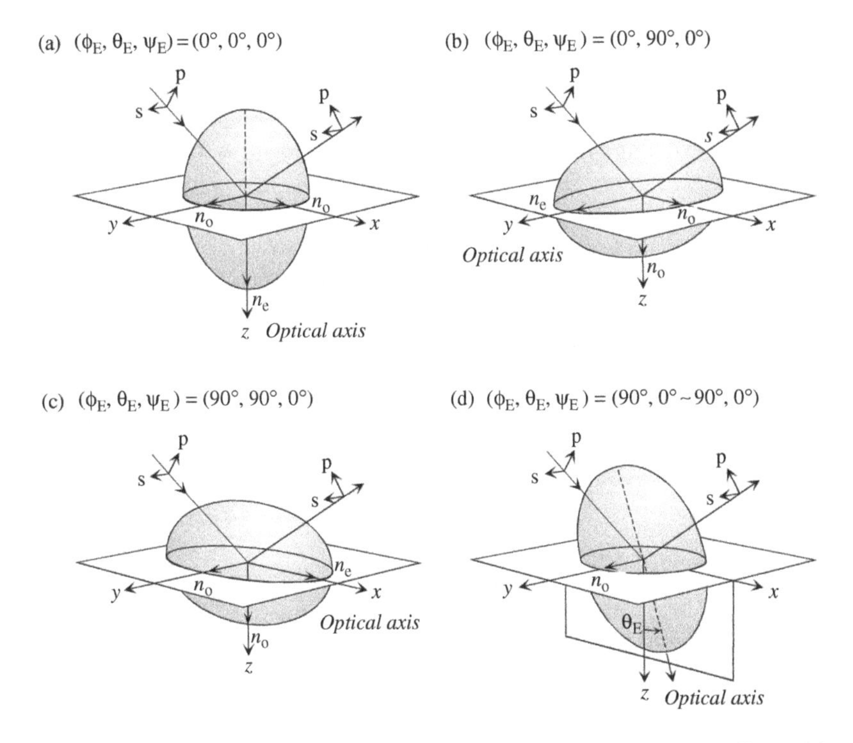

Figure 6.10 Orientations of the index ellipsoid when the Jones matrix becomes diagonal ($r_{ps} = r_{sp} = 0$). In these figures, the index ellipsoid of a positive uniaxial material is rotated using the Euler angles (ϕ_E, θ_E, ψ_E). n_o and n_e represent the refractive indices of the ordinary and extraordinary rays, respectively.

s-polarization defined in Fig. 2.15, the z axis is selected as downward, as shown in Fig. 6.10. The Jones matrix is diagonal when the principal axes of the index ellipsoid coincide with the (x, y, z) coordinates. If the optical axis lies parallel to the plane of incidence [Fig. 6.10(d)], the Jones matrix is still diagonal. As we will see in the next section, the diagonal elements of the Jones matrix can be calculated rather easily in the case of Figs. 6.10(a)–(c).

6.2 FRESNEL EQUATIONS FOR ANISOTROPIC MATERIALS

When the principal axes of an index ellipsoid coincide with the (x, y, z) coordinates, the light reflection can be described from the Fresnel equations. Thus, if the optical axis of an anisotropic sample is parallel or perpendicular to the sample surface, we may align the sample so that these equations can be used. This section will explain the Fresnel equations that can be applied for anisotropic substrates and anisotropic thin films formed on isotropic substrates.

6.2.1 ANISOTROPIC SUBSTRATE

In this section, we will treat the simplest case where a sample is composed of an anisotropic substrate only. Figure 6.11 illustrates the light reflection on a positive uniaxial substrate. Here, the ambient is air and the principal axes of the index ellipsoid are parallel to the (x, y, z) coordinates. In Fig. 6.11, $\boldsymbol{K}_{tp}$ and $\boldsymbol{K}_{ts}$ show the propagation vectors of transmitted p- and s-polarizations, respectively. $\boldsymbol{K}_{inc}$, $\boldsymbol{K}_{ref}$, $\boldsymbol{K}_{tp}$, and $\boldsymbol{K}_{ts}$ are parallel to the plane of incidence, and the following equations hold [2]:

$$K_{inc} \sin\theta_i = K_{ref} \sin\theta_r = K_{tp} \sin\theta_{tp} = K_{ts} \sin\theta_{ts} \tag{6.18}$$

where θ_{tp} and θ_{ts} show the transmission angles of p- and s-polarizations, respectively (see Fig. 6.11). In anisotropic materials, however, K_{tp} and K_{ts} generally vary with K_{inc}. In order to determine θ_{tp} and θ_{ts}, graphical methods can be applied [2,12]. Since air is an isotropic ambient, we obtain $\theta_i = \theta_r$ from Eq. (6.18). Thus, the law of reflection [Eq. (2.51)] is also valid for light reflection on anisotropic samples. Now recall from Fig. 6.5 that the refractive indices of an anisotropic sample are determined from the cross-section of the index ellipsoid. In particular, the refractive indices for the transmitted p- and s-polarizations are obtained from the directions of $\boldsymbol{D}_{tp}$ and $\boldsymbol{D}_{ts}$, respectively. It is evident that the refractive index of the s-polarized light (n_s) is given by $n_s = n_o$. In the case of Fig. 6.11, the refractive index for the p-polarization (n_p) becomes larger than n_s, and thus we observe $\theta_{tp} < \theta_{ts}$.

From Fig. 6.2(a), it can be seen that the transmitted p- and s-polarizations in Fig. 6.11 correspond to the extraordinary and ordinary rays, respectively. With respect to s-polarization, therefore, $\boldsymbol{D}_{ts}$ and $\boldsymbol{E}_{ts}$ are parallel and Snell's law holds.

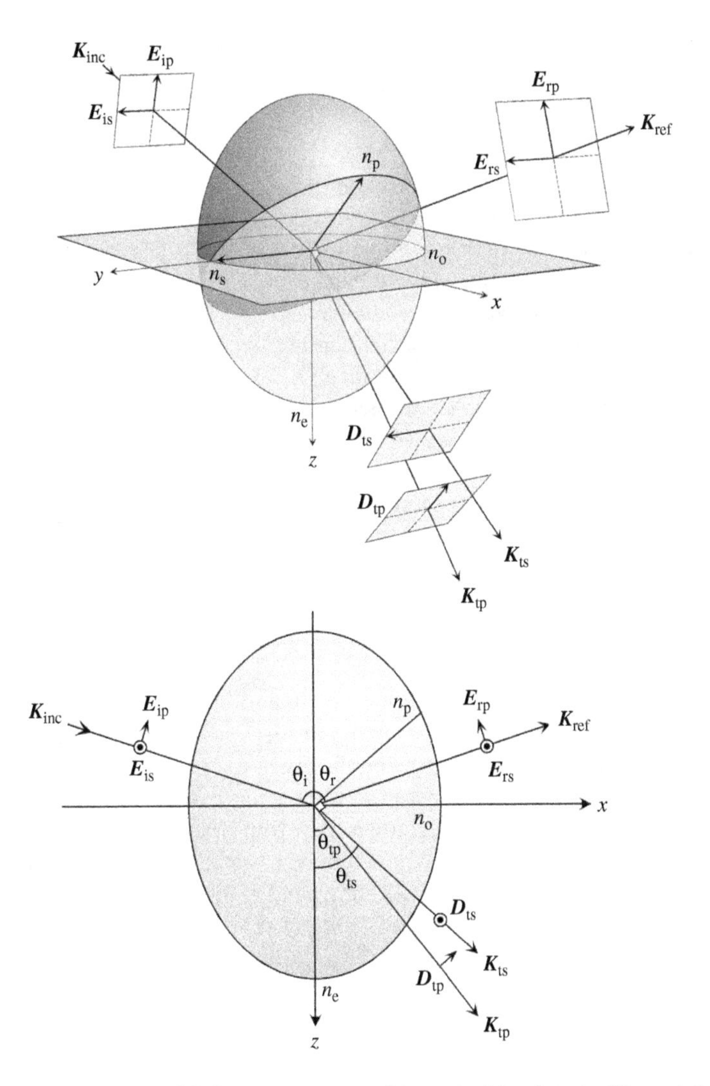

Figure 6.11 Reflection of light waves on a positive uniaxial substrate. The notation of the electric field $\boldsymbol{E}$ follows Fig. 2.15. The subscripts tp and ts show the transmitted p- and s-polarizations, respectively. n_p and n_s represent the refractive indices for $\boldsymbol{D}_{tp}$ and $\boldsymbol{D}_{ts}$.

Consequently, the amplitude reflection coefficient of the s-polarization is expressed by the Fresnel equation shown in Eq. (2.64):

$$r_{ss} = \frac{n_i \cos\theta_i - (n_y^2 - n_i^2 \sin^2\theta_i)^{1/2}}{n_i \cos\theta_i + (n_y^2 - n_i^2 \sin^2\theta_i)^{1/2}} \tag{6.19}$$

where n_i and n_y show the refractive indices of the ambient and y axis ($n_y = n_o = n_s$ in Fig. 6.11). As mentioned earlier, the transmitted p-polarization in Fig. 6.11 is the extraordinary ray, and $\boldsymbol{D}_{tp}$ and $\boldsymbol{E}_{tp}$ are not parallel [see also Fig. 6.3(e)]. In this case, r_{pp} is described by the following equation [9,13]:

$$r_{pp} = \frac{n_x n_z \cos\theta_i - n_i(n_z^2 - n_i^2 \sin^2\theta_i)^{1/2}}{n_x n_z \cos\theta_i + n_i(n_z^2 - n_i^2 \sin^2\theta_i)^{1/2}} \quad (6.20)$$

where n_x and n_z represent the refractive indices of the x and z axes, respectively. It follows from Fig. 6.11 that $n_x = n_o$ and $n_z = n_e$. Eq. (6.20) shows that r_{pp} is determined by the refractive index components of n_x and n_z. If $n_x = n_z$, Eq. (6.20) reduces to Eq. (2.64). r_{pp} and r_{ss} of the configurations shown in Figs. 6.10(b) and (c) are obtained by simply substituting their (n_x, n_y, n_z) values into Eqs. (6.19) and (6.20). We can use the complex refractive indices of N_x, N_y, and N_z, instead of n_x, n_y, and n_z, in Eqs. (6.19) and (6.20). These equations can also be applied for biaxial materials ($N_x \neq N_y \neq N_z$). Since $r_{ps} = r_{sp} = 0$ in the above configurations, the Jones matrix of the anisotropic samples is obtained rather easily from Eqs. (6.19) and (6.20).

6.2.2 ANISOTROPIC THIN FILM ON ISOTROPIC SUBSTRATE

When the principal axes of an anisotropic thin film are aligned to the (x, y, z) coordinates, r_{pp} and r_{ss} of the anisotropic thin film on an isotropic substrate can be calculated in a similar manner. However, the effect of optical interference within the anisotropic thin film must be taken into account in this case. Figure 6.12 illustrates an optical model for an ambient/(anisotropic thin film)/(isotropic substrate) structure. Here, the anisotropic thin film is assumed to be a biaxial material with the complex refractive indices of (N_{1x}, N_{1y}, N_{1z}). As shown in Fig. 6.12, the reflection (incidence) angle θ_0 and transmission angle θ_2 of multiply reflected waves are constants and

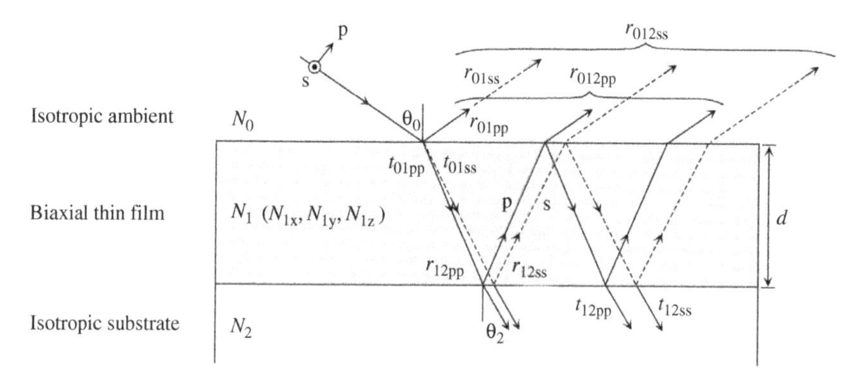

Figure 6.12 Optical model for an ambient/(biaxial thin film)/(isotropic substrate) structure. The principal axes of the biaxial thin film are assumed to be parallel to the (x, y, z) coordinates.

are related by $N_0 \sin\theta_0 = N_2 \sin\theta_2$, where N_0 and N_2 show the complex refractive indices of the ambient and substrate, respectively. In Fig. 6.12, the transmitted waves are separated into p- and s-polarizations, similar to Fig. 6.11. The light reflection and transmission at the ambient/film interface can be expressed as follows [13]:

$$r_{01pp} = \frac{N_{1x}N_{1z}\cos\theta_0 - N_0(N_{1z}^2 - N_0^2\sin^2\theta_0)^{1/2}}{N_{1x}N_{1z}\cos\theta_0 + N_0(N_{1z}^2 - N_0^2\sin^2\theta_0)^{1/2}} \tag{6.21a}$$

$$r_{01ss} = \frac{N_0\cos\theta_0 - (N_{1y}^2 - N_0^2\sin^2\theta_0)^{1/2}}{N_0\cos\theta_0 + (N_{1y}^2 - N_0^2\sin^2\theta_0)^{1/2}} \tag{6.21b}$$

$$t_{01pp} = \frac{2N_0\cos\theta_0\left[N_0^2\sin^2\theta_0(N_x^2 - N_z^2) + N_z^4\right]^{1/2}}{N_z\left[N_zN_x\cos\theta_0 + N_0(N_z^2 - N_0^2\sin^2\theta_0)^{1/2}\right]} \tag{6.21c}$$

$$t_{01ss} = \frac{2N_0\cos\theta_0}{N_0\cos\theta_0 + (N_{1y}^2 - N_0^2\sin^2\theta_0)^{1/2}} \tag{6.21d}$$

The reflection and transmission coefficients at the film/substrate interface are given by the following equations [13]:

$$r_{12pp} = \frac{N_2(N_{1z}^2 - N_2^2\sin^2\theta_2)^{1/2} - N_{1x}N_{1z}\cos\theta_2}{N_2(N_{1z}^2 - N_2^2\sin^2\theta_2)^{1/2} + N_{1x}N_{1z}\cos\theta_2} \tag{6.22a}$$

$$r_{12ss} = \frac{(N_{1y}^2 - N_2^2\sin^2\theta_2)^{1/2} - N_2\cos\theta_2}{(N_{1y}^2 - N_2^2\sin^2\theta_2)^{1/2} + N_2\cos\theta_2} \tag{6.22b}$$

$$t_{12pp} = \frac{2N_z^2N_x(N_z^2 - N_0^2\sin^2\theta_0)^{1/2}}{\left[N_0^2\sin^2\theta_0(N_x^2 - N_z^2) + N_z^4\right]^{1/2}\left[N_zN_x\cos\theta_2 + N_2(N_z^2 - N_0^2\sin^2\theta_0)^{1/2}\right]} \tag{6.22c}$$

$$t_{12ss} = \frac{2(N_{1y}^2 - N_2^2\sin^2\theta_2)^{1/2}}{(N_{1y}^2 - N_2^2\sin^2\theta_2)^{1/2} + N_2\cos\theta_2} \tag{6.22d}$$

As shown in Eqs. (6.21c) and (6.22c), the formulas for the transmission of p-polarization become complicated, since $\boldsymbol{D}_{tp}$ and $\boldsymbol{E}_{tp}$ are not parallel. Using Eqs. (2.83) and (2.85), we obtain the total reflection and transmission coefficients for p- and s-polarizations [9,13]:

$$r_{012pp} = \frac{r_{01pp} + r_{12pp}\exp(-i2\beta_p)}{1 + r_{01pp}r_{12pp}\exp(-i2\beta_p)} \tag{6.23a}$$

$$r_{012ss} = \frac{r_{01ss} + r_{12ss}\exp(-i2\beta_s)}{1 + r_{01ss}r_{12ss}\exp(-i2\beta_s)} \tag{6.23b}$$

$$t_{012pp} = \frac{t_{01pp}t_{12pp}\exp(-i\beta_p)}{1 + r_{01pp}r_{12pp}\exp(-i2\beta_p)} \tag{6.23c}$$

$$t_{012ss} = \frac{t_{01ss}t_{12ss}\exp(-\mathrm{i}\beta_s)}{1 + r_{01ss}r_{12ss}\exp(-\mathrm{i}2\beta_s)} \tag{6.23d}$$

where β_p and β_s are the phase variations for p- and s-polarized waves:

$$\beta_p = \frac{2\pi d}{\lambda}\left(\frac{N_{1x}}{N_{1z}}\right)\left(N_{1z}^2 - N_0^2\sin^2\theta_0\right)^{1/2} \tag{6.24a}$$

$$\beta_s = \frac{2\pi d}{\lambda}\left(N_{1y}^2 - N_0^2\sin^2\theta_0\right)^{1/2} \tag{6.24b}$$

It follows from the above equations that $\tan\psi_{pp}\exp(\mathrm{i}\Delta_{pp}) = r_{012pp}/r_{012ss}$. Accordingly, the calculation procedure of the (anisotropic thin film)/(isotropic substrate) structure is essentially the same as that of isotropic thin films described in Section 2.4.

6.3 4 × 4 MATRIX METHOD

The Fresnel approach discussed in Section 6.2 allows ellipsometry data analysis only for anisotropic samples oriented in specific directions. Such restrictions can be avoided if we perform data analysis using a method referred to as the 4 × 4 matrix method. The 4 × 4 matrix method provides the most general approach for the data analysis of anisotropic samples and thus has been employed widely. In this section, we will address the principles and actual calculation procedures of the 4 × 4 matrix method in detail. Due to the requirements of the 4 × 4 matrix method, however, the definition of $N \equiv n + \mathrm{i}k$, instead of $N \equiv n - \mathrm{i}k$, will bc used throughout this section.

6.3.1 PRINCIPLES OF THE 4 × 4 MATRIX METHOD

In 1972, Berreman derived a first-order differential equation from Maxwell's equations, known as Berreman's equation [14]:

$$\frac{\partial\Psi}{\partial z} = \mathrm{i}\frac{\omega}{c}\Delta_B\Psi \tag{6.25}$$

where ω and c show the angular frequency and speed of light, respectively. In Eq. (6.25), Ψ represents a vector whose elements are the electric field E and magnetic field H:

$$\Psi = \begin{bmatrix} E_x \\ E_y \\ H_x \\ H_y \end{bmatrix} \tag{6.26}$$

The E and H components in Eq. (6.26) are defined using the (x, y, z) coordinates shown in Fig. 6.13 and represent the tangential components parallel to the sample surface. It can be seen that the (x, y, z) coordinates shown in Fig. 6.13 are identical to those in Figs. 6.10 and 6.11. The matrix Δ_B in Eq. (6.25) is a 4×4 matrix expressed as [14–16]:

$$\Delta_B = \begin{bmatrix} -K_{xx}\dfrac{\varepsilon_{zx}}{\varepsilon_{zz}} & -K_{xx}\dfrac{\varepsilon_{zy}}{\varepsilon_{zz}} & 0 & 1-\dfrac{K_{xx}^2}{\varepsilon_{zz}} \\ 0 & 0 & -1 & 0 \\ \varepsilon_{yz}\dfrac{\varepsilon_{zx}}{\varepsilon_{zz}}-\varepsilon_{yx} & K_{xx}^2-\varepsilon_{yy}+\varepsilon_{yz}\dfrac{\varepsilon_{zy}}{\varepsilon_{zz}} & 0 & K_{xx}\dfrac{\varepsilon_{yz}}{\varepsilon_{zz}} \\ \varepsilon_{xx}-\varepsilon_{xz}\dfrac{\varepsilon_{zx}}{\varepsilon_{zz}} & \varepsilon_{xy}-\varepsilon_{xz}\dfrac{\varepsilon_{zy}}{\varepsilon_{zz}} & 0 & -K_{xx}\dfrac{\varepsilon_{xz}}{\varepsilon_{zz}} \end{bmatrix} \tag{6.27}$$

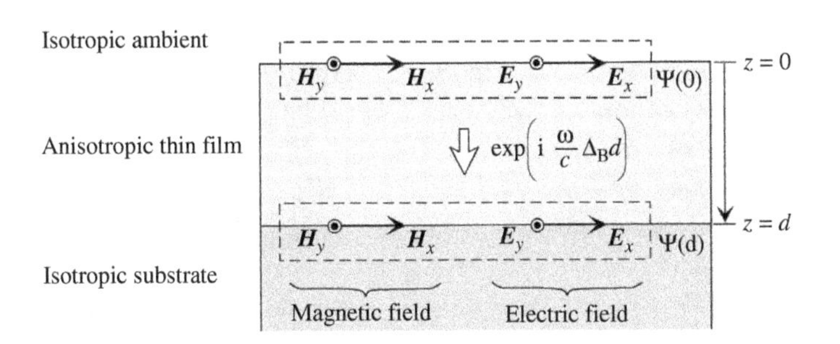

Figure 6.13 Representation of the electric field $\boldsymbol{E}$ and magnetic field $\boldsymbol{H}$ in an ambient/(anisotropic thin film)/(isotropic substrate) structure. In this figure, the x and y axes are parallel to the interfaces and the z axis is selected to be downward. Ψ represents a vector consisting of $E_{x,y}$ and $H_{x,y}$.

In Eq. (6.27), each ε corresponds to the element of the dielectric tensor shown in Eq. (6.9), and K_{xx} can be related to the component of the propagation vector $\boldsymbol{K}$ in the x direction. If Eq. (2.29) is applied, the x component of $\boldsymbol{K}$ is given by

$$K_x = \frac{\omega}{c} n_i \sin\theta_i \tag{6.28}$$

where n_i is the refractive index of air (or an ambient), and θ_i shows the angle of incidence. From Eq. (6.28), K_{xx} is defined as follows [16]:

$$K_{xx} \equiv \frac{c}{\omega} K_x = n_i \sin\theta_i \tag{6.29}$$

It should be emphasized that Berreman's equation is exact and can be derived directly from Maxwell's equations without any restricting assumptions [14]. Thus, optical interference caused by multiple light reflections can also be expressed explicitly from Berreman's equation. However, Berreman's equation has been derived from Maxwell's equations in Gaussian units in which length is expressed in centimeters [cm]. Thus, in Gaussian units, the speed of light is given by $c = 2.99792 \times 10^{10}\,\text{cm/s}$. Except for the length, there are no changes for the values of ε, n and ω. When we apply Gaussian units, the magnetic field H in Eq. (6.26) is equivalent to the magnetic induction B (see Section 2.1.2) for nonmagnetic materials. Notice that, although $\boldsymbol{E}_{x,y}$ and $\boldsymbol{H}_{x,y}$ are illustrated separately in Fig. 6.13, $(\boldsymbol{E}_x, \boldsymbol{H}_y)$ and $(\boldsymbol{E}_y, \boldsymbol{H}_x)$ are paired in actual electromagnetic waves, as shown in Fig. 2.15.

The solution of Eq. (6.25) has the form

$$\Psi(d) = \exp\left(\mathrm{i}\frac{\omega}{c}\Delta_{\mathrm{B}} d\right)\Psi(0) \tag{6.30}$$

where $\Psi(d)$ and $\Psi(0)$ represent the tangential components of E and H at $z = d$ and $z = 0$, respectively [14]. Now suppose a sample structure in which an anisotropic thin film with a thickness of d is formed on an isotropic substrate, as shown in Fig. 6.13. Eq. (6.30) shows that the x and y components of the electric and magnetic fields can be transferred by the exponential term in Eq. (6.30). When the (E, H) components are transferred in the opposite direction (i.e., from $z = d$ to $z = 0$), the solution is expressed by

$$\Psi(0) = \exp\left[\mathrm{i}\frac{\omega}{c}\Delta_{\mathrm{B}}(-d)\right]\Psi(d) \tag{6.31}$$

If we calculate $\partial\Psi(0)/\partial(-d)$ from Eq. (6.31), we obtain

$$\frac{\partial\Psi(0)}{\partial(-d)} = \mathrm{i}\frac{\omega}{c}\Delta_{\mathrm{B}}\exp\left[\mathrm{i}\frac{\omega}{c}\Delta_{\mathrm{B}}(-d)\right]\Psi(d) = \mathrm{i}\frac{\omega}{c}\Delta_{\mathrm{B}}\Psi(0) \tag{6.32}$$

Thus, it is obvious that Eq. (6.31) satisfies Eq. (6.25). By substituting Eq. (6.30) into Eq. (6.25), we can also confirm that Eq. (6.30) is a solution of Berreman's equation.

In 1996, Schubert established a new 4×4 matrix formalism based on the above equations, which can be employed for the calculation of the Jones matrix [16–18]. In particular, Schubert has introduced the incident matrix $\boldsymbol{L}_{\mathrm{i}}$, exit matrix $\boldsymbol{L}_{\mathrm{t}}$, and partial transfer matrix $\boldsymbol{T}_{\mathrm{p}}(-d)$ defined by

$$\boldsymbol{L}_{\mathrm{i}}\begin{bmatrix} E_{\mathrm{is}} \\ E_{\mathrm{rs}} \\ E_{\mathrm{ip}} \\ E_{\mathrm{rp}} \end{bmatrix} \equiv \Psi(0) \tag{6.33}$$

$$\boldsymbol{L}_{\mathrm{t}}\begin{bmatrix} E_{\mathrm{ts}} \\ 0 \\ E_{\mathrm{tp}} \\ 0 \end{bmatrix} \equiv \Psi(d) \tag{6.34}$$

$$\boldsymbol{T}_{\mathrm{p}}(-d) \equiv \exp\left[\mathrm{i}\frac{\omega}{c}\Delta_{\mathrm{B}}(-d)\right] \tag{6.35}$$

In Eqs. (6.33) and (6.34), the E components represent the electric fields defined by Fig. 6.14. Notice that these definitions are identical to those shown in Fig. 2.15. As indicated in Fig. 6.14, the incident matrix $\boldsymbol{L}_{\mathrm{i}}$ projects the electric fields of the incident waves ($E_{\mathrm{ip}}, E_{\mathrm{is}}$) and reflected waves ($E_{\mathrm{rp}}, E_{\mathrm{rs}}$) onto the surface of the anisotropic thin film, while $\boldsymbol{L}_{\mathrm{i}}^{-1}$ converts the in-plane electric fields to the original configuration. On the other hand, the exit matrix $\boldsymbol{L}_{\mathrm{t}}$ projects the electric fields of the transmitted waves ($E_{\mathrm{tp}}, E_{\mathrm{ts}}$) onto the thin film/substrate interface. In Fig. 6.14, we assume that the thickness of the isotropic substrate is infinite and there is no backside reflection. When there is backside reflection, the vector of Eq. (6.34) should be represented by $[E_{\mathrm{ts}}, E_{\mathrm{bs}}, E_{\mathrm{tp}}, E_{\mathrm{bp}}]^{\mathrm{T}}$, where E_{bs} and E_{bp} denote the electric fields of s- and p-polarized waves reflected on the backside, respectively [16–18].

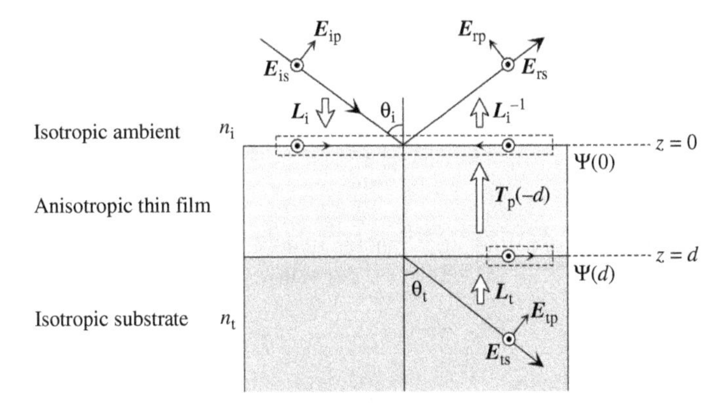

Figure 6.14 Representation of the electric fields for incident, reflected, and transmitted waves in an ambient/(anisotropic thin film)/(isotropic substrate) structure. The incident matrix $\boldsymbol{L}_{\mathrm{i}}$ projects the electric fields of the incident and reflected waves onto the thin film surface ($z = 0$), while the exit matrix $\boldsymbol{L}_{\mathrm{t}}$ projects the electric fields of the transmitted waves onto the thin film/substrate interface ($z = d$). The partial transfer matrix $\boldsymbol{T}_{\mathrm{p}}(-d)$ transfers the field components from $z = d$ to $z = 0$, and $\boldsymbol{L}_{\mathrm{i}}^{-1}$ converts the tangential components into the incident and reflected waves.

By substituting Eqs. (6.33)–(6.35) into Eq. (6.31), we get

$$\begin{bmatrix} E_{\mathrm{is}} \\ E_{\mathrm{rs}} \\ E_{\mathrm{ip}} \\ E_{\mathrm{rp}} \end{bmatrix} = \boldsymbol{L}_{\mathrm{i}}^{-1}\boldsymbol{T}_{\mathrm{p}}(-d)\boldsymbol{L}_{\mathrm{t}} \begin{bmatrix} E_{\mathrm{ts}} \\ 0 \\ E_{\mathrm{tp}} \\ 0 \end{bmatrix} \tag{6.36}$$

This equation can be obtained by multiplying $\boldsymbol{L}_{\mathrm{i}}^{-1}$ on both sides and using $\boldsymbol{L}_{\mathrm{i}}^{-1}\boldsymbol{L}_{\mathrm{i}} = 1$. Eq. (6.36) shows that the p- and s-polarized waves transmitting through the isotropic substrate (E_{ts} and E_{tp}) are first projected onto the thin film/substrate interface by $\boldsymbol{L}_{\mathrm{t}}$. These in-plane components are then transferred to the ambient/thin film interface by the partial transfer matrix $\boldsymbol{T}_{\mathrm{p}}(-d)$. Recall from Eq. (6.31) that $\boldsymbol{T}_{\mathrm{p}}(-d)$ transfers the tangential components in the opposite direction along the z axis. Finally, $\Psi(0)$

is converted by $\boldsymbol{L}_{\mathrm{i}}^{-1}$ to express the incident and reflected waves. In this 4×4 matrix method, therefore, the field components are transferred by the matrices from bottom to top. In this sense, this calculation procedure is rather similar to that of optical interference in multilayer structures shown in Fig. 2.24. From Eq. (6.36), we define a new matrix called the transfer matrix $\boldsymbol{T}$:

$$\boldsymbol{T} \equiv \boldsymbol{L}_{\mathrm{i}}^{-1}\boldsymbol{T}_{\mathrm{p}}(-d)\boldsymbol{L}_{\mathrm{t}} \tag{6.37}$$

The above procedure can be extended easily to express light propagation in a multilayer structure. When there are two layers on an isotropic substrate, for example, the transfer matrix is expressed by $\boldsymbol{L}_{\mathrm{i}}^{-1}\boldsymbol{T}_{1\mathrm{p}}(-d_1)\boldsymbol{T}_{2\mathrm{p}}(-d_2)\boldsymbol{L}_{\mathrm{t}}$, where $\boldsymbol{T}_{1\mathrm{p}}(-d_1)$ and $\boldsymbol{T}_{2\mathrm{p}}(-d_2)$ denote the partial transfer matrices of the two layers with thicknesses of d_1 and d_2, respectively. Thus, we can rewrite Eq. (6.37) in the more general form [16,17]:

$$\boldsymbol{T} \equiv \boldsymbol{L}_{\mathrm{i}}^{-1}\prod_{j=1}^{N}\boldsymbol{T}_{j\mathrm{p}}(-d_j)\boldsymbol{L}_{\mathrm{t}} \tag{6.38}$$

It is clear from Eq. (6.36) that the transfer matrix $\boldsymbol{T}$ is a 4×4 matrix. In matrix form, the transfer matrix is expressed by the following matrix [5,6,16]:

$$\begin{bmatrix} E_{\mathrm{is}} \\ E_{\mathrm{rs}} \\ E_{\mathrm{ip}} \\ E_{\mathrm{rp}} \end{bmatrix} = \boldsymbol{T}\begin{bmatrix} E_{\mathrm{ts}} \\ 0 \\ E_{\mathrm{tp}} \\ 0 \end{bmatrix} = \begin{bmatrix} T_{11} & T_{12} & T_{13} & T_{14} \\ T_{21} & T_{22} & T_{23} & T_{24} \\ T_{31} & T_{32} & T_{33} & T_{34} \\ T_{41} & T_{42} & T_{43} & T_{44} \end{bmatrix}\begin{bmatrix} E_{\mathrm{ts}} \\ 0 \\ E_{\mathrm{tp}} \\ 0 \end{bmatrix} \tag{6.39}$$

From the elements of the transfer matrix shown in Eq. (6.39), the elements of the Jones matrices in Eqs. (6.12) and (6.13) are described by the following equations [5,6]:

$$r_{\mathrm{pp}} = \left(\frac{E_{\mathrm{rp}}}{E_{\mathrm{ip}}}\right)_{E_{\mathrm{is}}=0} = \frac{T_{11}T_{43} - T_{13}T_{41}}{T_{11}T_{33} - T_{13}T_{31}} \tag{6.40a}$$

$$r_{\mathrm{sp}} = \left(\frac{E_{\mathrm{rs}}}{E_{\mathrm{ip}}}\right)_{E_{\mathrm{is}}=0} = \frac{T_{11}T_{23} - T_{13}T_{21}}{T_{11}T_{33} - T_{13}T_{31}} \tag{6.40b}$$

$$r_{\mathrm{ss}} = \left(\frac{E_{\mathrm{rs}}}{E_{\mathrm{is}}}\right)_{E_{\mathrm{ip}}=0} = \frac{T_{21}T_{33} - T_{23}T_{31}}{T_{11}T_{33} - T_{13}T_{31}} \tag{6.40c}$$

$$r_{\mathrm{ps}} = \left(\frac{E_{\mathrm{rp}}}{E_{\mathrm{is}}}\right)_{E_{\mathrm{ip}}=0} = \frac{T_{33}T_{41} - T_{31}T_{43}}{T_{11}T_{33} - T_{13}T_{31}} \tag{6.40d}$$

$$t_{\mathrm{pp}} = \left(\frac{E_{\mathrm{tp}}}{E_{\mathrm{ip}}}\right)_{E_{\mathrm{is}}=0} = \frac{T_{11}}{T_{11}T_{33} - T_{13}T_{31}} \tag{6.40e}$$

$$t_{sp} = \left(\frac{E_{ts}}{E_{ip}}\right)_{E_{is}=0} = \frac{-T_{13}}{T_{11}T_{33} - T_{13}T_{31}} \tag{6.40f}$$

$$t_{ss} = \left(\frac{E_{ts}}{E_{is}}\right)_{E_{ip}=0} = \frac{T_{33}}{T_{11}T_{33} - T_{13}T_{31}} \tag{6.40g}$$

$$t_{ps} = \left(\frac{E_{tp}}{E_{is}}\right)_{E_{ip}=0} = \frac{-T_{31}}{T_{11}T_{33} - T_{13}T_{31}} \tag{6.40h}$$

For example, r_{sp} represents reflected s-polarization (E_{rs}) generated by incident p-polarization (E_{ip}) when there is no incident s-polarization ($E_{is} = 0$), as confirmed from Fig. 6.9(a). Thus, r_{sp} is described as $r_{sp} = (E_{rs}/E_{ip})_{E_{is}=0}$ in Eq. (6.40b). We can derive Eq. (6.40) rather easily from Eq. (6.39). By expanding Eq. (6.39), we first obtain

$$\begin{bmatrix} E_{is} \\ E_{rs} \\ E_{ip} \\ E_{rp} \end{bmatrix} = \begin{bmatrix} T_{11}E_{ts} + T_{13}E_{tp} \\ T_{21}E_{ts} + T_{23}E_{tp} \\ T_{31}E_{ts} + T_{33}E_{tp} \\ T_{41}E_{ts} + T_{43}E_{tp} \end{bmatrix} \tag{6.41}$$

In the case of r_{ss}, for example, we obtain $E_{ip} = T_{31}E_{ts} + T_{33}E_{tp} = 0$ from the condition $E_{ip} = 0$. Thus, using $E_{tp}/E_{ts} = -T_{31}/T_{33}$ and Eq. (6.41), we get

$$r_{ss} = \left(\frac{E_{rs}}{E_{is}}\right)_{E_{ip}=0} = \frac{T_{21} + T_{23}(E_{tp}/E_{ts})}{T_{11} + T_{13}(E_{tp}/E_{ts})} = \frac{T_{21}T_{33} - T_{23}T_{31}}{T_{11}T_{33} - T_{13}T_{31}} \tag{6.42}$$

All the other equations shown in Eq. (6.40) can be derived from a similar procedure. By applying Eq. (6.40), we obtain the ellipsometry parameters:

$$\rho_{pp} = r_{pp}/r_{ss} = \tan\psi_{pp}\exp(-i\Delta_{pp}) \tag{6.43a}$$

$$\rho_{ps} = r_{ps}/r_{ss} = \tan\psi_{ps}\exp(-i\Delta_{ps}) \tag{6.43b}$$

$$\rho_{sp} = r_{sp}/r_{ss} = \tan\psi_{sp}\exp(-i\Delta_{sp}) \tag{6.43c}$$

Recall that the 4×4 matrix method described in this section uses the convention of $N \equiv n + ik$ ($\varepsilon \equiv \varepsilon_1 + i\varepsilon_2$). Thus, Δ in Eq. (6.43) has a minus sign (see Appendix 2). In this case, $\Delta = -\arg(\rho)$ must be used for the calculation of Δ. When substrates are optically isotropic, the definition of $N \equiv n - ik$ can also be used if we replace i with $-$i in the above equations. Nevertheless, we cannot use this procedure in the calculation of anisotropic substrates (see Section 6.3.3) and, therefore, the convention of $N \equiv n + ik$ has been used in Section 6.3.

6.3.2 CALCULATION METHOD OF PARTIAL TRANSFER MATRIX

In the 4 × 4 matrix method, light reflection and transmission are described by $\boldsymbol{L}_\mathrm{i}^{-1}$, $\boldsymbol{T}_\mathrm{p}(-d)$, and $\boldsymbol{L}_\mathrm{t}$. Here, we will see the calculation method of the partial transfer matrix $\boldsymbol{T}_\mathrm{p}(-d)$. It has been shown that there are four solutions for Berreman's equation in Eq. (6.25) [14]:

$$\Psi_j(d) = \exp\left(\mathrm{i}\frac{\omega}{c} q_j d\right) \Psi_j(0) \quad j = 1, 2, 3, 4 \tag{6.44}$$

Here, q_j represent the eigenvalues of the matrix Δ_B shown in Eq. (6.27). Since Δ_B is a 4 × 4 matrix, there are four eigenvalues (q_{1-4}). Notice that Eq. (6.44) is quite similar to Eq. (6.30). The term $\omega q/c$ in Eq. (6.44) corresponds to $K = \omega n/c$ [Eq. (2.29)] and thus q represents the phase part (refractive index component) of a wave traveling along the z axis. Among four eigenvalues, two eigenvalues have positive real parts, while the other two eigenvalues have negative real parts. The positive eigenvalues show waves moving in the positive direction of the z axis, while the negative eigenvalues represent waves traveling in the negative direction. In the case of isotropic materials, the eigenvalues in Eq. (6.44) are described as follows [16]:

$$|q_j| = n_\mathrm{t} \cos\theta_\mathrm{t} \tag{6.45}$$

where n_t and θ_t show the refractive index and transmission angle of an isotropic material. Thus, at normal incidence ($\theta_\mathrm{i} = \theta_\mathrm{t} = 0°$), we obtain $q_1 = n_\mathrm{t}$, while $q_1 = 0$ at $\theta_\mathrm{t} = 90°$. This result shows clearly that q_j represent the refractive index component of waves traveling along the z axis.

From Eq. (6.44) and Eq. (6.25), we obtain

$$\frac{\partial \Psi_j(d)}{\partial d} = \mathrm{i}\frac{\omega}{c} q_j \exp\left(\mathrm{i}\frac{\omega}{c} q_j d\right) \Psi_j(0) \tag{6.46a}$$

$$\mathrm{i}\frac{\omega}{c} \Delta_\mathrm{B} \Psi_j(d) = \mathrm{i}\frac{\omega}{c} q_j \Psi_j(d) \tag{6.46b}$$

$$\Delta_\mathrm{B} \Psi_j(d) = q_j \Psi_j(d) \tag{6.46c}$$

Accordingly, the eigenvalues q_j can be obtained from the eigenvalue equation shown in Eq. (6.46c) [14]:

$$\det(\Delta_\mathrm{B} - q\boldsymbol{I}) = 0 \tag{6.47}$$

where det denotes the determinant of a matrix and $\boldsymbol{I}$ shows the identity matrix, similar to Eq. (6.6). If we perform numerical analysis using a computer, the eigenvalues can be obtained directly from Eq. (6.47) [19].

On the other hand, Wöhler *et al.* have shown that the partial transfer matrix can be expressed from a finite series [20,21]:

$$\boldsymbol{T}_{\mathrm{p}}(-d) \equiv \exp\left[\mathrm{i}\frac{\omega}{c}\Delta_{\mathrm{B}}(-d)\right] = \beta_0 \boldsymbol{I} + \beta_1 \Delta_{\mathrm{B}} + \beta_2 \Delta_{\mathrm{B}}^2 + \beta_3 \Delta_{\mathrm{B}}^3 \tag{6.48}$$

Using the eigenvalues q_{1-4} determined by Eq. (6.47), we obtain β_{0-3} values as follows [20,21]:

$$\beta_0 = -\sum_{j=1}^{4} q_k q_l q_m \frac{\exp[\mathrm{i}\omega q_j(-d)/c]}{(q_j - q_k)(q_j - q_l)(q_j - q_m)} \tag{6.49a}$$

$$\beta_1 = \sum_{j=1}^{4} (q_k q_l + q_k q_m + q_l q_m) \frac{\exp[\mathrm{i}\omega q_j(-d)/c]}{(q_j - q_k)(q_j - q_l)(q_j - q_m)} \tag{6.49b}$$

$$\beta_2 = -\sum_{j=1}^{4} (q_k + q_l + q_m) \frac{\exp[\mathrm{i}\omega q_j(-d)/c]}{(q_j - q_k)(q_j - q_l)(q_j - q_m)} \tag{6.49c}$$

$$\beta_3 = \sum_{j=1}^{4} \frac{\exp[\mathrm{i}\omega q_j(-d)/c]}{(q_j - q_k)(q_j - q_l)(q_j - q_m)} \tag{6.49d}$$

In the summation of Eq. (6.49), $(k, l, m) = (2, 3, 4)$ for $j = 1$, $(k, l, m) = (1, 3, 4)$ for $j = 2$, $(k, l, m) = (1, 2, 4)$ for $j = 3$, and $(k, l, m) = (1, 2, 3)$ for $j = 4$. It should be emphasized that $\boldsymbol{T}_{\mathrm{p}}(-d)$ calculated from Eq. (6.48) includes the optical interference effect. Thus, when the light absorption in a thin film is small, the interference effect will appear in spectra calculated from the 4×4 matrix method.

6.3.3 CALCULATION METHODS OF INCIDENT AND EXIT MATRICES

We can express the exit matrix $\boldsymbol{L}_{\mathrm{t}}$ from the electric and magnetic fields at $z = d$ in Fig. 6.14. Since $B = H$ in Gaussian units (see Section 6.3.1), the tangential components at $z = d$ can be obtained directly from Fig. 2.15 [16]:

$$\Psi(d) = \begin{bmatrix} E_x \\ E_y \\ H_x \\ H_y \end{bmatrix}_{z=d} = \begin{bmatrix} E_{\mathrm{tp}} \cos\theta_{\mathrm{t}} \\ E_{\mathrm{ts}} \\ -B_{\mathrm{ts}} \cos\theta_{\mathrm{t}} \\ B_{\mathrm{tp}} \end{bmatrix} = \begin{bmatrix} E_{\mathrm{tp}} \cos\theta_{\mathrm{t}} \\ E_{\mathrm{ts}} \\ -n_{\mathrm{t}} E_{\mathrm{ts}} \cos\theta_{\mathrm{t}} \\ n_{\mathrm{t}} E_{\mathrm{tp}} \end{bmatrix} \tag{6.50}$$

In Eq. (6.50), n_{t} and θ_{t} represent the refractive index and transmission angle for the isotropic substrate shown in Fig. 6.14, and $B = nE$ is used for the transformation (see also Section 2.3.2). By combining Eqs. (6.34) and (6.50), we get

$$\boldsymbol{L}_{\mathrm{t}} \begin{bmatrix} E_{\mathrm{ts}} \\ 0 \\ E_{\mathrm{tp}} \\ 0 \end{bmatrix} \equiv \Psi(d) = \begin{bmatrix} E_{\mathrm{tp}} \cos\theta_{\mathrm{t}} \\ E_{\mathrm{ts}} \\ -n_{\mathrm{t}} E_{\mathrm{ts}} \cos\theta_{\mathrm{t}} \\ n_{\mathrm{t}} E_{\mathrm{tp}} \end{bmatrix} \tag{6.51}$$

From Eq. (6.51), we obtain $\boldsymbol{L}_\mathrm{t}$ defined by the following matrix [16–18]:

$$\boldsymbol{L}_\mathrm{t} = \begin{bmatrix} 0 & 0 & \cos\theta_\mathrm{t} & 0 \\ 1 & 0 & 0 & 0 \\ -n_\mathrm{t}\cos\theta_\mathrm{t} & 0 & 0 & 0 \\ 0 & 0 & n_\mathrm{t} & 0 \end{bmatrix} \quad (6.52)$$

As we have seen in Fig. 6.12, $n_\mathrm{i}\sin\theta_\mathrm{i} = n_\mathrm{t}\sin\theta_\mathrm{t}$ holds in Fig. 6.14. Thus, $\cos\theta_\mathrm{t}$ in Eq. (6.52) is given by

$$\cos\theta_\mathrm{t} = [1 - (n_\mathrm{i}/n_\mathrm{t})^2 \sin^2\theta_\mathrm{i}]^{1/2} \quad (6.53)$$

In the matrix shown in Eq. (6.52), the backside reflection on a substrate is not taken into account. If backside reflection is present, the second and fourth columns of $\boldsymbol{L}_\mathrm{t}$ have finite values [17,18].

Similarly, the incident matrix $\boldsymbol{L}_\mathrm{i}$ can be derived from the tangential components at $z = 0$ in Fig. 6.14 [16]:

$$\Psi(0) = \begin{bmatrix} E_x \\ E_y \\ H_x \\ H_y \end{bmatrix}_{z=0} = \begin{bmatrix} E_\mathrm{ip}\cos\theta_\mathrm{i} \\ E_\mathrm{is} \\ -n_\mathrm{i}E_\mathrm{is}\cos\theta_\mathrm{i} \\ n_\mathrm{i}E_\mathrm{ip} \end{bmatrix}_\mathrm{inc} + \begin{bmatrix} -E_\mathrm{rp}\cos\theta_\mathrm{i} \\ E_\mathrm{rs} \\ n_\mathrm{i}E_\mathrm{rs}\cos\theta_\mathrm{i} \\ n_\mathrm{i}E_\mathrm{rp} \end{bmatrix}_\mathrm{ref} \quad (6.54)$$

where the subscripts inc and ref indicate the incident and reflected waves. By combining Eqs. (6.33) and (6.54), we obtain

$$\boldsymbol{L}_\mathrm{i} \begin{bmatrix} E_\mathrm{is} \\ E_\mathrm{rs} \\ E_\mathrm{ip} \\ E_\mathrm{rp} \end{bmatrix} \equiv \Psi(0) = \begin{bmatrix} E_\mathrm{ip}\cos\theta_\mathrm{i} - E_\mathrm{rp}\cos\theta_\mathrm{i} \\ E_\mathrm{is} + E_\mathrm{rs} \\ -n_\mathrm{i}E_\mathrm{is}\cos\theta_\mathrm{i} + n_\mathrm{i}E_\mathrm{rs}\cos\theta_\mathrm{i} \\ n_\mathrm{i}E_\mathrm{ip} + n_\mathrm{i}E_\mathrm{rp} \end{bmatrix} \quad (6.55)$$

It can be confirmed from Eq. (6.55) that the inverse matrix $\boldsymbol{L}_\mathrm{i}^{-1}$ is expressed by the following matrix [16–18]:

$$\boldsymbol{L}_\mathrm{i}^{-1} = \frac{1}{2} \begin{bmatrix} 0 & 1 & -1/(n_\mathrm{i}\cos\theta_\mathrm{i}) & 0 \\ 0 & 1 & 1/(n_\mathrm{i}\cos\theta_\mathrm{i}) & 0 \\ 1/\cos\theta_\mathrm{i} & 0 & 0 & 1/n_\mathrm{i} \\ -1/\cos\theta_\mathrm{i} & 0 & 0 & 1/n_\mathrm{i} \end{bmatrix} \quad (6.56)$$

In order to confirm the above results, we now consider an ambient/(isotropic substrate) structure. If we express $\Psi(0)$ in this structure using Eqs. (6.51) and (6.55), we obtain $\Psi(0) = \boldsymbol{L}_\mathrm{i}[E_\mathrm{is}, E_\mathrm{rs}, E_\mathrm{ip}, E_\mathrm{rp}]^\mathrm{T} = \boldsymbol{L}_\mathrm{t}[E_\mathrm{ts}, 0, E_\mathrm{tp}, 0]^\mathrm{T}$. Thus, it follows from Eqs. (6.52) and (6.56) that

$$\begin{bmatrix} E_\mathrm{is} \\ E_\mathrm{rs} \\ E_\mathrm{ip} \\ E_\mathrm{rp} \end{bmatrix} = \boldsymbol{L}_\mathrm{i}^{-1}\boldsymbol{L}_\mathrm{t} \begin{bmatrix} E_\mathrm{ts} \\ 0 \\ E_\mathrm{tp} \\ 0 \end{bmatrix} = \frac{1}{2} \begin{bmatrix} [1 + n_\mathrm{t}\cos\theta_\mathrm{t}/(n_\mathrm{i}\cos\theta_\mathrm{i})]E_\mathrm{ts} \\ [1 - n_\mathrm{t}\cos\theta_\mathrm{t}/(n_\mathrm{i}\cos\theta_\mathrm{i})]E_\mathrm{ts} \\ (\cos\theta_\mathrm{t}/\cos\theta_\mathrm{i} + n_\mathrm{t}/n_\mathrm{i})E_\mathrm{tp} \\ (-\cos\theta_\mathrm{t}/\cos\theta_\mathrm{i} + n_\mathrm{t}/n_\mathrm{i})E_\mathrm{tp} \end{bmatrix} \quad (6.57)$$

From Eq. (6.57), the amplitude reflection coefficient for s-polarization is given by

$$r_s \equiv \frac{E_{rs}}{E_{is}} = \frac{1 - n_t \cos\theta_t/(n_i \cos\theta_i)}{1 + n_t \cos\theta_t/(n_i \cos\theta_i)} = \frac{n_i \cos\theta_i - n_t \cos\theta_t}{n_i \cos\theta_i + n_t \cos\theta_t} \tag{6.58}$$

Accordingly, the Fresnel equation shown in Eq. (2.61) can be derived directly from Eq. (6.57). In other words, we have determined $\boldsymbol{L}_i^{-1}$ and $\boldsymbol{L}_t$ so that the boundary conditions are satisfied. If we include the partial transfer matrix, the optical response of thin films can be incorporated systematically. Although refractive indices have been used for $\boldsymbol{L}_i^{-1}$ and $\boldsymbol{L}_t$, complex refractive indices can also be employed in these matrices.

It should be emphasized that Eq. (6.52) represents the exit matrix for optically isotropic substrates. When a substrate shows optical anisotropy, the exit matrix for anisotropic substrates must be used. This matrix can be obtained from the following procedure. Now consider a sample composed of an anisotropic substrate only. In this case, we need to determine the in-plane field components at $z = 0$ [i.e., $\Psi(0)$ in Fig. 6.14]. If we modify Eq. (6.46c) slightly, we obtain

$$\Delta_B \Psi_j(0) = q_j \Psi_j(0) \tag{6.59}$$

The eigenvectors $\Psi_j(0)$ and eigenvalues q_j can be obtained numerically from Eq. (6.47). When there is no backside reflection in an ambient/(anisotropic substrate) structure, no waves travel in the negative direction of the z axis. Thus, we can neglect the two eigenvalues that have negative real parts. Let q_a and q_b be the two eigenvalues whose real parts are positive. In this case, their eigenvectors are described by

$$\Psi_a(0) = \begin{bmatrix} \Xi_{1a} \\ \Xi_{2a} \\ \Xi_{3a} \\ \Xi_{4a} \end{bmatrix} \qquad \Psi_b(0) = \begin{bmatrix} \Xi_{1b} \\ \Xi_{2b} \\ \Xi_{3b} \\ \Xi_{4b} \end{bmatrix} \tag{6.60}$$

Using Eq. (6.60), we can express the exit matrix for anisotropic substrates as follows [16]:

$$\boldsymbol{L}_{t,\mathrm{ani}} \equiv \begin{bmatrix} \Xi_{1a} & 0 & \Xi_{1b} & 0 \\ \Xi_{2a} & 0 & \Xi_{2b} & 0 \\ \Xi_{3a} & 0 & \Xi_{3b} & 0 \\ \Xi_{4a} & 0 & \Xi_{4b} & 0 \end{bmatrix} \tag{6.61}$$

Since there is no backside reflection, the 2nd and 4th columns in Eq. (6.61) are assumed to be zero, similar to Eq. (6.52). The transfer matrix of the ambient/(anisotropic substrate) structure is then expressed by

$$\boldsymbol{T} = \boldsymbol{L}_i^{-1} \boldsymbol{L}_{t,\mathrm{ani}} \tag{6.62}$$

If there is a thin layer on this anisotropic substrate, the transfer matrix is obtained from $\boldsymbol{T} = \boldsymbol{L}_{\mathrm{i}}^{-1}\boldsymbol{T}_{\mathrm{p}}(-d)\boldsymbol{L}_{\mathrm{t,ani}}$. It should be noted that (ψ, Δ) values remain the same even if Ξ_{ja} and Ξ_{jb} in Eq. (6.61) are exchanged. As mentioned earlier, it is rather difficult to use the convention of $N \equiv n - ik$ for the calculation of $\boldsymbol{L}_{\mathrm{t,ani}}$, since there is no proper i that can be replaced with −i.

6.3.4 CALCULATION PROCEDURE OF THE 4 × 4 MATRIX METHOD

Figure 6.15 summarizes the calculation procedure of the 4 × 4 matrix method developed by Schubert [16–18]. As shown in this figure, we first select the principal dielectric constants of an index ellipsoid. Keep in mind that $\varepsilon \equiv \varepsilon_1 + \mathrm{i}\varepsilon_2$ in the 4 × 4 matrix method. If the index ellipsoid is tilted relative to the (x, y, z) coordinates, we calculate the dielectric tensor in measurement configuration using Eq. (6.9).

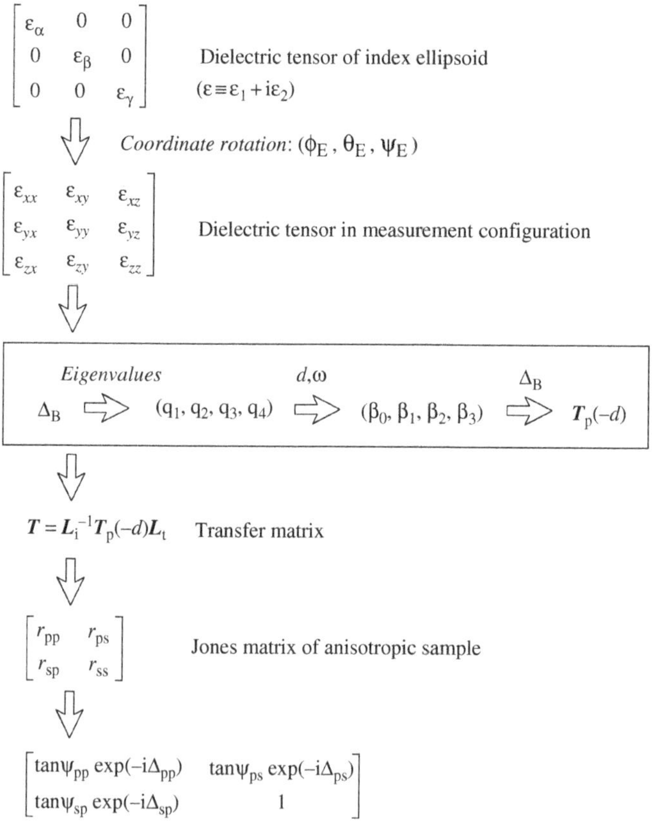

$$\begin{bmatrix} \varepsilon_\alpha & 0 & 0 \\ 0 & \varepsilon_\beta & 0 \\ 0 & 0 & \varepsilon_\gamma \end{bmatrix}$$ Dielectric tensor of index ellipsoid ($\varepsilon \equiv \varepsilon_1 + \mathrm{i}\varepsilon_2$)

⇩ *Coordinate rotation*: $(\phi_E, \theta_E, \psi_E)$

$$\begin{bmatrix} \varepsilon_{xx} & \varepsilon_{xy} & \varepsilon_{xz} \\ \varepsilon_{yx} & \varepsilon_{yy} & \varepsilon_{yz} \\ \varepsilon_{zx} & \varepsilon_{zy} & \varepsilon_{zz} \end{bmatrix}$$ Dielectric tensor in measurement configuration

⇩

Δ_B ⇨ (*Eigenvalues*) (q_1, q_2, q_3, q_4) ⇨ (d, ω) $(\beta_0, \beta_1, \beta_2, \beta_3)$ ⇨ (Δ_B) $\boldsymbol{T}_{\mathrm{p}}(-d)$

⇩

$\boldsymbol{T} = \boldsymbol{L}_{\mathrm{i}}^{-1}\boldsymbol{T}_{\mathrm{p}}(-d)\boldsymbol{L}_{\mathrm{t}}$ Transfer matrix

⇩

$$\begin{bmatrix} r_{pp} & r_{ps} \\ r_{sp} & r_{ss} \end{bmatrix}$$ Jones matrix of anisotropic sample

⇩

$$\begin{bmatrix} \tan\psi_{pp}\exp(-\mathrm{i}\Delta_{pp}) & \tan\psi_{ps}\exp(-\mathrm{i}\Delta_{ps}) \\ \tan\psi_{sp}\exp(-\mathrm{i}\Delta_{sp}) & 1 \end{bmatrix}$$

Figure 6.15 Calculation procedure of the 4 × 4 matrix method.

By substituting the values of the dielectric tensor into Eq. (6.27), we obtain Δ_B. The eigenvalues of this matrix (q_{1-4}) is then determined from Eq. (6.47). Using the values of q_{1-4}, d, and ω (angular frequency of measurement light), we obtain β_{0-3} from Eq. (6.49). Substituting β_{0-3} and Δ_B into Eq. (6.48) gives the partial transfer matrix $\boldsymbol{T}_p(-d)$. It should be noted that $\boldsymbol{T}_p(-d)$ can be calculated even when a layer is optically isotropic. In this case, $\boldsymbol{T}_p(-d)$ is obtained by simply setting $\varepsilon = \varepsilon_\alpha = \varepsilon_\beta = \varepsilon_\gamma$ and $\phi_E = \theta_E = \psi_E = 0°$ in the above calculation. The transfer matrix $\boldsymbol{T}$ of the structure shown in Fig. 6.14 is obtained from $\boldsymbol{T} = \boldsymbol{L}_i^{-1}\boldsymbol{T}_p(-d)\boldsymbol{L}_t$ using Eqs. (6.52) and (6.56). In multilayer structures, Eq. (6.38) should be applied for the calculation of $\boldsymbol{T}$. If a substrate is optically anisotropic, $\boldsymbol{L}_{t,ani}$ [Eq. (6.61)] is used, instead of $\boldsymbol{L}_t$ [Eq. (6.52)]. The Jones matrix of anisotropic samples can be determined from Eq. (6.40) and then converted to ρ using Eq. (6.43). Finally, the (ψ, Δ) values are obtained from ρ using Eq. (4.5) and $\Delta = -\arg(\rho)$ [see Eq. (4.6)]. Although the calculation procedure of the 4×4 matrix method is rather complicated, this method is quite general and can be applied to all the possible configurations of the index ellipsoid and multilayer structures consisting of anisotropic and isotropic layers. If we calculate $(\varepsilon_\alpha, \varepsilon_\beta, \varepsilon_\gamma)$ from the effective medium theories described in Section 5.3, the surface roughness layer of anisotropic materials can also be expressed.

6.4 INTERPRETATION OF (ψ, Δ) FOR ANISOTROPIC MATERIALS

In anisotropic samples, (ψ, Δ) values vary according to the orientation of the index ellipsoid. In this section, we will see actual changes in (ψ, Δ) calculated from the 4×4 matrix method. In particular, we will discuss the variations of (ψ, Δ) in transparent anisotropic substrates, followed by those in anisotropic thin films formed on isotropic substrates.

6.4.1 VARIATIONS OF (ψ, Δ) IN ANISOTROPIC SUBSTRATES

Here, as the simplest example, we will examine anisotropic samples consisting of a substrate only. Figure 6.16 shows ψ_{pp} values of three anisotropic substrates with different orientations, plotted as a function of the incidence angle θ_i. The ambient is air and the anisotropic substrates are assumed to be transparent uniaxial crystals with refractive indices of $n_o = 2$ and $n_e = 2.5$. Since $n_o < n_e$, these are positive uniaxial crystals shown in Fig. 6.4(b). In the case of uniaxial crystals, coordinate rotation is described completely by (ϕ_E, θ_E), as confirmed from Fig. 6.6. In Fig. 6.16, the orientations of the anisotropic substrates are changed by varying θ_E over the range 0° to 90° with a constant ϕ_E of 90°. In these orientations, the optical axes of the index ellipsoids are parallel to the plane of incidence. Thus, the Jones matrices of these samples are diagonal (i.e., $r_{sp} = r_{ps} = 0$), as we have seen in Fig. 6.10.

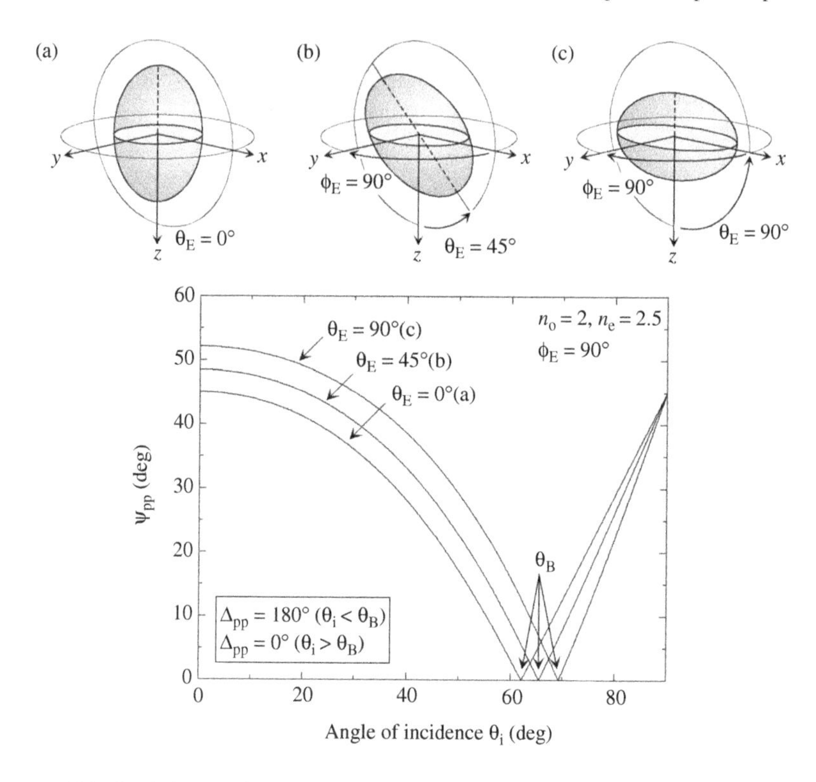

Figure 6.16 Variations of ψ_{pp} with the incidence angle θ_i in three anisotropic substrates: (a) $\theta_E = 0°$, (b) $\theta_E = 45°$ and (c) $\theta_E = 90°$. In the calculation, $n_o = 2$, $n_e = 2.5$ and $\phi_E = 90°$ are assumed. θ_B in the figure denotes the Brewster angle for each orientation.

The result shown in Fig. 6.16 was calculated from the transfer matrix described in Eq. (6.62).

In Fig. 6.16, ψ_{pp} becomes zero at the Brewster angle θ_B (see Section 2.3.4) and θ_B increases as the Euler angle θ_E increases. It should be noted that we observe $\Delta_{pp} = 180°$ at $\theta_i < \theta_B$ and $\Delta_{pp} = 0°$ at $\theta_i > \theta_B$. This variation is quite similar to that observed in the air/c-Si structure shown in Fig. 5.1(c). In the case of Fig. 6.16, θ_B can be estimated roughly from n_x of the index ellipsoids. When $\theta_E = 0°$, for example, $n_x = n_o = 2$ and we obtain $\theta_B \sim \tan^{-1}(2) = 63.4°$ by applying Eq. (2.73). The θ_B observed in Fig. 6.16 is 62.2° and the approximated angle slightly differs from the actual value. When $\theta_E = 90°$, on the other hand, n_x becomes 2.5 (n_e) and we obtain $\tan^{-1}(2.5) = 68.2°(\theta_B = 69.3°$ in Fig. 6.16). Accordingly, with increasing the Euler angle θ_E, the refractive index n_x increases, which in turn shifts θ_B toward higher angles. It can be seen from Fig. 6.16 that the ψ_{pp} value at $\theta_i = 0°$ also increases with θ_E. This behavior can be explained from n_x and n_y of the index ellipsoid. As we discussed in Section 6.2.1 (see also Fig. 6.11), r_{pp} can be related to n_x and n_z, while r_{ss} is determined by n_y. At $\theta_i = 0°$, however, r_{pp} is characterized completely by n_x. When $\theta_E = 0°$, we observe $\psi = 45°(|r_{pp}| = |r_{ss}|)$ since $n_x = n_y = n_o$, as confirmed from Fig. 6.16(a). In the case of $\theta_E = 90°$, the refractive index in the x direction is larger than that in the y direction [i.e., $n_e(n_x) > n_o(n_y)$]. Recall from Eq. (2.72) that reflectance increases with increasing refractive index. Consequently, ψ_{pp} becomes larger than 45° at $\theta_E = 90°$ [see Eq. (4.5)]. Thus, the ellipsometry measurement at small incidence angles is quite helpful to find the in-plane anisotropy of samples. If $\psi_{pp} > 45°$ in the measurement, samples are obviously anisotropic.

Figure 6.17 shows ψ_{pp} of three anisotropic substrates with different orientations ($\theta_E = 0°, 45°, 90°$), plotted as a function of the Euler angle ϕ_E. The optical constants of the index ellipsoids in this figure are the same as those used in Fig. 6.16. In Fig. 6.17, however, the substrates are rotated in the x–y plane using ϕ_E. The incidence angle used for the calculation is 70°. In this case, we always observe $\Delta_{pp} = 0°$, independent of ϕ_E, since $\theta_i = 70°$ is larger than the maximum Brewster angle observed for $\theta_E = 90°$ ($\theta_B = 69.3°$). As shown in Fig. 6.17, ψ_{pp} of $\theta_E = 0°$ shows a constant value because the dielectric tensor does not change with ϕ_E when $\theta_E = 0°$. In contrast, ψ_{pp} of $\theta_E = 45°$ and 90° varies with ϕ_E and its amplitude increases with θ_E. It can be seen from Fig. 6.17 that ψ_{pp} values show minima at $\phi_E = 90°$ and 270°. Notice that ψ_{pp} at $\phi_E = 90°$ corresponds to ψ at $\theta_i = 70°$ in Fig. 6.16. As mentioned earlier, θ_B shifts toward $\theta_i = 70°$ with increasing θ_E. Thus, $|r_{pp}|$ decreases with increasing θ_E, leading to a reduction in ψ_{pp} since $\tan\psi_{pp} = |r_{pp}|/|r_{ss}|$. In contrast, when $\phi_E = 0°$ and 180°, we obtain $n_x = 2$, independent of θ_E, and r_{pp} basically becomes larger. However, n_y and r_{ss} increase with θ_E and consequently ψ_{pp} at $\phi_E = 0°$ decreases slightly with increasing θ_E. Strictly speaking, r_{pp} at $\phi_E = 0°$ also decreases slightly with θ_E due to the reduction in n_z.

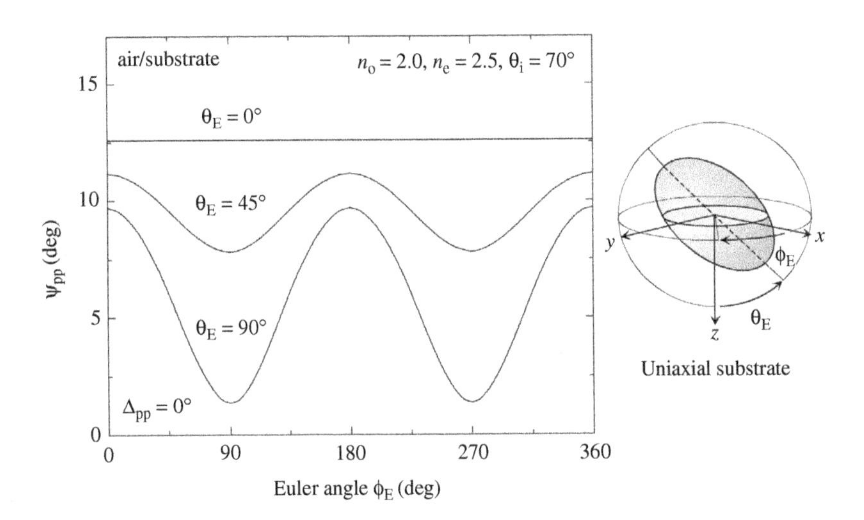

Figure 6.17 ψ_{pp} of three anisotropic substrates with different orientations ($\theta_E = 0°, 45°, 90°$), plotted as a function of the Euler angle ϕ_E. In this calculation, $n_o = 2.0$, $n_e = 2.5$, and $\theta_i = 70°$ are assumed.

Figure 6.18 shows ψ_{ps}, ψ_{sp}, Δ_{ps} and Δ_{sp} for $\theta_E = 90°$ and 45°, plotted as a function of the Euler angle ϕ_E. The results shown in Fig. 6.18 have been obtained from the same calculation as Fig. 6.17. It can be seen from Fig. 6.18 that the Jones matrix is not diagonal anymore, except for the specific orientations in Fig. 6.10. In Fig. 6.18(a), $\psi_{ps} = \psi_{sp}$ shows a maximum value at around $\phi_E = 45°$. Thus, the measurement sensitivity for optical anisotropy is maximized at $\phi_E = 45°$, as we have seen in Section 6.1.4 [see Eq. (6.10)]. Notice that the value of ψ_{ps} (ψ_{sp}) is quite small due to small r_{ps} (r_{sp}). Since $\tan\psi_{ps} = (R_{ps}/R_{ss})^{1/2}$ [Eq. (4.5)], we obtain $R_{ps}/R_{ss} \sim 0.4\%$ from $\psi_{ps} = 3.5$. Accordingly, in the case of Fig. 6.18(a), the actual reflectance of R_{ps} (R_{sp}) is only 0.4 % of R_{ss}. In ellipsometry, we can still characterize these small values, as we will see in Chapter 7.5. As shown in Figs. 6.18(b) and (c), Δ values are either 0° or 180°, since $\varepsilon_2 = k = 0$ in this case.

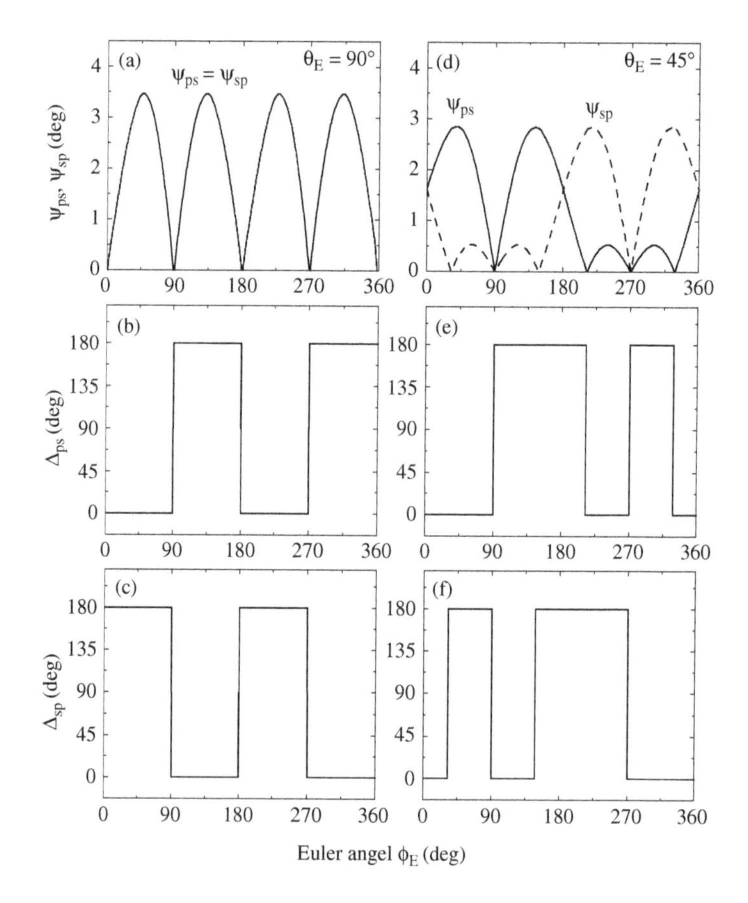

Figure 6.18 (a) ψ_{ps}, ψ_{sp}, (b) Δ_{ps}, and (c) Δ_{sp} for $\theta_E = 90°$ and (d) ψ_{ps}, ψ_{sp}, (e) Δ_{ps}, and (f) Δ_{sp} for $\theta_E = 45°$, plotted as a function of the Euler angle ϕ_E. The result has been obtained from the same calculation as Fig. 6.17.

When $\theta_E = 45°$, we observe $\psi_{ps} \neq \psi_{sp}$, although their values are symmetric in terms of ϕ_E. In Fig. 6.18(d), ψ_{ps} is maximized around 45°, similar to the case of $\theta_E = 90°$. The Jones matrix of this sample becomes diagonal only when $\phi_E = 90°$ and 270° [see Fig. 6.10(d)]. It is evident from Figs. 6.16–6.18 that we can confirm in-plane optical anisotropy from in-plane rotation or measurement at low incidence angles. Furthermore, the result shown in Fig. 6.18 implies that the tilting of the optical axis relative to the plane of incidence can be estimated roughly from ψ_{ps} and ψ_{sp} values; i.e., we observe $\psi_{ps} = \psi_{sp} = 0°$ for $\theta_E = 0°$, $\psi_{ps} \neq \psi_{sp}$ for $0° < \theta_E < 90°$, and $\psi_{ps} = \psi_{sp}$ for $\theta_E = 90°$ when $\phi_E = 45°$.

6.4.2 VARIATIONS OF (ψ, Δ) IN ANISOTROPIC THIN FILMS

In this section, we will examine the variations of (ψ, Δ) in an air/(anisotropic thin film)/(isotropic substrate) structure. Since the calculation procedure of the 4×4 matrix method is rather complicated, a numerical example for this sample will also be shown in this section. Here, the sample is an anisotropic thin film formed on a crystalline silicon (c-Si) substrate, and the anisotropic thin film is a transparent positive-uniaxial crystal with refractive indices of $n_o = 2.0$ and $n_e = 2.5$. The incidence angle, thin-film thickness, and angular frequency of the probe light (En = 2 eV) are assumed to be $\theta_i = 70°$, $d = 1000 \times 10^{-8}$ cm and $\omega = 3.04 \times 10^{15}$ rad/s, respectively. Recall from Section 6.3.1 that the 4×4 matrix method uses Gaussian units in which length is defined in centimeters. When the orientation of the anisotropic thin film is $\phi_E = \theta_E = 45°$, we obtain the dielectric tensor from Eq. (6.9) using $\varepsilon_\alpha = \varepsilon_\beta = n_o^2$ and $\varepsilon_\gamma = n_e^2$:

$$\begin{bmatrix} \varepsilon_{xx} & \varepsilon_{xy} & \varepsilon_{xz} \\ \varepsilon_{yx} & \varepsilon_{yy} & \varepsilon_{yz} \\ \varepsilon_{zx} & \varepsilon_{zy} & \varepsilon_{zz} \end{bmatrix} = \begin{bmatrix} 4.563 & -0.563 & 0.795 \\ -0.563 & 4.563 & -0.795 \\ 0.795 & -0.795 & 5.125 \end{bmatrix} \tag{6.63}$$

As confirmed from Eq. (6.63), the dielectric tensor is symmetric. From $\theta_i = 70°$, we obtain $K_{xx} = 0.94$ using Eq. (6.29). By substituting the value $K_{xx} = 0.94$ and Eq. (6.63) into Eq. (6.27), Δ_B is determined. The eigenvalues of this anisotropic thin film are then determined to be

$$\begin{bmatrix} q_1 \\ q_2 \\ q_3 \\ q_4 \end{bmatrix} = \begin{bmatrix} -2.174 \\ 1.765 \\ -1.765 \\ 1.882 \end{bmatrix} \tag{6.64}$$

By inserting d, ω, q_{1-4}, and $c = 2.998 \times 10^{10}$ cm/s into Eq. (6.49), we obtain

$$\begin{bmatrix} \beta_0 \\ \beta_1 \\ \beta_2 \\ \beta_3 \end{bmatrix} = \begin{bmatrix} 0.566 - i0.025 \\ 0.026 - i0.917 \\ -0.251 + i0.008 \\ -0.008 + i0.117 \end{bmatrix} \tag{6.65}$$

From Δ_B and the values of Eq. (6.65), $\boldsymbol{T}_p(-d)$ can be calculated:

$$\boldsymbol{T}_p(-d) = \begin{bmatrix} -0.352 - i0.057 & 0.091 - i0.001 & 0.033 + i0.044 & 0.056 - i0.397 \\ 0.104 + i0.08 & -0.326 - i0.011 & 0.004 + i0.502 & -0.033 - i0.044 \\ 0.162 - i0.014 & -0.017 + i1.753 & -0.326 - i0.011 & -0.091 + i0.001 \\ 0.3 - i2.12 & -0.162 + i0.014 & -0.104 - i0.08 & -0.352 - i0.057 \end{bmatrix} \quad (6.66)$$

If we assume that the complex refractive index of c-Si at 2 eV is given by $N_{sub} = 3.898 + i0.016$, $\boldsymbol{L}_t$ is obtained using $n_t = N_{sub}$ and $\cos\theta_t = 0.971 + i2.4 \times 10^{-4}$ [Eq. (6.53)]. The incident matrix $\boldsymbol{L}_i^{-1}$ can also be calculated easily using $n_i = 1$ and $\cos\theta_i = 0.342$. From $\boldsymbol{L}_i^{-1}\boldsymbol{T}_p(-d)\boldsymbol{L}_t$, we obtain $\boldsymbol{T}$ expressed by

$$\boldsymbol{T} = \begin{bmatrix} -1.946 - i3.588 & 1.668 - i1.548 & 0.273 - i0.03 & 0.63 - i0.148 \\ 1.614 + i1.679 & -1.989 + i3.435 & -0.301 - i0.064 & -0.861 - i0.101 \\ 0.065 - i0.086 & 0.039 + i0.097 & -0.71 - i3.486 & -0.003 - i1.266 \\ 0.167 + i0.403 & -0.593 - i0.386 & -0.369 + i1.199 & -1.66 + i3.094 \end{bmatrix} \quad (6.67)$$

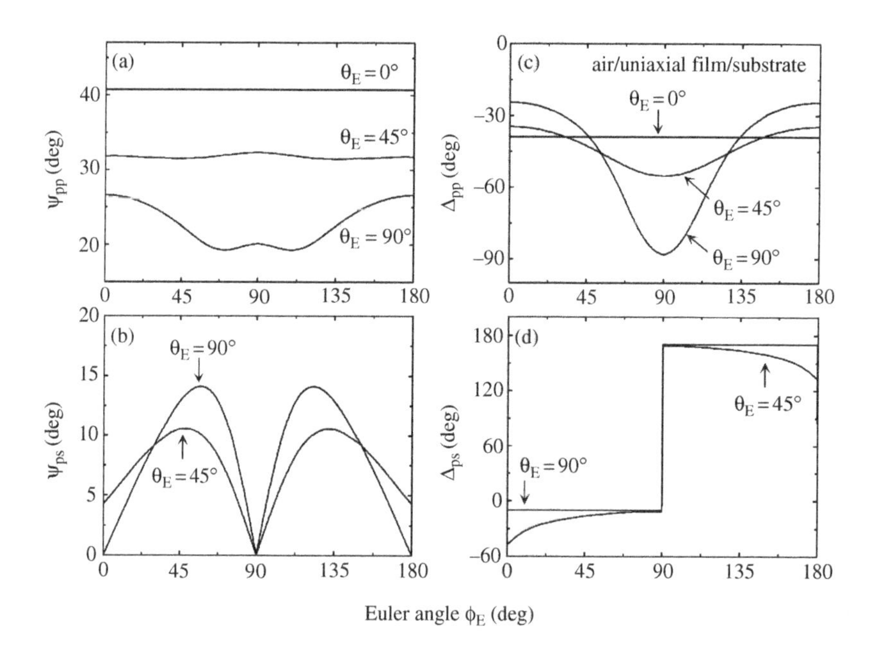

Figure 6.19 (a) ψ_{pp}, (b) ψ_{ps}, (c) Δ_{pp}, and (d) Δ_{ps} for three different orientations of uniaxial thin films ($\theta_E = 0°, 45°, 90°$) in an air/(uniaxial thin film)/(isotropic substrate) structure, plotted as a function of the Euler angle ϕ_E. The refractive indices of the thin films are $n_o = 2.0$ and $n_e = 2.5$, and the complex refractive index of the substrate (c-Si) is $3.898 + i0.016$ (2 eV). The film thickness and incidence angle are assumed to be 1000 Å and 70°, respectively.

Using the values of the elements in Eq. (6.67), we get $r_{pp} = -0.311 - i0.161$, $r_{ps} = -0.107 - i0.002$, $r_{ss} = -0.551 + i0.151$, and $r_{sp} = 0.042 - i0.036$. Finally, (ψ, Δ) values are determined to be $\psi_{pp} = 31.46°$, $\Delta_{pp} = -42.67°$, $\psi_{ps} = 10.57°$, $\Delta_{ps} = -16.35°$, $\psi_{sp} = 5.54°$, and $\Delta_{sp} = -154.93°$.

Figure 6.19 shows the (ψ, Δ) values of the above model with three different orientations of the anisotropic thin films ($\theta_E = 0°, 45°, 90°$), plotted as a function of the Euler angle ϕ_E. As shown in Fig. 6.19, the Jones matrix of this structure is diagonal when $\theta_E = 0°$ and $\phi_E = 90°$. It can be seen that the variations of ψ_{pp} and Δ_{pp} with ϕ_E become larger with increasing θ_E. When $\theta_E = 0°$, on the other hand, ψ_{pp} and Δ_{pp} are independent of ϕ_E. Furthermore, ψ_{ps} is maximized around $\phi_E = 45°$ and its value increases with θ_E. Accordingly, although the variations of (ψ, Δ) in this structure are highly complicated, the trends of these changes are essentially similar to those of the anisotropic substrates discussed in the previous section. Thus, the alignment of anisotropic samples can also be performed by referring to (ψ, Δ) values.

6.5 MEASUREMENT AND DATA ANALYSIS OF ANISOTROPIC MATERIALS

As we have seen in Section 6.1.4, the Jones matrix of anisotropic samples has six independent parameters ($\psi_{pp}, \Delta_{pp}, \psi_{ps}, \Delta_{ps}, \psi_{sp}, \Delta_{sp}$). In the ellipsometry of anisotropic samples, measurement and data analysis are performed for these six parameters, instead of the two parameters for isotropic samples (i.e., ψ and Δ). Unfortunately, the data analysis for anisotropic samples generally becomes more complicated due to the increase in the number of analytical parameters, although the measurement and data analysis procedures for anisotropic samples are essentially the same as those for isotropic samples. In this section, we will review the measurement and data analysis methods for anisotropic samples.

6.5.1 MEASUREMENT METHODS

In a complete Mueller-matrix ellipsometer, the six independent parameters of the Jones matrix are characterized from a single measurement [22]. However, several measurements are necessary for more conventional instruments including RAE, RCE, and PME. In RAE, for example, the two Fourier coefficients (α, β) are measured [Eq. (4.19)] and thus at least three measurements are required to characterize the Jones matrix completely [11].

The measurement of anisotropic samples by RAE can be expressed by replacing $\boldsymbol{S}$ in Eq. (4.14) with the Jones matrix of anisotropic samples $\boldsymbol{S}_{ani}$ [Eq. (6.14)]:

$$\boldsymbol{L}_{out} = \boldsymbol{A}\boldsymbol{R}(A)\boldsymbol{S}_{ani}\boldsymbol{R}(-P)\boldsymbol{P}\boldsymbol{L}_{in} \tag{6.68}$$

In Eq. (6.68), the term $\boldsymbol{R}(-P)\boldsymbol{P}\boldsymbol{L}_{\text{in}}$ can be simplified to

$$\boldsymbol{R}(-P)\boldsymbol{P}\boldsymbol{L}_{\text{in}} = \begin{bmatrix} \cos P & -\sin P \\ \sin P & \cos P \end{bmatrix} \begin{bmatrix} 1 & 0 \\ 0 & 0 \end{bmatrix} \begin{bmatrix} 1 \\ 0 \end{bmatrix} = \begin{bmatrix} \cos P \\ \sin P \end{bmatrix} = \cos P \begin{bmatrix} 1 \\ \tan P \end{bmatrix} \tag{6.69}$$

By substituting Eq. (6.69) into Eq. (6.68), we get

$$\begin{bmatrix} E_A \\ 0 \end{bmatrix} = \begin{bmatrix} 1 & 0 \\ 0 & 0 \end{bmatrix} \begin{bmatrix} \cos A & \sin A \\ -\sin A & \cos A \end{bmatrix} \begin{bmatrix} \rho_{\text{pp}} & \rho_{\text{ps}} \\ \rho_{\text{sp}} & 1 \end{bmatrix} \begin{bmatrix} 1 \\ \tan P \end{bmatrix} \tag{6.70}$$

In the above equation, proportional constants are neglected. Expansion of Eq. (6.70) yields

$$E_A = (\rho_{\text{pp}} + \rho_{\text{ps}} \tan P) \cos A + (\rho_{\text{sp}} + \tan P) \sin A \tag{6.71}$$

From Eq. (6.71), we obtain

$$I = |E_A|^2 = I_0 (1 + \alpha \cos 2A + \beta \sin 2A) \tag{6.72}$$

where (α, β) are the normalized Fourier coefficients expressed by the following equations [11]:

$$\alpha = \frac{|\rho_{\text{pp}} + \rho_{\text{ps}} \tan P|^2 - |\rho_{\text{sp}} + \tan P|^2}{|\rho_{\text{pp}} + \rho_{\text{ps}} \tan P|^2 + |\rho_{\text{sp}} + \tan P|^2} \tag{6.73a}$$

$$\beta = \frac{2\text{Re}\left[(\rho_{\text{pp}} + \rho_{\text{ps}} \tan P)(\rho_{\text{sp}} + \tan P)\right]}{|\rho_{\text{pp}} + \rho_{\text{ps}} \tan P|^2 + |\rho_{\text{sp}} + \tan P|^2} \tag{6.73b}$$

The above equations can be derived rather easily using the formulas described in Appendix 1(e) and (f). However, Eq. (6.73) has been modified slightly so that (α, β) become real numbers. When $P = 0°$ in Eq. (6.73) [i.e., $\tan P = 0$], (α, β) are characterized by ρ_{pp} and ρ_{sp} since only p-polarization illuminates a sample in this condition.

Eq. (6.73) shows clearly that (α, β) vary according to the setting of the polarizer angle P. We can determine each element of the Jones matrix by utilizing this characteristic. Let $(\alpha_{\text{ex}}, \beta_{\text{ex}})$ and $(\alpha_{\text{cal}}, \beta_{\text{cal}})$ be experimental and calculated Fourier coefficients. In the case of RAE, the experimental and calculated values are related by

$$\alpha_{\text{ex}}(P) = \alpha_{\text{cal}}(P, \rho_{\text{pp}}, \rho_{\text{ps}}, \rho_{\text{sp}}) \tag{6.74a}$$

$$\beta_{\text{ex}}(P) = \beta_{\text{cal}}(P, \rho_{\text{pp}}, \rho_{\text{ps}}, \rho_{\text{sp}}) \tag{6.74b}$$

The actual values of $(\alpha_{\text{cal}}, \beta_{\text{cal}})$ can be obtained from Eq. (6.73). The number of unknown parameters in Eq. (6.74) is six (i.e., $\psi_{\text{pp}}, \Delta_{\text{pp}}, \psi_{\text{ps}}, \Delta_{\text{ps}}, \psi_{\text{sp}}, \Delta_{\text{sp}}$). Thus, if there are at least three data sets of $(\alpha_{ex}, \beta_{ex})$ obtained at different polarizer angles, we can extract the six independent parameters from linear regression analysis of Eq. (6.74) [11]. This method has been employed widely to measure the Jones matrix of anisotropic samples. In particular, when ellipsometry measures all the elements of the Jones matrix, such measurements are often referred to as generalized ellipsometry [23,24]. The above technique can also be extended easily to RCE and RAE with compensator. If we use Eq. (6.70) and Table 3.2, RAE with compensator ($PCSA_R$) can be expressed as

$$\begin{bmatrix} E_A \\ 0 \end{bmatrix} = \begin{bmatrix} 1 & 0 \\ 0 & 0 \end{bmatrix} \begin{bmatrix} \cos A & \sin A \\ -\sin A & \cos A \end{bmatrix} \begin{bmatrix} \rho_{pp} & \rho_{ps} \\ \rho_{sp} & 1 \end{bmatrix} \begin{bmatrix} 1 & 0 \\ 0 & \exp(-i\delta) \end{bmatrix} \begin{bmatrix} 1 \\ \tan P \end{bmatrix} \quad (6.75)$$

Accordingly, (α, β) values of this instrument are given simply by replacing $\tan P$ in Eq. (6.73) with $\tan P \exp(-i\delta)$ [25].

For the characterization of anisotropic samples, transmission-type ellipsometry has also been employed [26–29]. This measurement is particularly sensitive to in-plane optical anisotropy, as we have seen in Fig. 6.16. In this measurement, however, a refractive index difference, rather than the absolute values of optical constants, is determined [26,27]. If we combine reflection and transmission ellipsometry, more reliable results can be obtained [27–29]. When an anisotropic sample depolarizes incident light, however, the characterization should be performed by using Mueller matrix ellipsometry (see Section 4.2.7).

6.5.2 DATA ANALYSIS METHODS

In principle, the data analysis procedure for anisotropic samples is exactly the same as that for isotropic samples, and only measurement and calculation methods are different. As mentioned above, the calculation of (ψ, Δ) for anisotropic samples is generally performed using the Fresnel equations described in Section 6.2 or the 4×4 matrix method explained in Section 6.3. In the case of anisotropic samples, however, there are many measurement and analytical parameters, compared with isotropic samples. With respect to uniaxial materials, for example, the unknown parameters for optical constants are $(N_o, N_e, \phi_E, \theta_E)$, where N_o and N_e represent the complex refractive indices of the ordinary and extraordinary rays, respectively. If a sample is a biaxial crystal, the number of unknown parameters increases to six (i.e., $N_x, N_y, N_z, \phi_E, \theta_E, \psi_E$). If we count refractive index and extinction coefficient separately, the total number of unknown parameters becomes nine. In addition, although the Euler angles are generally independent of wavelength, optical constants change with wavelength. Furthermore, if anisotropic layers are thin films, the thickness of these layers has to be determined. Accordingly, in order to characterize anisotropic samples properly, we need additional data sets in the data analysis.

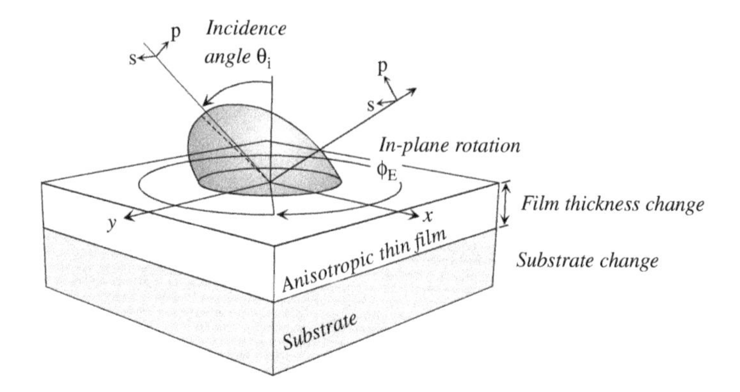

Figure 6.20 Ellipsometry measurements for an air/(anisotropic thin film)/(isotropic substrate) structure.

Figure 6.20 illustrates variable parameters that can be used as additional data sets in the data analysis of anisotropic samples. In the characterization of anisotropic samples, the incident angle θ_i or rotation angle ϕ_E is often varied to obtain sufficient data for the analysis[11,27–31]. In the case of anisotropic thin films, data may be obtained from samples with different film thicknesses [30]. If possible, the substrate on which an anisotropic layer is formed is changed [31]. It is helpful to employ transmission ellipsometry at normal or oblique incidence if samples are transparent [27–29]. In order to obtain reliable results, at least one or two independent variables should be changed. In linear regression analysis, measured parameters (i.e., ψ_{pp}, Δ_{pp}, ψ_{ps}, Δ_{ps}, ψ_{sp}, Δ_{sp}) are fitted simultaneously. Thus, recent advances in computer technologies have been quite beneficial, particularly to the data analysis of anisotropic samples. Examples for the data analysis of anisotropic samples will be shown in Chapter 7.5.

REFERENCES

[1] E. Hecht, *Optics*, 4th edition, Addison Wesley, San Francisco (2002).
[2] P. Yeh, *Optical Waves in Layered Media*, Wiley-Interscience Publication, New York (1988).
[3] R. W. Collins and A. S. Ferlauto, Optical Physics of Materials, in *Handbook of Ellipsometry*, edited by H. G. Tompkins and E. A. Irene, Chapter 2, 93–235, William Andrew, New York (2005).
[4] M. Born and E. Wolf, *Principles of Optics*, 7th edition, Cambridge University Press, Cambridge (1999).
[5] P. Yeh, Electromagnetic propagation in birefringent layered media, *J. Opt. Soc. Am.*, **69** (1979) 742–756.
[6] P. Yeh, Optics of anisotropic layered media: a new 4 × 4 matrix algebra, *Surf. Sci.*, **96** (1980) 41–53.
[7] G. B. Arfken and H. J. Weber, *Mathematical Methods for Physicists*, 4th edition, Academic Press, San Diego (1995).

[8] S. Teitler and B. W. Henvis, Refraction in stratified, anisotropic media, *J. Opt. Soc. Am.*, **60** (1970) 830–834.
[9] R. M. A. Azzam and N. M. Bashara, *Ellipsometry and Polarized Light*, North-Holland, Amsterdam (1977).
[10] G. E. Jellison, Jr, Data analysis for spectroscopic ellipsometry, in *Handbook of Ellipsometry*, edited by H. G. Tompkins and E. A. Irene, Chapter 3, 237–296, William Andrew, New York (2005).
[11] M. Schubert, B. Rheinländer, J. A. Woollam, B. Johs, and C. M. Herzinger, Extension of rotating-analyzer ellipsometry to generalized ellipsometry: determination of the dielectric function tensor from uniaxial TiO_2, *J. Opt. Soc. Am. A*, **13** (1996) 875–883.
[12] S. Huard, *Polarization of Light*, John Wiley & Sons, Ltd, Chichester (1997).
[13] D. den Engelsen, Ellipsometry of anisotropic films, *J. Opt. Soc. Am.*, **61** (1971) 1460–1466.
[14] D. W. Berreman, Optics in stratified and anisotropic media: 4×4-matrix formulation, *J. Opt. Soc. Am*, **62** (1972) 502–510.
[15] K. Eidner, Light propagation in stratified anisotropic media: orthogonality and symmetry properties of the 4×4 matrix formalisms, *J. Opt. Soc. Am. A*, **6** (1989) 1657–1660.
[16] M. Schubert, Polarization-dependent optical parameters of arbitrarily anisotropic homogeneous layered systems, *Phys. Rev. B*, **53** (1996) 4265–4274.
[17] M. Schubert, Theory and application of generalized ellipsometry, in *Handbook of Ellipsometry*, edited by H. G. Tompkins and E. A. Irene, Chapter 9, 637–717, William Andrew, New York (2005).
[18] M. Schubert, *Infrared Ellipsometry on Semiconductor Layer Structures: Phonons, Plasmons, and Polaritons*, Springer, Heidelberg (2004).
[19] Although the explicit equations for the calculation of the eigenvalues are described in Ref. [16], it appears that there are some mistakes in the equations of this reference.
[20] H. Wöhler, G. Haas, M. Fritsch, and D. A. Mlynski, Faster 4×4 matrix method for uniaxial inhomogeneous media, *J. Opt. Soc. Am. A*, **5** (1988) 1554–1557.
[21] H. Wöhler, M. Fritsch, G. Haas, and D. A. Mlynski, Characteristic matrix method for stratified anisotropic media: optical properties of special configurations, *J. Opt. Soc. Am. A*, **8** (1991) 536–540.
[22] For a review, see C. Chen, I. An, G. M. Ferreira, N. J. Podraza, J. A. Zapien, and R. W. Collins, Multichannel Mueller matrix ellipsometer based on the dual rotating compensator principle, *Thin Solid Films*, **455–456** (2004) 14–23.
[23] R. M. A. Azzam and N. M. Bashara, Generalized ellipsometry for surfaces with directional preference: application to diffraction gratings, *J. Opt. Soc. Am.*, **62** (1972) 1521–1523.
[24] R. M. A. Azzam and N. M. Bashara, Application of generalized ellipsometry to anisotropic crystals, *J. Opt. Soc. Am.*, **64** (1974) 128–133.
[25] M. Schubert, Generalized ellipsometry and complex optical systems, *Thin Solid Films*, **313–314** (1998) 323–332.
[26] R. A. Yarussi, A. R. Heyd, H. V. Nguyen, and R. W. Collins, Multichannel transmission ellipsometer for characterization of anisotropic optical materials, *J. Opt. Soc. Am. A*, **11** (1994) 2320–2330.
[27] A. Sassella, A. Borghesi, Th. Wagner, and J. Hilfiker, Generalized anisotropic ellipsometry applied to an organic single crystal: potassium acid phthalate, *J. Appl. Phys.*, **90** (2001) 3838–3842.
[28] T. Wagner, J. N. Hilfiker, T. E. Tiwald, C. L. Bungay, and S. Zollner, Materials characterization in the vacuum unltraviolet with variable angle spectroscopic ellipsometry, *Phys. Stat. Sol. A*, **188** (2001) 1553–1562.
[29] C. M. Ramsdale and N. C. Greenham, Ellipsometric determination of anisotropic optical constants in electroluminescent conjugated polymers, *Adv. Mater.*, **14** (2002) 212–215.

[30] T. U. Kampen, A. M. Paraian, U. Rossow, S. Park, G. Salvan, Th. Wagner, M. Friedrich and D. R. T. Zahn, Optical anisotropy of organic layers deposited on semiconductor surfaces, *Phys. Stat. Sol. A*, **188** (2001) 1307–1317.
[31] L. A. A. Pettersson, F. Carlsson, O. Inganäs, and H. Arwin, Spectroscopic ellipsometry studies of the optical properties of doped poly(3,4-ethylenedioxythiophene): an anisotropic metal, *Thin Solid Films*, **313–314** (1998) 356–361.

7 Data Analysis Examples

This chapter will introduce data analysis examples for insulators, semiconductors, metals, and organic materials. Here, we will treat the analysis of (ψ, Δ) spectra obtained from *ex situ* measurements performed after sample preparation. When the dielectric functions of samples are known, film thickness can be estimated quite easily. In this chapter, therefore, data analysis examples that include modeling of dielectric functions will be explained. Analysis examples for anisotropic materials will also be given in this chapter. Since spectroscopic ellipsometry has been applied extensively for semiconductor characterization, we will address the dielectric function modeling and analysis of semiconductors in more detail.

7.1 INSULATORS

In the UV/visible region, the data analysis of insulators can be performed rather easily, as the dielectric functions commonly exhibit simple structures that can be represented by the Cauchy and Sellmeier models. In thermal oxides, surfaces are very flat and the influence of overlayers can be neglected. Thus, a simple optical model (oxide/substrate structure) can be employed for the data analysis. In this section, we will look at the analysis of a SiO_2 thermal oxide formed on crystalline Si (c-Si), as the simplest example of ellipsometry data analysis. This section will also introduce the analysis of SiO_2/c-Si interface structures, as a more advanced analysis.

7.1.1 ANALYSIS EXAMPLES

Figure 7.1 shows an optical model for a SiO_2 thermal oxide formed on a c-Si substrate. This optical model corresponds to the one shown in Fig. 5.32(b). With respect to the substrate, when the substrate thickness (d_{sub}) is sufficiently larger

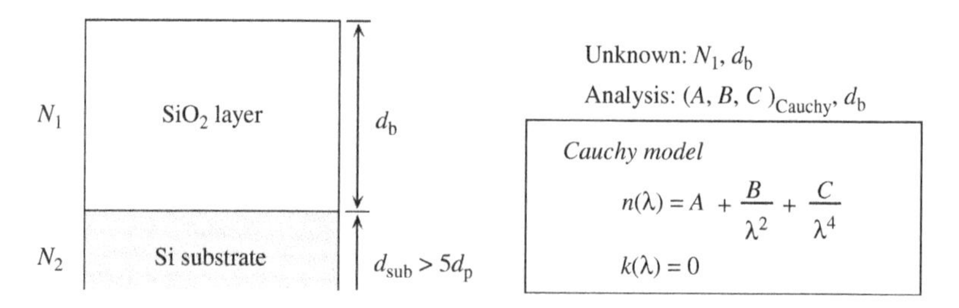

Figure 7.1 Optical model for a SiO_2 thermal oxide formed on a crystalline Si substrate. N_1 and N_2 represent the complex refractive indices of the SiO_2 bulk layer and Si substrate, respectively. The thicknesses of the bulk layer and substrate are denoted as d_b and d_{sub}, respectively. d_p indicates the penetration depth of light.

than the penetration depth (d_p) of light (i.e., $d_{sub} > 5d_p$), only light reflection at the SiO_2/c-Si interface is taken into account, since there is no backside reflection (see Section 5.1.3). The (ψ, Δ) of this optical model can be calculated from Eq. (5.2), and the parameters in this equation are described by

$$\tan\psi \exp(i\Delta) = \rho(N_0, N_1, N_2, d_b, \theta_0, \theta_1) \tag{7.1}$$

where N_0, N_1, and N_2 represent the complex refractive indices of the ambient (air), SiO_2 bulk layer, and c-Si substrate, respectively. Here, we assume that the angle of incidence θ_0 is known, although θ_0 can also be employed as an analysis parameter. The transmission angle in the SiO_2 layer (θ_1) is obtained by applying Snell's law (see Section 2.3.1). If we use $N_0 = 1 - i0$ and the reported value for N_2, the unknown parameters in this optical model are N_1 and the bulk layer thickness d_b. The resonant frequency for the electric polarization of SiO_2 is in the UV region (En $\sim$ 10 eV) [1]. Thus, in the region En = 1–5 eV, dielectric function modeling can be performed from the Cauchy and Sellmeier models (see Fig. 5.10). In the Cauchy model, N_1 is expressed by three parameters (A, B, C) [Eq. (5.17)]. Accordingly, when we apply the Cauchy model, the final analysis parameters become (A, B, C, d_b).

Figure 7.2(a) shows the (ψ, Δ) spectra obtained from a sample having a SiO_2/c-Si structure. This measurement was performed using a PC_RSA instrument using $\theta_0 = 75°$. The solid lines in the figure show the calculated spectra obtained from the linear regression analysis (see Section 5.5.1). From the analysis, $A = 1.4573$, $B = 2.383 \times 10^{-3}\,\mu m^2$, $C = 9.750 \times 10^{-5}\,\mu m^4$, and $d_b = 1049.5 \pm 0.5$ Å are estimated. Figure 7.2(b) shows the refractive index spectrum of the SiO_2 layer calculated from these (A, B, C) values. From a similar analysis using the Sellmeier model, $A = 1$, $B = 1.1082$, $\lambda_0 = 0.096\,\mu m$ can be obtained, as the parameter values of Eq. (5.16). The dotted line in Fig. 7.2(b) represents the result estimated from the Sellmeier model. The analysis using the Sellmeier model provides $d_b = 1051.5 \pm 0.5$ Å, which is quite similar to the one determined from the Cauchy model.

As we have seen in Fig. 5.39, the analysis of spectroscopic ellipsometry is performed by minimizing the fitting error. The fitting error is evaluated by the

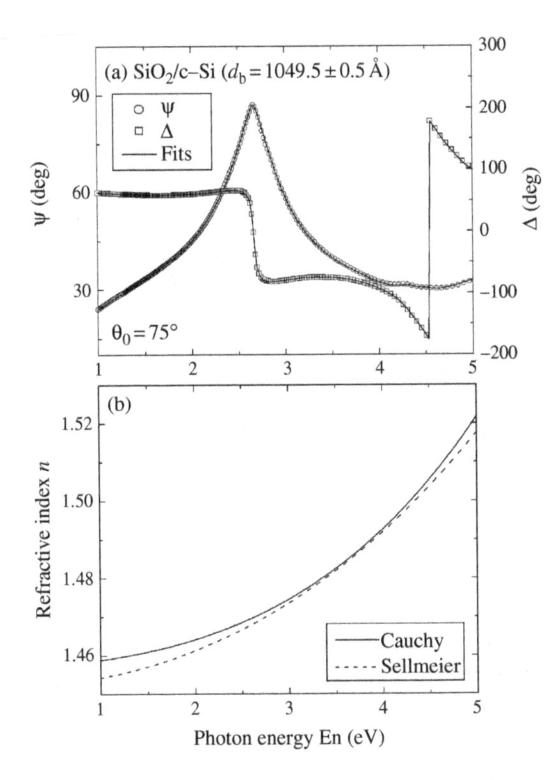

Figure 7.2 (a) (ψ, Δ) spectra obtained from a SiO_2 thermal oxide/crystalline Si (c-Si) substrate structure and (b) refractive index spectra of the SiO_2 thermal oxide layer deduced from the Cauchy and Sellmeier models. In (a), the incidence angle of the measurement is $\theta_0 = 75°$.

unbiased estimator σ or biased estimator χ (see Section 5.5.2). In the analysis example shown in Fig. 7.2, σ of the Cauchy model is smaller than that of the Sellmeier model by 10 %, and we choose the result obtained from the Cauchy model as the final result. Furthermore, since the refractive index spectrum of the thermal oxide determined from the Cauchy model is quite similar to the one reported previously [1–3], we can judge that the result obtained is appropriate. In spectroscopic ellipsometry, the optical constants and thickness of a sample are determined from such an analysis. In particular, since the measurement of spectroscopic ellipsometry is performed over a wide wavelength region, even if there are several analysis parameters, these values can be estimated from linear regression analysis. In contrast, if the number of analysis parameters exceeds the number of measured values, we cannot perform the analysis. Accordingly, more reliable analysis can be performed from (ψ, Δ) spectra measured over a wider wavelength region with a larger number of data points.

Figure 7.3 shows the calculated spectra obtained from the Cauchy model shown in Fig. 7.2(b) when the SiO_2 thickness is varied intentionally. At En of $\psi \sim 90°$, the phase shift between the primary and secondary beams is half a wavelength,

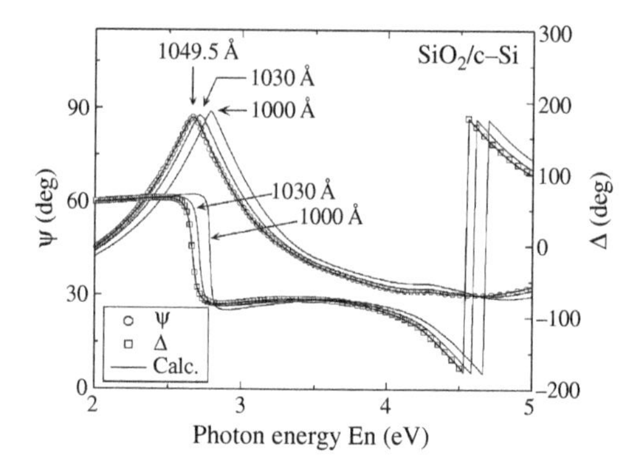

Figure 7.3 Variation in (ψ, Δ) spectra with the SiO_2 layer thickness in a SiO_2/c-Si structure.

and the thickness of a thin film determines this position (see Section 5.1.2). In Fig. 7.3, therefore, the position of $\psi \sim 90°$ changes with d_b. Even when d_b varies slightly, the (ψ, Δ) spectra change over the whole energy region, and the fitting error changes significantly. As a result, the confidence limit of d_b shows a very small value of ± 0.5 Å in this analysis (see Fig. 5.40). This example shows clearly that we can determine film thickness very precisely from spectroscopic ellipsometry. It has been reported that SiO_2 thicknesses estimated from ellipsometry show excellent agreement with those characterized by transmission electron microscope (TEM) [4], X-ray photoemission spectroscopy (XPS) [5,6], and capacitance voltage (CV) measurements [5,7].

When we characterize the refractive index spectrum and d_b of SiO_2 layers, the above analysis is generally sufficient. If we employ the reported dielectric function for SiO_2 [1–3], d_b is determined more easily. The fitting in the analysis can be improved by increasing the number of parameters in the Cauchy or Sellmeier models. In the case of the Sellmeier model, for example, the number of j in Eq. (5.16) is increased. Nevertheless, for the accurate characterization of SiO_2 layers, it is necessary to take the influence of the interface layer into account, as will be discussed in Section 7.1.2.

7.1.2 ADVANCED ANALYSIS

In this section, we will see analysis examples that include the effect of an interface layer formed at a SiO_2/c-Si interface [2,3]. Figure 7.4 shows (a) optical model without the interface layer and (b) optical model with the interface layer. When the dielectric function of SiO_2 is expressed using the Sellmeier model, the analysis parameters are (B, λ_0), if $A = 1$ is assumed. Here, we calculate the complex refractive index of the interface layer in Fig. 7.4(b) from the effective medium approximation using the complex refractive indices of SiO_2 and c-Si. The analysis

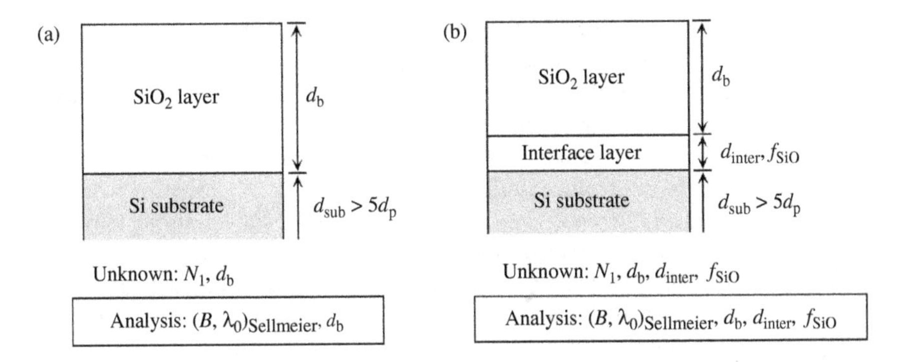

Figure 7.4 (a) Optical model without interface layer and (b) optical model with interface layer in a SiO_2/c-Si structure. In (b), f_{SiO} indicates the volume fraction of the SiO_2 component in the interface layer.

parameters of this interface layer are the thickness of the interface layer (d_{inter}) and the volume fraction of SiO_2 within the interface layer (f_{SiO}), similar to Fig. 5.27. As shown in Fig. 7.4(b), there are five analysis parameters for the optical model with the interface layer.

Table 7.1 shows the result obtained from the analysis using the optical models shown in Fig. 7.4 [2]. In Table 7.1, the numerical values in italic form indicate the fixed parameters in the analysis, and thus these parameter values were not changed during the linear regression analysis. Fits 1–3 in Table 7.1 represent the analyses performed without the interface layer, while the interface layer is included in the analyses shown in fits 4–9. In fit 1, (B, λ_0) values obtained from fused silica were used, but the value of χ^2 is quite large. This indicates that the refractive indices of thermally grown SiO_2 and fused silica are different even though these materials have the same chemical composition of SiO_2. For fit 3, (B, λ_0) values were employed as fitting parameters, and the χ^2 value is close to the ideal value

Table 7.1 Analysis of SiO_2/c-Si structure

Fit	d_b (nm)	B	λ_0 (nm)	d_{inter}(nm)	f_{SiO} (vol.%)	χ^2 [a]
1	59.43 ± 0.02	*1.099*[b]	*92.27*	*0*	*0*	18.53
2	59.79 ± 0.02	1.127 ± 0.001	*92.27*	*0*	*0*	2.34
3	58.81 ± 0.03	1.137 ± 0.002	87.4 ± 0.7	*0*	*0*	1.15
4	58.80 ± 0.02	*1.099*	*92.27*	1.09 ± 0.03	*50*	2.07
5	58.74 ± 0.03	1.114 ± 0.002	*92.27*	0.61 ± 0.08	*50*	1.04
6	58.76 ± 0.04	1.122 ± 0.007	90.4 ± 1.5	0.41 ± 0.18	*50*	1.01
7	58.70 ± 0.33	*1.099*	*92.27*	1.04 ± 0.12	37 ± 47	1.87
8	58.72 ± 0.19	1.112 ± 0.004	*92.27*	0.62 ± 0.09	43 ± 94	1.03
9	58.75 ± 0.02	1.118 ± 0.008	91.0 ± 1.8	0.49 ± 0.22	45 ± 141	1.01

[a] biased estimator, [b] numerical values in italic form show the fixed parameters in the analysis. Data from Jellison, Jr (Ref.[2])

of $\chi^2 = 1$. In the analysis of fits 4–6, f_{SiO} of the interface layer was fixed to 0.5. As confirmed from fits 4–6, when the number of analysis parameters increases, the confidence limits of the analysis parameters also increase. In other words, the precision of the analysis degrades. In general, this phenomenon is observed when there are strong correlations between analysis parameters. If there are no such correlations, the confidence limits remain unchanged. In particular, when SiO_2 layers are characterized in the UV/visible region, n and d_b show a strong correlation described by the nd product (see Section 5.1.2), as SiO_2 exhibits no light absorption in the region. Accordingly, even if we introduce an interface layer that has a similar refractive index, d_{inter} and d_b cannot be distinguished well, and the confidence limit of d_{inter} increases. In fits 7–9, f_{SiO} was used as an analysis parameter, but the confidence limits of f_{SiO} are quite large. Thus, we cannot determine f_{SiO}, at least from these analyses. Among the analyses shown in Table 7.1, we can choose fit 3 or fit 5 as the final result, since these analyses show the smaller values for χ^2 and the confidence limit. From the value of χ^2, we can judge that fit 5 is slightly better than fit 3 [2]. It can be understood from the above discussion that we need to minimize χ (or σ) and the confidence limits by selecting appropriate analysis parameters that show weak correlations with other analysis parameters.

Although the influence of the SiO_2 interface layer is not clear in the above example, it has been reported that the characterization of SiO_2 bulk layers is rather strongly affected by the presence of the SiO_2 interface layer [3,4]. Figure 7.5 shows the refractive index spectra extracted from thermally grown SiO_2 layers

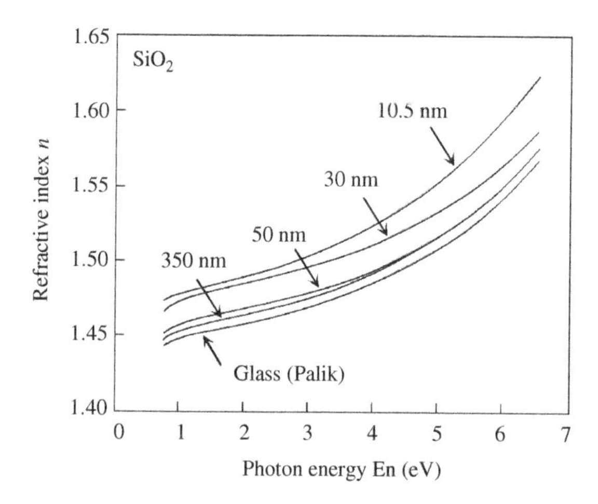

Figure 7.5 Refractive index spectra extracted from thermally grown SiO_2 layers with different thicknesses. In the figure, the bulk layer thickness (d_b) of each layer is indicated. The spectrum denoted as 'Glass' represents the one obtained from a thick glass substrate. Reprinted with permission from C. M. Herzinger, B. Johs, W. A. McGahan, J. A. Woollam, and W. Paulson, Ellipsometric determination of optical constants for silicon and thermally grown silicon dioxide via a multi-sample, multi-wavelength, multi-angle investigation, *Journal of Applied Physics*, **83**, 3323–3336 (1998). Copyright 1998, American Institute of Physics.

with different thicknesses [3]. This analysis was performed using the optical model without the interface layer [see Fig. 7.4(a)], and each d_b estimated from the analysis is indicated in Fig. 7.5. The spectrum denoted as 'Glass' represents the one obtained from a thick glass substrate [1]. The refractive index spectra shown in Fig. 7.5 can be calculated from the model expressed as follows [3]:

$$\varepsilon_1 = n^2 = A + \frac{B\lambda^2}{\lambda^2 - \lambda_0^2} - C\lambda^2 \quad \varepsilon_2 = 0 \tag{7.2}$$

In Eq. (7.2), the term $-C\lambda^2$ is added to the Sellmeier model to express the rapid reduction in n below 1.0 eV (see Fig. 7.5). This reduction is caused by the light absorption of SiO_2 in the infrared region (see Fig. 2.11). It can be seen from Fig. 7.5 that the overall refractive index decreases with increasing d_b. This implies that there exists an interface layer with a high refractive index at the SiO_2/c-Si interface [3,4,8–10]. In other words, the effect of the interface layer becomes insignificant as d_b increases, since the high refractive index of the interface layer is averaged out by the thicker bulk layer. As confirmed from Fig. 7.5, the influence of the interface layer is rather strong when $d_b < 500$ Å. Although not conclusive [5,11], the presence of the interface layer has been attributed to the formation of suboxide [$SiO_x (x < 2)$] at the SiO_2/c-Si interface [8,9]. Since the refractive index in the UV/visible region is essentially determined by electric polarization (see Section 2.2.3), the refractive index of SiO_x is expected to be intermediate between c-Si [$\varepsilon_s = \varepsilon_\infty = 11.6\,(n = 3.41)$] and SiO_2 [$\varepsilon_\infty = 2.1\,(n = 1.46)$]. Consequently, we observe a higher refractive index in SiO_x than in SiO_2. The formation of high stress at the SiO_2/c-Si interface has also been reported to change the refractive index in the interface region [12].

Unfortunately, the characterization of the interface layer has been rather difficult due to a strong correlation between d_{inter} and n_{inter} (n of the interface layer) in the analysis [3]. Furthermore, when d_b<100 Å, we cannot even evaluate d_b and n (bulk layer) independently [see Fig. 5.7]. In this case, it is necessary to estimate d_b from other measurement techniques, in order to characterize the average n in the SiO_2 bulk layer [10]. So far, with respect to the interface layer formed in SiO_2/c-Si structures, $d_{inter} = 7-10$ Å and $n_{inter} = 2.0$–3.2 (λ = 5461Å) have been reported [3,8,9]. In the study shown in Fig. 7.5, n_{inter} was determined by analyzing the samples shown in Fig. 7.5 simultaneously. From this study, the parameters in Eq. (7.2), $A = 3.5$, $B = 0.69417$, $\lambda_0 = 0.115\,\mu$m, and $C = 0.010591\,\mu\text{m}^{-2}$, have been obtained for the interface layer, while A = 1.30, $B = 0.81996$, $\lambda_0 = 0.10396\,\mu$m, and C = $0.01082\,\mu\text{m}^{-2}$ have been reported for the SiO_2 bulk layer [3]. Similar interface layer formation has also been observed for HfO_2 [13,14], which is now being studied intensively as an alternative gate oxide in ULSI devices. In a HfO_2/c-Si structure, however, the average refractive index of HfO_2 layers *increases* with d_b, since the refractive index of the interface layer is *lower* than that of the HfO_2 bulk layer [13]. It has been suggested that the low refractive index in the interface layer is caused by the formation of SiO_2 or $HfSiO_x$ [13,14]. The above examples show that we can examine interface structures of insulators from the dependence of optical properties on film thickness.

7.2 SEMICONDUCTORS

The data analysis of semiconductors can be performed in the same manner as the above examples, and the thickness of semiconductor layers is estimated easily if the dielectric functions of samples are known. The optical constants of various semiconductors have been tabulated in several books [15,16], and we can perform data analysis by utilizing these optical constants. In addition, tabulated optical constants for major semiconductors have been reported in several articles; (Si, Ge, GaP, GaAs, GaSb, InP, InAs, InSb) [17], $Al_xGa_{1-x}As$ [18], $Si_{1-x}Ge_x$ [19]. When the dielectric functions of samples are unknown, however, the analysis often becomes complicated, since the dielectric functions of semiconductors generally show complicated structures. In this case, the mathematical inversion described in Section 5.5.3 can be employed to extract the dielectric function of a sample. From the analysis of the extracted dielectric function, we can estimate the temperature, strain, and grain size of samples. Conversely, when the temperature or grain size of a sample differs, it may become difficult to use a reported dielectric function for the sample analysis. In this section, we will address the dielectric function modeling, analysis examples, and dielectric function analysis for semiconductors.

7.2.1 OPTICAL TRANSITIONS IN SEMICONDUCTORS

Figure 7.6(a) illustrates the optical transition process in a semiconductor in En–$\boldsymbol{K}$ space. In this figure, $\boldsymbol{K}$ represents the momentum (wave vector) of an electron. In a semiconductor, conduction and valence bands, separated by certain energy, are formed and each band shows a parabolic dependence on $\boldsymbol{K}$. In Fig. 7.6(a), $\boldsymbol{K}$ at the bottom of the conduction band coincides with $\boldsymbol{K}$ at the top of the valence

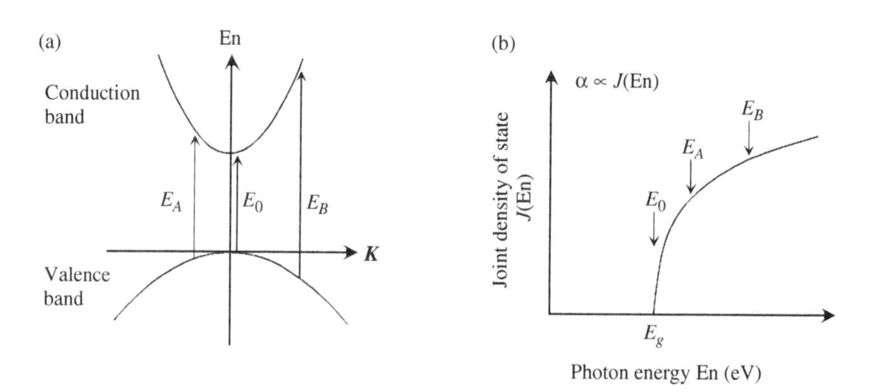

Figure 7.6 (a) Optical transitions in a direct bandgap semiconductor in En–$\boldsymbol{K}$ space and (b) joint density of state J(En) versus En. The optical transitions (E_0, E_A, E_B) shown in (a) correspond to those in (b). In (b), E_g represents the bandgap of the semiconductor.

band, and semiconductors of this type are called direct bandgap semiconductors. On the other hand, when the positions of $\boldsymbol{K}$ differ, such semiconductors are called indirect bandgap semiconductors. As shown in Fig. 7.6(a), if light illuminates a direct bandgap semiconductor, the electrons in the semiconductor are excited from the valence band to the conduction band. This optical transition occurs vertically in En–$\boldsymbol{K}$ space so that the transition satisfies the momentum conservation law. In indirect bandgap materials, however, optical transitions occur at the different positions of $\boldsymbol{K}$ with the aid of lattice vibration (phonon).

Figure 7.6(b) illustrates the joint density of state [J(En)] in the direct bandgap semiconductor shown in Fig. 7.6(a). The joint density of state represents the density of paired initial–final states that can participate in the optical transition with a certain En. Here, the initial and final states indicate the valence and conduction bands, respectively. As shown in Fig. 7.6(b), the optical transition in the semiconductor begins at En corresponding to the bandgap E_g, and J(En) increases with En at En > E_g. E_0 in Fig. 7.6 is particularly known as the critical point. Since the absorption coefficient α is proportional to J(En), we can estimate E_g from α(En). With respect to direct bandgap semiconductors, α(En) is expressed by the following equation [20]:

$$\alpha(\mathrm{En}) = A(\mathrm{En} - E_g)^{1/2} \tag{7.3}$$

On the other hand, the following equation holds in indirect bandgap semiconductors [20]:

$$\alpha(\mathrm{En}) = A(\mathrm{En} - E_g)^2 \tag{7.4}$$

In Eqs. (7.3) and (7.4), A shows the proportional constant. In the case of direct bandgap semiconductors, if we plot α^2 versus En, E_g is determined from the intercept at $\alpha^2 = 0$ (see Fig. 7.25). Similarly, E_g of indirect bandgap materials is estimated by plotting $\alpha^{1/2}$ versus En. The bandgap of amorphous semiconductors can be expressed from the Tauc gap [21] or Cody gap [22]. The Tauc gap has been given by Eq. (5.18) and the Cody gap is expressed as follows [22]:

$$\varepsilon_2 = A(\mathrm{En} - E_g)^2 \tag{7.5}$$

Thus, the Cody gap can be determined from the energy position of $(\varepsilon_2)^{1/2} = 0$ in the plot of $(\varepsilon_2)^{1/2}$ versus En. It has been known that the Cody model provides superior fitting to experimental spectra, compared with the Tauc model [22,23].

Figure 7.7 shows the band structures of (a) GaAs crystal [24] and (b) Si crystal [25]. The GaAs crystal is a direct bandgap material and the structure around the Γ point can be approximated by the parabolic bands shown in Fig. 7.6(a). As indicated by the arrows in Fig. 7.7(a), there exist various optical transitions in GaAs. Energy positions at which these optical transitions occur are also referred to as critical

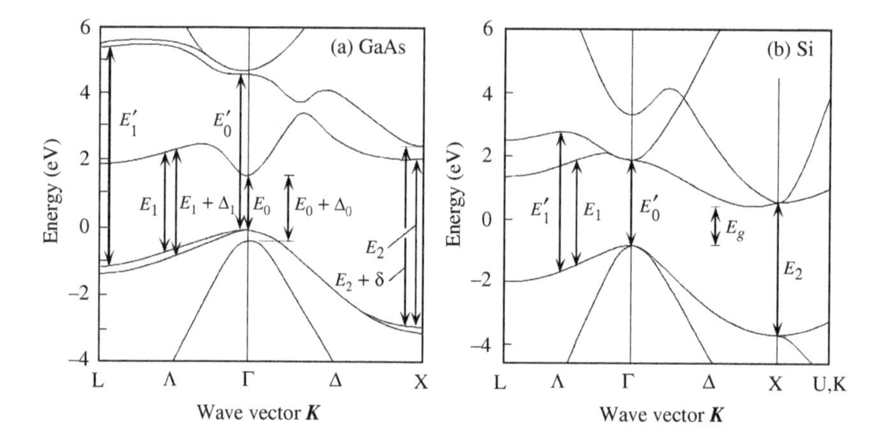

Figure 7.7 Band structures of (a) GaAs and (b) Si. Drawing (a): Adapted with permission from J. R. Chelikowsky and M. L. Cohen, Nonlocal pseudopotential calculations for the electronic structure of eleven diamond and zinc-blend semiconductors, *Phys. Rev. B*, **14** (1976) 556–582. Copyright 1976, the American Physical Society. Drawing (b): Adapted with permission from U. Schmid, N. E. Christensen, and M. Cardona, Relativistic band structure of Si, Ge, and GeSi: Inversion-asymmetry effects, *Phys. Rev. B*, **41** (1990) 5919–5930. Copyright 1990, the American Physical Society.

points, and each critical point shows specific $J(\mathrm{En})$ according to the types of optical transitions [1,20,26]. In photoluminescence measurement, sample characterization is performed only in the region of E_0 at the Γ point. In contrast, spectroscopic ellipsometry enables us to characterize various interband transitions shown in the figure. On the other hand, c-Si is an indirect bandgap material, and the positions at the top of the valence band and bottom of the conduction band are different. In the ε_2 spectrum of c-Si [Fig. 5.11(b)], the optical transitions of E_0' and E_1 in Fig. 7.7(b) can be seen at 3.4 eV, while the optical transitions of E_2 and E_1' appear at 4.25 eV and 5.3 eV, respectively.

7.2.2 MODELING OF DIELECTRIC FUNCTIONS

In dielectric function modeling of semiconductors, the optical transition at each critical point is modeled. As shown in Fig. 7.7, there are many critical points in a semiconductor crystal, and dielectric function modeling generally becomes complicated. In a wavelength region where light absorption is small, the dielectric function of a semiconductor changes smoothly, and we can often model the dielectric function by

$$n(\mathrm{En}) = A + B\mathrm{En}^2 + C\mathrm{En}^4 \tag{7.6a}$$

$$k(\mathrm{En}) = D\exp[(\mathrm{En} - E_g)/E_u]. \tag{7.6b}$$

Eq. (7.6a) represents the Cauchy model expressed by En, and Eq. (7.6b) shows the light absorption by the Urbach tail [20,26]. The Urbach tail is generated by the imperfection or doping of semiconductor crystals, and the light absorption by the tail state near E_g is described by the Urbach energy E_u. In some cases, Eq. (7.6b) can be utilized to express optical transitions that have no direct relation to the Urbach tail. For example, it is possible to perform modeling of c-Si using the parameters (A, B, C, D, E_g, E_u) in Eq. (7.6) in an energy region of En = 1–3 eV to some extent. Such modeling provides an effective way for estimating layer thickness that can be used for mathematical inversion (see Section 5.5.3). For Eq. (7.6a), the Sellmeier model can also be applied [27,28]. However, keep in mind that Eq. (7.6) does not satisfy the Kramers–Kronig relations (see Section 5.2.6).

There is a more realistic model called the harmonic oscillator approximation (HOA) in which quantum mechanical interpretation is included [29,30]:

$$\varepsilon(\mathrm{En}) = 1 + \sum_j \left(\frac{A_j}{\mathrm{En} + \mathrm{En}_{0j} - \mathrm{i}\Gamma_j} - \frac{A_j}{\mathrm{En} - \mathrm{En}_{0j} - \mathrm{i}\Gamma_j} \right) \tag{7.7}$$

The above equation shows the case for $\varepsilon \equiv \varepsilon_1 - \mathrm{i}\varepsilon_2$, and we must replace $-\mathrm{i}$ in Eq. (7.7) with $+\mathrm{i}$ when $\varepsilon \equiv \varepsilon_1 + \mathrm{i}\varepsilon_2$ is used [29,30]. By rearranging Eq. (7.7), we get

$$\varepsilon(\mathrm{En}) = 1 + \sum_j \left(\frac{2A_j \mathrm{En}_{0j}}{\mathrm{En}_{0j}^2 - \mathrm{En}^2 + \mathrm{i}2\Gamma_j \mathrm{En} + \Gamma_j^2} \right) \tag{7.8}$$

When the dielectric function of GaAs is expressed by HOA, we obtain $\Gamma < 1$ [30]. In particular, if $\Gamma \ll 1$, we get $\Gamma^2 \to 0$ in Eq. (7.8). In this case, Eq. (7.8) is reduced to the Lorentz model described in Eqs. (5.9) and (5.10). In the quantum mechanical model, however, the physical meaning of Γ becomes clearer, and Γ basically represents the average lifetime of electrons excited from the valence band to the conduction band [23,31]. If the average lifetime of electrons is given by $\langle \tau \rangle$, we obtain $\Gamma = \langle \tau \rangle^{-1}$, similar to the Drude model (see Section 5.2.5). Accordingly, the half-width of a ε_2 peak becomes narrower as the lifetime of electrons in the excited state increases.

Figure 7.8 shows the dielectric function of GaAs expressed from HOA. Table 7.2 shows the parameter values of HOA used in the calculation of Fig. 7.8 [30]. It can be seen from Fig. 7.8 that HOA exhibits a shape similar to the Lorentz model. The dielectric function of GaAs shown in Fig. 7.8 is calculated from seven oscillators using a total of 21 parameters. Naturally, the fitting to experimental spectra improves as the number of parameters increases. A model in which the phase factor ϕ is added to Eq. (7.8) has also been proposed [32]:

$$\varepsilon(\mathrm{En}) = 1 + \sum_j \left[\frac{A_j \exp(-\mathrm{i}\phi_j)}{\mathrm{En} + \mathrm{En}_{0j} - \mathrm{i}\Gamma_j} - \frac{A_j \exp(-\mathrm{i}\phi_j)}{\mathrm{En} - \mathrm{En}_{0j} - \mathrm{i}\Gamma_j} \right] \tag{7.9}$$

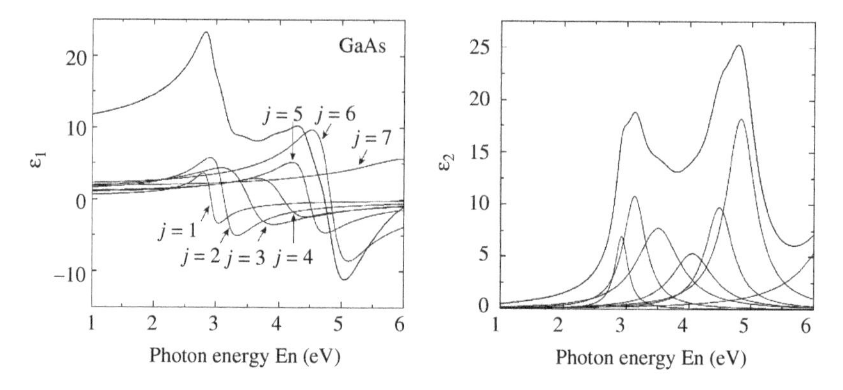

Figure 7.8 Dielectric function of GaAs expressed from harmonic oscillator approximation (HOA). This dielectric function was calculated by using parameter values shown in Table 7.2.

Table 7.2 Modeling of GaAs by HOA

j	A[a]	Γ[a]	En_0[a]	CP[b]
1	0.8688	0.1249	2.9207	2.91 (E_1)
2	2.3366	0.2160	3.1267	3.14 ($E_1+\Delta_1$)
3	3.2010	0.4114	3.5036	—
4	2.1293	0.4015	4.0500	—
5	2.6305	0.2693	4.4790	4.45 (E_0')
6	5.3206	0.2918	4.8210	5.00 (E_2)
7	6.0531	0.5828	6.5595	6.60 (E_1')

[a] Data from Erman *et al.* (Ref.[30].)
[b] critical point energies of GaAs at room temperature. Data from Adachi (Ref. [26].)

For the definition of $\varepsilon \equiv \varepsilon_1 + i\varepsilon_2$, $-i$ in the above equation is replaced with $+i$. If we employ Eqs. (7.7) and (7.9), the variation of dielectric function with temperature [33] or composition [32] can also be modeled.

As confirmed from Table 7.2, dielectric function modeling by HOA is performed using hypothetical oscillators that do not correspond to the critical points of GaAs [26], in order to obtain better fitting. This implies that the simple HOA model cannot describe $J(\mathrm{En})$ at each critical point accurately. On the other hand, Adachi proposed model dielectric function (MDF) from more explicit quantum calculation [26,34,35]. In this model, the optical transitions at various critical points are expressed from different theoretical expressions. Although the detail of this model is beyond the scope of this book, if we employ MDF, the dielectric function of c-Si can be expressed from the following equations [35]:

1) E_0', E_1' and $E_2(1)$

$$\varepsilon(\mathrm{En}) = \frac{C}{[1-(\mathrm{En}/E_\mathrm{H})^2] - \mathrm{i}(\mathrm{En}/E_\mathrm{H})\gamma} \tag{7.10a}$$

2) $E_1(1)$

$$\varepsilon(\mathrm{En}) = \frac{B_{1x}}{E_1 - \mathrm{En} - \mathrm{i}\Gamma_1} \tag{7.10b}$$

3) $E_1(2)$

$$\varepsilon(\mathrm{En}) = -B_1\left(\frac{E_1}{\mathrm{En}+\mathrm{i}\Gamma}\right)^2 \ln\left[1-\left(\frac{\mathrm{En}+\mathrm{i}\Gamma}{E_1}\right)^2\right] \tag{7.10c}$$

4) $E_2(2)$

$$\varepsilon(\mathrm{En}) = -F\left(\frac{E_2}{\mathrm{En}+\mathrm{i}\Gamma}\right)^2 \ln\left\{\frac{1-[(\mathrm{En}+\mathrm{i}\Gamma)/E_1]^2}{1-[(\mathrm{En}+\mathrm{i}\Gamma)/E_2]^2}\right\} \tag{7.10d}$$

The dielectric function of c-Si can be obtained by summing Eqs. (7.10a)–(7.10d). With respect to the E_1 and E_2 transitions, each transition is expressed from two models $[E_1(1,2), E_2(1,2)]$. It should be noted that the convention of $\varepsilon \equiv \varepsilon_1 + \mathrm{i}\varepsilon_2$ is used in Eq. (7.10). For the definition of $\varepsilon \equiv \varepsilon_1 - \mathrm{i}\varepsilon_2$, we first determine $\mathrm{Re}(\varepsilon)$ and $\mathrm{Im}(\varepsilon)$ from Eq. (7.10), and then convert by $\varepsilon = \mathrm{Re}(\varepsilon) - \mathrm{i}\mathrm{Im}(\varepsilon)$.

In the case of c-Si, the actual values of Eq. (7.10) are $E_\mathrm{H}(E_0') = 3.35$, $C(E_0') = 0.07$, $\gamma(E_0') = 0.09$, $E_\mathrm{H}(E_1') = 5.33$, $C(E_1') = 0.30$, $\gamma(E_1') = 0.12$, $E_\mathrm{H}(E_2) = E_2 = 4.28$, $C(E_2) = 3.08$, $\gamma(E_2) = 0.10$, $E_1 = 3.39$, $B_1 = 5.22$, $\Gamma = 0.10$, $B_{1x} = 1.44$, $\Gamma_1 = 0.08$, and $F = 3.82$ [35]. In MDF, therefore, the dielectric function of c-Si is expressed by a total of 15 parameters. Figure 7.9 shows the dielectric function of c-Si calculated from Eq. (7.10) using the above parameter values.

There are several other models that can be employed for the dielectric function modeling of semiconductors. Recently, a dielectric function model in which the HOA model is broadened by the Gaussian function has been proposed [36,37]. More recently, Johs *et al.* have proposed the parametric semiconductor model based on a similar function [38]. A simple model developed from Eq. (7.9) has also been proposed [39,40]. For the modeling of amorphous semiconductors, various models described in Section 5.2.4 have been used. In general, the dielectric function of a semiconductor is described by using 10–50 parameters. Accordingly, it is advisable to perform data analysis as follows:

(a) estimate thickness by applying a simple model such as the Cauchy and Sellmeier models,

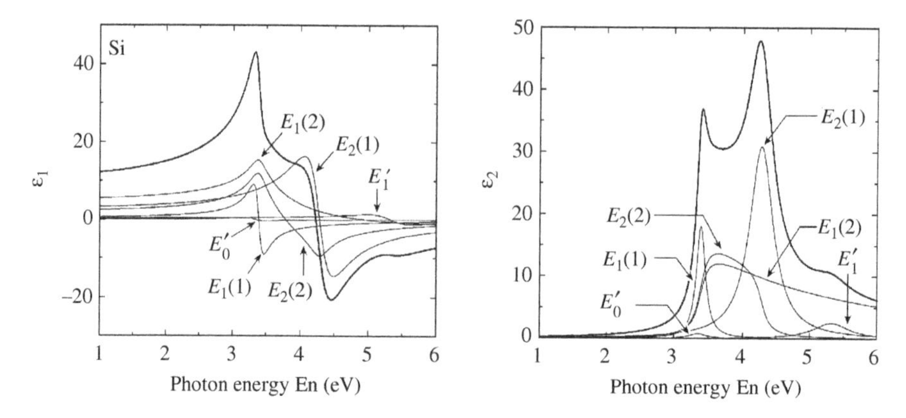

Figure 7.9 Dielectric function of Si expressed from model dielectric function (MDF). This dielectric function was calculated by using reported parameter values [35].

(b) determine the dielectric function from the mathematical inversion,
(c) analyze the inverted dielectric function using a dielectric function model,
(d) finally, perform fitting to (ψ, Δ) spectra again to optimize all the parameters including dielectric function and thickness parameters.

The above procedure is quite effective to separate structural parameters from dielectric function parameters. Keep in mind that we often need to fix several parameters if there are too many parameters in the analysis.

7.2.3 ANALYSIS EXAMPLES

Here, as analysis examples of semiconductors, we will treat analysis of a semiconductor layer formed on a substrate [41], analysis of a semiconductor layer using the effective medium approximation (EMA) [42], and compositional analysis of a semiconductor alloy [32], as described below.

Analysis of a Semiconductor Layer Formed on a Substrate

Figure 7.10(a) illustrates the optical model for a GaN layer formed on a sapphire (α-Al_2O_3) substrate. The sapphire shows strong optical anisotropy in the infrared region [43,44], but its optical properties are rather isotropic in the UV/visible region [45]. Thus, conventional analysis can be performed in this region. In this example, the back surface of the transparent sapphire substrate was roughened and coated with black paint to eliminate backside reflection [see Fig. 5.35(b)]. In this case, the optical model in Fig. 7.10(a) corresponds to the one shown in Fig. 5.32(c). We can calculate N_1 from N_2 using EMA, if $f_{void} = 0.5$ is assumed for the surface roughness layer (see Fig. 5.27). When the optical constants of the substrate (N_3) are known,

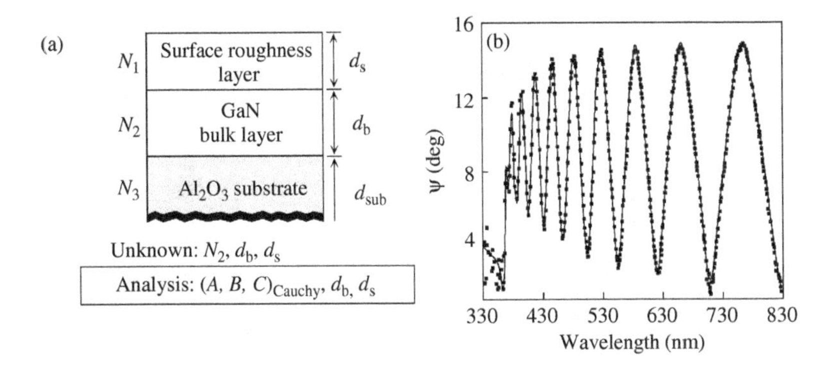

Figure 7.10 (a) Optical model for a GaN layer formed on a sapphire (α-Al_2O_3) substrate and (b) ψ spectrum obtained from the structure shown in (a) at $\theta_0 = 70°$. In (b), the closed circles represent an experimental spectrum, and the solid line indicates a calculated spectrum. Drawing (b): Reprinted with permission from *Japanese Journal of Applied Physics*, **37**, L1105–1108, Part 2, T. Yang, S. Goto, M. Kawata, K. Uchida, A. Niwa, and J. Gotoh (1998). Copyright 1998, The Institute of Pure and Applied Physics.

the unknown parameters in the optical model become N_2, d_b, and d_s, as shown in Fig. 7.10(a). In this analysis example, the dielectric function modeling of GaN was performed from the Cauchy and parametric semiconductor models. When the Cauchy model is applied, the analysis parameters are $(A, B, C)_{Cauchy}$, d_b, and d_s, similar to Fig. 7.1. Thus, when surface roughness is present, we can simply add d_s as an additional analysis parameter. Figure 7.10(b) shows the ψ spectrum obtained from the structure shown in Fig. 7.10(a) at $\theta_0 = 70°$ [41]. In this example, (ψ, Δ) spectra were measured at three angles of incidence ($\theta_0 = 60°, 65°, 70°$) using a RAE instrument. The solid line in Fig. 7.10(b) represents the fitting result calculated from the parametric semiconductor model. From this analysis, $\chi = 5.732$ is obtained as the value of the biased estimator. Recall from Section 5.5.2 that the large noise appearing in the low wavelength region in Fig. 7.10(b) does not increase the fitting error significantly if the biased estimator is employed. In Fig. 7.10(b), the oscillation observed above 360 nm represents the optical interference within the GaN layer, and the spectral feature observed at $\lambda < 360$ nm shows the strong light absorption in GaN. From the analysis of Fig. 7.10(b), $d_s = 6.6$ Å and $d_b = 1.0817\,\mu$m are obtained. On the other hand, from an AFM measurement with a scan size of $5 \times 5\,\mu m^2$, a root mean square roughness of $d_{rms} = 3$ Å was estimated for this sample. Although sample and scan areas are different, the relation between d_s and d_{rms} shows good agreement with $d_s = 1.5d_{rms} + 4$ Å obtained from Fig. 5.30. Moreover, the value of χ increases from 5.732 to 6.257 if the analysis is performed without the surface roughness layer. A bulk layer thickness determined from SEM (1.06 μm) also agrees well with the ellipsometry result. In this way, we can judge whether ellipsometry results determined from analyses are appropriate.

Figure 7.11 shows the extinction coefficient and refractive index of GaN obtained from the fitting analyses [41]. As shown in Fig. 7.11, the extinction coefficient of GaN increases rapidly at En higher than the bandgap of GaN [$E_g = 3.42$ eV ($\lambda = 363$ nm)]. It is clear that the refractive index of GaN at $\lambda > 430$ nm can be represented

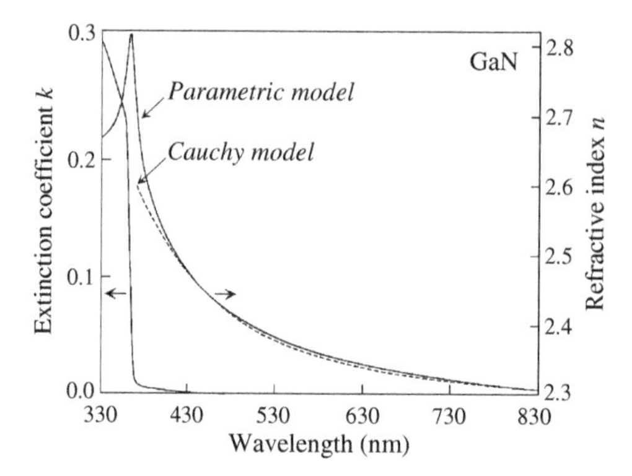

Figure 7.11 Extinction coefficient and refractive index of GaN obtained from analyses using the Cauchy model (dotted line) and parametric semiconductor model (solid line). Reprinted with permission from *Japanese Journal of Applied Physics*, **37**, L1105–1108, Part 2, T. Yang, S. Goto, M. Kawata, K. Uchida, A. Niwa, and J. Gotoh (1998). Copyright 1998, The Institute of Pure and Applied Physics.

by the Cauchy model (dotted line). In fact, the layer thicknesses estimated from the Cauchy model are quite similar to those determined from the semiconductor parametric model [41]. Thus, the dielectric function of GaN can also be obtained from mathematical inversion (see Fig. 5.41). From the parameter A of the Cauchy model, we can evaluate the high-frequency dielectric constant as $\varepsilon_\infty = A^2$, since the terms B/λ^2 and C/λ^4 vanish at $\lambda \rightarrow \infty$ (or En $\rightarrow$ 0 eV) [see Eq. (5.17)]. In this analysis example, $\varepsilon_\infty = A^2 = 5.20$ has been obtained as ε_∞ of GaN. Finally, it should be noted that epitaxial GaN layers grown on sapphire substrates show small optical anisotropy with a refractive index difference below 3% [46], although this effect was neglected in this analysis example.

Analysis of a Semiconductor Layer Using EMA

Figure 7.12(a) shows the optical model for a polycrystalline Si (poly-Si) layer formed on an oxide-covered c-Si substrate. If we assume that the optical constants of the SiO_2 layer (N_3) and c-Si substrate (N_4) are known, the unknown parameters in this optical model become N_1, N_2, d_s, d_b, and d_{SiO}, as shown in Fig. 7.12(a). In this example [42], the complex refractive indices of the surface roughness layer (N_1) and poly-Si layer (N_2) were calculated by applying EMA. In particular, a mixture of poly-Si and void was assumed for the surface roughness layer, while the three phase mixture (c-Si, poly-Si, void) was employed to express the poly-Si bulk layer. To simplify the analysis, reported dielectric functions were used for c-Si [1] and poly-Si [47]. In this case, if the volume fractions of c-Si ($f_{c\text{-}Si}$), poly-Si ($f_{poly\text{-}Si}$), and void (f_{void}) are known, N_1 and N_2 can be calculated from Eq. (5.44). As a result, the analysis parameters of the optical model shown in Fig. 7.12(a) become ($f_{poly\text{-}Si}$, f_{void}) in the surface roughness layer, ($f_{c\text{-}Si}$, $f_{poly\text{-}Si}$, f_{void}) in the bulk layer, d_s, d_b, and d_{SiO}.

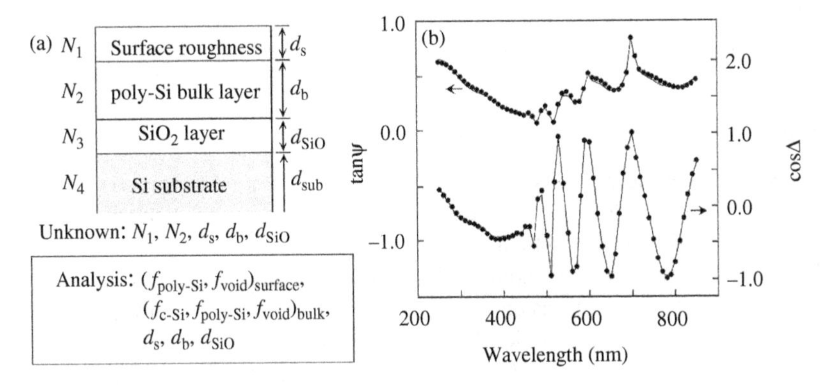

Figure 7.12 (a) Optical model for a polycrystalline Si (poly-Si) layer formed on a SiO_2/c-Si substrate and (b) $(\tan\psi, \cos\Delta)$ spectra obtained from the structure shown in (a) at $\theta_0 = 75°$. In (a), $f_{c\text{-}Si}$, $f_{poly\text{-}Si}$, and f_{void} represent the volume fractions of c-Si, poly-Si and void components, respectively. In (b), the closed circles represent experimental spectra, and the solid lines indicate calculated spectra. Drawing (b): Reprinted from *Thin Solid Films*, **313–314**, P. Petrik, M. Fried, T. Lohner, R. Berger, L. P. Bíro, C. Schneider, J. Gyulai, H. Ryssel, Comparative study of polysilicon-on-oxide using spectroscopic ellipsometry, atomic force microscopy, and transmission electron microscopy, 259–263, Copyright (1998), with permission from Elsevier.

Figure 7.12(b) shows the $(\tan\psi, \cos\Delta)$ spectra obtained from the structure shown in Fig. 7.12(a) [42]. These spectra were measured from a RAE-type instrument at $\theta_0 = 75°$ and the poly-Si layer was formed by low-pressure chemical vapor deposition (LPCVD). The oscillation observed at $\lambda > 450$ nm shows the interference effect in the multilayer structure and the spectral region below 450 nm represents the optical response of the poly-Si layer. Recall from Fig. 2.6 that the penetration depth of light reduces rapidly with increasing light absorption. In absorbing films, therefore, spectral features at higher energies (or shorter wavelengths) are generally governed by the optical properties of the top layer due to the small penetration depth of light. Notice that the $\tan\psi$ spectrum in Fig. 7.12(b) is similar to the ψ spectrum shown in Fig. 5.3(a) and two small peaks observed in the $\tan\psi$ spectrum at 360 nm and 290 nm represent the (E_0', E_1) and E_2 transitions, respectively (see Fig. 7.9). The solid lines in Fig. 7.12(b) indicate the calculation result and we can see good fitting to the experimental spectra. From the analysis, $(f_{poly\text{-}Si}, f_{void})_{surface} = (0.61 \pm 0.02, 0.39 \pm 0.02)$, $(f_{c\text{-}Si}, f_{poly\text{-}Si}, f_{void})_{bulk} = (0.08 \pm 0.04, 0.84 \pm 0.05, 0.08 \pm 0.01)$, $d_s = 77 \pm 7$ Å, $d_b = 4030 \pm 50$ Å, and $d_{SiO} = 1120 \pm 60$ Å are obtained.

It should be emphasized that the optical properties of poly-Si generally vary with process conditions. In particular, the peak widths of the interband transitions change with the grain size of poly-Si (see Section 7.2.4). Moreover, amorphous components may exist in poly-Si layers [42]. For the analysis of poly-Si, therefore, we need to find analysis parameters that minimize fitting errors. In the analysis of Fig. 7.12(b), the introduction of a thin interface layer ($d_{inter} \sim 300$ Å) improves the fitting quality. This layer has a higher f_{void} of ~ 0.3 and represents a transition layer formed by the initial nucleation of the poly-Si layer on the SiO_2 layer. The structural parameters obtained from the analysis have been reported to show excellent agreement with those estimated by TEM and AFM [42].

Compositional Analysis of a Semiconductor Alloy

Here, we will see the analysis of the Al composition in an $Al_xGa_{1-x}As$ multilayer, as an analysis example that includes the parameterization of dielectric functions. Figure 7.13(a) shows the optical model for the $Al_xGa_{1-x}As$ multilayer structure characterized in this analysis example [32]. In this sample, two $Al_xGa_{1-x}As$ layers with different Al compositions and one GaAs layer are formed on a GaAs substrate. In addition, there is a native oxide layer on the sample surface. With respect to the native oxides of III-V semiconductors, we can use reported optical constants [48]. Since the dielectric function of GaAs is known well, the analysis parameters in the optical model become the optical constants of the two $Al_xGa_{1-x}As$ layers (N_3 and N_4) and the thickness of each layer, as illustrated in Fig. 7.13(a). Figure 7.13(b) shows the dielectric functions of $Al_xGa_{1-x}As$ with different Al compositions [18]. As confirmed from this figure, the dielectric function of $Al_xGa_{1-x}As$ shows a strong dependence on the Al composition x. In this example, the parameterization of the $Al_xGa_{1-x}As$ dielectric function was performed from nine oscillators ($j = 9$) using Eq. (7.9). If the dielectric function of $Al_xGa_{1-x}As$ is parameterized as a function of the Al composition x, we can calculate the dielectric functions at various Al compositions. For example, if we obtain the oscillator strength A_1 at several Al compositions and perform fitting by $A_1(x) = a_0 + a_1x + a_2x^2 + a_3x^3$, the coefficients (a_0, a_1, a_2, a_3) can be determined from the analysis. Although a total of $9 \times 4 =$

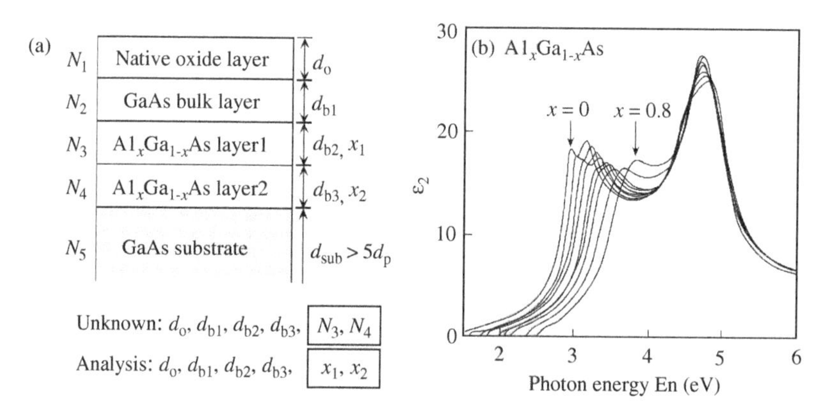

Figure 7.13 (a) Optical model for an Al_xGa_{1-x} As multilayer structure and (b) ε_2 spectra of $Al_xGa_{1-x}As$. In (a), x_1 and x_2 show the Al compositions in the two $Al_xGa_{1-x}As$ layers. Drawing (a): Reprinted with permission from *Journal of Applied Physics*, **70**, F. L. Terry, Jr, A modified harmonic oscillator approximation scheme for the dielectric constants of $Al_xGa_{1-x}As$, 409–417 (1991). Copyright 1991, American Institute of Physics. Drawing (b): Reprinted with permission from *Journal of Applied Physics*, **60**, D. E. Aspnes, S. M. Kelso, R. A. Logan and R. Bhat, Optical properties of $Al_xGa_{1-x}As$, 754–767 (1986). Copyright 1986, American Institute of Physics.

36 parameters are required for the parameterization of one dielectric function, if we obtain (a_0, a_1, a_2, a_3) for all these parameters, we can calculate the dielectric function of $Al_xGa_{1-x}As$ for arbitrary x. In the case of $Al_xGa_{1-x}As$, reasonable results can still be obtained even if each parameter is assumed to change linearly with x [32].

As shown in Fig. 7.13(a), when the dielectric function of $Al_xGa_{1-x}As$ is parameterized as a function of the composition, the Al compositions x_1 and x_2 in the $Al_xGa_{1-x}As$ layers can be used as analysis parameters. Figure 7.14 shows the $\cos\Delta$ spectrum of this sample [32], and the calculated values (dotted line) show quite good agreement with the experimental values (solid line). However, $\cos\Delta$ in the high energy region is mainly determined by the light reflection at the native oxide/GaAs interface, as the thickness of the GaAs top layer is rather thick in this sample ($d_{b1} \sim 2000\,\text{Å}$). In the analysis, therefore, the Al compositions x_1 and x_2 are evaluated from the optical interference appearing at En < 2.5 eV.

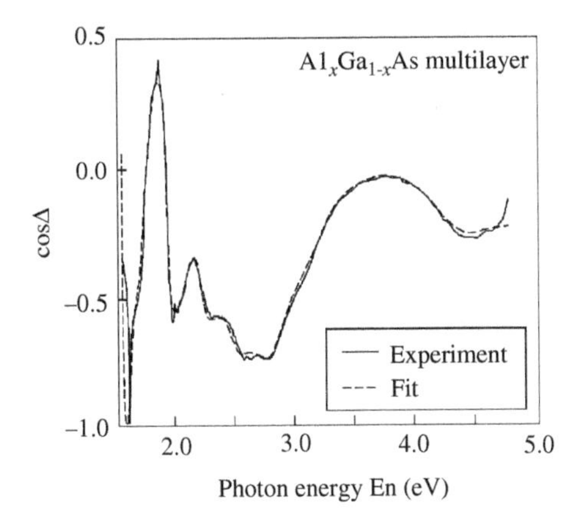

Figure 7.14 Experimental and fitted $\cos\Delta$ spectra obtained from the structure shown in Fig. 7.13(a). Reprinted with permission from *Journal of Applied Physics*, **70**, F. L. Terry, Jr, A modified harmonic oscillator approximation scheme for the dielectric constants of $Al_xGa_{1-x}As$, 409–417 (1991). Copyright 1991, American Institute of Physics.

Table 7.3 shows the results obtained from the linear regression analysis of Fig. 7.14 [32]. In this table, the first layer is the native oxide layer shown in Fig. 7.13(a). From this analysis, rather small confidence limits for the layer thicknesses and compositions are obtained. Accordingly, the thickness and composition of each layer can be evaluated from spectroscopic ellipsometry with high precision. Nevertheless, since ellipsometry is an indirect measurement technique, we need to justify ellipsometry results from other measurement techniques. In particular, if there is a large error in dielectric function modeling, all the results obtained from this modeling may include this error.

Table 7.3 Analysis of an $Al_xGa_{1-x}As$ multilayer structure

Layer	Nominal	Measurement result
1		$d_o = 18.3 \pm 0.6$ Å
2	2000 Å	$d_{b1} = 2015.8 \pm 22.3$ Å
3	500 Å	$d_{b2} = 479.5 \pm 11.2$ Å
	$x_1 = 0.25$	$x_1 = 0.199 \pm 0.009$
4	1000 Å	$d_{b3} = 861.0 \pm 18.5$ Å
	$x_2 = 0.65$	$x_2 = 0.675 \pm 0.023$
unbiased estimator		$\sigma = 1.50 \times 10^{-2}$

In the above parameterization scheme, however, a large number of parameters are required for dielectric function modeling. A simpler model in which this disadvantage is improved has also been proposed [49]. In this model, the dielectric function of an alloy is determined from the average of two dielectric functions that have similar compositions. For example, the dielectric function of an alloy ($x = 0.35$) is calculated as the average of $x = 0.3$ and $x = 0.4$ using this model. In particular, these dielectric functions are synthesized by sliding each dielectric function horizontally so that critical point energies match with experimental values. If we employ this method, the dielectric function for arbitrary compositions can be calculated directly from several dielectric functions with different alloy compositions.

7.2.4 ANALYSIS OF DIELECTRIC FUNCTIONS

When the dielectric functions of semiconductors are extracted by using the mathematical inversion, we can perform a critical point (CP) analysis to characterize the structures of critical points. From this analysis, we can evaluate the band structures of semiconductors relatively easily. Since the band structures vary according to temperature, strain, grain size, etc., these physical parameters can also be deduced from CP analysis. Here, we will see examples of dielectric function analysis using CP analysis.

Theoretical expressions used for CP analysis can be derived from a dielectric function given by the following equation [50,51]:

$$\varepsilon(\mathrm{En}) = C - A\exp(i\phi)(\mathrm{En} - \mathrm{En}_0 + i\Gamma)^n \tag{7.11}$$

Notice that Eq. (7.11) is rather similar to Eq. (7.9). As mentioned earlier, the shape of $J(\mathrm{En})$ around the critical point depends on the band structure, and each critical point is classified into one, two, and three dimensions according to the dimensions of the wave vectors $(\boldsymbol{K}_x, \boldsymbol{K}_y, \boldsymbol{K}_z)$ that participate in the optical transition [1,26]. n in Eq. (7.11) represents the dimension of $\boldsymbol{K}$, and the actual values of n are $n = 1/2$ (three-dimensional), $n = 0$ (two-dimensional), and $n = -1/2$ (one-dimensional). When $n = 0$, the term $(\mathrm{En} - \mathrm{En}_0 + \mathrm{i}\Gamma)^n$ in Eq. (7.11) is rewritten as $\ln(\mathrm{En} - \mathrm{En}_0 + \mathrm{i}\Gamma)$. Furthermore, when an optical transition exhibits excitonic behavior, we get $n = -1$. Since Eq. (7.11) expresses the variation in ε only around the critical point, good fitting may not be obtained from dielectric function modeling using Eq. (7.11) [34].

Critical point analysis is performed using second derivative spectra of $\varepsilon(\mathrm{En})$ or $\varepsilon(\omega)$. From Eq. (7.11), the theoretical formulas for second derivative spectra are expressed as follows [50,51]:

$$\frac{\mathrm{d}^2\varepsilon}{\mathrm{dEn}^2} = n(n-1)A\exp(\mathrm{i}\phi)(\mathrm{En} - \mathrm{En}_0 + \mathrm{i}\Gamma)^{n-2} \quad (n \neq 0) \tag{7.12a}$$

$$\frac{\mathrm{d}^2\varepsilon}{\mathrm{dEn}^2} = A\exp(\mathrm{i}\phi)(\mathrm{En} - \mathrm{En}_0 + \mathrm{i}\Gamma)^{-2} \quad (n = 0). \tag{7.12b}$$

Eqs. (7.11) and (7.12) have been derived using the definition $\varepsilon \equiv \varepsilon_1 + \mathrm{i}\varepsilon_2$. However, since the second derivative spectra for ε_1 and ε_2 are analyzed in CP analysis, it is not necessary to convert these equations to those of $\varepsilon \equiv \varepsilon_1 - \mathrm{i}\varepsilon_2$. From CP analysis using second derivative spectra, the specific features of critical points can be characterized more distinctively. In particular, since the second derivative spectra become zero in a region where the dielectric function varies smoothly, we can evaluate critical points efficiently from CP analysis. In this analysis, therefore, there is no need to model the entire dielectric function, and the analysis can be performed more easily.

It has been known that the dielectric functions of semiconductors show significant changes by the following effects:

(a) surface temperature of semiconductors [50,51],
(b) doping of semiconductors [52–56],
(c) strain in semiconductors [57–59],
(d) grain size of polycrystalline semiconductors [35,60–63],
(e) alloy composition of alloy semiconductors [18],
(f) quantum effect in semiconductor layers [59,64–66].

In the above examples, the band structures of semiconductors change largely. Thus, we can characterize these effects indirectly from CP analyses. Conversely, when the conditions in (a)–(f) above are different, it becomes difficult to use reported dielectric functions in data analysis. In the following part, we will see examples of CP analyses reported for (a)–(f) above.

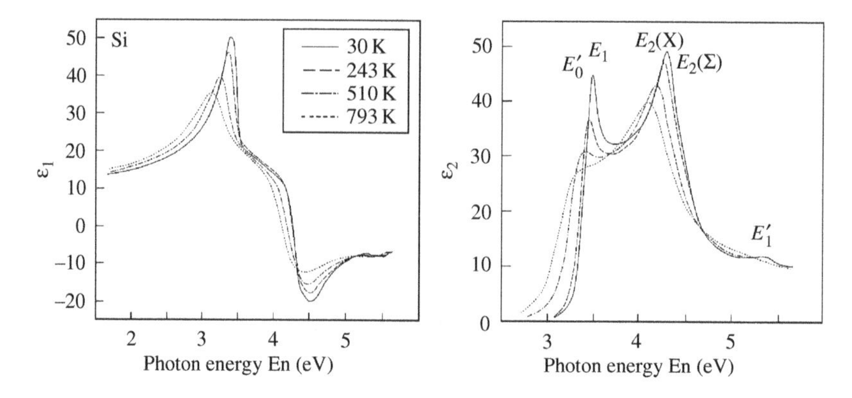

Figure 7.15 Variation of c-Si dielectric function with temperature. Adapted with permission from P. Lautenschlager, M. Garriga, L. Viña, and M. Cardona, Temperature dependence of the dielectric function and interband critical points in silicon, *Phys. Rev. B*, **36** (1987) 4821–4830. Copyright 1987, the American Physical Society.

Surface Temperature of Semiconductors

Figure 7.15 shows the variation of c-Si dielectric function with temperature [50]. These dielectric functions were obtained by removing native oxide (17.5 Å) from the c-Si surface using mathematical inversion (see also Section 8.1.1). As confirmed from Fig. 7.15, sample temperature has a large effect on the dielectric function of c-Si. Moreover, with increasing temperature, the ε_2 peaks broaden and shift toward lower energies. It should be noted that the $E_2(\Sigma)$ transition in Fig. 7.15 represents the interband transition at the Σ point [25], although this transition has not been shown in Fig. 7.7(b).

Figure 7.16(a) shows the second derivative spectrum for ε_1 obtained from c-Si at 30 K [50]. The open circles in the figure show the experimental spectra, and the solid line represents the fitting result calculated from Eq. (7.12). In this CP analysis, $(A, \phi, \mathrm{En}_0, \Gamma)$ are used as analysis parameters, and the linear regression analysis is performed for each value of $n = -1, -1/2, 0$, and $1/2$. In this case, from the n value that minimizes the fitting error, the final result for $(A, \phi, \mathrm{En}_0, \Gamma)$ is determined. From this analysis, the E_1 transition has been reported to show excitonic behavior $(n = -1)$ [50]. This result implies that electrons excited by the E_1 transition are highly localized. For other critical points (E_0', E_2, E_1'), the two-dimensional critical point provides the best fit to the experimental spectra. It has been reported that optical transitions become excitonic, as the ionicity of the atoms comprising a crystal increases [50].

Figure 7.16(b) shows the temperature dependence of the peak positions obtained from the CP analysis [50]. As shown in this figure, the E_0' and E_1 peaks become a single peak (excitonic) at $T > 350$ K. If we use an empirical relation developed by Varshni, the peak position versus the temperature T is expressed from the following equation [20,26]:

$$\mathrm{En}(T) = \mathrm{En}(0) - \alpha T^2/(T + \beta) \tag{7.13}$$

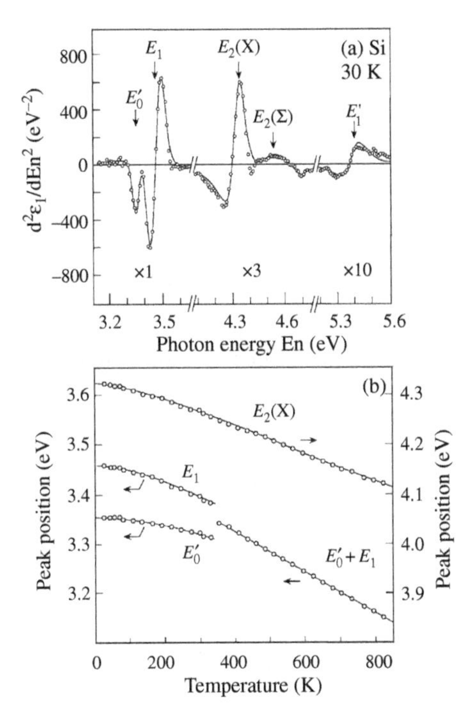

Figure 7.16 (a) Second derivative spectrum for ε_1 obtained from c-Si at 30 K and (b) temperature dependence of peak positions estimated from the critical point (CP) analysis. In (a), the open circles represent an experimental spectrum, and the solid line indicates a calculated spectrum. Adapted with permission from P. Lautenschlager, M. Garriga, L. Viña, and M. Cardona, Temperature dependence of the dielectric function and interband critical points in silicon, *Phys. Rev. B*, **36** (1987) 4821–4830. Copyright 1987, the American Physical Society.

where En(0) represents the peak position at $T = 0$ K, and (α, β) are the constants. From the analysis shown in Fig. 7.16(b), En(0) = 4.324 eV, $\alpha = 2.87 \times 10^{-4}$ eV/K, and $\beta = 124$ K are obtained for the E_2(X) transition. When the peak positions of critical points are parameterized in this manner, we can estimate sample temperatures from the energy positions of critical points. In particular, we can estimate near-surface temperature from this method, since the absorption coefficient of these transition peaks is quite large (see Fig. 5.3) and the resulting penetration depth of light becomes small. Accordingly, for the characterization of processing temperatures, this technique is quite effective.

As confirmed from Fig. 7.15, the ε_2 peaks broaden with increasing sample temperature. Thus, the Γ values in Eq. (7.12) become larger, similar to the Lorentz model. As mentioned earlier, Γ is inversely proportional to the average lifetime of excited electrons (see Section 7.2.2). Accordingly, the peak broadening implies that the lifetime of excited electrons becomes shorter at higher temperatures due to increasing lattice vibration. The variation of Γ with temperature can be expressed from the following equation [50]:

$$\Gamma(T) = \Gamma_0\left[1 + \frac{2}{\exp(\theta/T) - 1}\right] + \Gamma_1 \tag{7.14}$$

For the $E_2(X)$ transition, $\Gamma_0 = 24\,\text{meV}$, $\Gamma_1 = 39\,\text{meV}$, and $\theta = 326\,\text{K}$ have been reported [50]. It can be seen from Eq. (7.14) that $\Gamma(T)$ increases rapidly at $T > \theta$.

Doping of Semiconductors

Figure 7.17(a) shows the dielectric functions of c-Si ($N_f = 2.1 \times 10^{14}\,\text{cm}^{-3}$) and P-doped poly-Si ($N_f = 3.3 \times 10^{20}\,\text{cm}^{-3}$) [53]. Here, N_f indicates the free-carrier concentration defined in Eq. (5.27). The dielectric function of the poly-Si has been obtained from the mathematical inversion shown in Fig. 5.43. As confirmed from Fig. 7.17(a), when the doping concentration is very high, the dielectric function in the high energy region (En > 3 eV) changes drastically. In particular, the E_1 transition peak broadens significantly, compared with the E_2 transition peak. This originates from the fact that the E_1 peak is excitonic and its character is different from the E_2 peak. The broadening of the E_1 transition implies that the metastable bound exciton created by the E_1 transition is screened by high-concentration free carriers [53]. In other words, since an exciton is originally composed of an electron–hole pair, the exciton becomes unstable if the Coulomb attraction of the exciton is screened by free electrons surrounding it. As a result, the lifetime of the E_1 excitons becomes shorter as the electron concentration increases, and the Γ value of the E_1 peak increases. In contrast, since the E_2 transition is not excitonic, there

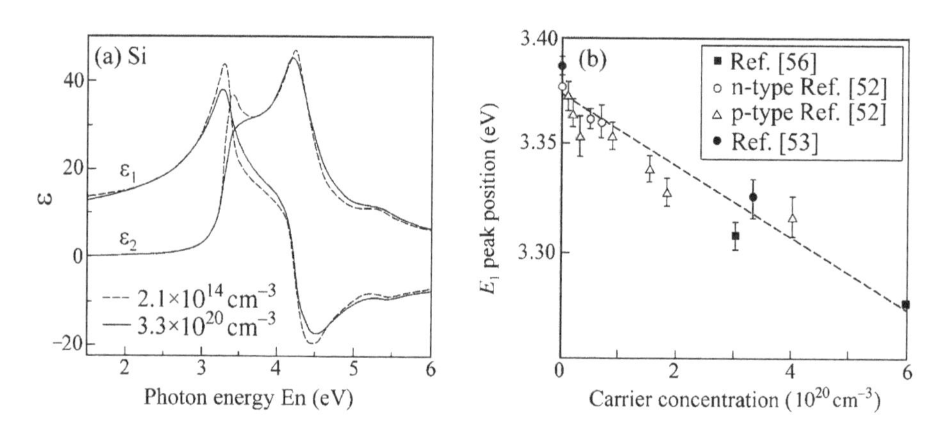

Figure 7.17 (a) Dielectric functions of c-Si ($N_f = 2.1 \times 10^{14}\,\text{cm}^{-3}$) and P-doped poly-Si ($N_f = 3.3 \times 10^{20}\,\text{cm}^{-3}$) and (b) E_1 peak position versus carrier concentration. Adapted with permission from D. E. Aspnes, A. A. Studna, and E. Kinsbron, Dielectric properties of heavily doped crystalline and amorphous silicon from 1.5 to 6.0 eV, *Phys. Rev. B*, **29** (1984) 768–779. Copyright 1984, the American Physical Society.

is little influence of free carriers on the E_2 transition. Notice from Fig. 7.17(a) that ε_1 of the poly-Si reduces slightly at lower energies (En < 2 eV) due to the effect of free-carrier absorption [see Fig. 5.23(a)]. The dielectric functions of doped poly-Si and c-Si are almost the same in the region En = 2–3 eV. Thus, if we perform a structural analysis using this region, the dielectric functions of doped Si films can be obtained by applying the mathematical inversion.

Figure 7.17(b) shows the E_1 peak position, plotted as a function of the carrier concentration in Si [53]. It can be seen that the E_1 peak shifts almost linearly as the carrier concentration increases. The E_2 peak shows a similar trend [53]. At lower carrier concentrations ($N_f < 5 \times 10^{19}\,\mathrm{cm}^{-3}$), however, the estimation of the carrier concentration from the peak shift becomes difficult due to the small peak shift in this region. The carrier concentrations in semiconductors can be evaluated more explicitly from the analysis of free-carrier absorption (see Section 7.3.2). So far, the origin of the peak shift observed at high carrier concentration has also been discussed in detail [53,54,67].

Strain in Semiconductors

In the epitaxial growth of a semiconductor layer on a hetero substrate, high strain is often generated in the growing layer due to a lattice mismatch between the epitaxial layer and the substrate. In particular, when the thickness of the epitaxial layer is less than a thickness called the critical thickness, the lattice of the epitaxial layer deforms so that the lattice of the layer aligns with that of the substrate [68]. However, when the thickness exceeds the critical thickness, lattice relaxation occurs with the generation of lattice defects and dislocations, and the lattice structure of the epitaxial layer changes into the inherent configuration. In the GaAs–AlAs system, the lattice mismatch is exceptionally small. For other systems, however, the strain in semiconductor layers should be taken into account.

Figure 7.18(a) shows the dielectric functions of strained and relaxed $Si_{1-x}Ge_x$ layers formed on Si(100) substrates [19]. The Ge composition in these $Si_{1-x}Ge_x$ layers is $x = 0.195$. These dielectric functions were obtained from mathematical inversion, and the thicknesses used in the analyses were 520 Å (strained $Si_{1-x}Ge_x$) and 6100 Å (relaxed $Si_{1-x}Ge_x$). As confirmed from Fig. 7.18(a), the dielectric function varies according to the strain generated within the $Si_{1-x}Ge_x$ layer, even if the Ge composition is the same. Notice that the E_1 peak around En = 3.2 eV splits into two peaks in the case of the strained $Si_{1-x}Ge_x$ layer.

Figure 7.18(b) shows the peak positions obtained from CP analysis, plotted as a function of the Ge composition [58]. The change in band structure with strain can be analyzed theoretically [69], and the four dotted lines in the figure indicate the results obtained from the theoretical calculation. The dotted line located at lower energies shows the result when a deformation potential is reduced by 25 %. Since the $Si_{1-x}Ge_x$ layers are formed on Si(100) substrates, the strain inside the SiGe layer increases as the Ge composition increases. Thus, peak splitting between the E_1 and $E_1 + \Delta_1$ transitions also increases with Ge composition. As shown in

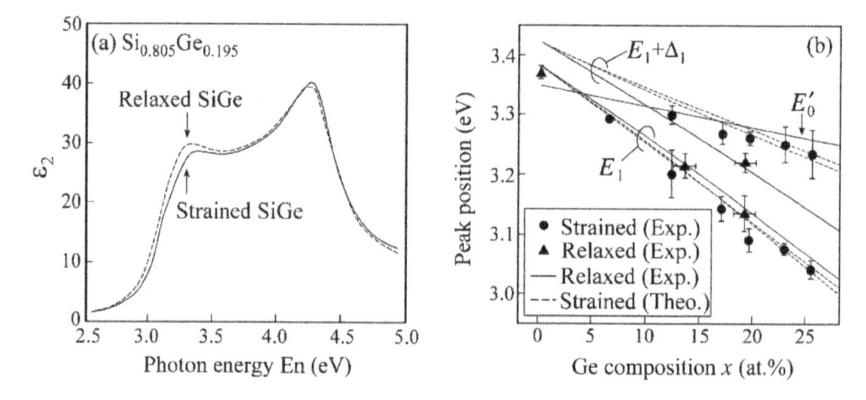

Figure 7.18 (a) ε_2 spectra for strained and relaxed $Si_{0.805}Ge_{0.195}$ layers and (b) peak positions of $Si_{1-x}Ge_x$ obtained from the CP analysis, plotted as a function of the Ge composition x. In (b), the dotted lines show results obtained from theoretical calculations. Drawing (a): Reprinted with permission from *Journal of Applied Physics*, **75**, C. Pickering and R. T. Carline, Dielectric function spectra of strained and relaxed $Si_{1-x}Ge_x$ alloys ($x = 0$–0.25), 4642–4647 (1994). Copyright 1994, American Institute of Physics. Drawing (b): Reprinted with permission from *Applied Physics Letters*, **64**, R. T. Carline, C. Pickering, D. J. Robbins, W. Y. Leong, A. D. Pitt, and A. G. Cullis, Spectroscopic ellipsometry of $Si_{1-x}Ge_x$ epilayers of arbitrary composition $0 \leq x \leq 0.255$, 1114–1116 (1994). Copyright 1994, American Institute of Physics.

Fig. 7.18(b), the calculated result agrees well with the experimental result. From the parameterization of $Si_{1-x}Ge_x$ dielectric functions, the composition analysis of Ge has been performed with a precision of $x = \pm 0.01$ [58].

Grain Size of Polycrystalline Semiconductors

In polycrystalline semiconductors, the dielectric function varies depending on grain size [35,60–63]. Figure 7.19(a) shows the change in the ε_2 spectrum when a hydrogenated amorphous silicon (a-Si:H) film is crystallized by thermal annealing at 1100 °C. The dielectric functions in Fig. 7.19(a) were calculated from MDF [Eq. (7.10)] using reported parameter values [35]. These ε_2 spectra represent those of the crystalline phase, and the contribution of the amorphous phase has been removed using EMA and mathematical inversion. When thermal annealing is performed for 30 min, the grain size is very small, and the E'_0, E_1, and E_2 peaks are broad, as shown in Fig. 7.19(a). When the sample is annealed for 400 min, the half width of the peaks becomes narrower, but the peaks are still broad, compared with Si single crystal (c-Si).

Figure 7.19(b) shows Γ estimated from CP analysis, plotted as a function of the inverse of the average gain size $\langle L \rangle$ (i.e., $\langle L \rangle^{-1}$) [62]. In this example, TEM was employed to determine $\langle L \rangle$ of poly-Si films fabricated by LPCVD and microcrystalline Si (μc-Si:H) films formed by sputtering. As confirmed from the

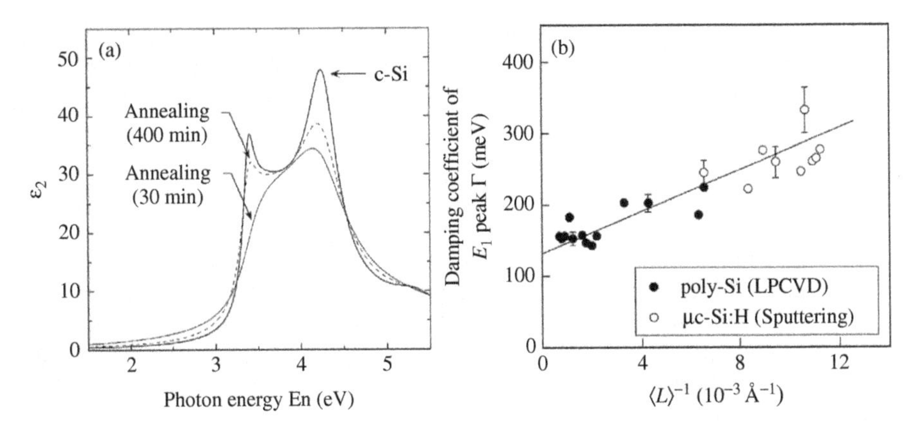

Figure 7.19 (a) Variation of the ε_2 spectrum (Si) with grain size and (b) damping coefficient of the E_1 peak (Γ), plotted as a function of the inverse of average grain size ($\langle L \rangle^{-1}$). The ε_2 spectra shown in (a) were calculated from MDF using reported parameter values [35]. In (b), the Γ values of poly-Si films fabricated by LPCVD and microcrystalline Si (μc-Si:H) films formed by sputtering are shown. Drawing (b): Reprinted with permission from *Journal of Applied Physics*, **73**, S. Boultadakis, S. Logothetidis, S. Ves, and J. Kircher, Optical properties of μc-Si:H/α-Si:H layered structures: Influence of the hydrogen bonds, crystallite size, and thickness, 914–925 (1993). Copyright 1993, American Institute of Physics.

figure, Γ of the E_1 peak changes linearly with $\langle L \rangle^{-1}$. Thus, the variation of Γ with $\langle L \rangle^{-1}$ can be expressed as follows [62]:

$$\Gamma = \Gamma_0 + Q\langle L \rangle^{-1} \tag{7.15}$$

where Γ_0 represents the Γ value when $\langle L \rangle$ is infinite and coincides with the Γ value of a single crystal. Q in Eq. (7.15) shows the proportional constant, and a physical model for Q has also been proposed [60]. Eq. (7.15) implies that, when grain sizes are small, the lifetime of excited electrons becomes shorter due to enhanced scattering at grain boundaries [60]. If we employ the result shown in Fig. 7.19(b), we can estimate the average grain size from the Γ value. In spectroscopic ellipsometry analysis, Eq. (7.15) has been used widely [60–63]. It has been confirmed that Eq. (7.15) also holds for GaAs [60].

Alloy Composition of Alloy Semiconductors

As we have seen in Fig. 7.13(b), critical point energy shows a strong dependence on alloy composition. In the case of $Al_xGa_{1-x}As$, the E_1 peak position changes with the Al composition as follows [18]:

$$E_1 = 2.924 + 0.965x + x(1-x)(-0.157 - 0.935x) \tag{7.16}$$

The numerical values in the above equation have been estimated from fitting to experimental values. When the positions of transition peaks are parameterized, we can characterize the alloy composition from CP analysis relatively easily. However, this analysis can be performed only for a sufficiently thick single layer. When we characterize a multilayer structure, the complete parameterization of dielectric functions is required (see Section 7.2.3).

Quantum Effect in Semiconductor Layers

When a thin semiconductor layer with a low bandgap is sandwiched by semiconductor layers having high bandgaps, electrons are confined in the low bandgap layer, and quantum levels will be formed in this layer [70]. In a AlGaAs/GaAs/AlGaAs quantum-well structure, quantum states are created in the GaAs layer and the optical transition peaks shift toward higher energies. If we use spectroscopic ellipsometry, the quantum effect on the E_1 transition (Λ point), for example, can be studied [64–66]. In the AlGaAs/GaAs/AlGaAs system, the dielectric function of a bulk layer can be employed for the AlGaAs layer since the quantum effect on this layer is negligible [66]. Thus, if the thicknesses of all the layers are known, the GaAs dielectric function that includes the quantum effect can be extracted using mathematical inversion [64]. From the dielectric function extracted, we can further analyze the quantum effects theoretically [64,65].

The quantum effect in semiconductors has been reported to change the dielectric functions greatly [59,64–66], and the dielectric functions generally shift toward the higher energy side. In the AlGaAs/GaAs/AlGaAs structure, the peak shift increases drastically when the thickness of the well layer (GaAs) becomes $d_b < 50$ Å [64]. In this case, we can improve the fit by sliding the dielectric function of the well layer toward higher energies [64]. This method is particularly useful when we estimate the layer thickness of the well layer.

7.3 METALS/SEMICONDUCTORS

The dielectric functions of metals and semiconductors with high carrier concentrations show a strong feature of free-carrier absorption that can be expressed from the Drude model (see Section 5.2.5). In particular, from analysis using the Drude model, the carrier concentration and mobility in semiconductors can be estimated quite accurately. In this section, we will address the dielectric functions of various metals and the analysis of free-carrier absorption in semiconductors.

7.3.1 DIELECTRIC FUNCTION OF METALS

Here, we will discuss the optical properties and dielectric function modeling of metals. Figure 7.20(a) shows the ε_1 spectrum of silver [71]. In this figure, ε_{exp} represents the ε_1 spectrum obtained from an experiment. This ε_{exp} can be decomposed into two contributions indicated as ε_{FCA} and ε_{bound} in Fig. 7.20(a).

ε_{FCA} and ε_{bound} represent the dielectric responses caused by free-carrier absorption and bound electrons, respectively. The formation of the bound electrons can be explained from the band structure of Ag shown in Fig. 7.20(b) [31]. It can be seen that the Fermi level (E_F) of Ag lies within the conduction band, and the conduction band is filled with electrons, similar to Fig. 5.22(a). In the case of Ag, however, the d band is located 4 eV below E_F, and an optical transition occurs from the d band to the Fermi surface, as denoted by the arrow in Fig. 7.20(b). The character of this optical transition is rather similar to the one observed in semiconductors, and the excited d electrons are highly localized on atomic sites. Accordingly, the bound electrons in the d state can be considered as electric dipoles.

In Fig. 7.20(a), E_p represents the plasma energy defined by Eq. (5.31). Recall from Section 5.2.5 that $\varepsilon_1 = 0$ at E_p. On the other hand, E_p(FCA) shows the plasma energy characterized by the free carrier contribution only. It is clear that the formation of the bound electrons increases the overall value of ε_1 and shifts E_p from 9.2 eV to 3.9 eV. As we have seen in Fig. 5.23(b), (n, k) values increase drastically at En $< E_p$, and reflectance becomes quite high in this region. In the case of Ag, therefore, the reflectance increases at En $<$ 3.9 eV. Since $\lambda = 320$ nm (En = 3.9 eV) is in the UV region, the light waves in the visible region are reflected on Ag surfaces [31], and thus the surface of Ag shows a whitish (silver) color. In the case of copper, on the other hand, the d band lies 2 eV below E_F, and strong light reflection (or absorption) begins below 2 eV ($\lambda > 600$ nm) [31]. Since this wavelength corresponds to red light, Cu has a reddish color. This is the reason why the colors of Ag and Cu look different.

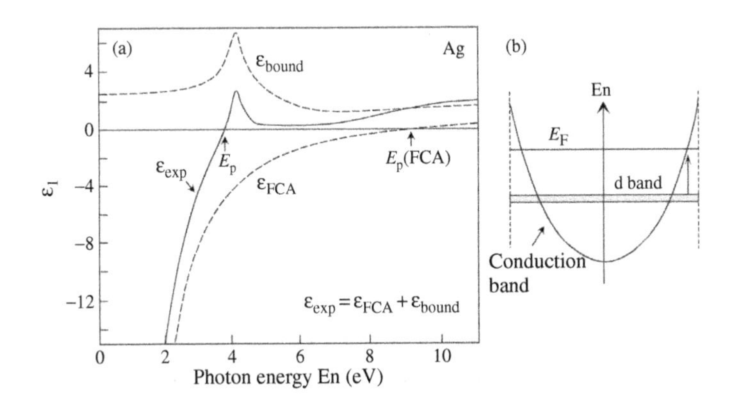

Figure 7.20 (a) Experimental ε_1 spectrum (ε_{exp}) of silver and (b) schematic band diagram for the noble metals. In (a), ε_{bound} and ε_{FCA} represent the contributions of bound electron and free-carrier absorption to ε_{exp}, and E_p indicates the plasma energy. In (b), E_F shows the Fermi level, and the arrow indicates the optical transition from the d band to E_F. Drawing (a): Adapted with permission from H. Ehrenreich and H. R. Philipp, Optical properties of Ag and Cu, *Phys. Rev.*, **128** (1962) 1622–1629. Copyright (1962), the American Physical Society. Drawing (b): Reprinted from *Optical Properties of Solids*, F. Wooten, Copyright (1972), with permission from Elsevier.

Table 7.4 Parameters of the Lorentz–Drude model

Parameter	Ag	Au	Cu	Al	Be	Cr	Ni	Pd	Pt	Ti	W
E_p(eV)	9.01	9.03	10.83	14.98	18.51	10.75	15.92	9.72	9.59	7.29	13.22
f_0	0.845	0.760	0.575	0.523	0.084	0.168	0.096	0.330	0.333	0.148	0.206
Γ_0(eV)	0.048	0.053	0.030	0.047	0.035	0.047	0.048	0.008	0.080	0.082	0.064
f_1	0.065	0.024	0.061	0.227	0.031	0.151	0.100	0.649	0.191	0.899	0.054
Γ_1(eV)	3.886	0.241	0.378	0.333	1.664	3.175	4.511	2.950	0.517	2.276	0.530
En_1(eV)	0.816	0.415	0.291	0.162	0.100	0.121	0.174	0.336	0.780	0.777	1.004
f_2	0.124	0.010	0.104	0.050	0.140	0.150	0.135	0.121	0.659	0.393	0.166
Γ_2(eV)	0.452	0.345	1.056	0.312	3.395	1.305	1.334	0.555	1.838	2.518	1.281
En_2(eV)	4.481	0.830	2.957	1.544	1.032	0.543	0.582	0.501	1.314	1.545	1.917
f_3	0.011	0.071	0.723	0.166	0.530	1.149	0.106	0.638	0.547	0.187	0.706
Γ_3(eV)	0.065	0.870	3.213	1.351	4.454	2.676	2.178	4.621	3.668	1.663	3.332
En_3(eV)	8.185	2.969	5.300	1.808	3.183	1.970	1.597	1.659	3.141	2.509	3.580
f_4	0.840	0.601	0.638	0.030	0.130	0.825	0.729	0.453	3.576	0.001	2.590
Γ_4(eV)	0.916	2.494	4.305	3.382	1.802	1.335	6.292	3.236	8.517	1.762	5.836
En_4(eV)	9.083	4.303	11.18	3.473	4.604	8.775	6.089	5.715	9.249	19.43	7.498
f_5	5.646	4.384	0.00	0.000	0.000	0.000	0.000	0.000	0.000	0.000	0.000
Γ_5(eV)	2.419	2.214	0.00	0.000	0.000	0.000	0.000	0.000	0.000	0.000	0.000
En_5(eV)	20.29	13.32	0.00	0.000	0.000	0.000	0.000	0.000	0.000	0.000	0.000

Data from Rakić *et al.* (Ref. [73].)

As evidenced from Fig. 7.20(a), the dielectric function of metals is expressed by

$$\varepsilon(\mathrm{En}) = \varepsilon_{\mathrm{FCA}}(\mathrm{En}) + \varepsilon_{\mathrm{bound}}(\mathrm{En}) \tag{7.17}$$

However, the dielectric function modeling of $\varepsilon_{\mathrm{bound}}(\mathrm{En})$ is generally difficult. We can express $\varepsilon_{\mathrm{bound}}(\mathrm{En})$ of Al by using the Ashcroft–Sturm model [72], but such models are not always available for other metals. In order to express the dielectric functions of various metals, the Lorentz–Drude model has been used [73]. In this model, $\varepsilon_{\mathrm{FCA}}(\mathrm{En})$ is expressed from the Drude model, while $\varepsilon_{\mathrm{bound}}(\mathrm{En})$ is described by several Lorentz oscillators [73]:

$$\varepsilon(\mathrm{En}) = \varepsilon_{\mathrm{Drude}}(\mathrm{En}) + \varepsilon_{\mathrm{Lorentz}}(\mathrm{En}) \tag{7.18}$$

where

$$\varepsilon_{\mathrm{Drude}}(\mathrm{En}) = 1 - \frac{f_0 E_{\mathrm{p}}^2}{\mathrm{En}^2 - \mathrm{i}\Gamma_0 \mathrm{En}} \tag{7.19a}$$

$$\varepsilon_{\mathrm{Lorentz}}(\mathrm{En}) = \sum_{j=1}^{5} \frac{f_j E_{\mathrm{p}}^2}{\mathrm{En}_j^2 - \mathrm{En}^2 + \mathrm{i}\Gamma_j \mathrm{En}} \tag{7.19b}$$

Notice that Eq. (7.19) is slightly different from Eqs. (5.9) and (5.29). Table 7.4 summarizes the parameters of Eq. (7.19) for 11 metals [73]. As shown in Table 7.4, $\varepsilon_{\mathrm{bound}}$ of each metal is expressed by 12–15 parameters. Although we cannot interpret these parameter values based on physical models, the Lorentz–Drude model provides an effective way for the construction of optical databases. For the noble metals (Ag, Au, Cu), fitting to experimental spectra can be improved by applying a model that utilizes a Gaussian line shape [73].

Figure 7.21 shows the dielectric function of Al calculated from the Lorentz–Drude model using the parameters shown in Table 7.4. The smooth varying backgrounds in Fig. 7.21 show the contribution of free-carrier absorption, and ε_1 reduces rapidly at lower energies, similar to Fig. 7.20(a). On the other hand, the ε_2 peak at 1.5 eV represents light absorption by the bound electrons in Al. This ε_2 peak has been expressed successfully by the Ashcroft–Sturm model [72]. From the dielectric function analysis using this model, the optical properties of Al can be studied in detail [74].

Unfortunately, the characterization of the dielectric functions of metals has been rather difficult due to the fact that:

(a) dielectric functions of metals vary according to process conditions or deposition methods [75,76],
(b) dielectric functions are sensitive to surface oxidation and surface roughness due to small penetration depths of light in metals [76–78],
(c) dielectric functions of metals show significant thickness dependences particularly when the thickness of metals is thin (typically < 100 Å) [74,79,80].

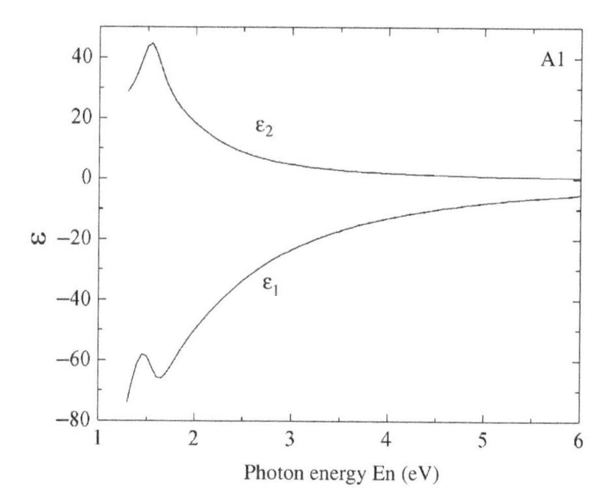

Figure 7.21 Dielectric function of Al. This dielectric function was calculated from the Lorentz–Drude model using the parameter values shown in Table 7.4.

The process dependence of metal dielectric functions can be attributed to the variations in void volume fraction, grain size, and surface roughness in metal films [76]. We can describe the surface roughness of metal films by using EMA [76] or the Maxwell Garnet model [74,81] (see Fig. 5.25). Since metals generally exhibit rapid surface oxidation [77], it is preferable to perform *in situ* (or real-time) ellipsometry measurements in a process chamber. It has been reported that the thickness dependence of Al dielectric function is caused by the variation in the electron scattering at defects and grain boundaries [74].

When a film is sufficiently thick and the effect of surface oxidation and roughness is negligible, the dielectric function is obtained simply from the pseudo-dielectric function (see Section 5.4.2). This is the most general method for obtaining the dielectric function of metals. However, keep in mind that metal films show thickness dependence of optical properties. Thus, one needs to confirm that film thickness is enough to represent bulk properties. For more explicit characterization, multi-sample analysis has been performed [79]. In this analysis, dielectric function is determined from several ellipsometry spectra obtained by varying film thickness. This analysis method is essentially similar to the global error-minimization method used in real-time data analysis (see Section 8.1.3). We may model dielectric functions obtained from the above procedure by using the Lorentz–Drude model. For some cases, the inclusion of void component using EMA may be required. However, the application of such modeled dielectric functions is limited due to the dependence of dielectric function on process conditions or deposition methods.

7.3.2 ANALYSIS OF FREE-CARRIER ABSORPTION

Here, we will see the analysis of free-carrier absorption in a Ga-doped polycrystalline ZnO that has a high carrier concentration ($N_f > 10^{19}\,cm^{-3}$) [82]. ZnO:Ga is a semiconductor widely used as a transparent conductive oxide (TCO). Figure 7.22(a) shows an optical model for the ZnO:Ga layer formed on a SiO_2/c-Si substrate. The rear surface of the c-Si substrate was roughened to eliminate backside reflection that occurs at $< 1.2\,eV$ in c-Si [83]. If we calculate N_1 from N_2 using EMA ($f_{void} = 0.5$) and determine the SiO_2 layer thickness ($d_{SiO} \sim 500\,Å$) prior to the ZnO:Ga deposition, the unknown parameters of this optical model are (N_2, d_s, d_b). It should be emphasized that the dielectric functions of conductive semiconductors can also be expressed from Eq. (7.17). In this analysis example, the dielectric function modeling of ZnO:Ga was performed by using the Drude model and Tauc–Lorentz (TL) model (see Section 5.2.4) [82]:

$$\varepsilon(En) = \varepsilon_D(En) + \varepsilon_{TL}(En) \tag{7.20}$$

where $\varepsilon_D(En)$ and $\varepsilon_{TL}(En)$ represent the dielectric functions calculated by the Drude and TL models, respectively. The Drude model in Eq. (7.20) is expressed by

$$\varepsilon_D(En) = -\frac{A_D}{En^2 - i\Gamma_D En} = \left(-\frac{A_D}{En^2 + \Gamma_D^2}\right) - i\left(\frac{A_D \Gamma_D}{En^3 + \Gamma_D^2 En}\right) \tag{7.21}$$

where (A_D, Γ_D) are the Drude parameters and correspond to (A, Γ) defined by Eq. (5.29). In Eq. (7.21), the equation is transformed by multiplying ($En^2 + i\Gamma_D En$)

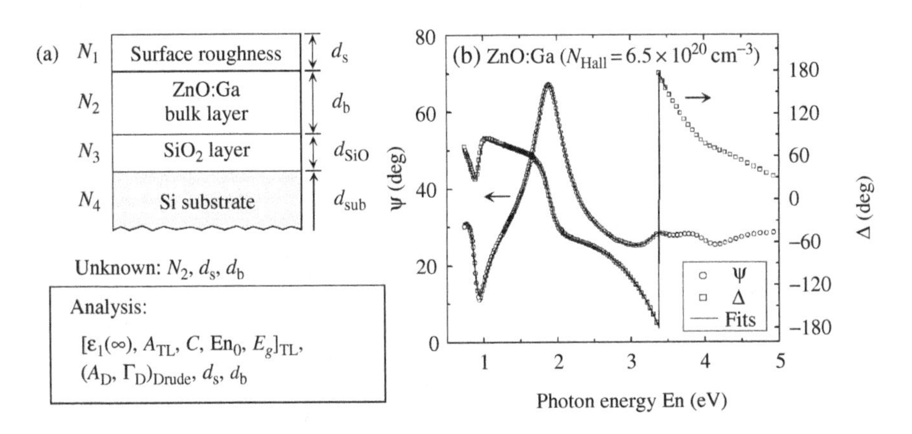

Figure 7.22 (a) Optical model for a Ga-doped polycrystalline ZnO layer formed on a SiO_2/c-Si substrate and (b) (ψ, Δ) spectra obtained from the structure shown in (a) at $\theta_0 = 70°$. The spectra shown in (b) were obtained from the ZnO:Ga layer with a Hall carrier concentration N_{Hall} of $6.5 \times 10^{20}\,cm^{-3}$. Drawing (b): Adapted with permission from H. Fujiwara and M. Kondo, Effects of carrier concentration on the dielectric function of ZnO:Ga and In_2O_3:Sn studied by spectroscopic ellipsometry: Analysis of free-carrier and band-edge absorption, *Phys. Rev. B*, **71** (2005) 075109. Copyright 2005, the American Physical Society.

by both numerator and denominator, in order to separate the real part from the imaginary part. Notice that ε_∞ in Eq. (5.29) is removed in Eq. (7.21) since the contribution of ε_∞ is calculated from $\varepsilon_{TL}(\text{En})$ in Eq. (7.20) (see Fig. 5.10). As confirmed from Eqs. (5.19)–(5.21), $\varepsilon_{TL}(\text{En})$ is calculated from five parameters [$\varepsilon_1(\infty)$, A_{TL}, C, En_0, E_g]. Here, A_{TL} shows the oscillator strength denoted as A in Eq. (5.19). As a result, the number of analysis parameters in the optical model becomes nine, as illustrated in Fig. 7.22(a). In this example, however, the number of analysis parameters was reduced to eight by assuming $\varepsilon_1(\infty) = 1$. The TL model has been reported to provide excellent fitting to various TCO materials [84,85]. The dielectric function modeling of TCO by the Lorentz model [86–88], Cauchy model [87,89], and Forouhi–Bloomer model [90] has also been reported.

Figure 7.22(b) shows the (ψ, Δ) spectra obtained from the sample shown in Fig. 7.22(a) [82]. The Hall carrier concentration N_{Hall} of the ZnO:Ga layer is $6.5 \times 10^{20}\,\text{cm}^{-3}$. The sample measurement was carried out at $\theta_0 = 70°$ using a RCE instrument ($\text{PC}_\text{R}\text{SA}$). In Fig. 7.22(b), the peak position of ψ observed at 2 eV represents the film thickness and shifts toward lower energies with increasing film thickness, as we have seen in Fig. 7.3. On the other hand, spectral features observed at $\text{En} < 1.5\,\text{eV}$ arise from free-carrier absorption. The solid lines in Fig. 7.22(b) show the fitting result obtained from linear regression analysis. In this analysis, the analyzed energy region is limited to $\text{En} < 3.5\,\text{eV}$, in order to avoid the complicated structures observed in the dielectric function of ZnO:Ga (see Fig. 7.23). From this analysis, the thickness parameters can be estimated ($d_s = 41.3 \pm 1.9\,\text{Å}$, $d_b = 605.6 \pm 1.4\,\text{Å}$). In this case, we can perform mathematical inversion to extract the dielectric function of ZnO:Ga over the whole measured region (see Fig. 5.41).

Figure 7.23 shows the dielectric functions of ZnO:Ga with different carrier concentrations [82]. In Fig. 7.23, N_{Hall} of each sample is indicated. These dielectric functions were extracted using mathematical inversion, and the dielectric function d ($6.5 \times 10^{20}\,\text{cm}^{-3}$) was obtained from the spectra shown in Fig. 7.22(b). At lower energies ($\text{En} < 3.0\,\text{eV}$), however, the dielectric functions calculated by Eq. (7.20) are shown to eliminate spectral noise. It can be seen that the dielectric function of ZnO:Ga is rather similar to that of Ag shown in Fig. 7.20(a). In Fig. 7.23, the ε_1 peak at ~3 eV shows a clear shift toward higher energies with increasing N_{Hall} due to the Burstein–Moss effect; namely, the conduction band filling with free electrons shifts the onset of band-edge absorption [20,26]. In other words, electrons cannot be excited to a conduction band already filled with electrons [see Fig. 5.22(a)]. It has been reported that the ε_1 peak in ZnO originates from excitonic transition [91,92]. Thus, the enhanced broadening of the ε_1 peak at high N_{Hall} is caused by the screening of excitons, discussed in Fig. 7.17(a). At lower energies, on the other hand, the ε_2 values increase with increasing N_{Hall} by the effect of free-carrier absorption. Notice that the slight increase in free-carrier absorption leads to a large reduction in ε_1. When the carrier concentration is high, the reduction in ε_1 is rather significant even in the visible region (~ 2 eV).

If (ε_∞, E_p, N_f,) in Eq. (5.31) are known, we can deduce the effective mass m^* of semiconductors directly from Eq. (5.31) [82,93]. The high-frequency dielectric

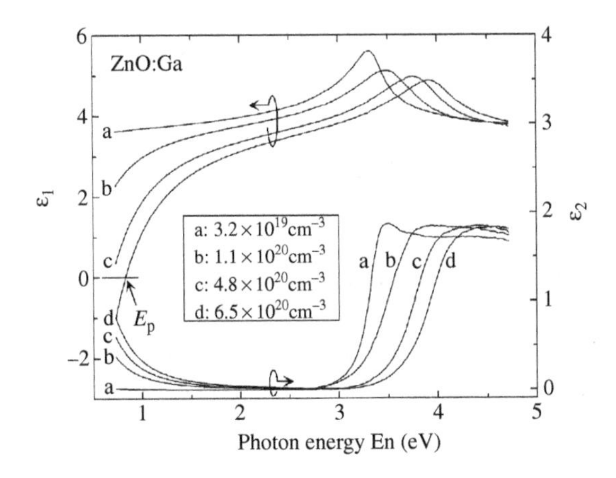

Figure 7.23 Dielectric functions of ZnO:Ga with different carrier concentrations. In the figure, the Hall carrier concentration N_{Hall} of each sample is indicated as a–d. The E_p represents the plasma energy. Adapted with permission from H. Fujiwara and M. Kondo, Effects of carrier concentration on the dielectric function of ZnO:Ga and In_2O_3:Sn studied by spectroscopic ellipsometry: Analysis of free-carrier and band-edge absorption, *Phys. Rev. B*, **71** (2005) 075109. Copyright 2005, the American Physical Society.

constant ε_∞ can be obtained from the analysis described below. At sufficiently low energies, the real part of $\varepsilon_{TL}(\mathrm{En})$ in Eq. (7.20) shows a constant value of ε_∞ [see Figs. 2.11 and 5.21(a)]. In this region, we can express $\varepsilon_1(\mathrm{En})$ from Eqs. (7.20) and (7.21) as follows [82,94]:

$$\varepsilon_1(\mathrm{En}) = \varepsilon_\infty - \frac{A_D}{\mathrm{En}^2 + \Gamma_D^2} \tag{7.22}$$

Thus, by plotting ε_1 versus $1/(\mathrm{En}^2 + \Gamma_D^2)$, we can determine ε_∞ from the intercept [82,94]. As shown in Fig. 7.23, the value of E_p can be obtained experimentally from the energy position of $\varepsilon_1(\mathrm{En}) = 0$. Accordingly, if we assume $N_f = N_{Hall}$, the effective mass can be estimated from the values of (ε_∞, E_p, N_{Hall}) using Eq. (5.31). Figure 7.24(a) shows m^*/m_0 of ZnO:Ga and In_2O_3:Sn (ITO) obtained from this procedure, plotted as a function of N_{Hall} [82]. Here, m_0 denotes the free-electron mass shown in Table 2.1. At $N_{Hall} = 0$, the reported values of $m^*(\mathrm{ZnO}) = 0.28m_0$ [95] and $m^*(In_2O_3) = 0.3m_0$ [96] are shown. The increase in m^* observed in heavily doped semiconductors has been explained by the nonparabolicity of the conduction band [94–96]. This nonparabolicity is caused by the conduction band filling with electrons, as shown in Fig. 5.22.

If m^* is known, we can determine N_f from the Drude parameter A_D by applying Eq. (5.30). Figure 7.24(b) shows N_f estimated from (A_D, m^*) using Eq. (5.30) [82]. In this calculation, m^* shown in Fig. 7.24(a) was used. In Fig. 7.24(b), N_f estimated from the ellipsometry analysis shows remarkable agreement with N_{Hall}. When the

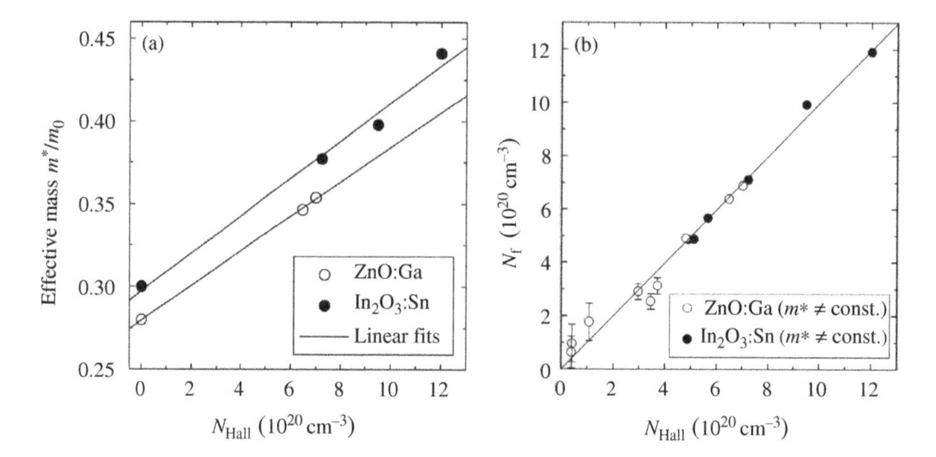

Figure 7.24 (a) Effective mass m^*/m_0 of ZnO:Ga and In_2O_3:Sn versus Hall carrier concentration N_{Hall} and (b) free-carrier concentration N_f estimated optically from spectroscopy ellipsometry versus N_{Hall}. In the analysis of N_f, m^*/m_0 shown in (a) was used. Adapted with permission from H. Fujiwara and M. Kondo, Effects of carrier concentration on the dielectric function of ZnO:Ga and In_2O_3:Sn studied by spectroscopic ellipsometry: Analysis of free-carrier and band-edge absorption, *Phys. Rev. B*, **71** (2005) 075109. Copyright 2005, the American Physical Society.

variation of m^* with carrier concentration is not taken into account, N_f shows poor agreement with N_{Hall} [82]. Thus, this result confirms the increase in m^* with carrier concentration. At lower carrier concentrations, however, the ellipsometry analysis becomes increasingly difficult due to low free-carrier absorption in the films. The sensitivity for free-carrier absorption can be improved by increasing the film thickness. The mobility μ can also be evaluated from (Γ_D, m^*) using Eq. (5.32). The electron mobilities estimated from the analysis, however, are constant at $\mu \sim 25\,\text{cm}^2/(\text{Vs})$ and show rather poor agreement with those evaluated from Hall measurement [82]. It should be emphasized that, in the case of polycrystalline materials, μ estimated from free-carrier absorption represents an average value within grains and does not include the effects of grain boundaries. In fact, mobilities estimated from free-carrier absorption generally show substantially higher values, compared with those evaluated by Hall measurement, when carrier transport is hindered considerably by grain boundaries [82,97]. Conversely, we can deduce the carrier transport properties at grain boundaries from the difference between mobilities evaluated from optical and electrical measurements.

Since ZnO is a direct bandgap material [98], the absorption coefficient can be approximated by $\alpha = A(\text{En} - E_g)^{1/2}$ [Eq. (7.3)]. Thus, the bandgap E_g of ZnO can be estimated by plotting α^2 versus En. Figure 7.25 shows α^2 versus photon energy for the ZnO:Ga [82]. The absorption coefficient can be obtained easily from the dielectric function (see Section 2.2.2). As shown in Fig. 7.25, the fundamental absorption edge shifts toward higher energies with increasing N_{Hall} by the Burstein–

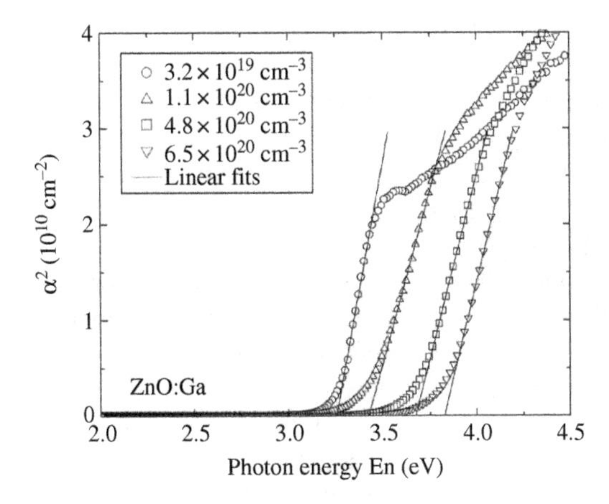

Figure 7.25 α^2 versus photon energy obtained from ZnO:Ga layers with different carrier concentrations. The bandgap is estimated from the intercept at $\alpha^2 = 0$. Adapted with permission from H. Fujiwara and M. Kondo, Effects of carrier concentration on the dielectric function of ZnO:Ga and In_2O_3:Sn studied by spectroscopic ellipsometry: Analysis of free-carrier and band-edge absorption, *Phys. Rev. B*, **71** (2005) 075109. Copyright 2005, the American Physical Society.

Moss effect. The solid lines show the linear fits to the experimental data, and E_g is estimated from the intercept at $\alpha^2 = 0$. As pointed out previously [99], however, this analysis is not applicable for heavily doped materials, since the conduction band is not parabolic anymore due to the conduction band filling with electrons. In order to determine E_g accurately, theoretical treatments are necessary [99]. Several theoretical studies have been reported with respect to the bandgap shift in TCO [99,100].

As evidenced from the above results, various information including optical constants (properties), thickness parameters, and electrical properties can be obtained from spectroscopic ellipsometry. In particular, from the analysis of free-carrier absorption, we can evaluate electrical properties accurately even without the requirement of forming electrodes on samples. As mentioned above, the drawbacks of the analysis using the Drude model are (a) the difficulty in characterizing samples with low carrier concentrations ($N_f < 10^{18}\,\mathrm{cm}^{-3}$) and (b) the requirement of m^* in the analysis. However, once m^* of a material is determined from the above procedure, we can evaluate the above properties relatively easily in a short time, when N_f is high enough. The measurement sensitivity for free-carrier absorption can be improved by using infrared spectroscopic ellipsometry, since free-carrier absorption increases in the lower energy region, as we have seen in Fig. 5.23. It should be noted that the electrical properties of TCO films generally show strong thickness dependence up to ~3000 Å [101,102]. In the above analysis, thin layers (~700 Å) were used to avoid such dependence. When strong thickness dependence is present, a multilayer structure should be used in the data analysis [103].

7.3.3 ADVANCED ANALYSIS

Here, as an advanced analysis, we will see the depth-profiling analysis of free-carrier absorption using a multilayer structure [104]. In this example, measurements were performed using an FTIR-PSC$_R$A instrument shown in Fig. 4.12(b) to improve measurement sensitivity. Figure 7.26(a) shows the ψ spectra of c-Si substrates implanted with As [104]. In order to activate the As atoms introduced into the c-Si substrates, the samples were annealed either in a furnace or rapid thermal annealing system. The peak at 1250 cm^{-1} in Fig. 7.26(a) represents atomic polarization induced by Si–O bonding within the native oxide. In general, when dopant atoms are introduced by ion implantation, their concentration is highly nonuniform in the thickness direction. Thus, the analysis of these samples was carried out by assuming a Gaussian distribution for the implanted As atoms. In this case, we can construct a optical model using the position of the Gaussian function (R), width of the Gaussian function (ΔR), and maximum carrier concentration at the peak position (N_{max}) as analysis parameters. From the analysis of the rapidly annealed sample, for example, $R = 44$ nm, $\Delta R = 64$ nm, and $N_{max} = 9.9 \times 10^{19}$ cm^{-3} can be obtained [Fig. 7.26(b)]. In other words, the carrier concentration becomes a maximum at a position of 44 nm from the surface and the width of this Gaussian distribution is 64 nm. If we perform the analysis assuming a Gaussian distribution, therefore, the number of analysis parameters can be reduced significantly. In the actual analysis, the Drude parameters ε_∞ and Γ (or mobility μ) are also required. Thus, we need a total of five parameters (N_{max}, Γ, ε_∞, R, ΔR) for this analysis.

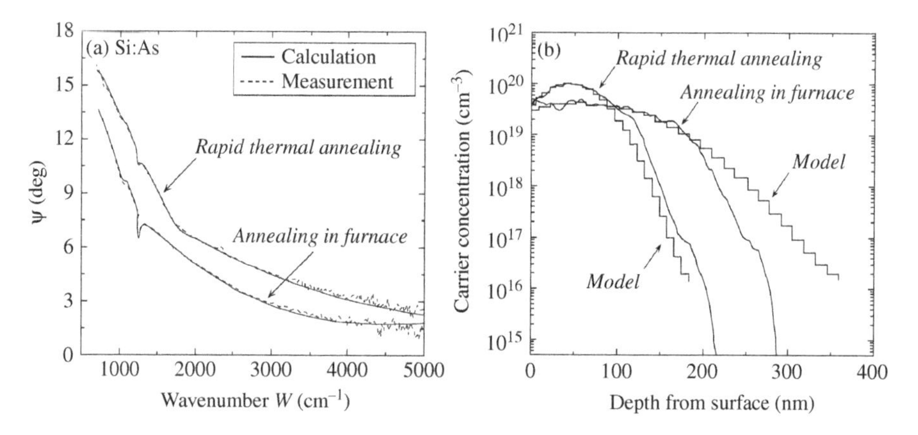

Figure 7.26 (a) ψ spectra of c-Si substrates implanted with As and (b) depth profiles of carrier concentration obtained from the samples shown in (a). In (b), the step structures denoted as 'Model' represent results determined from spectroscopic ellipsometry, and the solid lines show results characterized by spreading-resistance profile measurement. Reprinted from *Thin Solid Films*, **313–314**, T. E. Tiwald, D. W. Thompson, J. A. Woollam, W. Paulson, and R. Hance, Application of IR variable angle spectroscopic ellipsometry to the determination of free carrier concentration depth profiles, 661–666, Copyright (1998), with permission from Elsevier.

The solid lines in Fig. 7.26(a) show the fitting result obtained from the above analysis and the calculated spectra agree with the experimental spectra quite well. The step structures denoted as 'Model' in Fig. 7.26(b) represent the final result determined from the above analysis using the Gaussian distribution. In this analysis, the Gaussian distribution was described by a multilayer structure consisting of a total of 30 layers. As confirmed from Fig. 7.26(b), the carrier concentration profiles evaluated from infrared spectroscopic ellipsometry show excellent agreement with those characterized by spreading-resistance profile (SRP) measurement. This analysis example shows clearly that we can even characterize depth profiles of carrier concentration using spectroscopic ellipsometry. It has been reported that the analysis of depletion layers can also be performed if we employ a multilayer optical model [105,106].

7.4 ORGANIC MATERIALS/BIOMATERIALS

Since the late 1990s, spectroscopic ellipsometry has been applied extensively for the characterization of organic materials and biomaterials. The ellipsometry measurements for these materials can be classified into measurements in the UV/visible and infrared regions. In the UV/visible region, the structural characterization of thin organic films has been performed using spectroscopic ellipsometry. For this purpose, a single-wavelength ellipsometer, which usually employs a He–Ne laser as a light source, has also been used. From measurements in the infrared region, various atomic polarizations in organic materials have been studied.

7.4.1 ANALYSIS OF ORGANIC MATERIALS

With respect to organic materials, polymer thin films [107–116], self-assembled layers [117–119], Langmuir–Blodgett (LB) films [120–122], and liquid crystals [123–127] have been studied intensively using spectroscopic ellipsometry. Often, organic materials show strong optical anisotropy, and an analysis example of such materials will be shown in Section 7.5.3. Here, we will look at analysis examples for isotropic organic films including polymer, LB, and self-assembled films.

Figure 7.27 shows the dielectric function of a poly-pyrrole film formed in aqueous solution by using an electropolymerization process [107]. The polymer film ($\sim 500\,Å$) was deposited on a substrate covered with a Au film, and the measurement on this sample was performed in an aqueous solution using a P_RSA instrument. The dielectric function of the poly-pyrrole shown in Fig. 7.27 was obtained from mathematical inversion using an optical model consisting of an H_2O/polymer film/Au substrate structure. In general, the dielectric constants of organic films are low, and the small surface roughness on organic films has negligible effects on ellipsometry analysis (see Section 5.3.3). The solid lines in

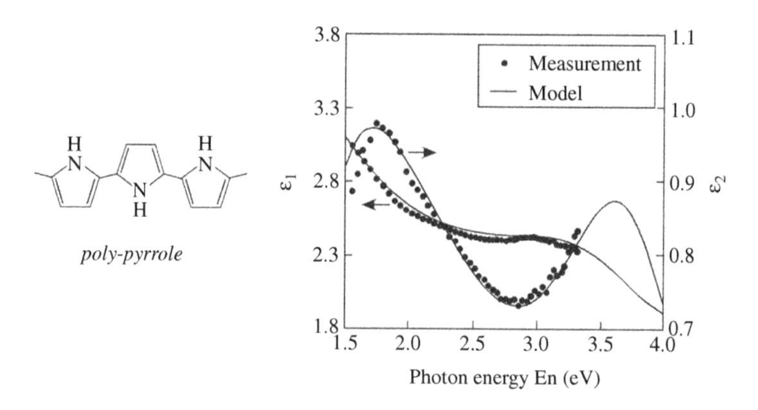

Figure 7.27 Dielectric function of poly-pyrrole (conducting polymer). The solid lines represent calculation results based on the Lorentz model. Reprinted from *Thin Solid Films*, **193/194**, Y.-T. Kim, D. L. Allara, R. W. Collins, K. Vedam, Real-time spectroscopic ellipsometry study of the electrochemical deposition of polypyrrole thin films, 350–360, Copyright (1990), with permission from Elsevier.

Fig. 7.27 represent the fitting result obtained from dielectric function modeling using the Lorentz model. The modeling was performed from three Lorentz peaks located at En = 1.65 eV, 2.3 eV, and 3.8 eV [107]. It should be noted that the poly-pyrrole is a conductive polymer and has a band structure similar to a semiconductor. The ε_2 peak at En = 3.8 eV represents the optical transition from the valence band to the conduction band, while the other peaks show the transitions from the valence band to defect states [107]. It is clear from this example that the film structures and optical properties of organic films can also be evaluated from spectroscopic ellipsometry.

Figure 7.28(a) shows the Δ spectra for a LB film and Au substrate obtained from the measurements in the infrared region at $\theta_0 = 70°$ [122]. For this measurement, an infrared spectroscopic ellipsometer (FTIR-P_RCSA) was used (see Section 4.2.6). The LB film is composed of nine double layers of the organic molecule depicted in the inset. As shown in Fig. 7.28(a), large changes appear in the Δ spectrum after the LB film formation on the Au substrate. The variation in the spectral slope is caused by the optical interference of the LB film and, from this change, we can determine the optical properties and film thickness of the LB film. In this example, however, several (ψ, Δ) spectra were measured at different angles of incidence, in order to separate the contributions of n and d correlated by the nd product (see Section 5.1.2). From the analysis, $d_b = 610$ Å and $\varepsilon_\infty = 2.04$ are obtained. Figure 7.28(b) shows the Δ and tan ψ spectra of this sample [122]. The Δ spectra in Fig. 7.28(b) correspond to the region 1300–1700 cm^{-1} in Fig. 7.28(a). The dotted lines in Fig. 7.28(b) represent the calculation results using the Lorentz model, and each peak shows the atomic polarization generated in the LB film. As confirmed from Fig. 7.28(b), the analysis of organic films in the infrared region is generally complicated, since a large number of peaks appear in the dielectric functions. In

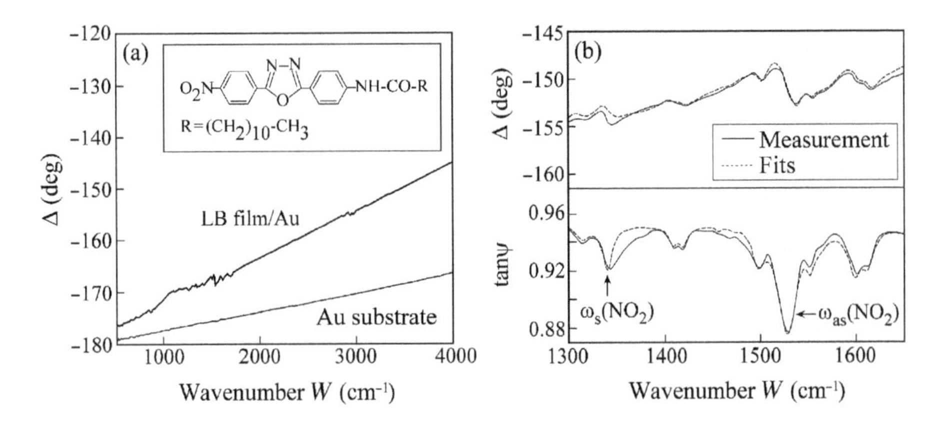

Figure 7.28 (a) Δ spectra obtained from a Langmuire–Blodgett (LB) film/Au substrate and a Au substrate at $\theta_0 = 70°$ and (b) (Δ, tan ψ) spectra of the LB film/Au in the range 1300–1700 cm^{-1}. In (b), $\omega_s(NO_2)$ and $\omega_{as}(NO_2)$ represent the symmetric and asymmetric vibration modes of NO_2 group, respectively. From *Physica Status Solidi A*, **188**, D. Tsankov, K. Hinrichs, A. Röseler, and E. H. Korte, FTIR ellipsometry as a tool for studying organic layers: from Langmuire–Blodgett films to can coatings, 1319–1329 (2001). Reproduced with permission of Wiley-VCH.

Fig. 7.28(b), we can see two peaks for the NO_2 group, assigned to symmetric vibration (ω_s) and asymmetric vibration (ω_{as}) modes. The light absorption of these vibration modes varies depending on the direction of polarized light. Accordingly, the orientation of the organic molecules can also be studied from the peak intensity ratio of these modes [122].

Figure 7.29 shows the dielectric function of a poly-arylene ether film (d_b = 407 nm) formed on a SiO_2/c-Si substrate by a spin-coating process [110]. This dielectric function was determined by using three measurements: (a) spectroscopic ellipsometry in the range En = 1.5–5.4 eV, (b) near-normal incidence reflectometry in the range En = 0.5– 6.5 eV, and (c) normal-incidence transmittance measurement using FTIR in the range En = 0.03– 0.52 eV. The ellipsometry measurement was carried out by a null ellipsometer using four-zone averaging (see Section 4.2.8). In this analysis example, the ellipsometry and reflectometry spectra were fitted simultaneously, in order to reduce the strong correlation between n and d described by the nd product. For the analysis, an optical model composed of an air/polymer film/SiO_2/c-Si substrate structure was used [109]. Here, the SiO_2 layer is a native oxide formed on the c-Si substrate. No surface roughness layer is provided in this optical model, probably due to the low sensitivity for roughness in the analysis. In this optical model, therefore, unknown parameters are the thickness and optical constants of the bulk layer (polymer film) only. The thickness of the film was determined by applying the Sellmeier model to the transparent region near En ~ 2 eV. From mathematical inversion using the thickness estimated from this analysis, the dielectric function over the whole measurement region (En = 1.5– 5.4 eV) was obtained [109]. In the region below 1.5 eV, however, there are only

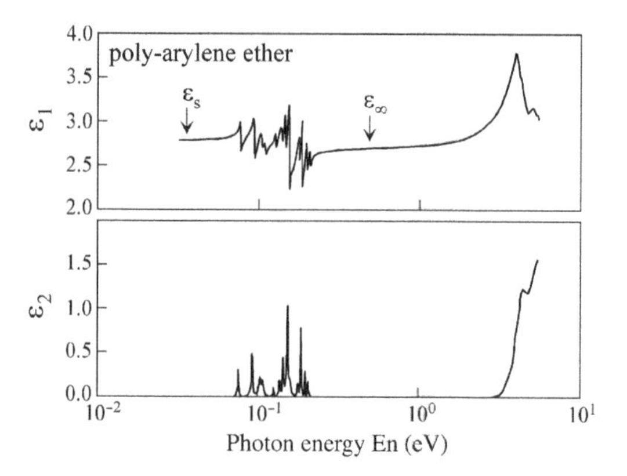

Figure 7.29 Dielectric function of a poly-arylene ether film formed on a SiO_2/c-Si substrate by a spin-coating process. In the figure, ε_s and ε_∞ represent the static and high-frequency dielectric constants, respectively. Reprinted with permission from *Applied Physics Letters*, **79**, K. Postava, T. Yamaguchi, and M. Horie, Estimation of the dielectric properties of low-k materials using optical spectroscopy, 2231–2233 (2001). Copyright 2001, American Institute of Physics.

reflectance and transmittance spectra. Thus, the dielectric function in this region was obtained using the Kramers–Kronig relations. The film thickness determined by spectroscopic ellipsometry was employed for this analysis. If we measure ellipsometry spectra in this region, this procedure can be avoided. In the infrared region, the dielectric function of the poly-arylene ether can be modeled from the Lorentz model. Thus, below the absorption edge in the UV region (En < 2.8 eV), the dielectric function is expressed by the following equation [110]:

$$\varepsilon(\mathrm{En}) = A + \frac{B\mathrm{En}_0^2}{\mathrm{En}_0^2 - \mathrm{En}^2} + \sum_j \frac{A_j \mathrm{En}_{0j}^2}{\mathrm{En}_{0j}^2 - \mathrm{En}^2 + \mathrm{i}\Gamma_j \mathrm{En}_{0j}\mathrm{En}} \tag{7.23}$$

In the above equation, the first and second terms on the right represent the Sellmeier model expressed by photon energy. The second term can be derived easily from Eq. (5.16) using the conversions $\lambda \to 1/\mathrm{En}$ and $\lambda_0 \to 1/\mathrm{En}_0$. On the other hand, the third term in Eq. (7.23) shows the Lorentz model, which is slightly different from Eq. (5.10). At the limit of $\mathrm{En} \to 0\,\mathrm{eV}$, Eq. (7.23) is transformed as follows [110]:

$$\varepsilon(\mathrm{En} \to 0) = A + B + \sum_j A_j \tag{7.24}$$

Notice that ε obtained from Eq. (7.24) represents the static dielectric constant ε_s. The high-frequency dielectric constant ε_∞ is also calculated from $(A+B)$ in Eq. (7.24). Recall from Fig. 2.11 that $\varepsilon_s > \varepsilon_\infty$. As a result, $\varepsilon_s = 2.78$ and $\varepsilon_\infty = 2.70$ have been reported from this analysis [110]. The value of ε_s can also be estimated from the constant ε_1 at lower energies.

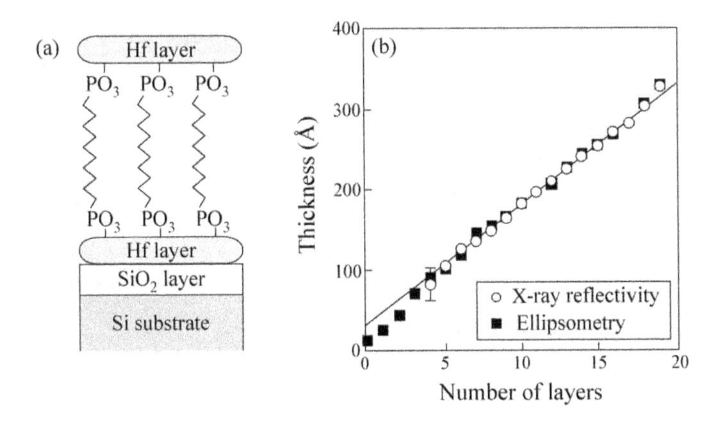

Figure 7.30 (a) Schematic structure of a self-assembled organic film consisting of a $Hf/O_3P(CH_2)_{10}PO_3$ layer and (b) thickness of the self-assembled film estimated by X-ray reflectivity and ellipsometry, plotted as a function of the number of the $Hf/O_3P(CH_2)_{10}PO_3$ layers. Drawing (b): Reprinted with permission from Zeppenfeld *et al.*, *J. Am. Chem. Soc.* 1994, **116**, 9158–9165. Copyright 1994, American Chemical Society.

Now, we will look at the ellipsometry analysis of the self-assembled organic film shown in Figure 7.30(a). In this sample, Hf layers were used as contact (binder) layers, and several $Hf/O_3P(CH_2)_{10}PO_3$ layers were prepared on a Hf/SiO_2/c-Si substrate by using wet chemical processing [118]. Figure 7.30(b) shows the thickness of the self-assembled film estimated by X-ray reflectivity and ellipsometry measurements, plotted as a function of the number of the $Hf/O_3P(CH_2)_{10}PO_3$ layers [118]. The ellipsometry measurement was performed using a single-wavelength ellipsometer using a wavelength of 6328 Å radiated from a tungsten–halogen lamp. The ellipsometry analysis was carried out by assuming an air/thin film/SiO_2/c-Si substrate structure. In this analysis example, however, the average refractive index of the self-assembled film was determined so that the layer thicknesses evaluated from X-ray reflectivity matched with those estimated from ellipsometry. From this analysis, the average refractive index of the organic layers was estimated to be $n = 1.500$. This refractive index is lower than that of the bulk material ($n = 1.544$) and indicates the lower density of the self-assembled layers. Furthermore, a layer spacing of 14.8 Å, characterized from the slope of Fig. 7.30(b), was found to be thinner than the expected value of 16.7 Å observed in the bulk material. The above results imply that the self-assembled layers are tilted in the actual samples and have lower densities (refractive indices), compared with the bulk material [118].

This example shows clearly that, if we use other characterization techniques, the film structures or optical constants of samples can be discussed in more detail. In particular, it is essential to separate the contributions of n and d, when we analyze samples in which correlation of the nd product is expected. In such cases, it is preferable to determine film thickness by other techniques. Unfortunately, the X-ray reflectivity technique cannot be applied to sample characterization when the surface roughness of a sample is larger than ~30 Å, as X-ray reflectivity reduces strongly with increasing surface roughness. Alternatively, we may measure ellipsometry spectra by varying the angle of incidence or sample thickness. We may perform the analysis of ellipsometry spectra together with reflectance and transmittance spectra to avoid the correlation between *n* and *d*. Keep in mind that we obtain more reliable ellipsometry results if an ellipsometry analysis is performed by referring to results evaluated from other techniques.

7.4.2 ANALYSIS OF BIOMATERIALS

The ellipsometry technique has also been applied to the characterization of biomaterials [128–134]. With respect to biomaterials, ellipsometry has been employed to determine the optical properties [131–134], layer structures [131–133], or adsorption processes [128–132] of biomaterials. In particular, ellipsometry enables us to monitor the adsorption processes of biomolecules in aqueous solution on a monolayer scale. Here, we will briefly overview biomaterial characterization using the ellipsometry technique.

Figure 7.31 shows the dielectric function of a protein called γ-globulin [131]. This dielectric function was extracted from one protein layer (60.5 Å) formed on a HgTe substrate using mathematical inversion. As confirmed from Fig. 7.31, the dielectric function of the protein layer shows low ε_1 values. In this sense, the optical properties of biomaterials are quite similar to those of the organic materials described in Section 7.4.1. It should be noted that, when we perform mathematical inversion of a thin layer (~ 50 Å), the optical properties of the inversion layer have to be different from those of the underlying layer (see Section 5.5.3). In the case of the above analysis, we can perform mathematical inversion of the thin protein

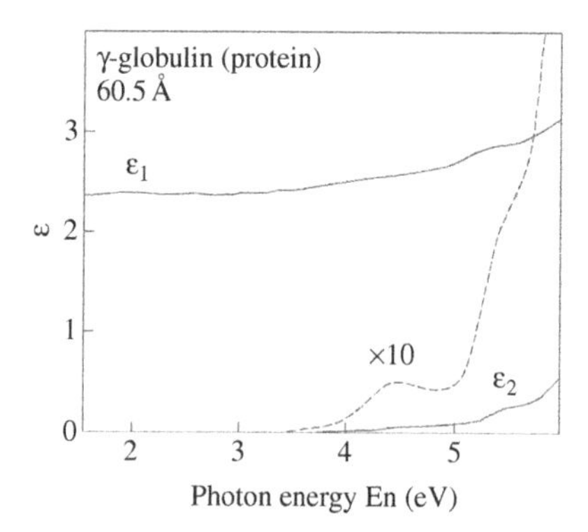

Figure 7.31 Dielectric function of γ-globulin (protein). The dielectric function was obtained from one protein layer (60.5 Å) formed on a HgTe substrate. Reprinted from *Thin Solid Films*, **313–314**, H. Arwin, Spectroscopic ellipsometry and biology: recent developments and challenges, 764–774, Copyright (1998), with permission from Elsevier.

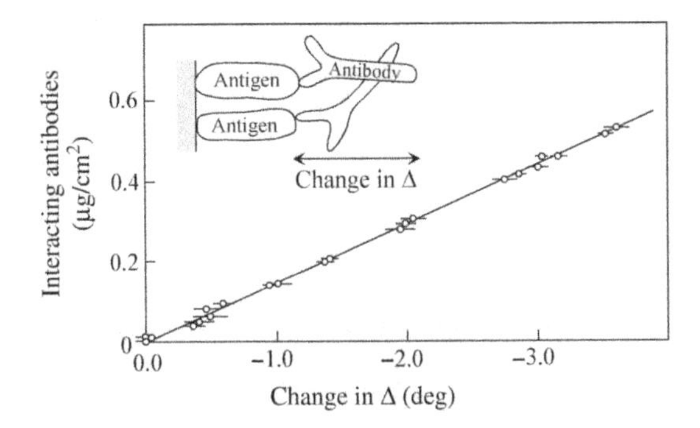

Figure 7.32 Weight density of rabbit antibodies interacted with human antigens formed on a c-Si substrate, plotted as a function of the change in Δ. Reprinted from *Journal of Colloid and Interface Science*, **103**, U. Jönsson, M. Malmqvist, and I. Rönnberg, Adsorption of immunoglobulin G, protein A, and fibronectin in the submonolayer region evaluated by a combined study of ellipsometry and radiotracer techniques, 360–372, Copyright (1985), with permission from Elsevier.

layer, since the optical constants of HgTe ($E_g = 0\,\text{eV}$) are rather similar to those of metals [16] and are quite different from those of γ-globulin.

Now we will look at the study of protein adsorption (antigen–antibody reaction) characterized by a single-wavelength ellipsometer [128]. Figure 7.32 shows the weight density of rabbit antibodies interacted with human antigens, plotted as a function of the change in Δ [128]. In this experiment, an n-type Si(111) substrate coated with the human antigen, immunoglobulin G (IgG), was used as a substrate, and the rabbit IgG antibodies were supplied to the substrate placed in a liquid cell to characterize the antigen–antibody reaction. The ellipsometry measurements were performed in the liquid cell using a null ellipsometer at a single wavelength of 6328 Å (He–Ne laser). The weight densities of the interacting antibodies were determined from a radiotracer technique that uses ^{14}C-labeled rabbit IgG as a probe. In this example, the amount of interacting antibodies was controlled by changing the concentration of the rabbit IgG antibodies. As shown in Fig. 7.32, the Δ value decreases linearly as the antibodies adsorbed on the antigens increase. Recall from Fig. 5.6(a) that the Δ value reduces linearly with increasing thickness when the film thickness on the c-Si substrate is thin (< 50Å). In this case, the adsorption process can be deduced directly from the absolute value of Δ [see Fig. 5.6(b)]. This characterization is particularly helpful for the real-time monitoring of protein adsorption processes.

Recently, the characterization of a DNA chip by ellipsometry has also been reported [133]. Figure 7.33 illustrates the structure of the DNA chip in which oligonucleotides (DNA molecules) are supported by linkers formed on a SiO_2/c-Si substrate [133]. The DNA molecule shown in Fig. 7.33 is a 14 base oligonucleotide with the sequence ATCATCTTTGGTGT and has a single-stranded DNA (ssDNA) structure. As known widely, DNA molecules have double-stranded structures and, if an unknown ssDNA binds with a known ssDNA to form a double-stranded DNA (dsDNA) structure, we can determine the sequence of the unknown DNA. The linker

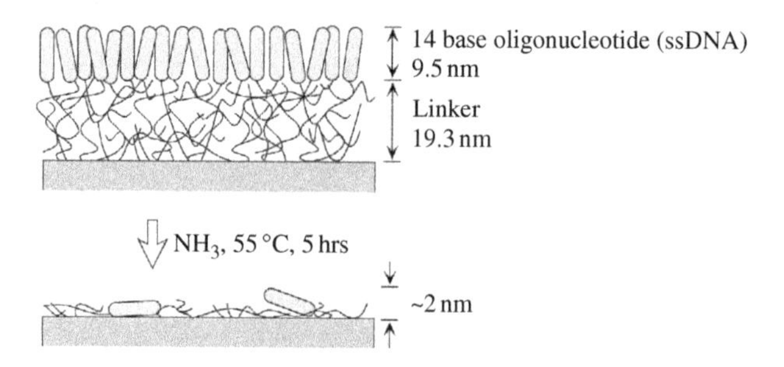

Figure 7.33 Schematic structure of a DNA chip deduced from ellipsometry analysis. In this figure, the linker is an organic layer and ssDNA indicates a single-stranded DNA molecule. Reprinted with permission from Lang *et al.*, *Langmuir* 1997, **13**, 2833–2842. Copyright 1997, American Chemical Society.

shown in Fig. 7.33 is an organic layer mainly composed of Si–O and O–$(CH_2)_n$ groups [133]. In this example, ellipsometry was also applied to characterize a sample structure treated by NH_3 at 55 °C for five hours.

The thickness of each layer shown in Fig. 7.33 was determined by a single-wavelength ellipsometer (6328 Å). From the analysis of the linker/SiO_2/c-Si structure, $d = 193$ Å and $n = 1.460$ were obtained as the thickness and refractive index of the linker layer, respectively. For SiO_2 and c-Si, reported optical constants were used. In this case, we can determine n and d of the linker directly from measured (ψ, Δ) values, provided that there are no surface roughness and interface layers. Recall from Fig. 5.7 that n and d of a film can be obtained independently when the film thickness is larger than 100 Å. Finally, by analyzing the DNA/linker/SiO_2/c-Si structure using $n = 1.460$ (linker), the thickness and refractive index of the ssDNA were estimated to be $d = 95$ Å and $n = 1.462$, respectively. This refractive index is consistent with one reported later [134]. From the study using spectroscopic ellipsometry, it has been reported that the optical constants of DNA can be expressed by the Lorentz model (single oscillator), and the refractive index of dsDNA is higher than that of ssDNA by $\sim$5% [134]. Thus, it is basically possible to detect the change from ssDNA to dsDNA by using the ellipsometry technique.

7.5 ANISOTROPIC MATERIALS

Recently, spectroscopic ellipsometry has become increasingly important due to its ability to characterize the optical constants of anisotropic materials. Historically, the application of ellipsometry to anisotropic sample characterization began as early as 1970s [135–140]. However, such characterization has become popular just recently, partly owing to the development of the new 4×4 matrix method in the late 1990s (see Section 6.3). At present, spectroscopic ellipsometry is employed widely to study the anisotropic properties of insulators, semiconductors, and organic materials. In this section, we will look at analysis examples for these materials in detail.

7.5.1 ANALYSIS OF ANISOTROPIC INSULATORS

Here, as the simplest analysis example, we will address the analysis of a uniaxial α-quartz (α-SiO_2) substrate that shows strong anisotropic behavior in the infrared region [43]. Figure 7.34(a) shows the optical model for the α-SiO_2 substrate. The unknown parameters in this optical model are the complex refractive indices for the ordinary ray (N_{1o}) and extraordinary ray (N_{1e}). Since $N_{1o} = n_{1o} - ik_{1o}$ and $N_{1e} = n_{1e} - ik_{1e}$, the total number of unknown parameters is four in this optical model. For this analysis, two uniaxial α-SiO_2 crystals with different orientations were used. In particular, these samples were prepared so that the optical axes of the two samples lay perpendicular and parallel to the sample surfaces. The ellipsometry measurements were performed for the orientations shown in Figs. 6.10(a)–(c) using FTIR-P_RCSA. In addition, for each orientation, the ellipsometry spectra were obtained at $\theta_0 = 65°, 75°, 80°$. In this example, therefore, a total of nine ellipsometry spectra were used in the analysis. Fig. 7.34(b) shows the Δ spectrum of the α-SiO_2 crystal measured at $\theta_0 = 75°$ with the orientation $\phi_E = \theta_E = 0°$ (see Fig. 6.10) [43].

Recall from Section 6.1.4 that the Jones matrix of anisotropic samples is diagonal in the above measurement configurations. Furthermore, the diagonal components of the Jones matrix (r_{ss} and r_{pp}) can be calculated simply from the Fresnel equations shown in Eqs. (6.19) and (6.20), since the principal axes of the index ellipsoid are parallel to the (x, y, z) coordinates in the measurements. In this case, the (ψ, Δ) values can be described as

$$\tan\psi_{pp}(\theta_0, \phi_E, \theta_E)\exp[i\Delta_{pp}(\theta_0, \phi_E, \theta_E)] = \rho_{pp}(\theta_0, \phi_E, \theta_E, N_0, N_{1o}, N_{1e}) \quad (7.25)$$

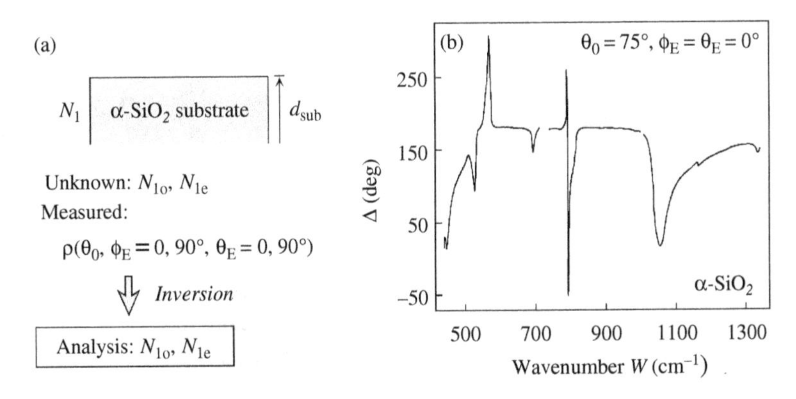

Figure 7.34 (a) Optical model for a uniaxial α-SiO_2 substrate and (b) Δ spectrum of the α-SiO_2 measured at $\theta_0 = 75°$. In (a), N_{1o} and N_{1e} represent the complex refractive indices for ordinary and extraordinary rays, respectively. $\rho(\theta_0, \phi_E, \theta_E)$ shows $\rho = \tan\psi_{pp}\exp(i\Delta_{pp})$ obtained at the incidence angle θ_0 with the Euler angles of ϕ_E and θ_E. In (b), the Euler angles of the uniaxial sample are $\phi_E = \theta_E = 0°$. Drawing (b): Reprinted from *Thin Solid Films*, **234**, J. Humlíček, and A. Röseler, IR ellipsometry of the highly anisotropic materials α-SiO_2 and α-Al_2O_3, 332–336, Copyright (1993), with permission from Elsevier.

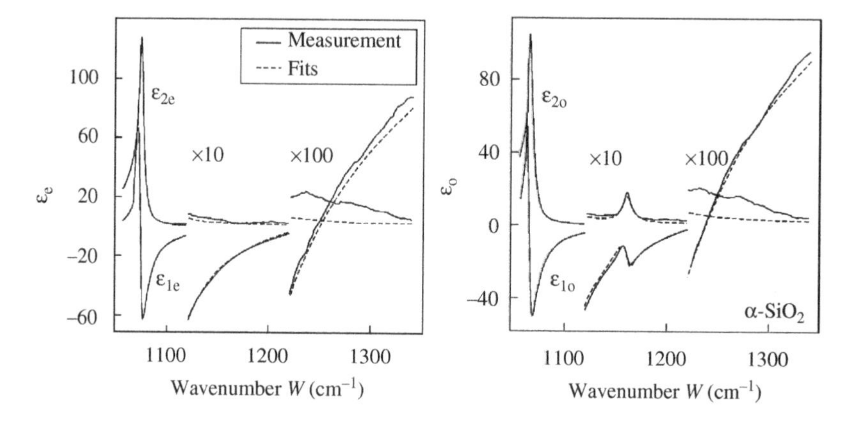

Figure 7.35 Dielectric functions of uniaxial α-SiO_2. In this figure, $\varepsilon_e = \varepsilon_{1e} - i\varepsilon_{2e}$ and $\varepsilon_o = \varepsilon_{1o} - i\varepsilon_{2o}$ represent the dielectric constants for the extraordinary and ordinary rays, respectively. The dotted lines show fits using the Lorentz model. Reprinted from *Thin Solid Films*, **234**, J. Humlíček, and A. Röseler, IR ellipsometry of the highly anisotropic materials α-SiO_2 and α-Al_2O_3, 332–336, Copyright (1993), with permission from Elsevier.

(N_{1o}, N_{1e}) in Eq. (7.25) can be extracted directly from the measured spectra by using mathematical inversion. To perform the mathematical inversion, however, the number of measured parameters has to be larger than the unknown parameters. Thus, we need at least two (ψ, Δ) spectra to obtain (n_{1o}, k_{1o}) and (n_{1e}, k_{1e}) separately. For this analysis, we will have two equations in the form of Eq. (7.25) with different values for $(\theta_0, \phi_E, \theta_E)$. However, it is preferable to perform the mathematical inversion using a larger number of (ψ, Δ) spectra.

Figure 7.35 shows the dielectric functions of the uniaxial α-SiO_2 crystal extracted from the mathematical inversion [43]. In this figure, the dielectric functions for the extraordinary and ordinary rays are denoted as $\varepsilon_e = \varepsilon_{1e} - i\varepsilon_{2e}$ and $\varepsilon_o = \varepsilon_{1o} - i\varepsilon_{2o}$, respectively. The sharp features in the dielectric functions originate from the transverse optical (TO) phonon and longitudinal optical (LO) phonon modes in α-SiO_2 [141]. The dotted lines indicate the result of dielectric function modeling using the Lorentz model. The Lorentz peaks for the extraordinary ray are shifted toward higher wavenumbers, compared with those of the ordinary ray. Notice that the peak at 1160 cm^{-1} is absent in the dielectric function for ε_e. So far, similar analyses have been performed for α-Al_2O_3 (sapphire) [43,44].

7.5.2 ANALYSIS OF ANISOTROPIC SEMICONDUCTORS

Various semiconductors having a zinc blend structure (or diamond structure) generally show isotropic optical character. Thus, most of group IV, III-V, and II-VI semiconductors, such as Si, GaAs, ZnSe, are optically isotropic. So far, optical anisotropy in semiconductors has been reported for wide-bandgap hexagonal (wurtzite) semiconductors including ZnO [91,92], GaN [16,46], 4H-SiC [16,142], 6H-SiC [16,143] and BN [16,144]. However, these semiconductors exhibit rather weak optical anisotropy in the visible region below the bandgaps. Here, we will look at ellipsometry data analyses of a uniaxial TiO_2 crystal [145] and 4H-SiC crystal [142] by using the 4×4 matrix method described in Section 6.3.

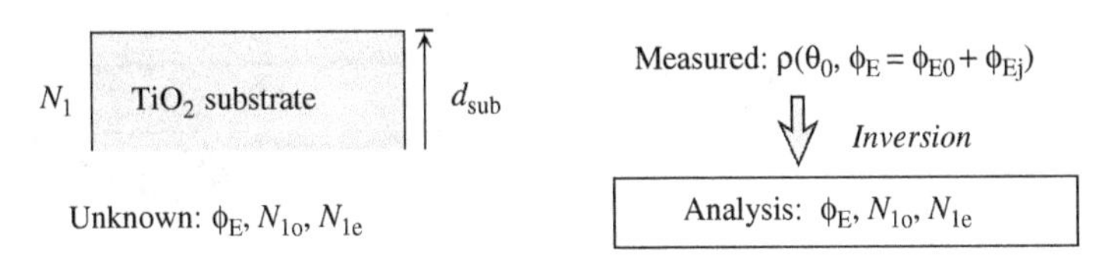

Figure 7.36 Optical model for a uniaxial TiO_2 substrate. In the figure, the Euler angle ϕ_E is expressed by $\phi_E = \phi_{E0} + \phi_{Ej}$, where ϕ_{Ej} shows the in-plane rotation angle of the sample in measurement.

Figure 7.36 shows the optical model for the uniaxial TiO_2 substrate studied in the analysis example. There are the two crystalline structures for TiO_2 (i.e., anatase and rutile), and both structures show strong anisotropy [146]. The sample used in this example is a rutile-type TiO_2 crystal with an orientation of (100), and the optical axis of this sample is parallel to the substrate surface (i.e., $\theta_E = 90°$). In this case, the unknown parameters in the optical model are N_{1o}, N_{1e}, and ϕ_E. When $\phi_E \neq 0°$ or 90° ($\theta_E = 90°$), the Jones matrix of this sample is not diagonal and shows finite values for r_{ps} and r_{sp} (see Section 6.1.4). In this example, therefore, the measurement of the sample was performed by the procedure described in Section 6.5.1 using an RAE instrument. From this measurement, the ellipsometry spectra for each (ψ_{pp}, Δ_{pp}, ψ_{ps}, Δ_{ps}, ψ_{sp}, Δ_{sp}) are obtained. Figure 7.37 shows the (ψ_{pp}, Δ_{pp}, ψ_{ps}, Δ_{ps}) spectra obtained by varying the in-plane rotation angle of the TiO_2 substrate (ϕ_{Ej}) [145]. If we use ϕ_{Ej}, the Euler angle of the sample (ϕ_E) is expressed as $\phi_E = \phi_{E0} + \phi_{Ej}$, where ϕ_{E0} represents the initial rotation angle between the optical axis and the x axis in the measurement configuration. When the sample is not rotated (i.e., $\phi_{Ej} = 0°$), it follows that $\phi_E = \phi_{E0}$. Recall that ϕ_E (or ϕ_{E0}) is the unknown parameter in this analysis.

Using Eq. (7.25), we can express the (ψ, Δ) values in this analysis as follows:

$$\tan\psi_{pp}(\theta_0, \phi_E, \theta_E)\exp[-i\Delta_{pp}(\theta_0, \phi_E, \theta_E)] = \rho_{pp}(\theta_0, \phi_E, \theta_E, N_0, N_{1o}, N_{1e}) \quad (7.26a)$$

$$\tan\psi_{ps}(\theta_0, \phi_E, \theta_E)\exp[-i\Delta_{ps}(\theta_0, \phi_E, \theta_E)] = \rho_{ps}(\theta_0, \phi_E, \theta_E, N_0, N_{1o}, N_{1e}) \quad (7.26b)$$

$$\tan\psi_{sp}(\theta_0, \phi_E, \theta_E)\exp[-i\Delta_{sp}(\theta_0, \phi_E, \theta_E)] = \rho_{sp}(\theta_0, \phi_E, \theta_E, N_0, N_{1o}, N_{1e}) \quad (7.26c)$$

The actual calculation of Eq. (7.26) can be performed using the 4×4 matrix method. Recall from Section 6.3 that the 4×4 matrix method uses the convention of $N \equiv n + ik$

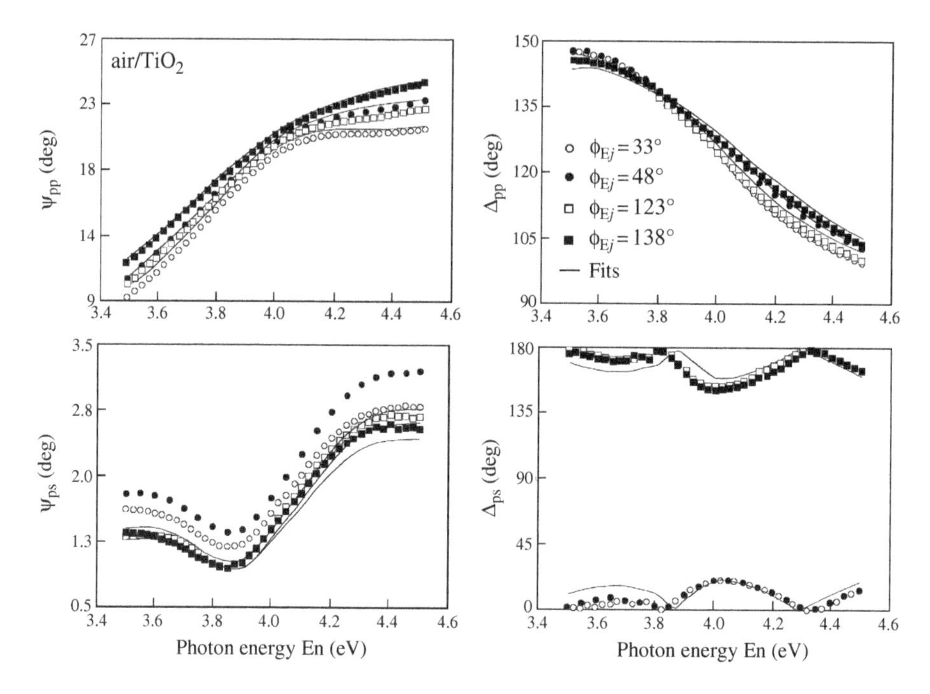

Figure 7.37 (ψ_{pp}, Δ_{pp}, ψ_{ps}, Δ_{ps}) spectra obtained from a rutile-type $TiO_2(100)$ substrate ($\theta_E = 90°$) at $\theta_0 = 70°$. In the measurement, the in-plane rotation angle of the TiO_2 substrate (ϕ_{Ej}) was changed. From *Journal of the Optical Society of America A*, **13**, M. Schubert, B. Rheinländer, J. A. Woollam, B. Johs and C. M. Herzinger, Extension of rotating-analyzer ellipsometry to generalized ellipsometry: determination of the dielectric function tensor from uniaxial TiO_2, 875–883 (1996). Reproduced with permission of the Optical Society of America.

and (ψ, Δ) are defined by $\tan\psi\exp(-i\Delta)$. Since θ_0, N_0, θ_E are known, the number of unknown parameters in Eq. (7.26) is five (i.e., ϕ_E, n_{1o}, k_{1o}, n_{1e}, k_{1e}). On the other hand, the total number of measured parameters is 24, since the 6 parameters for (ψ, Δ) were measured for each rotation angle indicated in Fig. 7.37. Accordingly, the unknown parameters (ϕ_E, N_{1o}, N_{1e}) can be determined by using mathematical inversion that employs Eq. (7.26). The solid lines in Fig. 7.37 represent the fitting result obtained from mathematical inversion. Although the result for (ψ_{sp}, Δ_{sp}) is not shown in Fig. 7.37, a similar fitting result has been reported for (ψ'_{sp}, Δ'_{sp}) defined by Eq. (6.16) [145]. Notice from Fig. 7.37 that the difference between the experimental and fitting spectra is larger in the Δ_{ps} spectra. This represents large measurement errors in RAE instruments in the regions near $\Delta = 0°$ and 180° (see Section 4.4.1).

Figure 7.38 shows the optical constants of the uniaxial TiO_2 obtained from the above analysis [145]. As shown in Fig. 7.38, (n_{1o}, n_{1e}) and (k_{1o}, k_{1e}) are quite different in TiO_2. From the analysis, the Euler angle ϕ_E was determined to be $35° \pm 3°$, and a similar value of $33.1° \pm 0.1°$ was also estimated from X-ray diffraction. The ϕ_E of TiO_2 was found to show a constant value versus wavelength. In this analysis example, the characterization of a $TiO_2(111)$ substrate was also

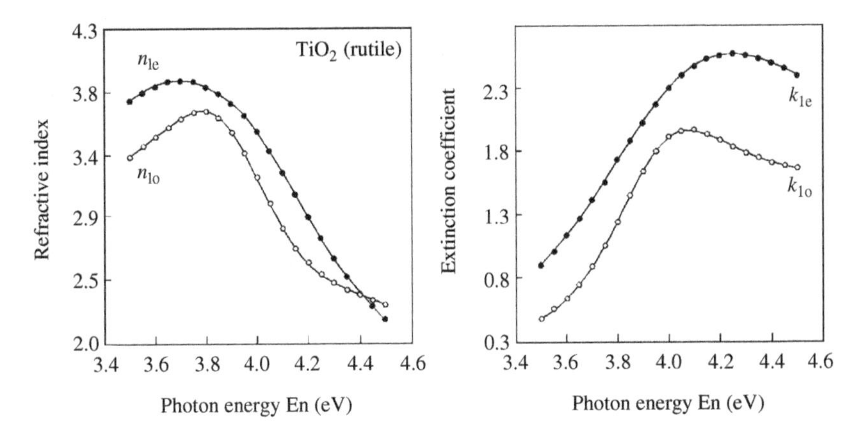

Figure 7.38 Refractive index and extinction coefficient of the uniaxial TiO_2 (rutile) crystal extracted from the analysis shown in Fig. 7.37. In the figure, the complex refractive indices for the ordinary and extraordinary rays are denoted as $N_{1o} = n_{1o} - ik_{1o}$ and $N_{1e} = n_{1e} - ik_{1e}$, respectively. From *Journal of the Optical Society of America A*, **13**, M. Schubert, B. Rheinländer, J. A. Woollam, B. Johs and C. M. Herzinger, Extension of rotating-analyzer ellipsometry to generalized ellipsometry: determination of the dielectric function tensor from uniaxial TiO_2, 875–883 (1996). Reproduced with permission of the Optical Society of America.

performed. In this case, the number of unknown parameters increases to six (i.e., ϕ_E, θ_E, n_{1o}, k_{1o}, n_{1e}, k_{1e}). For this analysis, ellipsometry spectra measured at different angles of incidence ($\theta_0 = 60°, 70°$) were also used to increase the number of measurement parameters. It has been confirmed from this analysis that we can determine ($\phi_E, \theta_E, N_{1o}, N_{1e}$) from a similar data analysis. With respect to TiO_2, the optical constants of the anataze structure have also been reported [146].

Figure 7.39 shows the dielectric functions obtained from a uniaxial 4H-SiC substrate [142]. In this figure, the dielectric constants for the ordinary and extraordinary rays are described using $\varepsilon_o = \varepsilon_{1o} - i\varepsilon_{2o}$ and $\varepsilon_e = \varepsilon_{1e} - i\varepsilon_{2e}$. The dielectric functions of the 4H-SiC were extracted from ellipsometry spectra obtained by varying the in-plane sample rotation and angle of incidence. In addition, ellipsometry spectra measured by a transmission mode were also used in the data analysis. For the measurements, an instrument that allows measurement up to the vacuum ultraviolet region was employed [142]. The optical anisotropy of 4H-SiC is rather weak below the bandgap (3.02 eV), but 4H-SiC exhibits quite strong anisotropy above 5 eV, as confirmed from Fig. 7.39. The ε_2 spectrum denoted as 'amorphous' in Fig. 7.39 shows one extracted from an amorphous SiC film.

7.5.3 ANALYSIS OF ANISOTROPIC ORGANIC MATERIALS

A large variety of organic materials including polymer films [111–116], LB films [120–122], organic crystals [147,148], and liquid crystals [123–127] show optical

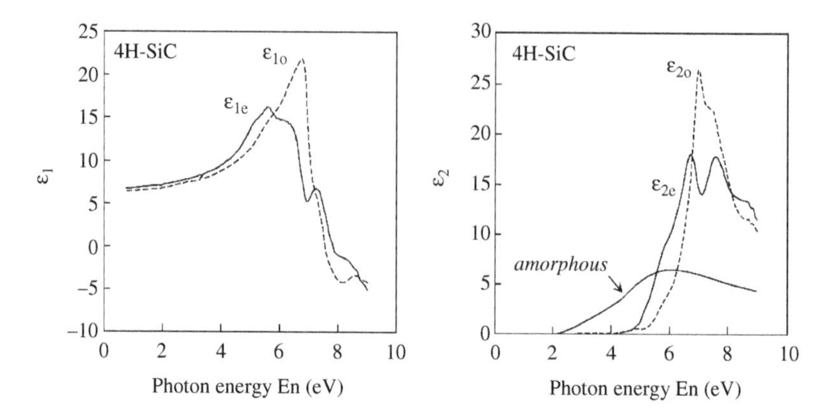

Figure 7.39 Dielectric functions of uniaxial 4H-SiC. The dielectric constants for the ordinary and extraordinary rays are represented by $\varepsilon_o = \varepsilon_{1o} - i\varepsilon_{2o}$ and $\varepsilon_e = \varepsilon_{1e} - i\varepsilon_{2e}$. The ε_2 spectrum denoted as 'amorphous' shows one extracted from an amorphous SiC film. From *Physica Status Solidi A*, **188**, T. Wagner, J. N. Hilfiker, T. E. Tiwald, C. L. Bungay, and S. Zollner, Materials characterization in the vacuum ultraviolet with variable angle spectroscopic ellipsometry, 1553–1562 (2001). Reproduced with permission of Wiley-VCH.

anisotropy, and numerous efforts have been made to characterize these materials. In particular, conjugated polymer films, applied extensively for organic light-emitting diodes, often exhibit quite strong anisotropy due to the preferential alignment of the polymer chains in the plane of the film [111–115]. Here, we will look at an analysis example of a conjugated polymer film that shows strong uniaxial anisotropy [113].

Figure 7.40 shows an optical model for a conjugated polymer film formed by a spin-coat process on an isotropic transparent substrate. Here, the conjugated polymer is a derivative of poly-phenylenevinylene (PPV) and its structure is depicted in the inset of Fig. 7.42. The substrate is a Spectrosil-B substrate that shows a constant refractive index and no light absorption in the visible/UV region (250–900 nm) [113]. The optical axis of the uniaxial polymer film is perpendicular to the surface (or parallel to the plane

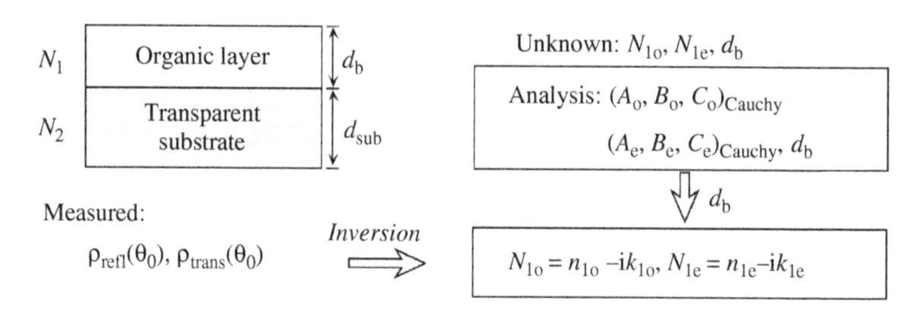

Figure 7.40 Optical model for a uniaxial organic layer ($\phi_E = \theta_E = 0°$) formed on an isotropic transparent substrate. In the figure, $\rho_{refl}(\theta_0)$ and $\rho_{trans}(\theta_0)$ represent $\rho = \tan\psi_{pp}\exp(i\Delta_{pp})$ measured at θ_0 by reflection and transmission modes, respectively. (A_o, B_o, C_o) and (A_e, B_e, C_e) show the Cauchy parameters for the ordinary and extraordinary rays, respectively.

of incidence), and the Jones matrix of this sample is diagonal, as confirmed from Fig. 6.10(a). Recall from Fig. 6.17 that ψ_{pp} of such samples does not vary with sample rotation. Thus, we can check the location of the optical axis from in-plane sample rotation. Since the principal axes of this sample are aligned to the (x, y, z) coordinates, we can perform the data analysis using the Fresnel equations described in Section 6.2.2. Consequently, the unknown parameters of the optical model shown in Fig. 7.40 become (N_{1o}, N_{1e}, d_b) because $\phi_E = \theta_E = 0°$.

In this example, the thickness of the polymer film (1200 Å) was estimated by applying the Cauchy model to the transparent region of the polymer film ($\lambda > 650$ nm). With respect to anisotropic materials, however, two sets of Cauchy parameters are required to describe the optical constants of the ordinary and extraordinary rays. (A_o, B_o, C_o) and (A_e, B_e, C_e) in Fig. 7.40 represent the Cauchy parameters for the ordinary and extraordinary rays, respectively. Furthermore, one backside reflection was also incorporated in the analysis, since a transparent substrate with flat surfaces was used in this example to perform ellipsometry measurements in both reflection and transmission modes. Although the detail of this calculation has not been given, we can express this backside reflection by the same procedure described in Section 5.4.4. In the case of this sample, t_1 in Fig. 5.36 corresponds to t_{012pp} and t_{012ss} shown in Eq. (6.23). t_1' and r_1' in Fig. 5.36 are calculated by converting Eqs. (6.21) and (6.22) using $N_0 \rightarrow N_2$ and $N_2 \rightarrow N_0$. The amplitude transmission coefficient can also be calculated from Eq. (6.40) by applying the 4×4 matrix method. Accordingly, even if a thin film shows optical anisotropy, we can incorporate the effect of backside reflection into the analysis. Several calculation methods have also been proposed for backside reflection in anisotropic substrates [149–152]. In this example, the validity of the analysis of the backside reflection has been confirmed from a similar measurement performed for a thick substrate that prevents the detection of the incoherent light reflected on the substrate back surface. As illustrated in Fig. 7.40, after the thickness of the polymer film is determined, we can perform mathematical inversion to extract the optical constants of the polymer film. Notice that the above analysis procedure is identical to the one illustrated in Fig. 5.41.

Figure 7.41 shows the (ψ_{pp}, Δ_{pp}) spectra of the sample obtained from reflection and transmission ellipsometry measurements using a RCE instrument [113]. For the reflection measurement, the incidence angle θ_0 was changed from 55° to 70° with a step of 5°, while the transmission measurements were performed at $\theta_0 = 40°$–$60°$ with a step of 5°. The dotted and solid lines in Fig. 7.41 represent the experimental spectra and calculated spectra obtained from the mathematical inversion, although these spectra overlap almost completely. It has been reported that mathematical inversion tends to diverge when a similar analysis is performed using the reflection spectra only [113]. In addition, the confidence limits of the extracted optical constants reduce significantly when both reflection and transmission spectra are used for the inversion [113]. The above results indicate that there are strong correlations among the analysis parameters $(n_{1o}, k_{1o}, n_{1e}, k_{1e})$ used in the analysis (see Section 7.1.2). In this example, therefore, the incorporation of the transmission data is quite important.

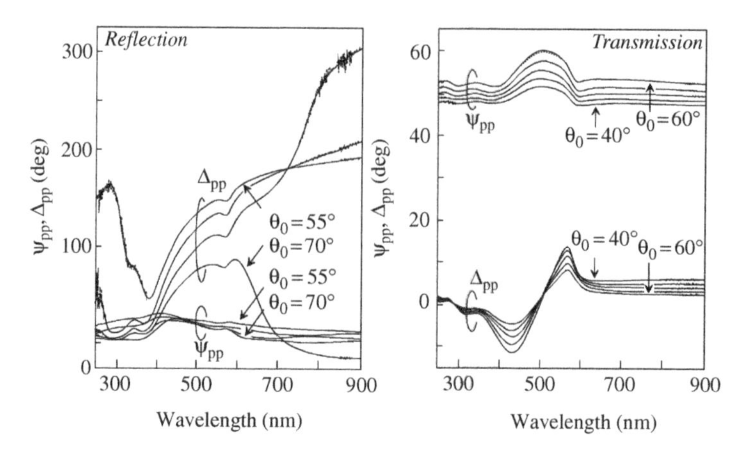

Figure 7.41 (ψ_{pp}, Δ_{pp}) spectra obtained from a uniaxial polymer film (see Fig. 7.42) formed on a transparent Spectrosil-B substrate using reflection and transmission ellipsometry measurements. In the measurements, the incidence angle θ_0 was changed with a step of 5°. The dotted and solid lines represent experimental and calculated spectra, respectively. From *Advanced Materials*, **14**, C. M. Ramsdale and N. C. Greenham, Ellipsometric determination of anisotropic optical constants in electroluminescent conjugated polymers, 212–215 (2002). Reproduced with permission of Wiley-VCH.

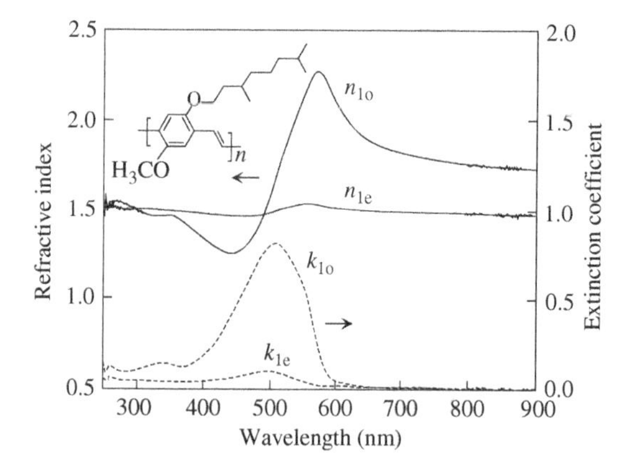

Figure 7.42 Refractive index and extinction coefficient of poly(2-methoxy-5-(3′, 7′-dimethyloctyloxy)-p-phenylenevinylene) extracted from the ellipsometry spectra shown in Fig. 7.41. In the figure, the complex refractive indices for the ordinary and extraordinary rays are denoted as $N_{1o} = n_{1o} - ik_{1o}$ and $N_{1e} = n_{1e} - ik_{1e}$, respectively. The optical axis of the polymer film is normal to the surface ($\phi_E = \theta_E = 0°$). From *Advanced Materials*, **14**, C. M. Ramsdale and N. C. Greenham, Ellipsometric determination of anisotropic optical constants in electroluminescent conjugated polymers, 212–215 (2002). Reproduced with permission of Wiley-VCH.

Figure 7.42 shows the optical constants of the polymer film extracted from the ellipsometry spectra shown in Fig. 7.41 [113]. As confirmed from this figure, this conjugated polymer film shows quite large optical anisotropy. Since $n_{1o} > n_{1e}$, this polymer is negative uniaxial (see Fig. 6.2). The strong absorption peak observed in the k_{1o} spectrum at 500 nm represents the $\pi \rightarrow \pi^*$ transition in the conjugated polymer [153]. It is clear from this result that the polymer chains are aligned parallel to the surface, and the optical transition occurs predominantly in this direction. In the k_{1e} spectrum, on the other hand, a slight peak shift toward shorter wavelength can be seen. This implies that the conjugation length of the polymer chains reduces in the direction perpendicular to the surface [113]. Thus, it is possible to deduce microscopic structures from the optical properties of polymer films. So far, similar studies have been performed for a large variety of conjugated polymers [111–115].

REFERENCES

[1] E. D. Palik (Ed.), *Handbook of Optical Constants of Solids*, Academic Press, San Diego (1985).

[2] G. E. Jellison, Jr, Use of the biased estimator in the interpretation of spectroscopic ellipsometry data, *Appl. Opt.*, **30** (1991) 3354–3360.

[3] C. M. Herzinger, B. Johs, W. A. McGahan, J. A. Woollam, and W. Paulson, Ellipsometric determination of optical constants for silicon and thermally grown silicon dioxide via a multi-sample, multi-wavelength, multi-angle investigation, *J. Appl. Phys.*, **83** (1998) 3323–3336.

[4] A. Kalnitsky, S. P. Tay, J. P. Ellul, S. Chongsawangvirod, J. W. Andrews, and E. A. Irene, Measurements and modeling of thin silicon dioxide films on silicon, *J. Electrochem. Soc.*, **137** (1990) 234–238.

[5] For a review, see M. L. Green, E. P. Gusev, R. Degraeve, and E. L. Garfunkel, ultrathin (< 4 nm) SiO_2 and Si-O-N gate dielectric layers for silicon microelectronics: understanding the processing, structure, and physical and electrical limits, *J. Appl. Phys.*, **90** (2001) 2057–2121.

[6] Z. H. Lu, J. P. McCaffrey, B. Brar, G. D. Wilk, R. M. Wallace, L. C. Feldman, and S. P. Tay, SiO_2 film thickness metrology by X-ray photoelectron spectroscopy, *Appl. Phys. Lett.*, **71** (1997) 2764–2766.

[7] S. -H. Lo, D. A. Buchanan, and Y. Taur, Modeling and characterization of quantization, polysilicon depletion, and direct tunneling effects in MOSFETs with ultrathin oxides, *IBM J. Res. Develop.*, **43** (1999) 327–337.

[8] E. Taft and L. Cordes, Optical evidence for a silicon-silicon oxide interlayer, *J. Electrochem. Soc.*, **126** (1979) 131–134.

[9] D. E. Aspnes and J. B. Theeten, Optical properties of the interface between Si and its thermally grown oxide, *Phys. Rev. Lett.*, **43** (1979) 1046–1050.

[10] Y. Wang and E. A. Irene, Consistent refractive index parameters for ultrathin SiO_2 films, *J. Vac. Sci. Technol. B*, **18** (2000) 279–282.

[11] D. A. Muller, T. Sorsch, S. Moccio, F. H. Baumann, K. Evans-Lutterodt and G. Timp, The electronic structure at the atomic scale of ultrathin gate oxides, *Nature*, **399** (1999) 758–761.

[12] N. V. Nguyen, D. Chandler-Horowitz, P. M. Amirtharaj, and J. G. Pellegrino, Spectroscopic ellipsometry determination of the properties of the thin underlying

strained Si layer and the roughness at SiO_2/Si interface, *Appl. Phys. Lett.*, **64** (1994) 2688–2690.
[13] Q. Fang, J. -Y. Zhang, Z. M. Wang, J. X. Wu, B. J. O'Sullivan, P. K. Hurley, T. L. Leedham, H. Davies, M. A. Audier, C. Jimenez, J. -P. Senateur, and I. W. Boyd, Characterisation of HfO_2 deposited by photo-induced chemical vapor deposition, *Thin Solid Films*, **427** (2003) 391–396.
[14] J. F. Conley, Jr, Y. Ono, D. J. Tweet, W. Zhuang, and R. Solanki, Atomic layer deposition of thin hafnium oxide films using a carbon free precursor, *J. Appl. Phys.*, **93** (2003) 712–718.
[15] E. D. Palik and G. Ghosh (Eds), *Handbook of Optical Constants of Solids*, Academic Press, San Diego (1997).
[16] S. Adachi, *Optical Constants of Crystalline and Amorphous Semiconductors, Numerical Data and Graphical Information*, Kluwer Academic Publishers, Norwell (1999).
[17] D. E. Aspnes and A. A. Studna, Dielectric functions and optical parameters of Si, Ge, GaP, GaAs, GaSb, InP, InAs and InSb from 1.5 to 6.0 eV, *Phys. Rev. B*, **27** (1983) 985–1009.
[18] D. E. Aspnes, S. M. Kelso, R. A. Logan and R. Bhat, Optical properties of $Al_xGa_{1-x}As$, *J. Appl. Phys.*, **60** (1986) 754–767.
[19] C. Pickering and R. T. Carline, Dielectric function spectra of strained and relaxed $Si_{1-x}Ge_x$ alloys ($x = 0$–0.25), *J. Appl. Phys.*, **75** (1994) 4642–4647.
[20] J. I. Pankove, *Optical Process in Semiconductors*, Dover Publications, New York (1971).
[21] J. Tauc, R. Grigorovici, and A. Vancu, Optical properties and electronic structure of amorphous germanium, *Phys. Stat. Sol.*, **15** (1966) 627–637.
[22] G. D. Cody, B. D. Brooks and B. Abeles, Optical absorption above the optical gap of amorphous silicon hydride, *Sol. Energy Mat.*, **8** (1982) 231–240.
[23] R. W. Collins and K. Vedam, Optical properties of solids, in *Encyclopedia of Applied Physics*, vol. 12 (1995) 285–336.
[24] J. R. Chelikowsky, and M. L. Cohen, Nonlocal pseudopotential calculations for the electronic structure of eleven diamond and zinc-blend semiconductors, *Phys. Rev. B*, **14** (1976) 556–582.
[25] U. Schmid, N. E. Christensen, and M. Cardona, Relativistic band structure of Si, Ge, and GeSi: Inversion-asymmetry effects, *Phys. Rev. B*, **41** (1990) 5919–5930.
[26] S. Adachi, *Optical Properties of Crystalline and Amorphous Semiconductors, Materials and Fundamental Principles*, Kluwer Academic Publishers, Norwell (1999).
[27] G. E. Jellison, Jr and F. A. Modine, Optical functions of silicon at elevated temperatures, *J. Appl. Phys.*, **76** (1994) 3758–3761.
[28] J. Lee, R. W. Collins, A. R. Heyd, F. Flack, and N. Samarth, Spectroellipsometry for characterization of $Zn_{1-x}Cd_xSe$ multilayered structures on GaAs, *Appl. Phys. Lett.*, **69** (1996) 2273–2275.
[29] D. E. Aspnes, Spectroscopic ellipsometry of solids, in *Optical Properties of Solids: New Developments*, edited by B. O. Seraphin, Chapter 15, 801–846, North-Holland, Amsterdam (1976).
[30] M. Erman, J. B. Theeten, P. Chambon, S. M. Kelso, and D. E. Aspnes, Optical properties and damage analysis of GaAs single crystals partly amorphized by ion implantation, *J. Appl. Phys.*, **56** (1984) 2664–2671.
[31] F. Wooten, *Optical Properties of Solids*, Academic Press, New York (1972).
[32] F. L. Terry, Jr, A modified harmonic oscillator approximation scheme for the dielectric constants of $Al_xGa_{1-x}As$, *J. Appl. Phys.*, **70** (1991) 409–417.
[33] H. Yao, P. G. Snyder, and J. A. Woollam, Temperature dependence of optical properties of GaAs, *J. Appl. Phys.*, **70** (1991) 3261–3267.

[34] T. Aoki and S. Adachi, Temperature dependence of the dielectric function of Si, *J. Appl. Phys.*, **69** (1991) 1574–1582.
[35] T. Suzuki and S. Adachi, Optical properties of amorphous Si partially crystallized by thermal annealing, *Jpn. J. Appl. Phys.*, **32** (1993) 4900–4906.
[36] C. C. Kim, J. W. Garland, H. Abad, and P. M. Raccah, Modeling the optical dielectric function of semiconductors: extension of the critical-point parabolic-band approximation, *Phys. Rev. B*, **45** (1992) 11749–11767.
[37] C. C. Kim, J. W. Garland, and P. M. Raccah, Modeling the optical dielectric function of the alloy system $Al_xGa_{1-x}As$, *Phys. Rev. B*, **47** (1993) 1876–1888.
[38] B. Johs, C. M. Herzinger, J. H. Dinan, A. Cornfeld, J. D. Benson, Development of a parametric optical constant model for $Hg_{1-x}Cd_xTe$ for control of composition by spectroscopic ellipsometry during MBE growth, *Thin Solid Films*, **313–314** (1998) 137–142.
[39] J. Leng, J. Opsal, H. Chu, M. Senko, and D. E. Aspnes, Analytic representations of the dielectric functions of materials for device and structural modeling, *Thin Solid Films*, **313–314** (1998) 132–136.
[40] J. Leng, J. Opsal, H. Chu, M. Senko, and D. E. Aspnes, Analytic representations of the dielectric functions of crystalline and amorphous Si and crystalline Ge for very large scale integrated device and structural modeling, *J. Vac. Sci. Technol. A*, **16** (1998) 1654–1657.
[41] T. Yang, S. Goto, M. Kawata, K. Uchida, A. Niwa, and J. Gotoh, Optical properties of GaN thin films on sapphire substrates characterized by variable-angle spectroscopic ellipsometry, *Jpn. J. Appl. Phys.*, **37** (1998) L1105–1108.
[42] P. Petrik, M. Fried, T. Lohner, R. Berger, L. P. Bíro, C. Schneider, J. Gyulai, and H. Ryssel, Comparative study of polysilicon-on-oxide using spectroscopic ellipsometry, atomic force microscopy, and transmission electron microscopy, *Thin Solid Films*, **313–314** (1998) 259–263.
[43] J. Humlíček and A. Röseler, IR ellipsometry of the highly anisotropic materials α-SiO_2 and α-Al_2O_3, *Thin Solid Films*, **234** (1993) 332–336.
[44] M. Schubert, T. E. Tiwald, and C. M. Herzinger, Infrared dielectric anisotropy and phonon modes of sapphire, *Phys. Rev. B*, **61** (2000) 8187–8201.
[45] H. Yao and C. H. Yan, Anisotropic optical responses of sapphire (α-Al_2O_3) single crystals, *J. Appl. Phys.*, **85** (1999) 6717–6722.
[46] G. Yu, H. Ishikawa, T. Egawa, T. Soga, J. Watanabe, T. Jimbo, and M. Umeno, Polarized reflectance spectroscopy and spectroscopic ellipsometry determination of the optical anisotropy of gallium nitride on sapphire, *Jpn. J. Appl. Phys.*, **36** (1997) L1029–1031.
[47] G. E. Jellison, Jr, M. F. Chisholm, and S. M. Gorbatkin, Optical functions of chemical vapor deposited thin-film silicon determined by spectroscopic ellipsometry, *Appl. Phys. Lett.*, **62** (1993) 3348–3350.
[48] S. Zollner, Model dielectric functions for native oxides on compound semiconductors, *Appl. Phys. Lett.*, **63** (1993) 2523–2524.
[49] P. G. Snyder, J. A. Woollam, S. A. Alterovitz, and B. Johs, Modeling $Al_xGa_{1-x}As$ optical constants as functions of composition, *J. Appl. Phys.*, **68** (1990) 5925–5926.
[50] P. Lautenschlager, M. Garriga, L. Viña, and M. Cardona, Temperature dependence of the dielectric function and interband critical points in silicon, *Phys. Rev. B*, **36** (1987) 4821–4830.
[51] P. Lautenschlager, M. Garriga, S. Logothetidis, and M. Cardona, Interband critical points of GaAs and their temperature dependence, *Phys. Rev. B*, **35** (1987) 9174–9189.
[52] L. Viña and M. Cardona, Optical constant of pure and heavily doped silicon and germanium: electronic interband transitions, *Physica*, **117**B and **118**B (1983) 356–358.
[53] D. E. Aspnes, A. A. Studna, and K. Kinsbron, Dielectric properties of heavily doped crystalline and amorphous silicon from 1.5 to 6.0 eV, *Phys. Rev. B*, **29** (1984) 768–779.

[54] L. Viña and M. Cardona, Optical properties of pure and ultraheavily doped germanium: theory and experiment, *Phys. Rev. B*, **34** (1986) 2586–2597.
[55] M. Erman, J. B. Theeten, N. Vodjdani, and Y. Demay, Chemical and structural analysis of the GaAs/AlGaAs heterojunctions by spectroscopic ellipsometry, *J. Vac. Sci. Technol. B*, **1** (1983) 328–333.
[56] D. E. Aspnes, G. K. Celler, J. M. Poate, G. A. Rozgonyi, and T. T. Sheng, in *Laser and Electron Beam Processing of Electronic Materials*, edited by C. L. Anderson, G. K. Celler, and G. A. Rozgonyi, vols. 80–1, Electrochemical Society, Princeton (1980).
[57] C. Pickering, R. T. Carline, M. T. Emeny, N. S. Garawal and L. K. Howard, Dielectric functions and critical points of strained $In_xGa_{1-x}As$ on GaAs, *Appl. Phys. Lett.*, **60** (1992) 2412–2414.
[58] R. T. Carline, C. Pickering, D. J. Robbins, W. Y. Leong, A. D. Pitt, and A. G. Cullis, Spectroscopic ellipsometry of $Si_{1-x}Ge_x$ epilayers of arbitrary composition $0 \leq x \leq 0.255$, *Appl. Phys. Lett.*, **64** (1994) 1114–1116.
[59] C. M. Herzinger, P. G. Snyder, F. G. Celii, Y.-C. Kao, D. Chow, B. Johs and J. A. Woollam, Studies of thin strained InAs, AlAs, and AlSb layers by spectroscopic ellipsometry, *J. Appl. Phys.*, **79** (1996) 2663–2674.
[60] G. F. Feng and R. Zallen, Optical properties of ion-implanted GaAs: the observation of finite-size effects in GaAs microcrystals, *Phys. Rev. B*, **40** (1989) 1064–1073.
[61] S. Boultadakis, S. Logothetidis, and S. Ves, Comparative study of thin poly-Si films grown by ion implantation and annealing with spectroscopic ellipsometry, Raman spectroscopy, and electron microscopy, *J. Appl. Phys.*, **72** (1992) 3648–3658.
[62] S. Boultadakis, S. Logothetidis, S. Ves, and J. Kircher, Optical properties of μc-Si:H/ α-Si:H layered structures: influence of the hydrogen bonds, crystallite size, and thickness, *J. Appl. Phys.*, **73** (1993) 914–925.
[63] H. V. Nguyen, and R. W. Collins, Finite-size effects on the optical functions of silicon microcrystallites: a real-time spectroscopic ellipsometry study, *Phys. Rev. B*, **47** (1993) 1911–1917.
[64] M. Erman, J. B. Theeten, P. Frijlink, S. Gaillard, F. J. Hia, and C. Alibert, Electronic states and thicknesses of GaAs/GaAlAs quantum wells as measured by electroreflectance and spectroscopic ellipsometry, *J. Appl. Phys.*, **56** (1984) 3241–3249.
[65] R. P. Vasquez, R. T. Kuroda, and A. Madhukar, Observation of quantum confinement effect away from the zone center in a spectroscopic ellipsometry study of the dielectric function of single $Al_{0.3}Ga_{0.7}As/GaAs/Al_{0.3}Ga_{0.7}As$ square quantum wells, *J. Appl. Phys.*, **61** (1987) 2973–2978.
[66] F. Lukeš and K. Ploog, Dielectric function of GaAs/AlAs surperlattices grown on GaAs substrates with different orientation, *Thin Solid Films*, **233** (1993) 162–165.
[67] K.-F. Berggren and B. E. Sernelius, Band-gap narrowing in heavily doped many-valley semiconductors, *Phys. Rev. B*, **24** (1981) 1971–1986.
[68] D. L. Smith, *Thin-Film Deposition: Principles and Practice*, McGraw-Hill, New York (1995).
[69] M. Chandrasekhar and F. H. Pollak, Effects of uniaxial stress on the electroreflectance spectrum of Ge and GaAs, *Phys. Rev. B*, **15** (1977) 2127–2144.
[70] K. J. Ebeling, *Integrated Optoelectronics: Waveguide Optics, Photonics, Semiconductors*, Springer-Verlag, Berlin Heidelberg (1993).
[71] H. Ehrenreich and H. R. Philipp, Optical properties of Ag and Cu, *Phys. Rev.*, **128** (1962) 1622–1629.
[72] N. W. Ashcroft and K. Sturm, Interband absorption and the optical properties of polyvalent metals, *Phys. Rev. B*, **3** (1971) 1898–1910.
[73] A. D. Rakić, A. B. Djurišić, J. M. Elazar, and M. L. Majewski, Optical properties of metallic films for vertical-cavity optoelectronic devices, *Appl. Opt.*, **37** (1998) 5271–5283.

[74] H. V. Nguyen, I. An, and R. W. Collins, Evolution of the optical functions of thin-film aluminum: a real-time spectroscopic ellipsometry study, *Phys. Rev. B*, **47** (1993) 3947–3965.

[75] H. G. Tompkins, T. Zhu, and E. Chen, Determining thickness of thin metal films with spectroscopic ellipsometry for applications in magnetic random-access memory, *J. Vac. Sci. Technol. A*, **16** (1998) 1297–1302.

[76] D. E. Aspnes, E. Kinsbron, and D. D. Bacon, Optical properties of Au: sample effects, *Phys. Rev. B*, **21** (1980) 3290–3299.

[77] X. Gao, J. Hale, S. Heckens, and J. A. Woollam, Studies of metallic multilayer structures, optical properties, and oxidation using *in situ* spectroscopic ellipsometry, *J. Vac. Sci. Technol. A*, **16** (1998) 429–435.

[78] For a review, see O. Hunderi, Optics of rough surfaces, discontinuous films and heterogeneous materials, *Surf. Sci.*, **96** (1980) 1–31.

[79] C. Liu, J. Erdmann, J. Maj, and A. Macrander, Thickness determination of metal thin films with spectroscopic ellipsometry for X-ray mirror and multilayer applications, *J. Vac. Sci. Technol. A*, **17** (1999) 2741–2748.

[80] G. K. Pribil, B. Johs, N. J. Ianno, Dielectric function of thin metal films by combined *in situ* transmission ellipsometry and intensity measurements, *Thin Solid Films*, **455–456** (2004) 443–449.

[81] S. Norrman, T. Andersson, C. G. Granqvist, and O. Hunderi, Optical properties of discontinuous gold films, *Phys. Rev. B*, **18** (1978) 674–695.

[82] H. Fujiwara and M. Kondo, Effects of carrier concentration on the dielectric function of ZnO:Ga and In_2O_3:Sn studied by spectroscopic ellipsometry: Analysis of free-carrier and band-edge absorption, *Phys. Rev. B*, **71** (2005) 075109.

[83] R. Joerger, K. Forcht, A. Gombert, M. Köhl, and W. Graf, Influence of incoherent superposition of light on ellipsometric coefficients, *Appl. Opt.*, **36** (1997) 319–327.

[84] P. I. Rovira and R. W. Collins, Analysis of specular and textured SnO_2:F films by high speed four-parameter Stokes vector spectroscopy, *J. Appl. Phys.*, **85** (1999) 2015–2025.

[85] H. Fujiwara, M. Kondo, and A. Matsuda, Interface-layer formation in microcrystalline Si:H growth on ZnO substrates studied by real-time spectroscopic ellipsometry and infrared spectroscopy, *J. Appl. Phys.*, **93** (2003) 2400–2409.

[86] T. Gerfin and M. Grätzel, Optical properties of tin-doped indium oxide determined by spectroscopic ellipsometry, *J. Appl. Phys.*, **79** (1996) 1722–1729.

[87] R. A. Synowicki, Spectroscopic ellipsometry characterization of indium tin oxide film microstructure and optical constants, *Thin Solid Films*, **313–314** (1998) 394–397.

[88] M. Losurdo, Relationships among surface processing at the nanometer scale, nanostructure and optical properties of thin oxide films, *Thin Solid Films*, **455–456** (2004) 301–312.

[89] H. E. Rhaleb, E. Benamar, M. Rami, J. P. Roger, A. Hakam, and A. Ennaoui, Spectroscopic ellipsometry studies of index profile of indium tin oxide films prepared by spray pyrolysis, *Appl. Surf. Sci.*, **201** (2002) 138–145.

[90] K. Zhang, A. R. Forouhi and I. Bloomer, Accurate and rapid determination of thickness, n and k spectra, and resistivity of indium-tin-oxide films, *J. Vac. Sci. Technol. A*, **17** (1999) 1843–1847.

[91] H. Yoshikawa and S. Adachi, Optical constants of ZnO, *Jpn. J. Appl. Phys.*, **36** (1997) 6237–6243.

[92] G. E. Jellison, Jr and L. A. Boatner, Optical functions of uniaxial ZnO determined by generalized ellipsometry, *Phys. Rev. B*, **58** (1998) 3586–3589; Erratum **65**, 049902 (2001).

[93] A. Kasic, M. Schubert, Y. Saito, Y. Nanishi, and G. Wagner, Effective electron mass and phonon modes in n-type hexagonal InN, *Phys. Rev. B*, **65** (2002) 115206.

[94] E. Shanthi, V. Dutta, A. Banerjee, and K. L. Chopra, Electrical and optical properties of undoped and antimony-doped tin oxide films, *J. Appl. Phys.*, **51** (1980) 6243–6251.

[95] K. Ellmer, Resistivity of polycrystalline zinc oxide films: current status and physical limit, *J. Phys. D: Appl. Phys.*, **34** (2001) 3097–3108.
[96] For a review, see Z. M. Jarzębski, Preparation and physical properties of transparent conducting oxide films, *Phys. Stat. Sol. A*, **71** (1982) 13–41.
[97] K. Peter, G. Willeke, K. Prasad, A. Shah, and E. Bucher, Free-carrier absorption in microcrystalline silicon thin films prepared by very-high-frequency glow discharge, *Philos. Mag. B*, **69** (1994) 197–207.
[98] Y. Mi, H. Odaka and S. Iwata, Electronic structures and optical properties of ZnO, SnO_2 and In_2O_3, *Jpn. J. Appl. Phys.*, **38** (1999) 3453–3458.
[99] For a review, see I. Hamberg and C. G. Granqvist, Evaporated Sn-doped In_2O_3 films: basic optical properties and applications to energy-efficient windows, *J. Appl. Phys.*, **60** (1986) R123–R159.
[100] B. E. Sernelius, K. -F. Berggren, Z. -C. Jin, I. Hamberg, and C. G. Granqvist, Band-gap tailoring of ZnO by means of heavy Al doping, *Phys. Rev. B*, **37** (1988) 10244–10248.
[101] T. Minami, H. Sato, H. Nanto, and S. Takata, Group III impurity doped zinc oxide thin films prepared by rf magnetron sputtering, *Jpn. J. Appl. Phys.*, **24** (1985) L781–784.
[102] Y. Shigesato, S. Takaki, and T. Haranou, Crystallinity and electrical properties of tin-doped indium oxide films deposited by DC magnetron sputtering, *Appl. Surf. Sci.*, **48/49** (1991) 269–275.
[103] J. A. Woollam, W. A. McGahan and B. Johs, Spectroscopic ellipsometry studies of indium tin oxide and other flat panel display multilayer materials, *Thin Solid Films*, **241** (1994) 44–46.
[104] T. E. Tiwald, D. W. Thompson, J. A. Woollam, W. Paulson, and R. Hance, Application of IR variable angle spectroscopic ellipsometry to the determination of free carrier concentration depth profiles, *Thin Solid Films*, **313–314** (1998) 661–666.
[105] J. Humlíček, R. Henn, and M. Cardona, Far-infared ellipsometry of depleted surface layer in heavily doped n-type GaAs, *Appl. Phys. Lett.*, **69** (1996) 2581–2583.
[106] M. Schubert, C. Bundesmann, H. v. Wenckstern, G. Jakopic, A. Haase, N. -K. Persson, F. Zhang, H. Arwin, and O. Inganäs, Carrier distribution in organic/inorganic (poly(3,4-ethylenedioxy thiophene/poly(styrenesulfonate)polymer)-Si) heterojunction determined from infrared ellipsometry, *Appl. Phys. Lett.*, **84** (2004) 1311–1313.
[107] Y.-T. Kim, D. L. Allara, R. W. Collins, K. Vedam, Real-time spectroscopic ellipsometry study of the electrochemical deposition of polypyrrole thin films, *Thin Solid Films*, **193/194** (1990) 350–360.
[108] C. L. Bungay, T. E. Tiwald, D. W. Thompson, M. J. DeVries, J. A. Woollam, and J. F. Elman, IR ellipsometry studies of polymers and oxygen plasma-treated polymers, *Thin Solid Films*, **313–314** (1998) 713–717.
[109] K. Postava and T. Yamaguchi, Optical functions of low-k materials for interlayer dielectrics, *J. Appl. Phys.*, **89** (2001) 2189–2193.
[110] K. Postava, T. Yamaguchi, and M. Horie, Estimation of the dielectric properties of low-k materials using optical spectroscopy, *Appl. Phys. Lett.*, **79** (2001) 2231–2233.
[111] J. Sturm, S. Tasch, A. Niko, G. Leising, E. Toussaere, J. Zyss, T. C. Kowalczyk, K. D. Singer, U. Scherf, and J. Huber, Optical anisotropy in thin films of a blue electroluminescent conjugated polymer, *Thin Solid Films*, **298** (1997) 138–142.
[112] L. A. A. Pettersson, F. Carlsson, O. Inganäs and H. Arwin, Spectroscopic ellipsometry studies of the optical properties of doped poly(3,4-ethylenedioxythiophene): an anisotropic metal, *Thin Solid Films*, **313–314** (1998) 356–361.
[113] C. M. Ramsdale and N. C. Greenham, Ellipsometric determination of anisotropic optical constants in electroluminescent conjugated polymers, *Adv. Mater.*, **14** (2002) 212–215.
[114] M. Losurdo, M. M. Giangregorio, P. Capezzuto, G. Bruno, F. Babudri, D. Colangiuli, G. M. Farinola, and F. Naso, Study of anisotropic optical properties of

poly(arylenephenylene) thin films: dependence on polymer backbone, *Macromolecules*, **36** (2003) 4492–4497.
[115] B. P. Lyons and A. P. Monkman, A comparison of the optical constants of aligned and unaligned thin polyfluorene films, *J. Appl. Phys.*, **96** (2004) 4735–4741.
[116] M. Schubert, C. Bundesmann, G. Jakopic, H. Maresch, H. Arwin, N. -C. Persson, F. Zhang, and O. Inganäs, Infrared ellipsometry characterization of conducting thin organic films, *Thin Solid Films*, **455–456** (2004) 295–300.
[117] J. P. Folkers, P. E. Laibinis and G. M. Whitesides, Self-assembled monolayers of alkanethiols on gold: comparisons of monolayers containing mixtures of short- and long-chain constituents with CH_3 and CH_2OH terminal groups, *Langmuir*, **8** (1992) 1330–1341.
[118] A. C. Zeppenfeld, S. L. Fiddler, W. K. Ham, B. J. Klopfenstein, and C. J. Page, Variation of layer spacing in self-assembled hafnium-1,10-decanediylbis(phosphonate) multilayers as determined by ellipsometry and grazing angle X-ray diffraction, *J. Am. Chem. Soc.*, **116** (1994) 9158–9165.
[119] Z. Li, S. Chang, and R. S. Williams, Self-assembly of alkanethiol molecules onto platinum and platinum oxide surfaces, *Langmuir*, **19** (2003) 6744–6749.
[120] A. Y. Tronin and A. F. Konstantinova, Ellipsometric study of the optical anisotropy of lead arachidate langumuire films, *Thin Solid Films*, **177** (1989) 305–314.
[121] B. Lecourt, D. Blaudez, and J. -M. Turlet, Specific approach of generalized ellipsometry for the determination of weak in-plane anisotropy: application to Langmuir–Blodgett ultrathin films, *J. Opt. Soc. Am. A*, **15** (1998) 2769–2782.
[122] D. Tsankov, K. Hinrichs, A. Röseler, and E. H. Korte, FTIR ellipsometry as a tool for studying organic layers: from Langmuir–Blodgett films to can coatings, *Phys. Stat. Sol. A*, **188** (2001) 1319–1329.
[123] Ch. Bahr and D. Fliegner, Behavior of a first-order smectic-A–smectic-C transition in free-standing liquid-crystal films, *Phys. Rev. A*, **46** (1992) 7657–7663.
[124] M. Schubert, B. Rheinländer, C. Cramer, H. Schmiedel, J. A. Woollam, C. M. Herzinger and B. Johs, Generalized transmission ellipsometry for twisted biaxial dielectric media: application to chiral liquid crystals, *J. Opt. Soc. Am. A*, **13** (1996) 1930–1940.
[125] P. M. Johnson, D. A. Olson, S. Pankratz, Ch. Bahr, J. W. Goodby and C. C. Huang, Ellipsometric studies of synclinic and anticlinic arrangements in liquid crystal films, *Phys. Rev. E*, **62** (2000) 8106–8113.
[126] T. Tadokoro, K. Akao, T. Yoshihara, S. Okutani, M. Kimura, T. Akahane and H. Toriumi, Dynamics of surface-stabilized ferroelectric liquid crystals at the alignment layer surface studied by total-reflection ellipsometry, *Jpn. J. Appl. Phys.*, **40** (2001) L453–455.
[127] J. N. Hilfiker, B. Johs, C. M. Herzinger, J. F. Elman, E. Montbach, D. Bryant, and P. J. Bos, Generalized spectroscopic ellipsometry and Mueller-matrix study of twisted nematic and super twisted nematic liquid crystals, *Thin Solid Films*, **455–456** (2004) 596–600.
[128] U. Jönsson, M. Malmqvist, and I. Rönnberg, Adsorption of immunoglobulin G, protein A, and fibronectin in the submonolayer region evaluated by a combined study of ellipsometry and radiotracer techniques, *J. Colloid. Interface Sci.*, **103** (1985) 360–372.
[129] F. Tiberg and M. Landgren, Characterization of thin nonionic surfactant films at the silica/wafer interface by means of ellipsometry, *Langmuir*, **9** (1993) 927–932.
[130] For a review, see P. Tengvall, I. Lundström, Bo Liedberg, Protein adsorption studies on model organic surfaces: an ellipsometric and infrared spectroscopic approach, *Biomaterials*, **19** (1998) 407–422.
[131] For a review, see H. Arwin, Spectroscopic ellipsometry and biology: recent developments and challenges, *Thin Solid Films*, **313–314** (1998) 764–774.
[132] For a review, see H. Arwin, Ellipsometry on thin organic layers of biological interest: characterization and applications, *Thin Solid Films*, **377–378** (2000) 48–56.

[133] D. E. Gray, S. C. Case-Green, T. S. Fell, P. J. Dobson, and E. M. Southern, Ellipsometric and interferometric characterization of DNA probes immobilized on a combinatorial array, *Langmuir*, **13** (1997) 2833–2842.
[134] S. Elhadj, G. Singh, and R. F. Saraf, Optical properties of an immobilized DNA monolayer from 255 to 700 nm, *Langmuir*, **20** (2004) 5539–5543.
[135] S. Teitler and B. W. Henvis, Refraction in stratified, anisotropic media, *J. Opt. Soc. Am.*, **60** (1970) 830–834.
[136] D. den Engelsen, Ellipsometry of anisotropic films, *J. Opt. Soc. Am.*, **61** (1971) 1460–1466.
[137] F. Meyer, E. E. de Kluizenaar, and D. den Engelsen, Ellipsometric determination of the optical anisotropy of gallium selenide, *J. Opt. Soc. Am.*, **63** (1973) 529–532.
[138] R. M. A. Azzam and N. M. Bashara, Application of generalized ellipsometry to anisotropic crystals, *J. Opt. Soc. Am.*, **64** (1974) 128–133.
[139] D. J. De Smet, Ellipsometry of anisotropic thin films, *J. Opt. Soc. Am.*, **64** (1974) 631–638.
[140] D. J. De Smet, Generalized ellipsometry and the 4×4 matrix formalism, *Surf. Sci.*, **56** (1976) 293–306.
[141] J. Humlíček, Transverse and longitudinal vibration modes in α-quartz, *Philos. Mag. B*, **70** (1994) 699–710.
[142] T. Wagner, J. N. Hilfiker, T. E. Tiwald, C. L. Bungay, and S. Zollner, Materials characterization in the vacuum ultraviolet with variable angle spectroscopic ellipsometry, *Phys. Stat. Sol. A*, **188** (2001) 1553–1562.
[143] S. Ninomiya and S. Adachi, Optical constants of 6H-SiC single crystals, *Jpn. J. Appl. Phys.*, **33** (1994) 2479–2482.
[144] M. Schubert, Generalized ellipsometry and complex optical systems, *Thin Solid Films*, **313–314** (1998) 323–332.
[145] M. Schubert, B. Rheinländer, J. A. Woollam, B. Johs, and C. M. Herzinger, Extension of rotating-analyzer ellipsometry to generalized ellipsometry: determination of the dielectric function tensor from uniaxial TiO_2, *J. Opt. Soc. Am. A*, **13** (1996) 875–883.
[146] G. E. Jellison, Jr, L. A. Boatner, J. D. Budai, B. -S. Jeong and D. P. Norton, Spectroscopic ellipsometry of thin film and bulk anatase (TiO_2), *J. Appl. Phys.*, **93** (2003) 9537–9541.
[147] M. I. Alonso, M. Garriga, J. O. Ossó, F. Schreiber, E. Barrena, and H. Dosch, Strong optical anisotropies of $F_{16}CuPc$ thin films studied by spectroscopic ellipsometry, *J. Chem. Phys.*, **119** (2003) 6335–6340.
[148] O. D. Gordan, M. Friedrich, and D. R. T. Zahn, Determination of the anisotropic dielectric function for metal free phthalocyanine thin films, *Thin Solid Films*, **455–456** (2004) 551–556.
[149] K. Forcht, A. Gombert, R. Joerger, and M. Köhl, Ellipsometric investigation of thick polymer films, *Thin Solid Films*, **313–314** (1998) 808–813.
[150] R. Ossikovski, M. Kildemo, M. Stchakovsky, and M. Mooney, Anisotropic incoherent reflection model for spectroscopic ellipsometry of a thick semitransparent anisotropic substrate, *Appl. Opt.*, **39** (2000) 2071–2077.
[151] K. Postava, T. Yamaguchi, and R. Kantor, Matrix description of coherent and incoherent light reflection and transmission by anisotropic multilayer structures, *Appl. Opt.*, **41** (2002) 2521–2531.
[152] H. Touir, M. Stchakovsky, R. Ossikovski, and M. Warenghem, Coherent and incoherent interference modeling and measurement of anisotropic multilayer stacks using conversional ellipsometry, *Thin Solid Films*, **455–456** (2004) 628–631.
[153] For example, see E. K. Miller, D. Yoshida, C. Y. Yang, and A. J. Heeger, Polarized ultraviolet absorption of highly oriented poly(2-methoxy, 5-(2′-ethyl)-hexyloxy) paraphenylene vinylene, *Phys. Rev. B*, **59** (1999) 4661–4664.

8 Real-Time Monitoring by Spectroscopic Ellipsometry

Real-time monitoring by spectroscopic ellipsometry allows various process diagnoses on the atomic scale. Furthermore, it is possible to perform real-time control of sample structures based on the ellipsometry technique. For the analysis of real-time ellipsometry spectra, we can employ global error minimization (GEM) and virtual substrate approximation (VSA), which are more powerful than the conventional analysis methods described so far. In this chapter, we will address data analysis methods used for real-time monitoring and look at analysis examples for each method. This chapter will also introduce examples of feedback process control by spectroscopic ellipsometry.

8.1 DATA ANALYSIS IN REAL-TIME MONITORING

From real-time monitoring by spectroscopic ellipsometry, a quite large number of (ψ, Δ) spectra are obtained, and we can perform very reliable ellipsometry analysis by utilizing these spectra. Table 8.1 summarizes analysis methods used frequently in the data analysis of real-time ellipsometry. The linear regression analysis (LRA) shown in Table 8.1 is identical to the one described in the previous chapters. In order to perform LRA, however, all the dielectric functions of a sample have to be known. When there are unknown dielectric functions in a sample, dielectric function modeling for layers or the substrate is required.

The global error minimization (GEM) in Table 8.1 is a data analysis method in which mathematical inversion is combined with LRA [1–3]. In data analysis by GEM, we assume that the dielectric function of an analyzed layer is independent of thickness. Since mathematical inversion is incorporated in GEM, it is possible to perform data analysis even when the dielectric function of a layer is completely unknown [1–10]. This is the greatest advantage of this method. However, the data analysis procedure of GEM is rather complicated, compared with LRA, and this analysis method cannot be employed for real-time control.

Spectroscopic Ellipsometry: Principles and Applications H. Fujiwara

Table 8.1 Comparisons of data analysis methods used for real-time ellipsometry

	Linear regression analysis (LRA)	Global error minimization (GEM)	Virtual substrate approximation (VSA)
Precondition	Dielectric functions are known	Dielectric function is thickness independent	A film and substrate are light absorbing
Difficulty of analysis	Very easy	Complicated	Moderate
Dielectric function of the sample	Necessary	Unnecessary	Necessary
Analysis of transparent materials	Yes	Yes	No
Analysis of graded layer	Difficult	Difficult	Yes
Real-time control	Yes	No	Yes

In the virtual substrate approximation (VSA) [11–13], the film and substrate are required to show relatively large light absorption. Basically, VSA is applied for the analysis of semiconductor layers formed on a semiconductor substrate [11–21]. If we employ VSA, in particular, we can perform the analysis of a compositionally graded layer in which the composition of an alloy varies continuously in the growth direction [13–19]. The change in crystalline volume fraction in the growth direction can also be characterized from VSA [21]. In contrast to LRA and GEM, however, VSA cannot be applied to samples that show low light absorption [22]. Recently, a new method that allows the characterization of transparent layers has also been proposed [23].

Each method shown in Table 8.1 has its own advantages and disadvantages, and we need to select a proper analysis method according to the situation. When the dielectric function of a layer is unknown, for example, we employ GEM to obtain the dielectric function of the layer. An optical database can be constructed from several dielectric functions determined from GEM. Based on such optical databases, we may carry out real-time control of film structures by applying LRA or VSA. This section will explain the principles of the above analysis methods and introduce analysis examples for each method.

8.1.1 PROCEDURES FOR REAL-TIME DATA ANALYSIS

The most important factor in the data analysis of real-time ellipsometry is the variation of dielectric function with temperature. As we have seen in Fig. 7.15, the dielectric functions of semiconductors change significantly with temperature. Thus, when a process temperature is not room temperature, it becomes difficult to use reported dielectric functions for the data analysis. The dielectric function of a substrate at process temperatures can be obtained from spectra measured before the processing. If there are no oxide and surface roughness layers on a substrate, the dielectric function at a process temperature is determined from the pseudo-dielectric function (see Section 5.4.2). When an oxide or a surface roughness layer is present, we obtain the dielectric function using mathematical inversion (see Section 5.5.3). For example, the optical model for a SiO_2/crystalline Si (c-Si) structure is expressed by Eq. (7.1), and we can describe the refractive index of the SiO_2 thermal oxide at a temperature T as $N_1(T) = 1.464 + 1.221 \times 10^{-5} T$ ($\lambda =$ 6328 Å) [24]. Accordingly, even if the temperature is increased to 1000 °C, the refractive index of SiO_2 shows a very small variation of ~ 0.01 [25]. If we assume that the optical constants and thickness of the SiO_2 layer estimated at room temperature do not change with temperature, the unknown parameter of the optical model is the complex refractive index of the c-Si substrate only. Thus, from (ψ, Δ) spectra measured at a process temperature, the dielectric function of c-Si at the process temperature can be extracted by using mathematical inversion. When several layers in a multilayer structure show temperature dependence, it is necessary to determine the dielectric function of each layer in advance using the above procedure (see Section 8.2.2). If the thickness of the top layer is sufficiently thicker than the penetration depth of light ($> 5d_p$), the dielectric function of the layer at the process temperature can be obtained by simply removing an overlayer using mathematical inversion [see Fig. 5.35(a)]. For some cases, we can perform the data analysis of real-time ellipsometry using dielectric functions obtained from such methods (see Section 8.1.2).

Since sample temperature often varies depending on the pressure or the source gas employed in a process, extra care is required for dielectric functions used in data analysis. Conversely, we can evaluate a process temperature from CP analysis (see Section 7.2.4). At process temperatures, the angle of incidence may change slightly because of the thermal expansion of a substrate heater. When a sample holder is rotated during film deposition, we need to suppress incidence angle variation caused by sample rotation [26,27].

Using real-time ellipsometry, on the other hand, dielectric functions at process temperatures can be evaluated. The dielectric functions of samples at room temperature can be obtained by the following procedure: we first estimate the thickness parameters of the sample at the process temperature, and then measure (ψ, Δ) spectra after bringing the temperature back to room temperature. Finally, to obtain the dielectric function of the sample at room temperature, we perform mathematical inversion of the room temperature spectra, assuming that the thickness parameters do not change after processing. It should be noted that this analysis becomes difficult when the dielectric function of the substrate changes after substrate heating or when the surface roughness of the sample varies after sample processing, for example.

8.1.2 LINEAR REGRESSION ANALYSIS (LRA)

From the analysis of real-time data using LRA, the time evolution of sample structures can be characterized. If the dielectric functions of samples at process temperatures are known, ellipsometry analysis can be performed easily by using the procedure described in the previous chapters. Here, as data analysis examples for LRA, we will look at the analyses for thermal oxidation of c-Si substrates [28] and polycrystalline Si (poly-Si) formation by the thermal crystallization of hydrogenated amorphous silicon (a-Si:H) [29].

Figure 8.1 shows an optical model for the SiO_2 layer formed on a c-Si substrate by thermal oxidation. This optical model is exactly the same as the one shown in Fig. 7.1. As mentioned earlier, the complex refractive index of the SiO_2 layer ($N_1 = n_1$) is temperature independent, and we can use the value at room temperature for the SiO_2 layer (see Section 8.1.1). With respect to the complex refractive index of c-Si (N_2), however, the temperature effect must be taken into account. Here, we employ the reported temperature variation for N_2 [25]. In this case, if the incidence angle θ_0 is known, the known parameters in the optical model are N_0, N_1, N_2, and θ_0. Thus, the unknown parameter is the time evolution of the SiO_2 layer thickness [$d_b(t)$] only, and $d_b(t)$ can be determined directly from real-time data [$\psi(t)$, $\Delta(t)$] using LRA. From the slope of $d_b(t)$ versus time t, we can further estimate instantaneous oxidation rates.

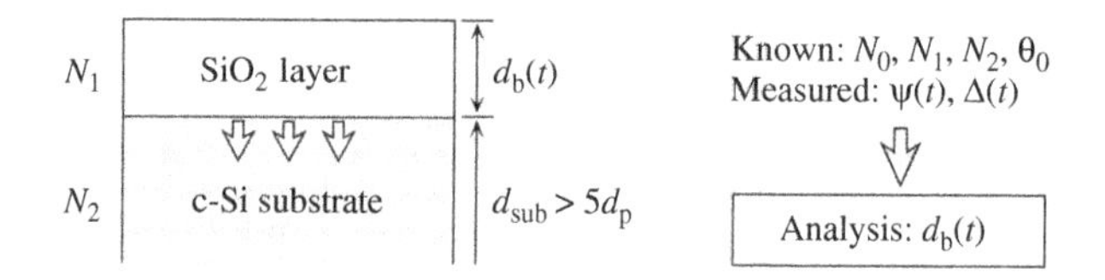

Figure 8.1 Optical model for a SiO_2 layer formed on a crystalline silicon (c-Si) substrate by thermal oxidation. N_0, N_1, and N_2 represent the complex refractive indices of the ambient, SiO_2 bulk layer, and c-Si substrate, respectively. The thicknesses of the bulk layer and substrate are denoted as d_b and d_{sub}, respectively. d_p indicates the penetration depth of light, and θ_0 shows the angle of incidence.

Figure 8.2 shows the instantaneous oxidation rate of Si(100) substrates, plotted as a function of the oxide thickness (d_b) [28]. In this example, the thermal oxidation was carried out in a quartz reactor by dry oxygen at different temperatures. In the analysis, the value of N_2 was changed depending on the process temperature. The ellipsometry measurement was performed using a single-wavelength ellipsometer with a He–Ne laser ($\lambda = 6328$ Å). It can be seen from Fig. 8.2 that the oxidation rate is 3–5 times faster when d_b is thin, and the overall oxidation rate increases with temperature. The solid lines show oxidation rates predicted by a linear-parabolic model in which the oxidation is determined by the diffusion of the oxidant through the SiO_2 bulk layer and its reaction at the SiO_2/c-Si interface [30]. The enhancement of oxidation at $d_b <$ 200 Å has been explained by the stress effects at the SiO_2/c-Si interface from first-principles calculations [31]. As evidenced from the above example, the ellipsometry technique is quite effective in characterizing such reaction processes. If we apply

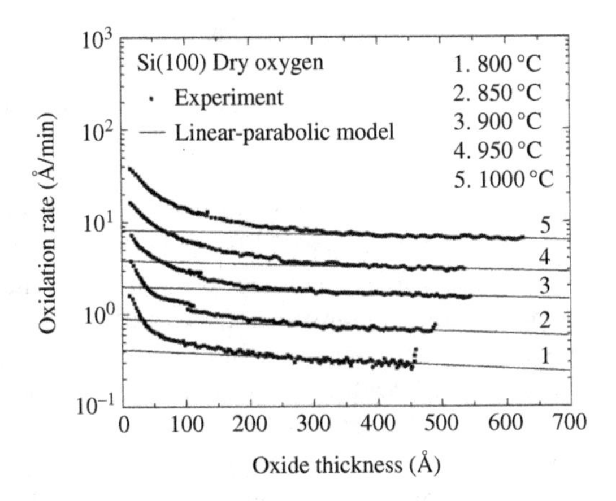

Figure 8.2 Oxidation rate of Si(100) substrates obtained from single-wavelength ellipsometry using LRA, plotted as a function of the oxide thickness. Reprinted with permission from Massoud *et al.*, *J. Electrochem. Soc.*, **132**, 2685 (1985). Copyright 1985, The Electrochemical Society.

spectroscopic ellipsometry, the data analysis discussed in Section 7.1 can also be performed.

Now, we will look at an analysis example for thermal crystallization of a-Si:H layers [29]. Figure 8.3 shows an optical model for a Si film/SiO_2 substrate structure. In this example, the a-Si:H layers were deposited on quartz substrates, and poly-Si layers were formed by thermal annealing of the a-Si:H layers. When d_b of the Si film is 1 μm, the condition $d_b > 5d_p$ is satisfied at En > 2.5 eV (see Section 5.1.3). In this region, (ψ, Δ) are characterized by light reflection at the ambient/Si film interface only, and we can neglect the optical effects of the substrate completely [see Fig. 5.35(a)]. In this analysis example, the complex refractive indices of the a-Si:H layer ($N_{\text{a-Si:H}}$) and poly-Si layer ($N_{\text{poly-Si}}$) were determined from pseudo-dielectric functions obtained before and after thermal annealing. When θ_0 is known, the unknown parameter in the optical model is the variation of N_1 with time. Here,

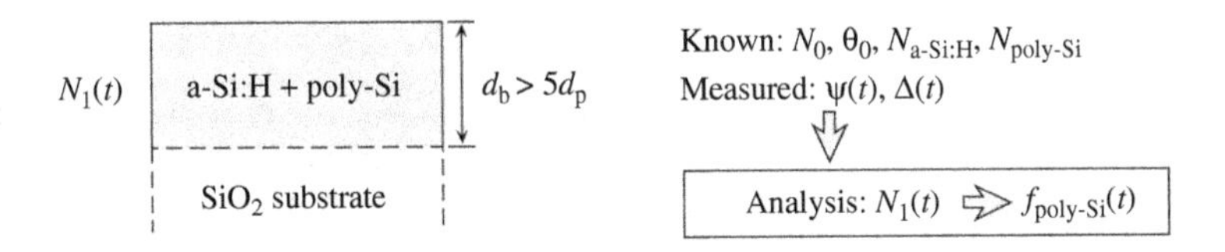

Figure 8.3 Optical model for a Si film formed on a quartz substrate. The Si film is composed of hydrogenated amorphous silicon (a-Si:H) and polycrystalline silicon (poly-Si), and the poly-Si is formed by crystallization of the a-Si:H. In this optical model, the thickness of the Si film is $d_b > 5d_p$. $N_{\text{a-Si:H}}$ and $N_{\text{poly-Si}}$ represent the complex refractive indices of the a-Si:H and poly-Si components, respectively, and $f_{\text{poly-Si}}$ shows the volume fraction of the poly-Si component in the Si film.

we assume a two phase mixture for the Si film and calculate $N_1(t)$ from $N_{\text{a-Si:H}}$ and $N_{\text{poly-Si}}$ by applying EMA [Eq. (5.43)]. If the volume fractions of a-Si:H ($f_{\text{a-Si:H}}$) and poly-Si ($f_{\text{poly-Si}}$) satisfy $f_{\text{a-Si:H}} + f_{\text{poly-Si}} = 1$, the analysis parameter in the optical model becomes $f_{\text{poly-Si}}$ (or $f_{\text{a-Si:H}}$) only. Accordingly, from LRA of real-time spectra $[\psi(t), \Delta(t)]$, we can determine $f_{\text{poly-Si}}(t)$.

Figure 8.4(a) shows the time evolution of the Δ spectrum during thermal crystallization of an a-Si:H layer at 615 °C [29]. The real-time measurement was performed using a P_RSA instrument at $\theta_0 = 70°$. As shown in Fig. 8.4(a), the spectral features of the E_1 and E_2 transitions appear after poly-Si formation (see Fig. 5.29), and the change in the Δ spectrum stops completely upon completion of the thermal crystallization ($t = 700$ s). In this example, $N_{\text{a-Si:H}}$ and $N_{\text{poly-Si}}$ at 615 °C were determined from the (ψ, Δ) spectra measured at $t = 0$ s and 700 s, respectively. In this calculation, we first obtain $(\langle\varepsilon_1\rangle, \langle\varepsilon_2\rangle)$ from Eq. (5.51) and then convert these using Eq. (2.48). The solid lines in Fig. 8.4(a) represent the fitting results calculated from the above model using EMA, and the calculated spectra agree well with the experimental spectra. Figure 8.4(b) shows the time evolution of $f_{\text{a-Si:H}}$ and $f_{\text{poly-Si}}$ obtained from this analysis [29]. As shown in Fig. 8.4(b), after the initiation of crystallization, $f_{\text{a-Si:H}}$ reduces rapidly, while $f_{\text{poly-Si}}$ increases. It can be seen that a slight increase in the process temperature leads to a large change in crystallization rate. As evidenced from Fig. 8.4(b), ellipsometry can also be applied to detect the endpoints of such processes.

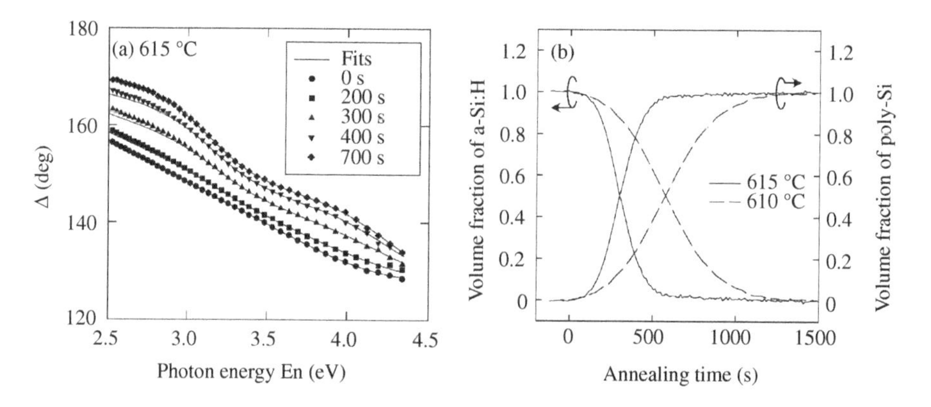

Figure 8.4 Time evolution of (a) Δ spectrum and (b) volume fractions of a-Si:H and poly-Si during thermal crystallization of an a-Si:H layer. Reprinted from *Thin Solid Films*, **313–314**, M. Wakagi, H. Fujiwara, and R. W. Collins, Real time spectroscopic ellipsometry for characterization of the crystallization of amorphous silicon by thermal annealing, 464–468, Copyright (1998), with permission from Elsevier.

It should be noted that the above Si films have a small surface roughness of $d_s = 10$ Å. Thus, $N_{\text{a-Si:H}}$ and $N_{\text{poly-Si}}$ in Fig. 8.3 are actually pseudo-complex refractive indices. However, since the change in the surface roughness is negligible in the above crystallization process, the pseudo-dielectric functions can be employed for ellipsometry analysis. The data analysis can also be performed by incorporating a surface roughness layer [29]. In this case, the optical model shown in Fig 5.27(b) is used and the dielectric functions of the a-Si:H and poly-Si layers are extracted using mathematical inversion. In this analysis, the complex refractive index of the bulk layer is first calculated by EMA and then the complex refractive index of the surface roughness layer is determined from the optical constants of the bulk layer using EMA ($f_{void} = 0.5$). With respect to this example, almost the same results are obtained even if the surface roughness layer is incorporated into the optical model. From CP analysis of the extracted dielectric function, the surface temperature and grain size of the poly-Si can be evaluated (see Section 7.2.4). The temperatures shown in Fig. 8.4(b) represent surface temperatures estimated from CP analyses.

8.1.3 GLOBAL ERROR MINIMIZATION (GEM)

A data analysis method called global error minimization (GEM) was developed by Collins's group [1–3], and this method enables us to determine the dielectric function and structures of a sample simultaneously [1–10]. Thus, when the dielectric function of a sample is unknown, GEM is a quite powerful analysis method. Here, we will examine the principle of GEM using an analysis example for a-Si:H growth.

Figure 8.5 shows pseudo-dielectric functions obtained from real-time monitoring of a-Si:H growth. The measurement was performed using a PC_RSA instrument at $\theta_0 = 70°$, and the a-Si:H layer was formed on a c-Si substrate covered with native oxide using a plasma process. The repetition time of this measurement was 7 s, and the growth rate of the a-Si:H layer was 0.4 Å/s. Thus, each spectrum in Fig. 8.5 shows the change in the film structure on the atomic scale. The two peaks in the $\langle\varepsilon_2\rangle$ spectrum at 3.4 eV and 4.25 eV ($t < 0.5$ min) represent the E_1 and E_2 transitions of the c-Si substrate (see Fig. 7.15).

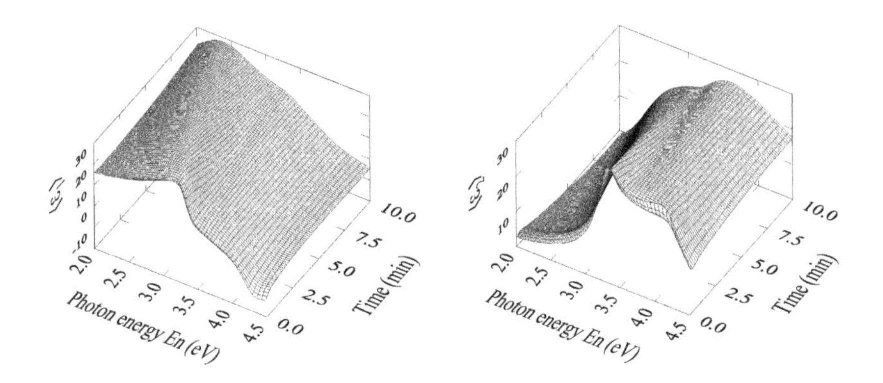

Figure 8.5 Pseudo-dielectric functions obtained from the real-time monitoring of a-Si:H growth on a SiO_2/c-Si substrate.

Figure 8.6 illustrates the data analysis procedure for GEM. In data analysis by GEM, first one pair of $(\langle\varepsilon_1\rangle, \langle\varepsilon_2\rangle)$ spectra measured at $t = x$ min is selected. Figure 8.6(a) shows the optical model for the spectra. In this model, ε_s, ε_b, and ε_{sub} denote the dielectric functions of the surface roughness layer, bulk layer, and the substrate, respectively. In this example, the bulk layer is the a-Si:H layer and the substrate is composed of SiO_2/c-Si. The dielectric function of the c-Si substrate at the process temperature can be extracted from (ψ, Δ) spectra measured before the film deposition using mathematical inversion (see Section 8.1.1). Here, ε_s is calculated from EMA assuming $f_{void} = 0.5$. As shown in Fig. 8.6(b), the unknown parameters in this optical model are the dielectric function of the bulk layer ($\varepsilon_b = \varepsilon_{1b} - i\varepsilon_{2b}$), surface roughness layer thickness d_s, and bulk layer thickness d_b. If d_s and d_b at time x are known, $(\varepsilon_{1b}, \varepsilon_{2b})$ can be obtained directly from the measured spectra $(\langle\varepsilon_1\rangle, \langle\varepsilon_2\rangle)$ using mathematical inversion. In GEM, the expected values of (d_s, d_b) at $t = x$ min are used and a tentative ε_b is obtained. In Fig. 8.6(b), the dielectric function of the a-Si:H layer extracted from the spectra at $t = 8.8$ min is shown.

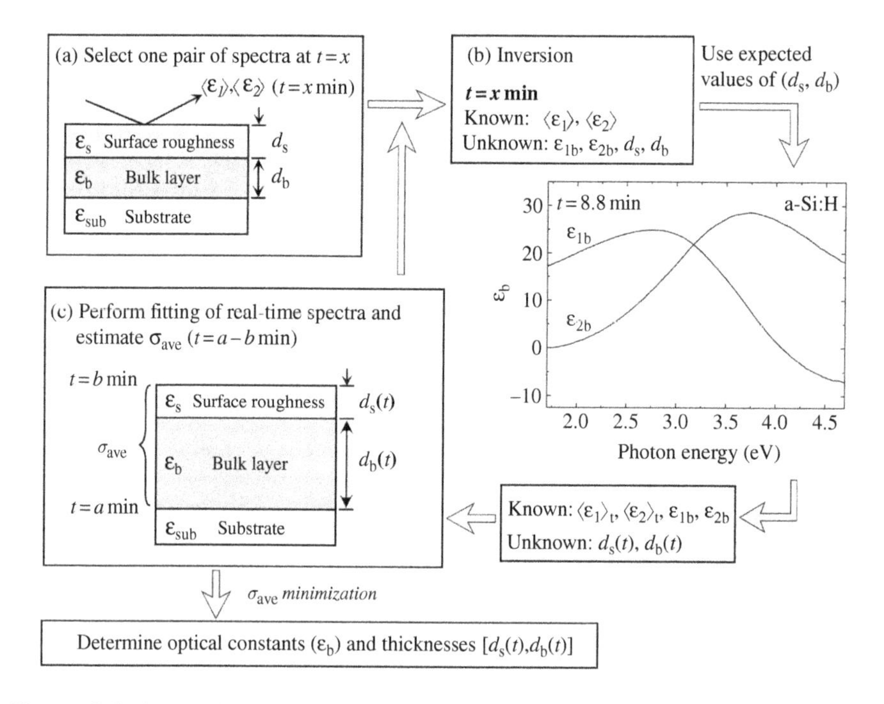

Figure 8.6 Data analysis procedure for global error minimization (GEM).

Linear regression analysis is then performed on all the spectra measured at $t = a$–b min using the $(\varepsilon_{1b}, \varepsilon_{2b})$ determined in Fig. 8.6(b). In this analysis, (d_s, d_b) are used as analysis parameters, and the unbiased estimator σ (or biased estimator χ)

is obtained for each of the spectra. From the analysis of the spectra measured at $t = a\text{–}b$ min, the average fitting error is calculated by

$$\sigma_{ave} = \frac{1}{F} \sum \sigma(t) \tag{8.1}$$

where F is the number of (ψ, Δ) spectra used in the analysis. As we have seen in Fig. 5.43, when (d_s, d_b) used in the mathematical inversion are inaccurate, anomalous structures appear in the extracted dielectric function. Thus, if this dielectric function is employed for LRA, the σ_{ave} value increases drastically. Conversely, when the extracted dielectric function is correct, the σ_{ave} value becomes smaller. Accordingly, ε_b can be determined from (d_s, d_b) that minimize σ_{ave}. By applying the ε_b obtained, $d_s(t)$ and $d_b(t)$ can also be evaluated. This is the basic principle of real-time data analysis by GEM.

Figure 8.7(a) shows the time evolution of σ obtained from LRA of the real-time spectra shown in Fig. 8.5. In Fig. 8.7(a), the results for two dielectric functions

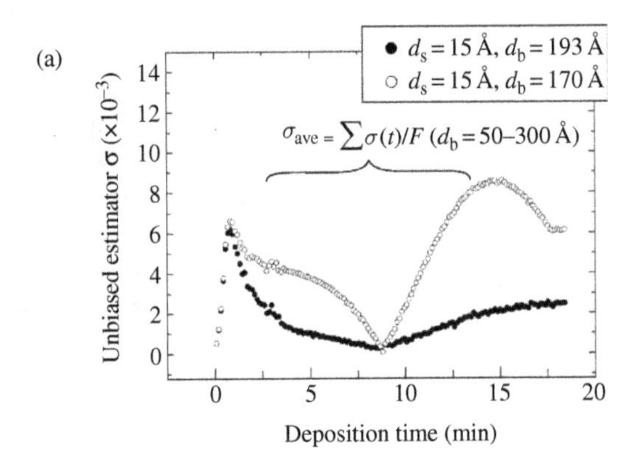

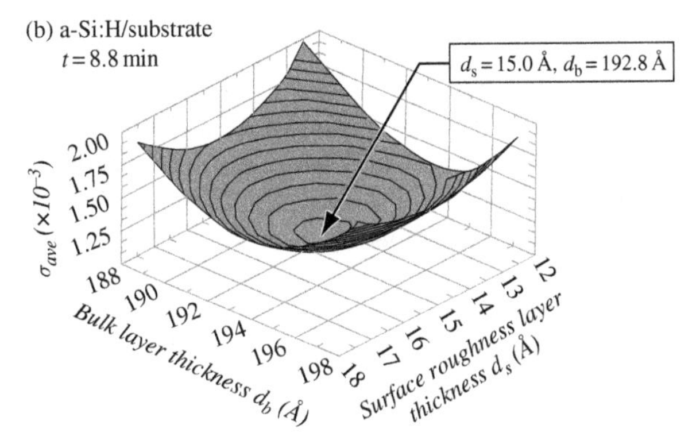

Figure 8.7 Analysis of real-time spectra by GEM: (a) time evolution of the unbiased estimator σ for the a-Si:H growth and (b) σ_{ave} obtained from each pair of (d_b, d_s) used in the mathematical inversion.

extracted from different values of (d_s, d_b) are shown. For the calculation of σ, Eq. (5.58) was used. As confirmed from Fig. 8.7(a), when the dielectric function is extracted assuming $(d_s, d_b) = (15\,\text{Å}, 193\,\text{Å})$ at 8.8 min, σ becomes small at $t > 2.5$ min. However, if we obtain the dielectric function using $(d_s, d_b) = (15\,\text{Å}, 170\,\text{Å})$, anomalous structures appear in the extracted dielectric function due to the deviation of the optical interference, and the σ values increase drastically except near $t = 8.8$ min. Figure 8.7(b) shows σ_{ave} for each pair of (d_s, d_b) used in the mathematical inversion. σ_{ave} in this figure was calculated from the region $t = 2.5$–13 min ($d_b = 50$–300 Å). As shown in Fig. 8.7(b), σ_{ave} is expressed using the two-dimensional plane of (d_s, d_b), and the thickness parameters at $t = 8.8$ min are determined from the values that minimize σ_{ave} ($d_s = 15.0\,\text{Å}$ and $d_b = 192.8\,\text{Å}$). The dielectric function shown in Fig. 8.6(b) represents the final ε_b obtained from the analysis. In data analysis by GEM, therefore, dielectric function modeling is unnecessary, and the dielectric function of a sample is obtained directly from real-time ellipsometry spectra.

As confirmed from the above procedure, GEM assumes that the dielectric function of the bulk layer (ε_b) is independent of the bulk layer thickness d_b. However, GEM can be employed if the dielectric function of a sample does not vary over a thickness of 100–200 Å. Thus, GEM can be employed to analyze a multilayer structure, if the thickness of each layer is 100–200 Å [9]. When ε_b varies strongly with d_b, however, GEM cannot be used, since σ_{ave} no longer shows a minimum in the analysis. Therefore, GEM is not a suitable analysis method for graded layers. In addition, GEM cannot be employed as a data analysis method for real-time control, because GEM requires all the real-time spectra for the analysis. On the other hand, GEM is a highly reliable method, as the data analysis is performed from many ellipsometry spectra obtained during film growth. Similarly, to improve the accuracy of the data analysis, the analysis of *ex situ* measurements is often performed from several (ψ, Δ) spectra obtained by varying film thickness [32].

Figure 8.8 shows the time evolution of Δ observed during the a-Si:H deposition in Fig. 8.5. The solid lines in this figure represent the fitting results calculated from ε_b, which shows a minimum σ_{ave}. As confirmed from Fig. 8.8, the calculated result agrees quite well with the experimental result over a wide energy range. Figure 8.9 shows the initial growth process of the a-Si:H layer obtained simultaneously from the above analysis. It can be seen that d_s increases rapidly in the initial stage of growth due to the island growth of the a-Si:H layer on the SiO_2/c-Si substrate. When the substrate surface is covered with a-Si:H islands ($t = 0.6$ min), d_s shows a maximum value, and the bulk layer is formed. At $t > 0.6$ min, d_s gradually reduces, and d_b increases linearly with a growth rate of 0.38 Å/s. Similar results have already been reported [2–5], and the island growth of a semiconductor crystal has also been observed [20]. As confirmed from the above example, it is possible to characterize thin-film growth on the atomic scale from real-time monitoring by spectroscopic ellipsometry.

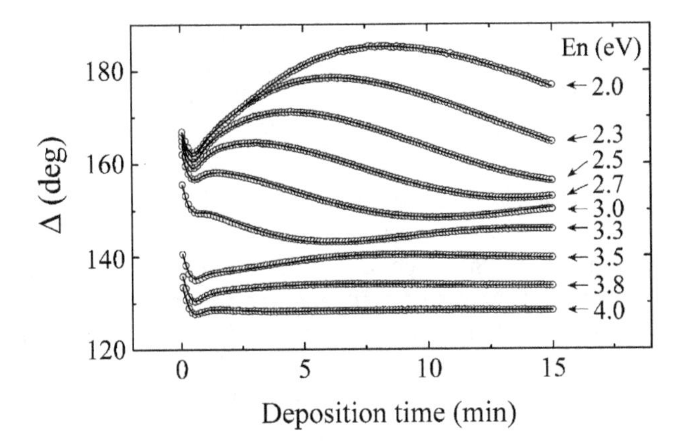

Figure 8.8 Time evolution of Δ for selected photon energies, observed during a-Si:H growth. The open circles show experimental results and the solid lines represent calculation results obtained from GEM.

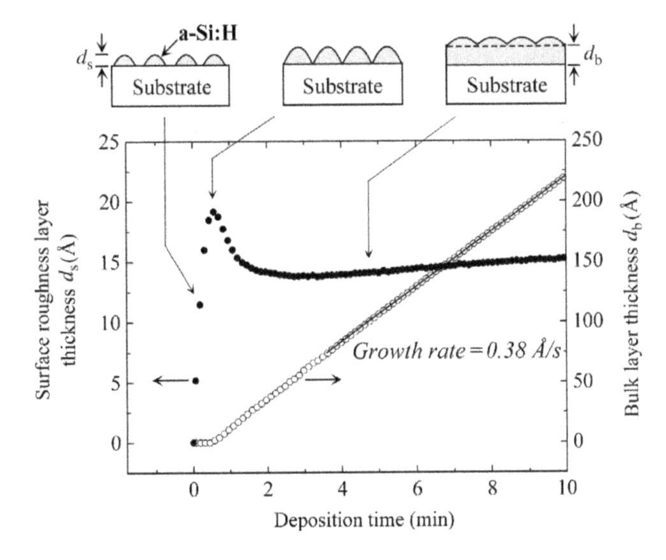

Figure 8.9 Initial growth process of an a-Si:H layer on a SiO_2/c-Si substrate characterized by real-time spectroscopic ellipsometry using GEM.

The variation in Δ observed at $t < 1$ min in Fig. 8.8 reflects the change in the surface roughness layer. This originates from the fact that a surface roughness layer with a small refractive index shifts Δ over the whole wavelength region (see Fig. 5.29). As confirmed from Fig. 8.8, the influence of optical interference is larger at lower energies due to smaller light absorption in the film. At higher energies, the interference effect vanishes as the film becomes thicker, since the penetration depth of light is smaller in this regime. Accordingly, we can roughly estimate the variation in the surface roughness from Δ at high energies.

The island growth of a-Si:H layers has been investigated in more detail by real-time spectroscopic ellipsometry combined with infrared attenuated total reflection spectroscopy (ATR) [8]. Figure 8.10 shows (a) time evolution of d_s and d_b determined by spectroscopic ellipsometry (SE) and (b) time evolution of the SiH_n ($n = 1$–3) bonds characterized by ATR, obtained from similar a-Si:H deposition on a SiO_2/c-Si substrate [8]. The ellipsometry result shown in Fig. 8.10(a) is almost the same as the one shown in Fig. 8.9. The ATR result in Fig. 8.10(b) represents the light absorption of the SiH_n species in the infrared region (stretching modes). In particular, the peak positions of SiH_n exposed to the film surface are different from those of SiH and SiH_2 present within the bulk layer, and we can separate the contributions of these species based on their peak positions [8]. In Fig. 8.10(b), the $SiH_{2,3}$ surface (open circles) represents the sum of the SiH_2 and SiH_3 surface modes. It can be seen that the variation in the $SiH_{2,3}$ surface mode reproduces the change in d_s shown in Fig. 8.10(a). This result confirms island formation and the following

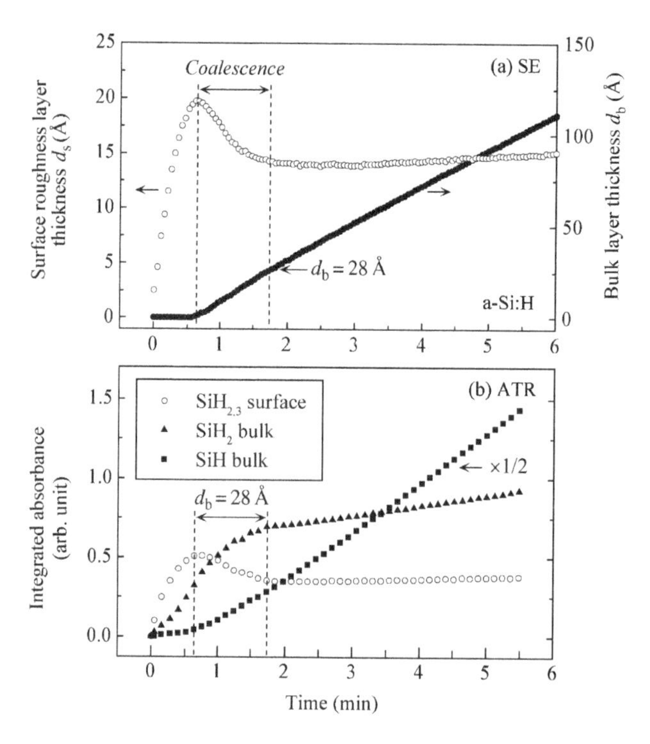

Figure 8.10 (a) Time evolution of the surface roughness layer thickness d_s and the bulk layer thickness d_b determined by spectroscopic ellipsometry (SE) and (b) time evolution of the integrated absorbances of the SiH_n ($n = 1$–3) bonds characterized by infrared attenuated total reflection spectroscopy (ATR) during a-Si:H growth on a SiO_2/c-Si substrate. Adapted with permission from H. Fujiwara, Y. Toyoshima, M. Kondo, and A. Matsuda, Interface-layer formation mechanism in a-Si:H thin-film growth studied by real-time spectroscopic ellipsometry and infrared spectroscopy, *Phys. Rev. B*, **60** (1999) 13598–13604. Copyright 1999, the American Physical Society.

coalescence of the a-Si:H islands on the substrate. Furthermore, both ellipsometry and ATR results show that the a-Si:H growth reaches a steady state after the end of the coalescence ($d_b > 28$ Å). Accordingly, the combination of ellipsometry with other techniques is quite effective for the detailed analysis of film growth.

From comparison between Fig. 8.7(a) and Fig. 8.9, it is evident that σ is quite large in the region where d_s varies rapidly ($t < 2$ min). The increase in σ implies that the dielectric function of the bulk layer differs when d_b is small. As confirmed from Fig. 8.10(b), SiH_2 bond formation is dominant when $d_b < 28$ Å, and the hydrogen content in this layer is higher than that in the thicker layer [8]. Thus, the increase in σ at $t < 2$ min can be attributed to the variation of dielectric function with thickness. Moreover, since $f_{void} > 0.5$ at the initial stage of island growth, σ increases if we perform the analysis assuming $f_{void} = 0.5$ [3]. When the analysis is performed by taking these effects into account, the increase in σ at $t < 2$ min is suppressed [3]. These results are consistent with the general fact that σ (or χ) increases when an optical model is oversimplified. In the case of Fig. 8.7(a), however, σ reduces as d_b increases, since the sensitivity for the interface region decreases gradually and f_{void} is approximated by 0.5. Accordingly, we can confirm the validity of ellipsometry analysis from the variation in σ (or χ). Conversely, we may try to find new structures in samples from the time evolution of σ.

8.1.4 VIRTUAL SUBSTRATE APPROXIMATION (VSA)

In 1993, Aspnes developed a data analysis method called the virtual substrate approximation (VSA) [11–13]. This analysis method is particularly powerful for the characterization of a graded layer in which dielectric function varies continuously in the thickness direction. If we apply VSA for the analysis of a compositionally graded layer, the composition of each layer can be determined [13–19]. Figure 8.11 shows optical models used for the analysis of graded layers. When the optical response of a graded layer is expressed by a multilayer structure, the thickness (d_j) and dielectric function ($\varepsilon_j = \varepsilon_{1j} - i\varepsilon_{2j}$) of each layer are required [Fig. 8.11(a)].

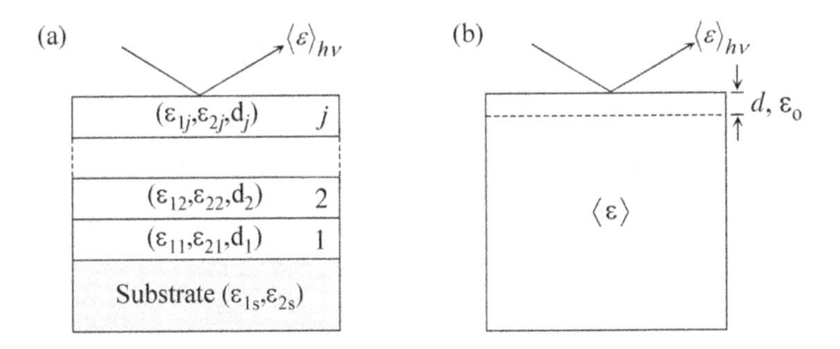

Figure 8.11 Optical models used for the analysis of graded layers: (a) a multilayer model and (b) virtual substrate approximation (VSA).

However, ellipsometry analysis using such optical models is generally difficult since there exist a large number of analysis parameters. Moreover, fitting errors in such analysis gradually increase toward the top layer by the propagation of analysis errors. In VSA, on the other hand, the complicated underlying structure is represented by a pseudo-dielectric function and only the thickness (d) and dielectric function (ε_o) of the top layer are taken into account [Fig. 8.11(b)]. Accordingly, analysis by VSA can be performed relatively easily even if the dielectric function of the sample varies continuously in the growth direction.

Figure 8.12 shows an optical model for VSA. In this figure, $\langle\varepsilon\rangle_{n-1}$ and $\langle\varepsilon\rangle_n$ represent the pseudo-dielectric functions calculated from Eq. (5.51) and n denotes the number of real-time spectra measured with a constant interval. The key feature of VSA is that the analysis is performed using the variation of pseudo-dielectric function with thickness [i.e. $\partial\langle\varepsilon\rangle/\partial d$]. In the analysis of $\langle\varepsilon\rangle_n$ shown in Fig. 8.12, for example, $\langle\varepsilon\rangle_{n-1}$ is employed as a virtual substrate and, from the variation in $\partial\langle\varepsilon\rangle/\partial d$, ε_o and d for the thin overlayer formed between $n-1$ and n are characterized.

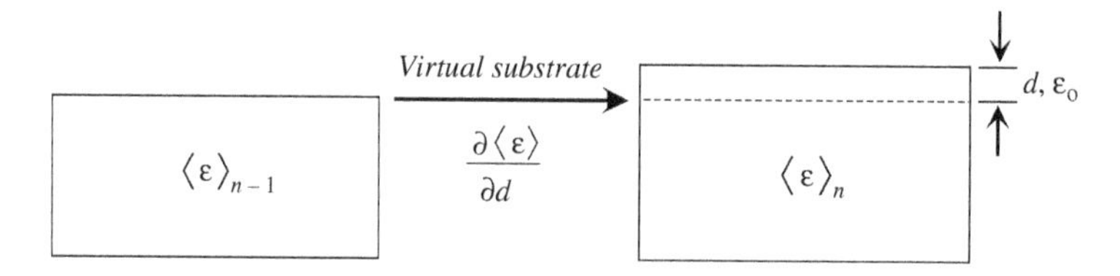

Figure 8.12 Optical model for virtual substrate approximation (VSA). $\langle\varepsilon\rangle_{n-1}$ and $\langle\varepsilon\rangle_n$ represent the pseudo-dielectric functions, and n denotes the spectral number. d and ε_o show the thickness and dielectric function of a thin overlayer formed between $n-1$ and n. In VSA, the variation of $\langle\varepsilon\rangle$ with thickness ($\partial\langle\varepsilon\rangle/\partial d$) is analyzed.

The theoretical expression for VSA can be derived from the optical model assuming an ambient/thin film/substrate structure [11–13]. In the convention of $N \equiv n + \mathrm{i}k$, ρ of this optical model is expressed as follows [see Eq. (5.2)]:

$$\rho = \left[\frac{r_{01,\mathrm{p}} + r_{12,\mathrm{p}}\exp(\mathrm{i}2\beta)}{1 + r_{01,\mathrm{p}}r_{12,\mathrm{p}}\exp(\mathrm{i}2\beta)}\right] \Big/ \left[\frac{r_{01,\mathrm{s}} + r_{12,\mathrm{s}}\exp(\mathrm{i}2\beta)}{1 + r_{01,\mathrm{s}}r_{12,\mathrm{s}}\exp(\mathrm{i}2\beta)}\right] \tag{8.2}$$

Here, r_{jk} in Eq. (8.2) is given by Eq. (2.65). By rearranging terms with respect to the thickness derivative $\partial\langle\varepsilon\rangle/\partial d$, we obtain the following equation [11–13]:

$$\frac{\partial\langle\varepsilon\rangle}{\partial d} = \frac{4\pi\mathrm{i}\langle\varepsilon\rangle_n\left(\langle\varepsilon\rangle_n - \varepsilon_o\right)(\varepsilon_o - 1)\left(\langle\varepsilon\rangle_n - \sin^2\theta_i\right)^{1/2}}{\lambda\varepsilon_o\left(\langle\varepsilon\rangle_n - 1\right)}, \tag{8.3}$$

where $\langle\varepsilon\rangle_n$ is the same as the one shown in Fig. 8.12 and λ is a wavelength. In Eq. (8.3), the dielectric constant of the ambient is assumed to be one. In the case of Fig. 8.12, $\partial\langle\varepsilon\rangle/\partial d$ is expressed simply by $\partial\langle\varepsilon\rangle/\partial d = \left(\langle\varepsilon\rangle_n - \langle\varepsilon\rangle_{n-1}\right)/d$.

The theoretical expression of VSA shown in Eq. (8.3) represents the first-order approximation of the optical model (ambient/thin film/substrate) in terms of the thickness derivative, and this equation holds only for a very thin film (~10 Å) formed on the virtual substrate. In a transparent film, however, optical interference is determined by the light reflection at the transparent film/substrate interface, and thus the virtual interface shown in Fig. 8.12 cannot be assumed [22]. Accordingly, VSA has been applied for the analysis of layers that show relatively large light absorption (mainly semiconductor layers) [13–21]. When an underlying layer changes continuously during processing, analysis by VSA becomes difficult. In addition, the analysis errors of VSA increase when the change in $\partial\langle\varepsilon\rangle/\partial d$ is small. In the compositional analysis of graded layers, dielectric function modeling for the alloy composition x is necessary (see Section 7.2.3). In this case, the dielectric function of an overlayer is expressed by $\varepsilon_o(x)$, and the analysis parameters in Eq. (8.3) become (x, d).

Although a surface roughness layer is not taken into account in the above analysis, we can incorporate the surface roughness layer into the VSA analysis [17]. Figure 8.13 shows the data analysis procedure for VSA that includes this surface roughness analysis. Here, the compositional analysis of a graded layer

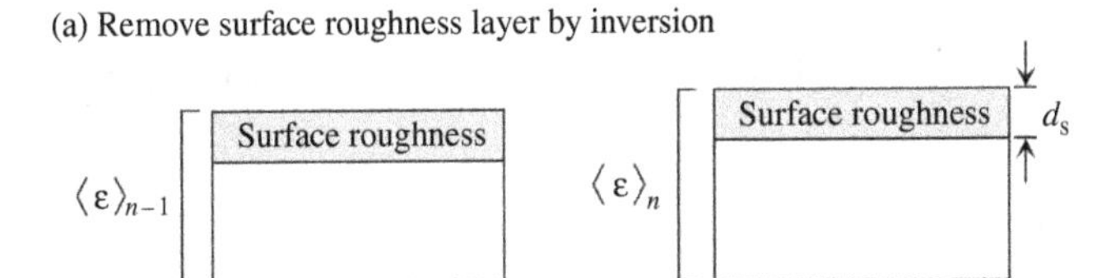

(b) Determine growth rate (r) and composition (x) by VSA

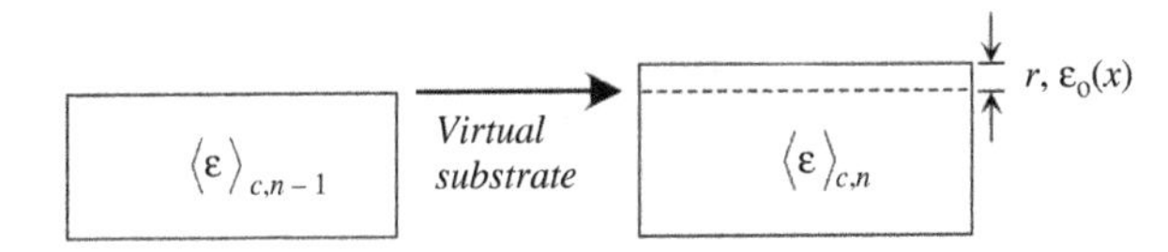

(c) Replace surface roughness layer using d_s in (a) and x estimated in (b), and compare with measured spectra

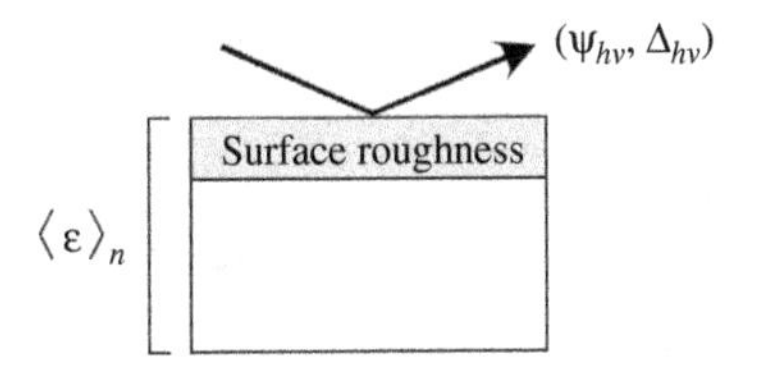

(d) By repeating (a)–(c), determine (d_s, r, x) that minimize σ (or χ)

Figure 8.13 Data analysis procedure for VSA that includes surface roughness analysis.

is considered and $\varepsilon(x)$ is the dielectric function for the alloy composition x. In this analysis, by applying mathematical inversion, the surface roughness layer is removed from $\langle\varepsilon\rangle_{n-1}$ and $\langle\varepsilon\rangle_n$ spectra using the same d_s and $\varepsilon(x)$ [Fig. 8.13(a)]. $\langle\varepsilon\rangle_{c,n-1}$ and $\langle\varepsilon\rangle_{c,n}$ in Fig. 8.13(b) represent the pseudo-dielectric functions obtained from this procedure. Secondly, the growth rate r (or thickness d) and composition x are estimated from $\langle\varepsilon\rangle_c$ using VSA [Fig. 8.13(b)]. In Fig. 8.13(c), the surface roughness layer is replaced using d_s employed in Fig. 8.13(a) and x estimated in Fig. 8.13(b), and the calculated (ψ, Δ) are compared with (ψ, Δ) obtained from actual measurement. Finally, by repeating Fig. 8.13(a)–(c), (d_s, r, x) are estimated from values that minimize σ (or χ). As we have seen in Fig. 5.34, surface roughness layers have a large influence on the pseudo-dielectric function. Thus, when surface roughness is present, it is necessary to use the above analysis method.

Figure 8.14 shows an analysis example of a compositionally graded layer by VSA [18]. In this example, the analysis of the surface roughness layer was performed using the procedure shown in Fig. 8.13. The analyzed layer is an a-$Si_{1-x}C_x$:H layer fabricated by a plasma process. The spectroscopic ellipsometry measurement was carried out using a P_RSA instrument. For the analysis, the dielectric function of a-$Si_{1-x}C_x$:H parameterized by the Tauc–Lorentz model (see Section 5.2.4) was employed [33]. As shown in Fig. 8.14(a), the carbon composition x was controlled by the flow ratio z of source gases. In Fig. 8.14(b)–(d), the carbon composition x, instantaneous growth rate r, and surface roughness layer thickness d_s estimated from the analysis are shown. In this example, the analysis was first performed using

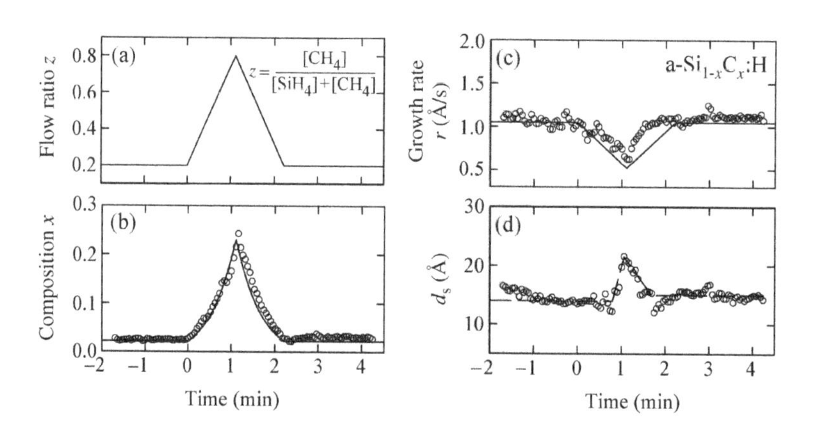

Figure 8.14 Growth process of a compositionally modulated a-$Si_{1-x}C_x$:H layer characterized from real-time spectroscopic ellipsometry using VSA: (a) gas flow ratio $z = [CH_4]/([SiH_4] + [CH_4])$, (b) carbon composition x, (c) growth rate r, and (d) surface roughness layer thickness d_s. In (b)–(d), the open circles show experimental results and the solid lines represent calculated results based on individual depositions with uniform compositions. Reprinted with permission from *Applied Physics Letters*, **70**, H. Fujiwara, J. Koh, C. R. Wronski, and R. W. Collins, Application of real time spectroscopic ellipsometry for high resolution depth profiling of compositionally graded amorphous silicon alloy thin films, 2150–2152 (1997). Copyright 1997, American Institute of Physics.

(x, r, d_s) as analysis parameters. The result shown in Fig. 8.14(d) represents d_s obtained from this analysis. In the following analysis, d_s was fixed to the values indicated by the dotted line in Fig. 8.14(d), in order to improve the stability in the analysis. The (x, r) determined from this analysis are shown in Figs. 8.14(b) and (c). The solid lines in Figs. 8.14(b) and (c) indicate calculation results predicted from individual depositions with uniform compositions. As confirmed from Fig. 8.14(b), the estimated carbon compositions show excellent agreement with the calculated values, but the peak position of the maximum composition deviates by 5 s due to the nonzero residence time of gases in the deposition system. From Figs. 8.14(c) and (d), it is evident that the growth rate increases when z is reduced ($t = 1.1$–2.2 min) and d_s increases when the carbon composition is high.

Since the instantaneous growth rate $r(t)$ is evaluated from the above analysis, the depth profile of the carbon composition can also be obtained from the integration of $r(t)$. Figure 8.15(a) shows the depth profile of the carbon composition x in an a-$Si_{1-x}C_x$:H layer determined from the above analysis using spectroscopic ellipsometry (SE) [19]. The dotted line shows calculated values based on individual depositions, and the solid line represents a depth profile obtained from secondary ion mass spectrometry (SIMS). In Fig. 8.15(a), the carbon profile estimated from ellipsometry shows poor agreement with that determined from SIMS, although the calculated result agrees well with the ellipsometry result. Figure 8.15(b) shows a simulation result obtained from the ellipsometry profile shown in Fig. 8.15(a) [19]. In this simulation, the ellipsometry profile was broadened intentionally assuming a Gaussian function. In particular, the half width of the Gaussian function was

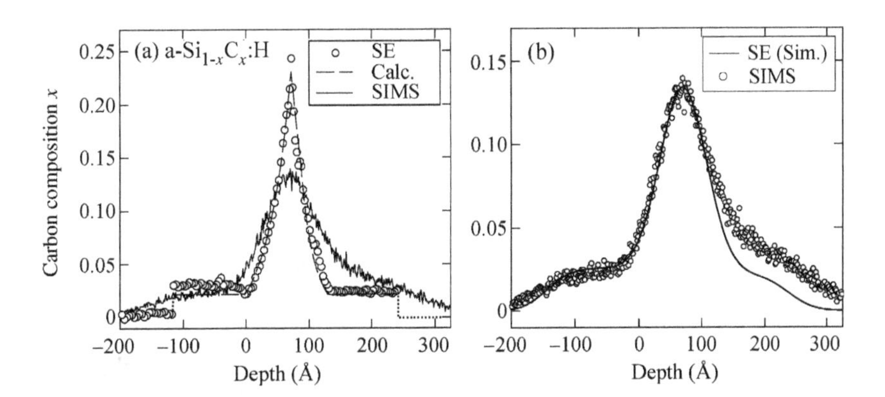

Figure 8.15 (a) Depth profiles of the carbon composition x in an a-$Si_{1-x}C_x$:H layer determined from spectroscopic ellipsometry (SE) and secondary ion mass spectrometry (SIMS) and (b) simulation result obtained from the SE profile shown in (a). In (a), the dotted line shows a calculated profile based on individual depositions. In (b), the SE profile in (a) was broadened intentionally assuming a Gaussian function. Reprinted with permission from *Applied Physics Letters*, **72**, H. Fujiwara, J. Koh, C. R. Wronski, R. W. Collins, and J. S. Burnham, Optical depth profiling of band gap engineered interfaces in amorphous silicon solar cells at monolayer resolution, 2993–2995 (1998). Copyright 1998, American Institute of Physics.

broadened by 44 Å and the overall carbon composition was increased by 20 %. As shown in Fig. 8.15(b), the simulated profile is in excellent agreement with the SIMS profile. This result shows that depth profiling analysis by spectroscopic ellipsometry has very high thickness resolution on the atomic scale.

8.2 OBSERVATION OF THIN-FILM GROWTH BY REAL-TIME MONITORING

If we apply spectroscopic ellipsometry to real-time monitoring of thin-film growth, thin-film formation processes can be investigated in detail. Thus, important information concerning device fabrication can be obtained from process diagnoses using spectroscopic ellipsometry. In particular, spectroscopic ellipsometry has been employed extensively for the characterization of Si thin films including a-Si:H and microcrystalline Si (μc-Si:H) [1–10,33–38]. This section will introduce the characterization of the Si thin-film growth, as analysis examples for real-time spectroscopic ellipsometry.

8.2.1 ANALYSIS EXAMPLES

Figure 8.16 shows the initial growth processes of (a) a-Si:H fabricated by plasma-enhanced chemical vapor deposition (PECVD) and (b) a-Si fabricated by magnetron sputtering [4]. In this example, the spectroscopic ellipsometry measurement was

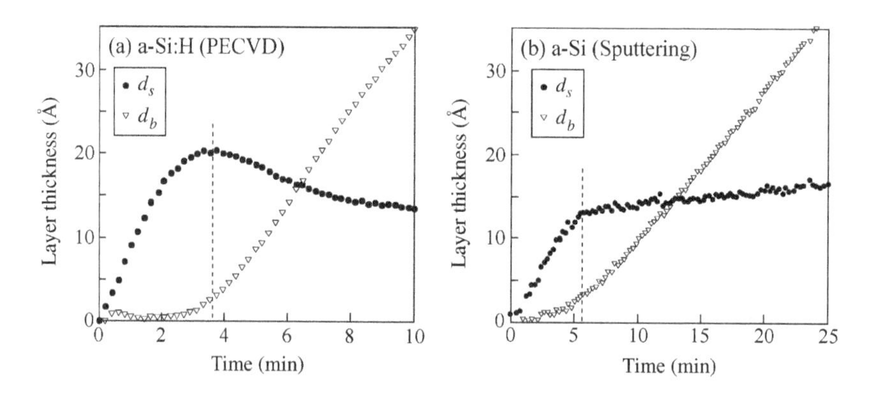

Figure 8.16 Initial growth processes of (a) a-Si:H fabricated by plasma-enhanced chemical vapor deposition (PECVD) and (b) a-Si fabricated by magnetron sputtering. d_s and d_b represent the surface roughness layer and bulk layer thicknesses, respectively. Reprinted from *Journal of Non-Crystalline Solids*, **137 & 138**, Y. M. Li, I. An, H. V. Nguyen, C. R. Wronski, and R. W. Collins, Real-time spectroscopic ellipsometry determination of the evolution of amorphous semiconductor optical functions, bandgap, and microstructure, 787–790, Copyright (1991), with permission from Elsevier.

performed using a P_RSA instrument. The growth temperatures were 250 °C (a-Si:H) and 300 °C (a-Si), and c-Si substrates covered with native oxide were used for the film deposition. For the analysis of real-time spectra, GEM described in Section 8.1.3 was employed. As shown in Fig. 8.16(a), the growth process of the a-Si:H layer by PECVD is similar to those shown in Figs. 8.9 and 8.10(a). In Fig. 8.16(a), a-Si:H islands uniformly cover the surface of the SiO_2/c-Si substrate by the time indicated by the dotted line, and the surface roughness layer thickness d_s gradually reduces during the coalescence of the islands. Similar island growth can be seen in the a-Si growth shown in Fig. 8.16(b). In the case of a-Si fabricated without hydrogen, however, the reduction in d_s is not observed during the coalescence of the a-Si islands, and d_s shows a constant value. This result indicates that the growth process in PECVD differs from that in magnetron sputtering. This difference has been attributed to variation in precursor diffusion on the growing surface [4]. In particular, the growing surface of a-Si:H layers is terminated with hydrogen [see Fig. 8.10(b)], and the diffusion length of the precursors is expected to become longer, compared with a-Si formed without surface hydrogen. Accordingly, the diffusion length of precursors can be deduced from the reduction in d_s observed during the coalescence. Furthermore, it has been reported that a-Si:H film properties improve as the reduction in d_s during the coalescence becomes larger [5].

If we assume that the islands formed on the substrate are perfectly hemispherical and oriented on a grid, the island density on the substrate can also be estimated from d_s when the substrate surface is covered with the islands. In this case, when the island density is small, d_s observed at the onset of the coalescence increases, since each island grows larger before coalescence. From the result shown in Fig. 8.16, island densities of $0.6 \times 10^{13}\,cm^{-2}$ (a-Si:H) and $1.2 \times 10^{13}\,cm^{-2}$ (a-Si) can be estimated [4].

Figure 8.17 shows the characterization of μc-Si:H growth by real-time spectroscopic ellipsometry [9]. The data analysis was performed using GEM shown in Fig. 8.6, and the μc-Si:H film was prepared by PECVD on a SiO_2/c-Si substrate. In PECVD, μc-Si:H films are formed when SiH_4 source gas is diluted by a large quantity of H_2 [38]. For the μc-Si:H deposition shown in Fig. 8.17, a hydrogen dilution ratio of $[H_2]/[SiH_4] = 20$ was used. The total film thickness was calculated from $d_b + 0.5d_s$, and the coefficient 0.5 for d_s corresponds to the void volume fraction assumed for the surface roughness layer ($f_{void} = 0.5$). In Fig. 8.17, after initiating the film deposition, d_s increases rapidly due to the island growth of the a-Si:H film on the substrate, and d_s decreases after the islands make contact on the substrate. In the following process, μc-Si:H nucleation occurs, and d_s increases again by the preferential growth of μc-Si:H grains. After the growing surface is covered by the μc-Si:H grains, columnar grain growth occurs, and d_s gradually decreases again. From this result, film thicknesses at which the μc-Si:H nucleates from the a-Si:H phase (d_{Nuclei}) and the μc-Si:H nuclei make contact on the surface ($d_{Contact}$) are estimated to be 110 Å and 950 Å, respectively. In this analysis, the dielectric function was extracted from the a-Si:H layer formed initially on the substrate, and the analysis of the μc-Si:H phase was performed by using this dielectric function. However, since the dielectric function of a-Si:H is different from that of μc-Si:H,

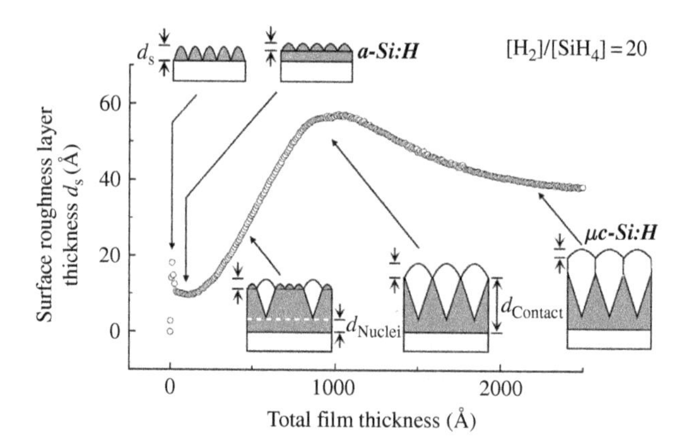

Figure 8.17 Characterization of microcrystalline silicon (μc-Si:H) growth by real-time spectroscopic ellipsometry. Adapted with permission from H. Fujiwara, M. Kondo and A. Matsuda, Real-time spectroscopic ellipsometry studies of the nucleation and grain growth processes in microcrystalline silicon thin films, *Phys. Rev. B*, **63** (2001) 115306-1–9. Copyright 2001, the American Physical Society.

fitting errors increase with increasing d_b. In addition, the analytical errors for d_s are relatively large during the phase transition from a-Si:H to μc-Si:H, since the void volume fraction in the surface roughness layer was fixed at 50 vol.% in this analysis.

In this example, the structural characterization of the above sample was also performed using transmission electron microscopy (TEM). Figure 8.18 shows

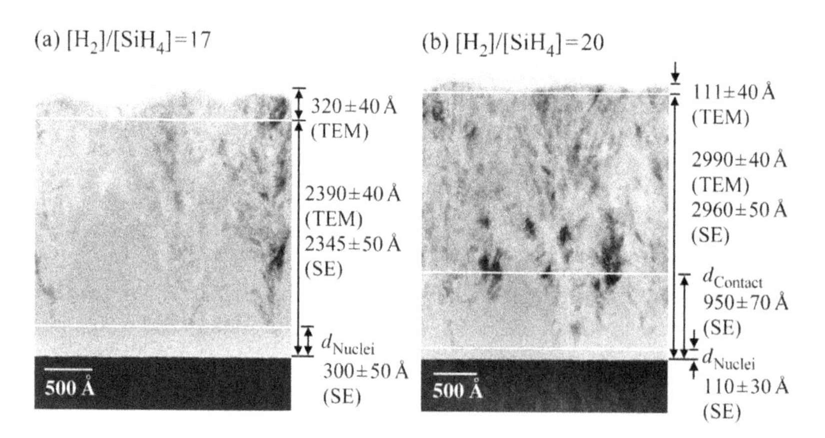

Figure 8.18 Cross-sectional TEM images of μc-Si:H films prepared using (a) $[H_2]/[SiH_4] = 17$ and (b) $[H_2]/[SiH_4] = 20$. d_{Nuclei} and $d_{Contact}$ estimated from spectroscopic ellipsometry (SE) represent film thicknesses at which the μc-Si:H nucleates from the a-Si:H phase and μc-Si:H nuclei make contact on the surface, respectively. In (b), these thicknesses were determined from the SE result shown in Fig. 8.17. Adapted with permission from H. Fujiwara, M. Kondo, and A. Matsuda, Real-time spectroscopic ellipsometry studies of the nucleation and grain growth processes in microcrystalline silicon thin films, *Phys. Rev. B*, **63** (2001) 115306-1–9. Copyright 2001, the American Physical Society.

cross-sectional TEM images of μc-Si:H films prepared using different hydrogen dilution ratios [9]. The TEM image in Fig. 8.18(b) was obtained from the same sample shown in Fig. 8.17. The white lines denoted as d_{Nuclei} and $d_{Contact}$ indicate the thicknesses evaluated from spectroscopic ellipsometry (SE). In Fig. 8.18, d_{Nuclei} and $d_{Contact}$ show remarkable agreement with the TEM results. The d_b values estimated from SE and TEM also agree quite well with errors less than 2 %. Nevertheless, the surface roughness of the sample in Fig. 8.18(a) is too large to apply the effective medium approximation (see Section 5.3.3). In this sample, therefore, the analysis errors for the surface roughness layer are quite large [9].

The above examples show that real-time spectroscopic ellipsometry allows the characterization of surface roughness layers on the atomic scale. From the time evolution of surface roughness, it is possible to discuss structural changes or reaction processes on growing surfaces. Although TEM is a highly reliable technique, TEM measurement is generally difficult, and only a limited number of samples can be measured. In contrast, spectroscopic ellipsometry measurement is rather easy and further enables us to characterize growth processes in real time. Accordingly, once data analysis procedures for specific materials or devices are established, spectroscopic ellipsometry provides a great ability to characterize thin-film growth.

8.2.2 ADVANCED ANALYSIS

Here, as an advanced analysis example, we will look at the analysis of an interface layer formed in μc-Si:H growth [10]. Figure 8.19(a) shows an optical model for a μc-Si:H film deposited on a ZnO:Ga layer [10]. As shown in this figure, the substrate of this sample is composed of ZnO:Ga/SiO_2/Si(100). The dielectric

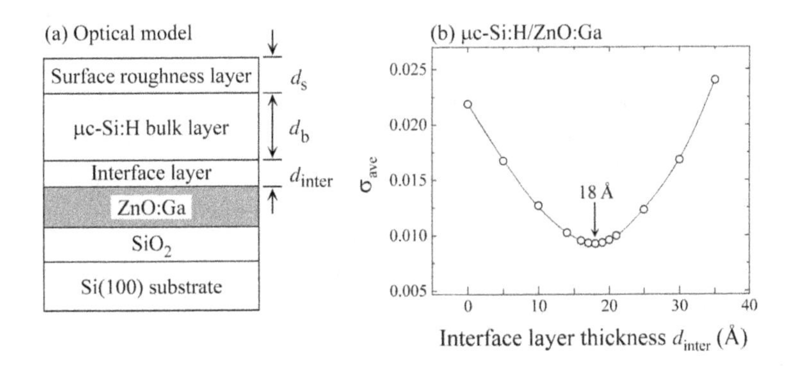

Figure 8.19 (a) Optical model for a μc-Si:H film deposited on a ZnO:Ga/SiO_2/Si(100) substrate and (b) variation of σ_{ave} with the interface layer thickness d_{inter} shown in (a). In (b), σ_{ave} represents the average fitting error calculated from Eq. (8.1). Reprinted with permission from H. Fujiwara, M. Kondo, and A. Matsuda, Interface-layer formation in microcrystalline Si:H growth on ZnO substrates studied by real-time spectroscopic ellipsometry and infrared spectroscopy, *Journal of Applied Physics*, **93**, 2400–2409 (2003). Copyright 2003, American Institute of Physics.

function of the ZnO:Ga layer at the process temperature (230 °C) can be obtained from the following procedure: first the thickness of the ZnO:Ga layer is estimated by the analysis described in Section 7.3.2 and then the sample (ZnO:Ga/SiO_2/c-Si) is measured at the process temperature. Finally, the dielectric function of the ZnO:Ga is extracted from the measured spectra using mathematical inversion. This analysis is performed assuming that the thickness of the ZnO:Ga layer estimated at room temperature does not vary at the process temperature. For this mathematical inversion, however, the dielectric function of c-Si measured in advance at the process temperature is necessary. For SiO_2, the dielectric function at room temperature can be employed (see Section 8.1.1). In this analysis, since the unknown parameter for the ZnO:Ga/SiO_2/c-Si is the dielectric function of the ZnO:Ga only, we can perform mathematical inversion. In this manner, even when a substrate has a multilayer structure, we can determine dielectric functions at process temperatures.

In the optical model shown in Fig. 8.19(a), an interface layer is introduced at the μc-Si:H/ZnO:Ga interface. Figure 8.19(b) shows σ_{ave} given by Eq. (8.1), plotted as a function of the interface layer thickness d_{inter} [10]. This σ_{ave} was estimated in a region where the thickness of the μc-Si:H layer is $d_b = 100$–200 Å. To simplify the analysis, however, the dielectric function of the SiO_2 layer was applied for the interface layer and, for the analysis of the μc-Si:H layer, the dielectric function of a μc-Si:H layer obtained from other analysis was employed. As shown in Fig. 8.19(b), σ_{ave} reduces rapidly with increasing d_{inter}, and σ_{ave} shows a minimum at $d_{inter} = 18$ Å. This analysis example shows clearly that σ_{ave} increases greatly if the interface layer is not taken into account, even when the thickness of the interface layer is quite small. However, when the μc-Si:H bulk layer is thick and probe light does not reach the interface region, the contribution of the interface layer to the measured spectra becomes negligible. Thus, the characterization of interface layers from *ex situ* measurements is generally difficult in absorbing films.

Figure 8.20(a) shows the growth process of the μc-Si:H obtained from the above analysis [10]. In Fig. 8.20(a), the interface layer is first formed on the substrate after starting the deposition, and the island growth of μc-Si:H occurs on this interface layer. Finally, after the growing surface is covered with μc-Si:H islands, the μc-Si:H bulk layer is formed, and the thickness of the bulk layer increases linearly with deposition time. In the analysis shown in Fig. 8.20(a), the thickness of the interface layer was fixed to the value estimated from the analysis of Fig. 8.19(b) (i.e., $d_{inter} = 18$ Å). Figure 8.20(b) shows the cross-sectional TEM image obtained from the same sample [10]. The TEM result shown in Fig. 8.20(b) supports the formation of the interface layer on the ZnO:Ga layer. Furthermore, the interface layer thickness estimated from TEM shows quite good agreement with the one obtained from the ellipsometry analysis. Although this analysis is rather complicated, the bulk layer thicknesses determined from these measurements also agree quite well. As evidenced from this example, thin-film growth processes can be characterized accurately from real-time spectroscopic ellipsometry. With respect to the Si films, more advanced characterization has also been reported [6,7].

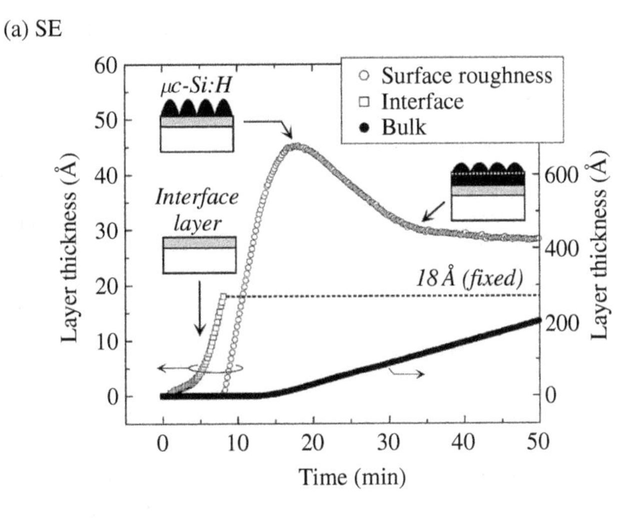

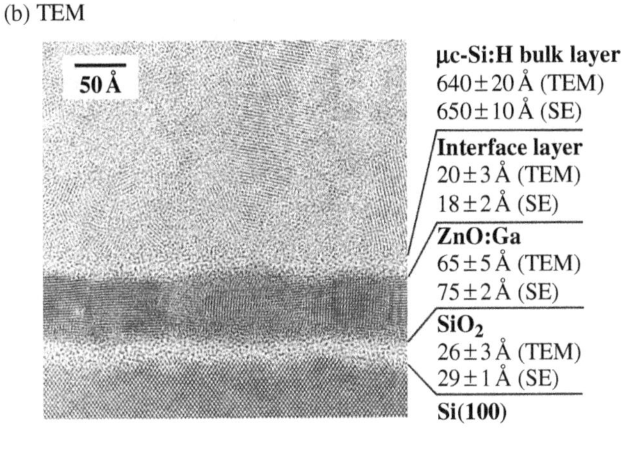

Figure 8.20 (a) Growth process of the μc-Si:H layer on the ZnO:Ga layer characterized from real-time spectroscopic ellipsometry (SE) and (b) cross-sectional TEM image obtained from the sample in (a). Reprinted with permission from H. Fujiwara, M. Kondo, and A. Matsuda, Interface-layer formation in microcrystalline Si:H growth on ZnO substrates studied by real-time spectroscopic ellipsometry and infrared spectroscopy, *Journal of Applied Physics*, **93**, 2400–2409 (2003). Copyright 2003, American Institute of Physics.

8.3 PROCESS CONTROL BY REAL-TIME MONITORING

From real-time monitoring by spectroscopic ellipsometry, it is possible to perform real-time control of film structures. In particular, real-time structural control by spectroscopic ellipsometry can be applied to various processes including solution processes. At present, process control has been carried out mainly in the semiconductor field. For data analysis of process control, LRA and VSA can be employed. In this section, we will address the structural control of thin films by LRA and VSA.

8.3.1 DATA ANALYSIS IN PROCESS CONTROL

With respect to process control by spectroscopic ellipsometry, various data analysis methods have been proposed:

(a) control of process conditions by LRA [24,26,27,39–42],
(b) control of process conditions by VSA [13–16,20],
(c) selecting a measurement wavelength where (ψ, Δ) change linearly with film thickness, and performing thickness control based on the measured (ψ, Δ) values [43],
(d) calculating (ψ, Δ) spectra of an intended structure in advance, and stopping the processing when the measured spectra match with the calculated spectra [22,44],
(e) determining the end of processing from changes in (ψ, Δ) [41].

As we will see later, from (a) and (b) above, feedback control for process temperature, growth rate, and alloy composition can be performed. (c) and (d) above have been applied for thickness control of a multilayer structure. On the other hand, (e) above has been employed for the control of a plasma etching process. In this case, we can judge the end of the etching process from the time evolution of (ψ, Δ) (see Fig. 8.24). Although LRA shows high stability and precision in data analysis, this method is not applicable to the analysis of graded layers (see Table 8.1). Thus, for the feedback control of alloy compositions, VSA has been utilized [13–16]. When we perform feedback control by LRA and VSA, however, the parameterization of dielectric functions is basically necessary.

8.3.2 PROCESS CONTROL BY LINEAR REGRESSION ANALYSIS (LRA)

Linear regression analysis has been employed widely as the data analysis method for real-time control. Here, we will look at real-time control by LRA applied for (a) a thermal oxidation process of c-Si [39], (b) a growth process of a-Si:H [40], and (c) an etching process of an LSI device [41].

Thermal Oxidation Process of c-Si

It has been reported that the process temperature and SiO_2 thickness during the thermal oxidation of c-Si can be controlled using single-wavelength ellipsometry [39]. For this analysis, we can use the optical model for the SiO_2/c-Si shown in Fig. 8.1. In order to perform feedback control of the process temperature, the parameterization of dielectric functions is required. In this example, the optical constants at $\lambda = 6328\,\text{Å}$ were modeled, since the measurement was carried out using a He–Ne laser ($\lambda = 6328\,\text{Å}$). If we parameterize the optical constants for the temperature T, the complex refractive indices of SiO_2 and c-Si are expressed by $N_1(T) = n_1(T)$ and $N_2(T) = n_2(T) - ik_2(T)$, respectively. In this case, the (ψ, Δ) of the sample are given by

$$\tan\psi\exp(i\Delta) = \rho\,[N_0, N_1(T), N_2(T), d_b, \theta_0] \tag{8.4}$$

The actual equation for $N_1(T)$ has been described in Section 8.1.1. In Eq. (8.4), the analysis parameters are (T, d_b) only, and we can characterize (T, d_b) independently of (ψ, Δ) measured at $\lambda = 6328\,\text{Å}$. In single-wavelength ellipsometry, however, the measurement sensitivity for (T, d_b) varies depending on the measurement wavelength and the optical constants of a sample.

Figure 8.21 shows (a) process temperature and (b) SiO_2 thickness obtained from real-time control using the above analysis [39]. The dotted lines in the figures represent the target values for the temperature ($T = 950\,°\text{C}$) and SiO_2 thickness ($d_b = 100\,\text{Å}$). In this example, a p-type Si(111) substrate was used, and the substrate heating was performed using a lamp heater. The feedback control of the process temperature was carried out by controlling the lamp heater output so that the temperature estimated from ellipsometry coincided with the target temperature. When the SiO_2 thickness, obtained simultaneously from the analysis, reached the target thickness, the lamp heating was terminated to stop the oxidation process. The reduction in the SiO_2 thickness observed after starting the substrate heating ($t \sim 1$ min) is due to the desorption of a contaminant layer by substrate heating. In this example, real-time control of the SiO_2 layer thickness with a precision of $\sim 2\,\text{Å}$

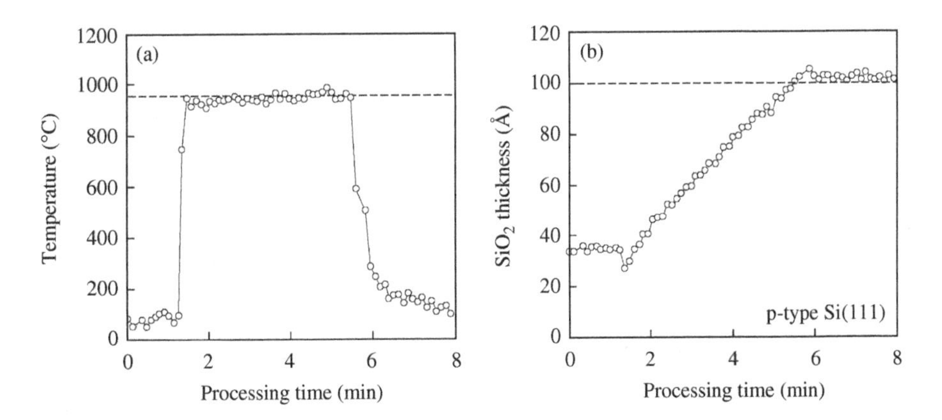

Figure 8.21 Real-time control of a thermal oxidation process by LRA using single-wavelength ellipsometry: (a) process temperature and (b) thickness of a SiO_2 layer formed on a p-type Si(111) substrate by thermal oxidation. The dotted lines represent target values used in the real-time control. Reprinted from *Thin Solid Films*, **233**, E. A. Irene, Applications of spectroscopic ellipsometry to microelectronics, 96–111, Copyright (1993), with permission from Elsevier.

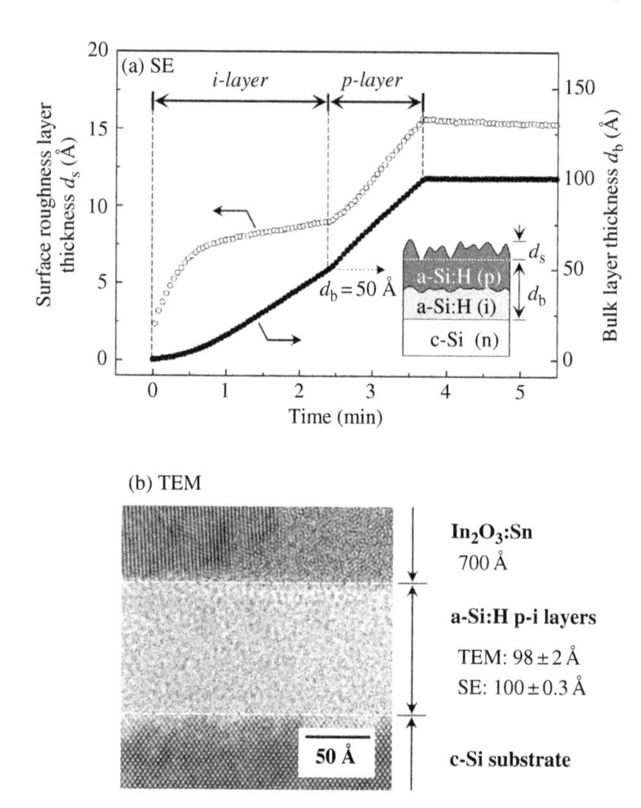

Figure 8.22 (a) Real-time thickness control of a-Si:H p-i layers applied for an a-Si:H/c-Si heterojunction solar cell and (b) cross-sectional TEM image of an a-Si:H/c-Si solar cell fabricated by the process shown in (a). These results were obtained by controlling the bulk layer thicknesses of the a-Si:H p-i layers to 50 Å. In (b), the thickness indicated by the white dotted lines represents the bulk layer thickness (d_b) of the a-Si:H p-i layers. Reprinted with permission from H. Fujiwara, and M. Kondo, Real-time monitoring and process control in amorphous/crystalline silicon heterojunction solar cells by spectroscopic ellipsometry and infrared spectroscopy, *Applied Physics Letters*, **86**, 032112 (2005). Copyright 2005, American Institute of Physics.

is demonstrated. As confirmed from Figs. 8.2 and 8.21, the oxidation rate is faster when the SiO_2 layer is thin. Thus, for the accurate control of oxide layer thickness, such real-time control is quite helpful.

Growth Process of a-Si:H

Figure 8.22(a) shows real-time thickness control of a-Si:H p-i layers applied for the fabrication of an a-Si:H/c-Si heterojunction solar cell [40]. In this figure, the time evolution of the surface roughness layer thickness d_s and bulk layer thickness d_b determined by spectroscopic ellipsometry is shown. In this example, a PC_RSA instrument was employed for the measurement, and the analysis was performed in real time by LRA using an optical model consisting of surface roughness layer/bulk layer/substrate, as shown in the inset of Fig. 8.22(a). For the surface roughness layer, $f_{\text{void}} = 0.5$ was assumed. The dielectric function of the a-Si:H layer was obtained in advance from an a-Si:H i-layer (200 Å) using GEM. The dielectric function of the i-layer was also used for the analysis of the p-layer, since the dielectric functions of the p- and i-layers are rather similar. In this example, the a-Si:H layers were formed on an n-type Si(100) substrate terminated with hydrogen. In this case, the dielectric function of the c-Si at the process temperature (130 °C) can be obtained from the pseudo-dielectric function [Eq. (5.51)], since there are no overlayers on the substrate. Consequently, the unknown parameters in this real-time analysis become d_s and d_b.

In the a-Si:H deposition shown in Fig. 8.22(a), the source gases for the p-layer were introduced into a PECVD reactor when d_b of the i-layer reached 50 Å at 2.4 min, and the plasma for the p-layer growth was terminated at $d_b = 100$ Å (3.7 min). As evidenced from Fig. 8.22(a), accurate control of the a-Si:H layer thicknesses with a precision better than ± 1 Å can be achieved by applying spectroscopic ellipsometry. In the early stages of the i-layer deposition, the island growth of the a-Si:H layer can be seen. However, the time evolution of d_s shown in Fig. 8.22(a) is different from the one shown in Fig. 8.10(a), since the a-Si:H layer in this example was formed on the H-terminated c-Si, instead of the SiO_2/c-Si in Fig. 8.10(a). In other words, when a-Si:H layers are deposited on SiO_2, a-Si:H islands becomes larger due to a lower island density of a-Si:H on the substrate. Accordingly, various growth modes on substrates can also be discussed from the time evolution of d_s (see Fig. 1.2). During the p-layer growth, surface roughening and higher deposition rate can be seen. These phenomena have been explained by higher dangling bond densities on the p-layer surface than on the i-layer surface [40]. Furthermore, it has been reported that a slight reduction in d_s observed after terminating the p-layer growth at $t > 3.7$ min is an artifact of the ellipsometry analysis and is induced by hydrogen effusion from the p-layer [45].

Figure 8.22(b) shows the cross-sectional TEM image of the a-Si:H/c-Si solar cell fabricated by the above process [40]. As shown in this figure, d_b estimated from the TEM image shows excellent agreement with the one determined from spectroscopic ellipsometry (SE). This result supports the validity of structural control by spectroscopic ellipsometry. TEM images obtained from this sample also indicate the presence of surface roughness on the a-Si:H layer with a thickness of $\sim$10 Å. In this example, however, the quantitative analysis of the surface roughness was not performed due to the weak contrast of the surface roughness on the TEM images.

Etching Process in an LSI Device

The real-time control of plasma etching processes can also be performed by using LRA [41,42]. Figure 8.23 shows a schematic diagram of an LSI structure used in a plasma etching process [41]. In Fig. 8.23, a TiN/poly-Si is formed on a gate oxide, and part of the TiN layer is covered by photoresist. In addition, field-oxide (FOX)

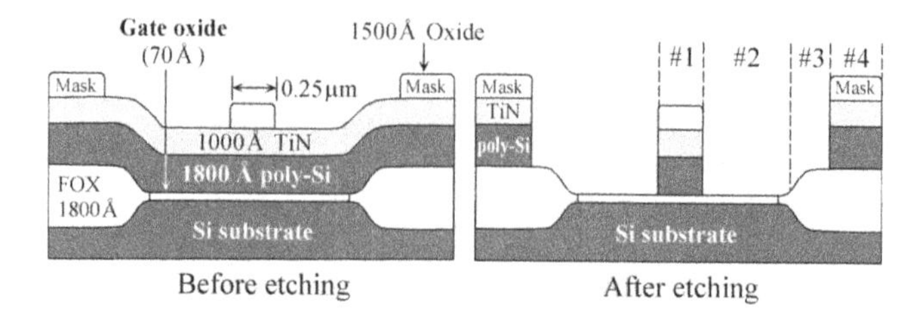

Figure 8.23 Schematic diagram of an LSI structure used in a plasma etching process. The area fractions #1–#4 were employed in quantitative analysis using the island-film model. Reprinted from *Thin Solid Films*, **313–314**, H. L. Maynard, N. Layadi, J. T. C. Lee, Plasma etching of submicron devices: in situ monitoring and control by multi-wavelength ellipsometry, 398–405, Copyright (1998), with permission from Elsevier.

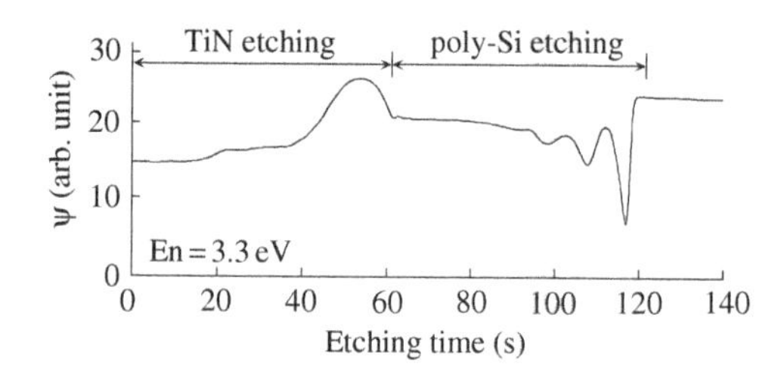

Figure 8.24 Time evolution of ψ observed during the plasma etching of the LSI structure shown in Fig. 8.23. Reprinted from *Thin Solid Films*, **313–314**, H. L. Maynard, N. Layadi, J. T. C. Lee, Plasma etching of submicron devices: in situ monitoring and control by multi-wavelength ellipsometry, 398–405, Copyright (1998), with permission from Elsevier.

is incorporated into this device to isolate transistors. Figure 8.24 shows the time evolution of ψ observed during the plasma etching of this device [41]. In Fig. 8.24, the sample is an LSI wafer consisting of the structure represented by Fig. 8.23. For this measurement, a PMSA instrument was used, and the probe light (En = 3.3 eV) was not intentionally aligned with any pattern on the wafer. Since a measured region ($2 \times 8\,\text{mm}^2$) is significantly larger than the scale of the device structure, ψ shown in Fig. 8.24 represents the averaged optical response of the etched structure.

In the etching process, the TiN layer is first etched, followed by etching of the poly-Si layer. As shown in Fig. 8.24, the value of ψ is almost constant up to 40 s, since the penetration depth of light in the TiN layer is quite small. However, ψ shows a large variation as the TiN thickness becomes smaller. Similarly, the change in ψ is negligible in the initial stages of poly-Si etching, but the probe light reaches the poly-Si/gate oxide interface at $t = 90$ s and, after that, optical interference effects appear. The variation in ψ stops completely after the removal of the poly-Si layer. The change in ψ observed during poly-Si etching is similar to that when the thickness direction is reversed in Fig. 5.8. In this way, etching processes can be characterized qualitatively from the time evolution of (ψ, Δ). If the gate oxide layer is exposed to the plasma, the plasma may damage the gate oxide layer. Thus, it is desirable to stop the plasma immediately after the complete removal of the poly-Si layer. As confirmed from Fig. 8.24, it is possible to perform such process control based on ellipsometry measurement.

If we employ the island-film model described in Section 5.3.3, quantitative analysis can also be performed. Recall that the amplitude reflection coefficient of the island-film model is calculated from area fractions (surface coverage ratios) [see Eq. (5.48)]. In this example, the analysis of the LSI structure was performed by using the area fractions #1–#4 indicated in Fig. 8.23. In this case, (ψ, Δ) is expressed by the following equation [41]:

$$\tan\psi \exp(i\Delta) = \left(\sum_{j=1}^{4} A_j r_{j,\mathrm{p}}\right) \Big/ \left(\sum_{j=1}^{4} A_j r_{j,\mathrm{s}}\right) \tag{8.5}$$

where A_j represent the area fractions of the area #1–#4 and $A_1 + A_2 + A_3 + A_4 = 1$. In Eq. (8.5), $r_{j,\mathrm{p}}$ and $r_{j,\mathrm{s}}$ show the amplitude reflection coefficients of the area j for p- and s-polarizations, respectively. When the dielectric functions of all the layers are known, the layer thicknesses and A_j of the LSI structure are estimated from LRA. Since A_j does not change during the etching, we can fix the parameter values for A_j. Consequently, the analysis parameters become the thicknesses of the TiN and poly-Si layers only. In this analysis, the influence of light diffraction, caused by the three-dimensional structure of the LSI device, was neglected.

Figure 8.25 shows the real-time characterization of the etching process based on the above analysis using the island-film model [41]. The control of the etching process was performed by simply terminating the plasma when the poly-Si layer thickness became zero. By employing this method, we can change etching conditions depending on the poly-Si layer thickness to suppress plasma damage. It has been reported that this analysis can still be performed even if the area fraction of the etched region is only 25 % [41]. Nevertheless, when the aspect ratio of device

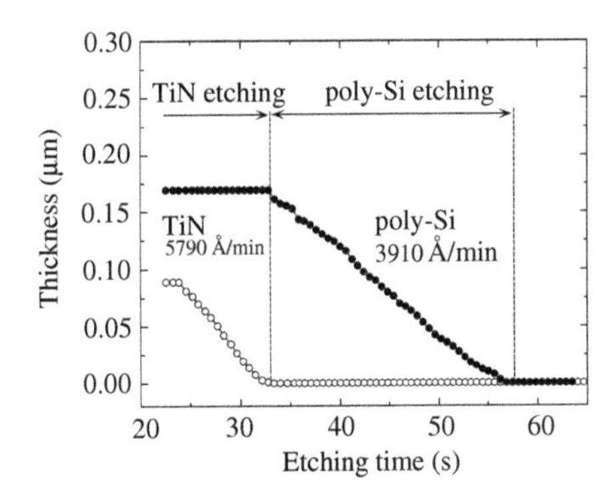

Figure 8.25 Time evolution of TiN and poly-Si layer thicknesses obtained from real-time characterization of the plasma etching process using LRA. Reprinted from *Thin Solid Films*, **313–314**, H. L. Maynard, N. Layadi, J. T. C. Lee, Plasma etching of submicron devices: in situ monitoring and control by multi-wavelength ellipsometry, 398–405, Copyright (1998), with permission from Elsevier.

structures is high, the characterization becomes difficult, since the probe light will not reach the etching surface due to the shadowing effect [42,46]. Recently, more advanced analysis that includes the effect of light scattering during etching has also been reported [47].

8.3.3 PROCESS CONTROL BY VIRTUAL SUBSTRATE APPROXIMATION (VSA)

Here, we will look at feedback control of $Al_xGa_{1-x}As$ composition [14] and CdTe growth rate [20] using VSA. As mentioned earlier, when VSA is applied for composition control, the parameterization of dielectric function is necessary. In the reported example, the composition control of the $Al_xGa_{1-x}As$ layer was performed using single-wavelength ellipsometry (En = 2.6 eV), and ε_2 of $Al_xGa_{1-x}As$ at the process temperature (~600° C) was parameterized from $\varepsilon_2(x) \sim \varepsilon_2(0) - 16.9x - 0.33x^2 (x < 0.3)$ [14]. Here, x shows the Al composition, and $\varepsilon_2(0)$ represents the value of GaAs (600 °C) measured before the growth. In this analysis, only the value of ε_2

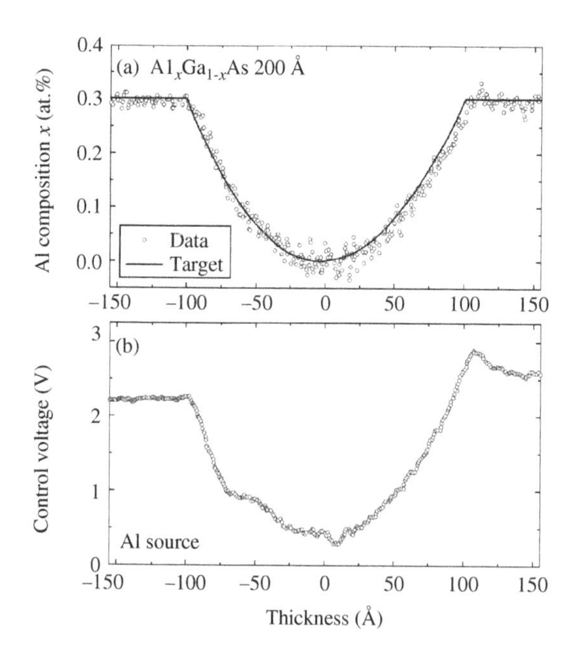

Figure 8.26 Feedback control of the Al composition in an $Al_xGa_{1-x}As$ layer by VSA using single-wavelength ellipsometry: (a) depth profile of the Al composition x and (b) control voltage of the flow meter for the Al source determined from real-time ellipsometry. Reprinted with permission from D. E. Aspnes, W. E. Quinn, M. C. Tamargo, M. A. A. Pudensi, S. A. Schwarz, M. J. S. P. Brasil, R. E. Nahory, and S. Gregory, Growth of $Al_xGa_{1-x}As$ parabolic quantum wells by real-time feedback control of composition, *Applied Physics Letters*, **60**, 1244–1246 (1992). Copyright 1992, American Institute of Physics.

was used. From single-wavelength ellipsometry, it is rather difficult to determine the growth rate and composition simultaneously. Thus, the growth rate of $Al_xGa_{1-x}As$ was approximated by $r = r_0/(1-x)$, where r_0 denotes the growth rate of GaAs. In this case, the analysis parameter of VSA is the Al composition x only. For the fabrication of the $Al_xGa_{1-x}As$ layer, chemical beam epitaxy (CBE) was employed. In the feedback control, the flow rate of the Al source was varied according to the difference between measured and target compositions.

Figure 8.26 shows (a) Al composition of $Al_xGa_{1-x}As$ and (b) control voltage of the flow meter determined from the ellipsometry measurement [14]. In Fig. 8.26(a), the open circles show the experimental data obtained from VSA and the solid line represents the target compositions. It can be seen that the measured compositions agree quite well with the target compositions. The growth rate of the GaAs was 0.95 Å/s, and the composition of $Al_xGa_{1-x}As$ with a thickness of ~3.1 Å was estimated from the analysis. In this example, the ellipsometry profile shows excellent agreement with the SIMS profile. In a quantum well structure, however, the dielectric function of a well layer changes (see Section 7.2.4). With respect to the $Al_xGa_{1-x}As$ quantum well structure shown in Fig. 8.26, the quantum effect can be neglected since the well layer is relatively thick [14].

Figure 8.27(a) shows the feedback control of a CdTe layer by VSA [20]. In this example, the CdTe layer was formed on a GaAs substrate by metal–organic vapor phase epitaxy (MOVPE), and the growth rate of the CdTe layer was controlled from the source gas flow. The measurement was performed with spectroscopic ellipsometry using a PSA_R instrument. When the dielectric functions of GaAs and CdTe at the process temperature (350 °C) are known, the analysis parameter in VSA is the growth rate of the CdTe layer only. In this example, the growth rate of the CdTe layer was initially adjusted to 4.8 Å/s and was changed to 2.8 Å/s at a process time of ~4 min. As shown in Fig. 8.27(a), the growth rates obtained from the

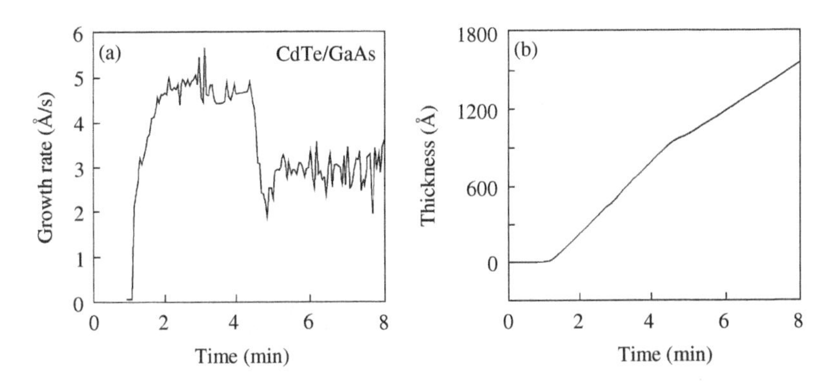

Figure 8.27 (a) Feedback control of the growth rate of a CdTe layer by VSA and (b) time evolution of the CdTe layer thickness estimated from LRA. In the feedback control, the CdTe growth rate was changed from 4.8 Å/s to 2.8 Å/s at a process time of ~4 min. Reprinted from *Thin Solid Films*, **233**, B. Johs, D. Doerr, S. Pittal, I. B. Bhat, and S. Dakshinamurthy, Real-time monitoring and control during MOVPE growth of CdTe using multiwavelength ellipsometry, 293–296, Copyright (1993), with permission from Elsevier.

analysis agree well with the target values. The repetition time of the measurement is 3 s in Fig. 8.27(a), and each measured value represents a growth rate of ~10 Å thick layer. Figure 8.27(b) shows the thickness of the CdTe layer obtained from LRA performed after the CdTe layer deposition [20]. In Fig. 8.27(b), the reduction in the growth rate after $t \sim 4$ min can be seen clearly.

As evidenced from the above examples, if we employ VSA, the composition and growth rate can be estimated relatively easily, and the feedback control of complicated structures can also be performed. However, the parameterization of dielectric function required for VSA is often complicated. This is the greatest drawback of real-time control by VSA. If the optical database for temperature and composition is perfected, real-time structural control based on ellipsometry technique will be widely applied.

REFERENCES

[1] I. An, Y. M. Li, C. R. Wronski, H. V. Nguyen, and R. W. Collins, *In situ* determination of dielectric functions and optical gap of ultrathin amorphous silicon by real time spectroscopic ellipsometry, *Appl. Phys. Lett.*, **59** (1991) 2543–2545.
[2] For a review, see R. W. Collins, I. An, H. V. Nguyen, Y. Li, and Y. Lu, Real-time spectroscopic ellipsometry studies of the nucleation, growth, and optical functions of thin films, Part I: tetrahedrally bonded materials, in *Physics of Thin Films*, edited by K. Vedam, vol. 19, 49–125 , Academic Press (1994).
[3] H. Fujiwara, J. Koh, P. I. Rovira, and R. W. Collins, Assessment of effective-medium theories in the analysis of nucleation and microscopic surface roughness evolution for semiconductor thin films, *Phys. Rev. B*, **61** (2000) 10832–10844.
[4] Y. M. Li, I. An, H. V. Nguyen, C. R. Wronski, and R. W. Collins, Real time spectroscopic ellipsometry determination of the evolution of amorphous semiconductor optical functions, bandgap, and microstructure, *J. Non Cryst. Solids*, **137 & 138** (1991) 787–790.
[5] R. W. Collins, J. S. Burnham, S. Kim, J. Koh, Y. Lu and C. R. Wronski, Insights into deposition processes for amorphous semiconductor materials and devices from real time spectroscopic ellipsometry, *J. Non-Cryst. Growth*, **198–200** (1996) 981–986.
[6] For a review, see R. W. Collins, I. An, H. Fujiwara, J. Lee, Y. Lu, J. Koh, and P. I. Rovira, Advances in multichannel spectroscopic ellipsometry, *Thin Solid Films*, **313–314** (1998) 18–32.
[7] For a review, see R. W. Collins, J. Koh, H. Fujiwara, P. I. Rovira, A. S. Ferlauto, J. A. Zapien, C. R. Wronski, and R. Messier, Recent progress in thin film growth analysis by multichannel spectroscopic ellipsometry, *Appl. Surf. Sci.*, **154–155** (2000) 217–228.
[8] H. Fujiwara, Y. Toyoshima, M. Kondo, and A. Matsuda, Interface-layer formation mechanism in a-Si:H thin-film growth studied by real-time spectroscopic ellipsometry and infrared spectroscopy, *Phys. Rev. B*, **60** (1999) 13598–13604.
[9] H. Fujiwara, M. Kondo, and A. Matsuda, Real-time spectroscopic ellipsometry studies of the nucleation and grain growth processes in microcrystalline silicon thin films, *Phys. Rev. B*, **63** (2001) 115306-1–9.
[10] H. Fujiwara, M. Kondo, and A. Matsuda, Interface-layer formation in microcrystalline Si:H growth on ZnO substrates studied by real-time spectroscopic ellipsometry and infrared spectroscopy, *J. Appl. Phys.*, **93** (2003) 2400–2409.
[11] D. E. Aspnes, Minimal-data approaches for determining outer-layer dielectric responses of films from kinetic reflectometric and ellipsometric measurements, *J. Opt. Soc. Am. A*, **10** (1993) 974–983.

[12] D. E. Aspnes, Optical approaches to determine near-surface compositions during epitaxy, *J. Vac. Sci. Technol. A*, **14** (1996) 960–966.

[13] For a review, see D. E. Aspnes, Real-time optical diagnostics for epitaxial growth, *Surf. Sci.*, **307–309** (1994) 1017–1027.

[14] D. E. Aspnes, W. E. Quinn, M. C. Tamargo, M. A. A. Pudensi, S. A. Schwarz, M. J. S. P. Brasil, R. E. Nahory, and S. Gregory, Growth of $Al_xGa_{1-x}As$ parabolic quantum wells by real-time feedback control of composition, *Appl. Phys. Lett.*, **60** (1992) 1244–1246.

[15] C. Pickering, Complementary in situ and post-deposition diagnostics of thin film semiconductor structures, *Thin Solid Films*, **313–314** (1998) 406–415.

[16] L. Mantese, K. Selinidis, P. T. Wilson, D. Lim, Y. Y. Jiang, J. G. Ekerdt, and M. C. Downer, In situ control and monitoring of doped and compositionally graded SiGe films using spectroscopic ellipsometry and second harmonic generation, *Appl. Surf. Sci.*, **154–155** (2000) 229–237.

[17] S. Kim, J. S. Burnham, J. Koh, L. Jiao, C. R. Wronski, and R. W. Collins, Real time spectroellipsometry characterization of optical gap profiles in compositionally-graded semiconductor structures: applications to bandgap engineering in amorphous silicon-carbon alloy solar cells, *J. Appl. Phys.*, **80** (1996) 2420–2429.

[18] H. Fujiwara, J. Koh, C. R. Wronski, and R. W. Collins, Application of real time spectroscopic ellipsometry for high resolution depth profiling of compositionally graded amorphous silicon alloy thin films, *Appl. Phys. Lett.*, **70** (1997) 2150–2152.

[19] H. Fujiwara, J. Koh, C. R. Wronski, R. W. Collins, and J. S. Burnham, Optical depth profiling of band gap engineered interfaces in amorphous silicon solar cells at monolayer resolution, *Appl. Phys. Lett.*, **72** (1998) 2993–2995.

[20] B. Johs, D. Doerr, S. Pittal, I. B. Bhat, and S. Dakshinamurthy, Real-time monitoring and control during MOVPE growth of CdTe using multiwavelength ellipsometry, *Thin Solid Films*, **233** (1993) 293–296.

[21] A. S. Ferlauto, G. M. Ferreira, R. J. Koval, J. M. Pearce, C. R. Wronski, R. W. Collins, M. M. Al-Jassim, and K. M. Jones, Evaluation of compositional depth profiles in mixed-phase (amorphous+crystalline) silicon films from real time spectroscopic ellipsometry, *Thin Solid Films*, **455–456** (2004) 665–669.

[22] M. Kildemo, S. Deniau, P. Bulkin, and B. Drévillon, Real-time control of the growth of silicon alloy multilayers by multiwavelength ellipsometry, *Thin Solid Films*, **290–291** (1996) 46–50.

[23] B. Johs, General virtual interface algorithm for *in situ* spectroscopic ellipsometric data analysis, *Thin Solid Films*, **455–456** (2004) 632–638.

[24] K. A. Conrad, R. K. Sampson, H. Z. Massoud, and E. A. Irene, Ellipsometric monitoring and control of the rapid thermal oxidation of silicon, *J. Vac. Sci. Technol. B*, **11** (1993) 2096–2101.

[25] Y. J. van der Meulen and N. C. Hien, Design and operation of an automated, high-temperature ellipsometer, *J. Opt. Soc. Am.*, **64** (1974) 804–811.

[26] G. N. Maracas, C. H. Kuo, S. Anand, R. Droopad, G. R. L. Sohie and T. Levola, Ellipsometry for III-V epitaxial growth diagnostics, *J. Vac. Sci. Technol. A*, **13** (1995) 727–732.

[27] B. Johs, C. Herzinger, J. H. Dinan, A. Cornfeld, J. D. Benson, D. Doctor, G. Olson, I. Ferguson, M. Pelczynski, P. Chow, C. H. Kuo, and S. Johnson, Real-time monitoring and control of epitaxial semiconductor growth in a production environment by *in situ* spectroscopic ellipsometry, *Thin Solid Films*, **313–314** (1998) 490–495.

[28] H. Z. Massoud, J. D. Plummer, and E. A. Irene, Thermal oxidation of silicon in dry oxygen growth-rate enhancement in the thin regime: I. Experimental results, *J. Electrochem. Soc.*, **132** (1985) 2685–2693.

[29] M. Wakagi, H. Fujiwara, and R. W. Collins, Real time spectroscopic ellipsometry for characterization of the crystallization of amorphous silicon by thermal annealing, *Thin Solid Films*, **313–314** (1998) 464–468.

[30] B. E. Deal and A. S. Grove, General relationship for the thermal oxidation of silicon, *J. Appl. Phys.*, **36** (1965) 3770–3778.
[31] H. Kageshima, K. Shiraishi and M. Uematsu, Universal theory of Si oxidation rate and importance of interfacial Si emission, *Jpn. J. Appl. Phys.*, **38** (1999) L971–974.
[32] C. M. Herzinger, B. Johs, W. A. McGahan, J. A. Woollam, and W. Paulson, Ellipsometric determination of optical constants for silicon and thermally grown silicon dioxide via a multi-sample, multi-wavelength, multi-angle investigation, *J. Appl. Phys.*, **83** (1998) 3323–3336.
[33] H. Fujiwara, J. Koh and R. W. Collins, Depth-profiles in compositionally graded amorphous silicon alloy thin films analyzed by real time spectroscopic ellipsometry, *Thin Solid Films*, **313–314** (1998) 474–478.
[34] T. Kamiya, K. Nakahata, A. Miida, C. M. Fortmann, and I. Shimizu, Control of orientation from random to (220) or (400) in polycrystalline silicon films, *Thin Solid Films*, **337** (1999) 18–22.
[35] H. Shirai, Surface morphology and crystalline size during growth of hydrogenated microcrystalline silicon by plasma-enhanced chemical vapor deposition, *Jpn. J. Appl. Phys.*, **34** (1995) 450–458.
[36] N. Layadi, P. Roca i Cabarrocas, B. Drévillon, and I. Solomon, Real-time spectroscopic ellipsometry study of the growth of amorphous and microcrystalline silicon thin films prepared by alternating silicon deposition and hydrogen plasma treatment, *Phys. Rev. B*, **52** (1995) 5136–5143.
[37] G. F. Feng, M. Katiyar, J. R. Abelson, and N. Maley, Dielectric functions and electronic band states of a-Si and a-Si:H, *Phys. Rev. B*, **45** (1992) 9103–9107.
[38] H. Fujiwara, M. Kondo and A. Matsuda, Stress-induced nucleation of microcrystalline silicon from amorphous phase, *Jpn. J. Appl. Phys.*, **41** (2002) 2821–2828.
[39] For a review, see E. A. Irene, Applications of spectroscopic ellipsometry to microelectronics, *Thin Solid Films*, **233** (1993) 96–111.
[40] H. Fujiwara and M. Kondo, Real-time monitoring and process control in amorphous/crystalline silicon heterojunction solar cells by spectroscopic ellipsometry and infrared spectroscopy, *Appl. Phys. Lett.*, **86** (2005) 032112-1–3.
[41] H. L. Maynard, N. Layadi, J. T. C. Lee, Plasma etching of submicron devices: in situ monitoring and control by multi-wavelength ellipsometry, *Thin Solid Films*, **313–314** (1998) 398–405.
[42] S. Cho, P. G. Snyder, C. M. Herzinger and B. Johs, Etch depth control in bulk GaAs using patterning and real time spectroscopic ellipsometry, *J. Vac. Sci. Technol. B*, **20** (2002) 197–202.
[43] M. Beaudoin, S. R. Johnson, M. D. Boonzaayer, Y.-H. Zhang, and B. Johs, Use of spectroscopic ellipsometry for feedback control during the growth of thin AlAs layers, *J. Vac. Sci. Technol. B*, **17** (1999) 1233–1236.
[44] M. Kildemo, P. Bulkin, B. Drévillon, and O. Hunderi, Real-time control by multiwavelength ellipsometry of plasma-deposited multilayers on glass by use of an incoherent-reflection model, *Appl. Opt.*, **36** (1997) 6352–6359.
[45] H. Fujiwara, J. Koh, Y. Lee, C. R. Wronski, and R. W. Collins, Real-time spectroscopic ellipsometry characterization of structural and thermal equilibration of amorphous silicon-carbon alloy p layers in p-i-n solar cell fabrication, *J. Appl. Phys.*, **84** (1998) 2278–2286.
[46] M. Haverlag and G. S. Oehrlein, In situ ellipsometry and reflectometry during etching of patterned surfaces: experiments and simulations, *J. Vac. Sci. Technol. B*, **10** (1992) 2412–2418.
[47] H. Huang and F. L. Terry, Jr, Spectroscopic ellipsometry and reflectometry from gratings (Scatterometry) for critical dimension measurement and in situ, real-time process monitoring, *Thin Solid Films*, **455–456** (2004) 828–836.

Appendix 1 Trigonometric Functions

a) Definitions of trigonometric functions

$$\sin A = c/a \qquad \cot A = b/c$$
$$\cos A = b/a \qquad \sec A = a/b$$
$$\tan A = c/b \qquad \csc A = a/c$$

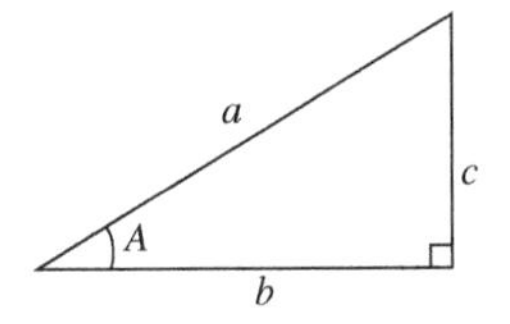

b) Basic formulas of trigonometric functions

$$\tan A = \sin A/\cos A \qquad \cot A = 1/\tan A$$
$$\cos^2 A + \sin^2 A = 1 \qquad \sec A = 1/\cos A$$

c)

Table A1.1 Conversion of trigonometric functions

	$-A$	$90° \pm A$ $\pi/2 \pm A$	$180° \pm A$ $\pi \pm A$
sin( ) =	$-\sin A$	$\cos A$	$\mp \sin A$
cos( ) =	$\cos A$	$\mp \sin A$	$-\cos A$
tan( ) =	$-\tan A$	$\mp \cot A$	$\pm \tan A$

d) Addition theorem

$$\sin(A \pm B) = \sin A \cos B \pm \cos A \sin B$$
$$\cos(A \pm B) = \cos A \cos B \mp \sin A \sin B$$
$$\tan(A \pm B) = \frac{\tan A \pm \tan B}{1 \mp \tan A \tan B}$$

Spectroscopic Ellipsometry: Principles and Applications H. Fujiwara

e) Double-angle formulas

$$\sin 2A = 2\sin A \cos A$$

$$\tan 2A = \frac{2\tan A}{1-\tan^2 A}$$

$$\cos 2A = \cos^2 A - \sin^2 A = \frac{1-\tan^2 A}{1+\tan^2 A}$$

f) Power of trigonometric functions

$$\sin^2 A = \frac{1}{2}(1-\cos 2A) \qquad \cos^2 A = \frac{1}{2}(1+\cos 2A)$$

Appendix 2 Definitions of Optical Constants

The optics and physics fields use different definitions (or conventions) for the optical constants. The definitions used in the optics and physics fields are summarized in Table A2.1. Unfortunately, since both definitions have been employed widely, these definitions are highly confusing [1,2]. Basically, the difference in these definitions originates from the definition for the phase of light waves. When the phase of a light

Table A2.1 Definitions of the optical constants used in the optics and physics fields

	Optics	Physics	Section
Phase of light	$(\omega t - Kx + \delta)$	$(Kx - \omega t + \delta)$	2.1.1
When the initial phase δ is positive	Wave advances	Wave lags	3.1.1
Complex refractive index	$N \equiv n - ik$	$N \equiv n + ik$	2.1.3
Complex dielectric constant	$\varepsilon \equiv \varepsilon_1 - i\varepsilon_2$	$\varepsilon \equiv \varepsilon_1 + i\varepsilon_2$	2.2.2
Optical interference	$r_{012} = \dfrac{r_{01} + r_{12}\exp(-i2\beta)}{1 + r_{01}r_{12}\exp(-i2\beta)}$	$r_{012} = \dfrac{r_{01} + r_{12}\exp(i2\beta)}{1 + r_{01}r_{12}\exp(i2\beta)}$	2.4.1
Right-circular polarization	$\dfrac{1}{\sqrt{2}}\begin{bmatrix} 1 \\ i \end{bmatrix}$	$\dfrac{1}{\sqrt{2}}\begin{bmatrix} 1 \\ -i \end{bmatrix}$	3.3.1
Left-circular polarization	$\dfrac{1}{\sqrt{2}}\begin{bmatrix} 1 \\ -i \end{bmatrix}$	$\dfrac{1}{\sqrt{2}}\begin{bmatrix} 1 \\ i \end{bmatrix}$	3.3.1
(ψ, Δ)	$\rho \equiv \tan\psi \exp(i\Delta)$	$\rho \equiv \tan\psi \exp(-i\Delta)$	4.1.1
Lorentz model	$\varepsilon = 1 + \dfrac{A}{\omega_0^2 - \omega^2 + i\Gamma\omega}$	$\varepsilon = 1 + \dfrac{A}{\omega_0^2 - \omega^2 - i\Gamma\omega}$	5.2.1
Drude model	$\varepsilon = \varepsilon_\infty\left(1 - \dfrac{\omega_p^2}{\omega^2 - i\omega\Gamma}\right)$	$\varepsilon = \varepsilon_\infty\left(1 - \dfrac{\omega_p^2}{\omega^2 + i\omega\Gamma}\right)$	5.2.5
Pseudo-dielectric function	$\langle\varepsilon\rangle \equiv \langle\varepsilon_1\rangle - i\langle\varepsilon_2\rangle$	$\langle\varepsilon\rangle \equiv \langle\varepsilon_1\rangle + i\langle\varepsilon_2\rangle$	5.4.2
Berreman's equation	$\dfrac{\partial\Psi}{\partial z} = -i\dfrac{\omega}{c}\Delta_B\Psi$	$\dfrac{\partial\Psi}{\partial z} = i\dfrac{\omega}{c}\Delta_B\Psi$	6.3.1

Spectroscopic Ellipsometry: Principles and Applications H. Fujiwara

wave is defined by $(\omega t + \delta)$, the wave advances forward with an increase in the initial phase δ. In this case, we can understand variations in polarized light waves more easily. This is the reason why the phase has been defined by $(\omega t - Kx + \delta)$ in this book. In the second international conference on ellipsometry, held at the University of Nebraska in 1968, the definition of the optics field shown in Table A2.1 was adopted as the convention for ellipsometry [1]. Thus, the convention used in ellipsometry studies is sometimes referred to as 'The Nebraska Convention'. Nevertheless, almost all dielectric function models, reported in physics journals including *Physical Review*, have been derived using the definition from the physics field. Often, these definitions can be changed by simply reversing the signs of imaginary numbers (i.e., $\pm i \rightarrow \mp i$). When this method cannot be employed, we obtain real and imaginary parts separately and then convert the definition. In the conversion of $\varepsilon = \varepsilon_1 + i\varepsilon_2 \rightarrow \varepsilon = \varepsilon_1 - i\varepsilon_2$, for example, the real and imaginary parts are expressed by $\mathrm{Re}(\varepsilon)$ and $\mathrm{Im}(\varepsilon)$, respectively, and thus the definition can be changed from $\varepsilon = \mathrm{Re}(\varepsilon) - i\mathrm{Im}(\varepsilon)$.

REFERENCES

[1] R. H. Muller, Definitions and conventions in ellipsometry, *Surf. Sci.*, **16** (1969) 14–33.
[2] R. T. Holm, Convention confusions, in *Handbook of Optical Constants of Solids II*, edited by E. D. Palik, Chapter 2, 21–55, Academic Press, San Diego (1991).

Appendix 3 Maxwell's Equations for Conductors

Maxwell's equations for conductors including metals and semiconductors are expressed by

$$\mathrm{div}\,\boldsymbol{E} = \rho/\varepsilon_{\mathrm{p}} \tag{A3.1}$$

$$\mathrm{div}\,\boldsymbol{B} = 0 \tag{A3.2}$$

$$\mathrm{rot}\,\boldsymbol{E} = -\frac{\partial \boldsymbol{B}}{\partial t} \tag{A3.3}$$

$$\mathrm{rot}\,\boldsymbol{B} = \mu_{\mathrm{p}}\left(\varepsilon_{\mathrm{p}}\frac{\partial \boldsymbol{E}}{\partial t} + \boldsymbol{J}\right) \tag{A3.4}$$

where $\boldsymbol{E}$ is the electric field and $\boldsymbol{B}$ shows the magnetic induction (see Fig. 2.4). ε_{p} and μ_{p} are the permittivity (see Section 2.2.2) and permeability of materials, respectively. ρ in Eq. (A3.1) shows the electric charge, and $\boldsymbol{J}$ in Eq. (A3.4) indicates the current density given by $\boldsymbol{J} = \sigma\boldsymbol{E}$, where σ is the conductivity. Eq. (A3.1) represents Gauss's law, and Eq. (A3.2) implies that the north and south poles of magnets do not exist separately. Eqs. (A3.3) and (A3.4) describe Faraday's induction law and Ampère's law modified by Maxwell, respectively. In order to derive an equation that expresses an electromagnetic wave propagating inside a conductor, we first multiply both sides of Eq. (A3.3) by rot:

$$\mathrm{rot}\ \mathrm{rot}\,\boldsymbol{E} = -\frac{\partial}{\partial t}\mathrm{rot}\,\boldsymbol{B} \tag{A3.5}$$

From the formula of vector calculation, it follows that $\mathrm{rot}\ \mathrm{rot}\,\boldsymbol{E} = \mathrm{grad}\ \mathrm{div}\,\boldsymbol{E} - \nabla^2\boldsymbol{E}$, where $\nabla^2 = \partial^2/\partial x^2 + \partial^2/\partial y^2 + \partial^2/\partial z^2$. By substituting Eq. (A3.4) into (A3.5) and assuming $\rho = 0$ ($\mathrm{div}\,\boldsymbol{E} = 0$), we get

$$\nabla^2\boldsymbol{E} = \varepsilon_{\mathrm{p}}\mu_{\mathrm{p}}\frac{\partial^2 \boldsymbol{E}}{\partial t^2} + \mu_{\mathrm{p}}\sigma\frac{\partial \boldsymbol{E}}{\partial t} \tag{A3.6}$$

Spectroscopic Ellipsometry: Principles and Applications H. Fujiwara

This shows the wave equation for an electromagnetic wave inside a conductor.

An electromagnetic wave propagating in vacuum can be expressed by inserting $\varepsilon_p = \varepsilon_0$, $\mu_p = \mu_0$ and $\sigma = 0$ into Eq. (A3.6):

$$\nabla^2 \boldsymbol{E} = \varepsilon_0 \mu_0 \frac{\partial^2 \boldsymbol{E}}{\partial t^2} \tag{A3.7}$$

where μ_0 shows the permeability of a vacuum. On the other hand, the wave equation for general waves is written as

$$\nabla^2 f = \frac{1}{s^2} \frac{\partial^2 f}{\partial t^2} \tag{A3.8}$$

Here, s represents the speed of the wave. With respect to a wave created by vibrating a string, its speed is described by $s = \sqrt{T/m}$, where T and m indicate the tension and line density, respectively. From comparison between Eqs. (A3.7) and (A3.8), it can be understood that the speed of electromagnetic waves propagating in a vacuum (light speed) is given by

$$c = \frac{1}{\sqrt{\varepsilon_0 \mu_0}} = 2.99792 \times 10^8 \text{ m/s} \tag{A3.9}$$

From Eqs. (2.43) and (A3.9), on the other hand, we obtain $\varepsilon = \varepsilon_p/\varepsilon_0$ and $\varepsilon_0 \mu_0 = 1/c^2$. Moreover, $\mu_p = \mu_0$ holds at the frequency of light ($\sim 10^{14}$ Hz). Thus, we get

$$\varepsilon_p \mu_p = \varepsilon_0 \mu_0 \varepsilon = \varepsilon / c^2 \tag{A3.10}$$

$$\mu_p \sigma = \mu_0 \sigma = \sigma / \left(\varepsilon_0 c^2 \right) \tag{A3.11}$$

By substituting Eqs. (A3.10) and (A3.11) into Eq. (A3.6), we obtain

$$\nabla^2 \boldsymbol{E} = \frac{\varepsilon}{c^2} \frac{\partial^2 \boldsymbol{E}}{\partial t^2} + \frac{\sigma}{\varepsilon_0 c^2} \frac{\partial \boldsymbol{E}}{\partial t} \tag{A3.12}$$

If we assume that the solution of this wave equation is expressed by

$$\boldsymbol{E} = \boldsymbol{E}_0 \exp[\mathrm{i}(\omega t - Kx)] \tag{A3.13}$$

it follows that $\nabla^2 \boldsymbol{E} = -K^2 \boldsymbol{E}$, $\partial^2 \boldsymbol{E}/\partial t^2 = -\omega^2 \boldsymbol{E}$ and $\partial \boldsymbol{E}/\partial t = \mathrm{i}\omega \boldsymbol{E}$. By inserting these results into Eq. (A3.12) and rearranging the terms, we get

$$K^2 = \frac{\omega^2}{c^2} \left(\varepsilon - \mathrm{i} \frac{\sigma}{\varepsilon_0 \omega} \right) = \frac{\omega^2}{c^2} (\varepsilon_1 - \mathrm{i}\varepsilon_2) \tag{A3.14}$$

Accordingly, the imaginary part of the complex dielectric constant $\varepsilon \equiv \varepsilon_1 - \mathrm{i}\varepsilon_2$ is defined by $\varepsilon_2 \equiv \sigma/(\varepsilon_0 \omega)$. In CGS units, $\varepsilon_0 = 1$ and Eq. (A3.14) derived using SI units is converted by $\varepsilon_0 \rightarrow 1/(4\pi)$. Thus, ε_2 is defined by $\varepsilon_2 \equiv 4\pi\sigma/\omega$ in CGS units. If we define the complex refractive index by

$$N^2 \equiv (\varepsilon_1 - i\varepsilon_2) \tag{A3.15}$$

and substitute this into Eq. (A3.14), the following equation is obtained:

$$K = \frac{\omega}{c}N = \frac{2\pi}{\lambda}N \tag{A3.16}$$

This is the same equation as the one shown in Eq. (2.29), if we replace n in Eq. (2.29) with N.

Appendix 4 Jones–Mueller Matrix Conversion

Here, we will look at conversion from a Jones matrix to a Mueller matrix using a procedure reviewed by Azzam and Bashara [1]. As we have seen in Chapter 3, the Jones vector basically describes the electric fields in the x and y directions (see Section 3.3.1), whereas the Stokes vector shows the light intensities of different polarization states (see Section 3.4.1). Thus, we can derive the formula for the Jones-to-Mueller matrix conversion by calculating the light intensities (the Stokes vector) from the electric fields (the Jones vector). However, the Jones matrix is defined by a 2×2 matrix, while a 4×4 matrix is used to describe the Mueller matrix. In the conversion, therefore, the 4×4 Mueller matrix is expressed from the direct product (or Kronecker product) of the 2×2 Jones matrices. If $\boldsymbol{A}$ and $\boldsymbol{B}$ are 2×2 matrices given by

$$\boldsymbol{A} = \begin{bmatrix} a_{11} & a_{12} \\ a_{21} & a_{22} \end{bmatrix} \quad \boldsymbol{B} = \begin{bmatrix} b_{11} & b_{12} \\ b_{21} & b_{22} \end{bmatrix} \tag{A4.1}$$

the direct product of $\boldsymbol{A}$ and $\boldsymbol{B}$ is described as follows [2]:

$$\boldsymbol{A} \otimes \boldsymbol{B} = \begin{bmatrix} a_{11}\boldsymbol{B} & a_{12}\boldsymbol{B} \\ a_{21}\boldsymbol{B} & a_{22}\boldsymbol{B} \end{bmatrix} = \begin{bmatrix} a_{11}b_{11} & a_{11}b_{12} & a_{12}b_{11} & a_{12}b_{12} \\ a_{11}b_{21} & a_{11}b_{22} & a_{12}b_{21} & a_{12}b_{22} \\ a_{21}b_{11} & a_{21}b_{12} & a_{22}b_{11} & a_{22}b_{12} \\ a_{21}b_{21} & a_{21}b_{22} & a_{22}b_{21} & a_{22}b_{22} \end{bmatrix} \tag{A4.2}$$

Now, let us consider the transformation of the Jones vector described by

$$\boldsymbol{E}_{\text{out}} = \boldsymbol{J}\boldsymbol{E}_{\text{in}} \tag{A4.3}$$

or, in matrix form,

$$\begin{bmatrix} E_x \\ E_y \end{bmatrix}_{\text{out}} = \begin{bmatrix} J_{11} & J_{12} \\ J_{21} & J_{22} \end{bmatrix} \begin{bmatrix} E_x \\ E_y \end{bmatrix}_{\text{in}} \tag{A4.4}$$

Spectroscopic Ellipsometry: Principles and Applications H. Fujiwara

where $\boldsymbol{E}_{\text{in}}$ and $\boldsymbol{E}_{\text{out}}$ represent the Jones vectors of the incoming and outgoing light waves, respectively, and $\boldsymbol{J}$ shows the Jones matrix. Eq. (A4.4) corresponds to Eqs. (4.8) and (6.12) in ellipsometry measurement. In order to express the light intensities of the Jones vectors $\boldsymbol{E}_{\text{in}}$ and $\boldsymbol{E}_{\text{out}}$, we calculate the direct product of Eq. (A4.3) and its complex conjugate:

$$\begin{aligned}\boldsymbol{E}_{\text{out}} \otimes \boldsymbol{E}_{\text{out}}^{*} &= \boldsymbol{J}\boldsymbol{E}_{\text{in}} \otimes \boldsymbol{J}^{*}\boldsymbol{E}_{\text{in}}^{*} \\ &= (\boldsymbol{J} \otimes \boldsymbol{J}^{*})(\boldsymbol{E}_{\text{in}} \otimes \boldsymbol{E}_{\text{in}}^{*})\end{aligned} \tag{A4.5}$$

If we apply Eq. (A4.2), the direct product $\boldsymbol{E} \otimes \boldsymbol{E}^{*}$ in Eq. (A4.5) produces a 4×1 vector, known as the coherency vector:

$$\boldsymbol{C} = \boldsymbol{E} \otimes \boldsymbol{E}^{*} = \begin{bmatrix} E_x \\ E_y \end{bmatrix} \otimes \begin{bmatrix} E_x^{*} \\ E_y^{*} \end{bmatrix} = \begin{bmatrix} E_x E_x^{*} \\ E_x E_y^{*} \\ E_y E_x^{*} \\ E_y E_y^{*} \end{bmatrix} \tag{A4.6}$$

Thus, from Eq. (A4.6), Eq. (A4.5) is rewritten as

$$\boldsymbol{C}_{\text{out}} = (\boldsymbol{J} \otimes \boldsymbol{J}^{*})\boldsymbol{C}_{\text{in}} \tag{A4.7}$$

If we use the Stokes parameters shown in Table 3.3 (Electric field A), the Stokes vector $\boldsymbol{S}$ can be expressed from the coherency vector $\boldsymbol{C}$ as follows:

$$\boldsymbol{S} = \boldsymbol{A}\boldsymbol{C} \tag{A4.8}$$

where

$$\boldsymbol{A} = \begin{bmatrix} 1 & 0 & 0 & 1 \\ 1 & 0 & 0 & -1 \\ 0 & 1 & 1 & 0 \\ 0 & \mathrm{i} & -\mathrm{i} & 0 \end{bmatrix} \tag{A4.9}$$

In matrix form, Eq. (A4.8) is expressed by

$$\begin{bmatrix} S_0 \\ S_1 \\ S_2 \\ S_3 \end{bmatrix} = \begin{bmatrix} 1 & 0 & 0 & 1 \\ 1 & 0 & 0 & -1 \\ 0 & 1 & 1 & 0 \\ 0 & \mathrm{i} & -\mathrm{i} & 0 \end{bmatrix} \begin{bmatrix} E_x E_x^{*} \\ E_x E_y^{*} \\ E_y E_x^{*} \\ E_y E_y^{*} \end{bmatrix} = \begin{bmatrix} E_x E_x^{*} + E_y E_y^{*} \\ E_x E_x^{*} - E_y E_y^{*} \\ E_x E_y^{*} + E_y E_x^{*} \\ \mathrm{i}\left(E_x E_y^{*} - E_y E_x^{*}\right) \end{bmatrix} \tag{A4.10}$$

Accordingly, Eq. (A4.10) describes the Stokes parameters shown in Table 3.3.

From Eq. (A4.8), it follows that $\boldsymbol{C}_{\text{in}} = \boldsymbol{A}^{-1}\boldsymbol{S}_{\text{in}}$ and $\boldsymbol{C}_{\text{out}} = \boldsymbol{A}^{-1}\boldsymbol{S}_{\text{out}}$. By substituting these into Eq. (A4.7), we obtain

$$\begin{aligned}\boldsymbol{S}_{\text{out}} &= \boldsymbol{A}(\boldsymbol{J} \otimes \boldsymbol{J}^{*})\boldsymbol{A}^{-1}\boldsymbol{S}_{\text{in}} \\ &= \boldsymbol{M}\boldsymbol{S}_{\text{in}}\end{aligned} \tag{A4.11}$$

where $\boldsymbol{M}$ represents the Mueller matrix given by

$$\boldsymbol{M} = \boldsymbol{A}(\boldsymbol{J} \otimes \boldsymbol{J}^*)\boldsymbol{A}^{-1} \tag{A4.12}$$

From Eq. (A4.12), we obtain the formula for the Jones-to-Mueller matrix conversion [1]:

$$\boldsymbol{M} = \begin{bmatrix} \frac{1}{2}(E_1+E_2+E_3+E_4) & \frac{1}{2}(E_1-E_2-E_3+E_4) & F_{13}+F_{42} & -G_{13}-G_{42} \\ \frac{1}{2}(E_1-E_2+E_3-E_4) & \frac{1}{2}(E_1+E_2-E_3-E_4) & F_{13}-F_{42} & -G_{13}+G_{42} \\ F_{14}+F_{32} & F_{14}-F_{32} & F_{12}+F_{34} & -G_{12}+G_{34} \\ G_{14}+G_{32} & G_{14}-G_{32} & G_{12}+G_{34} & F_{12}-F_{34} \end{bmatrix} \tag{A4.13}$$

where

$$\begin{aligned} E_k &= J_k J_k^* \quad k = 1, 2, 3, 4 \\ F_{kl} &= F_{lk} = \mathrm{Re}(J_k J_l^*) = \mathrm{Re}(J_k^* J_l) \quad k, l = 1, 2, 3, 4 \\ G_{kl} &= -G_{lk} = \mathrm{Im}(J_k^* J_l) = -\mathrm{Im}(J_k J_l^*) \quad k, l = 1, 2, 3, 4 \end{aligned} \tag{A4.14}$$

In Eq. (A4.14), J_1, J_2, J_3 and J_4 represent J_{11}, J_{22}, J_{12}, and J_{21} in Eq. (A4.4), respectively. By applying Eq. (A4.13), we can calculate the Muller matrices shown in Table 3.2 from the Jones matrices. It should be emphasized that we cannot perform the Jones-to-Muller matrix conversion when an optical system is depolarizing since the Jones matrix cannot describe partially polarized light (see Section 3.4.3).

REFERENCES

[1] R. M. A. Azzam and N. M. Bashara, *Ellipsometry and Polarized Light*, North-Holland, Amsterdam (1977).

[2] G. B. Arfken and H. J. Weber, *Mathematical Methods for Physicists*, 4th edition, Academic Press, San Diego (1995).

Appendix 5 Kramers–Kronig Relations

In 1926 and 1927, Kramers and Kronig independently derived equations, now known as the Kramers–Kronig relations. The Kramers–Kronig relations can be derived by considering the integral of the form

$$I = P\int_{-\infty}^{+\infty} \frac{f(x)}{x-a}\mathrm{d}x = \lim_{\delta\to 0}\left[\int_{-\infty}^{a-\delta} \frac{f(x)}{x-a}\mathrm{d}x + \int_{a+\delta}^{+\infty} \frac{f(x)}{x-a}\mathrm{d}x\right] \quad \text{(A5.1)}$$

P in Eq. (A5.1) shows the principal value of the integral [see Eq. (5.34)]. To obtain the Kramers–Kronig relations, we change the parameters in Eq. (A5.1) by $x \to \omega'$ and $f(x) \to \chi(\omega')$. Here, ω' represents the complex angular frequency ($\omega' = \omega'_1 + i\omega'_2$), and $\chi(\omega')$ shows the dielectric susceptibility expressed by $\varepsilon = 1 + \chi$ [see Eq. (2.44)]. We also replace a in Eq. (A5.1) with a constant ω ($\omega > 0$). It should be noted that the Kramers–Kronig relations are mathematically exact, and we will use the definition of $\varepsilon \equiv \varepsilon_1 + i\varepsilon_2$ in Appendix 5.

From the above conversion for Eq. (A5.1), we obtain

$$I = P\int_{-\infty}^{+\infty} \frac{\chi(\omega')}{\omega' - \omega}\mathrm{d}\omega' \quad \text{(A5.2)}$$

Figure A5.1(a) illustrates the integration of Eq. (A5.2) on the complex plane of ω'. Since the angular frequency of actual spectra is a real number (i.e., $\omega' = \omega'_1$), the integration of Eq. (A5.2) is taken along the real axis only. As shown in Fig. A5.1(a), however, there exists a pole at $\omega' = \omega$ since the denominator of Eq. (A5.2) becomes zero when $\omega' = \omega$. Now, let δ be the distance from the pole on the real axis. In this case, the integration shows a unique value only when δ approaches to zero equally from both sides of the pole. This integral value is obtained from the limit of $\delta \to 0$ and is referred to as the principal value of the integral. From the paths of integration shown in Fig. A5.1(b)–(d), it is clear that the integral of Fig. A5.1(a) is described by

$$I = I_A - I_B - I_C \quad \text{(A5.3)}$$

where I_A, I_B and I_C represent the integral values of the contours A, B, and C shown in Figs. A5.1(b), A5.1(c), and A5.1(d), respectively.

Spectroscopic Ellipsometry: Principles and Applications H. Fujiwara

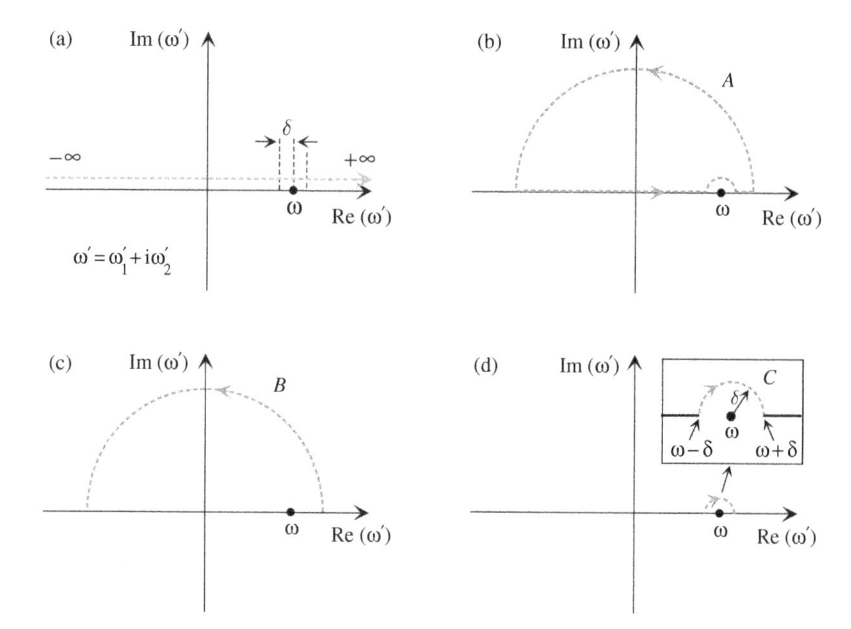

Figure A5.1 Integration of Eq. (A5.2) on the complex plane of ω'.

From Cauchy's theorem, it follows that $I_A = 0$. Now, in order to obtain I_B, we assume the Lorentz model [see Eq. (5.7)]. In this case, $\chi(\omega')$ is expressed by

$$\chi(\omega') = \frac{e^2 N_e}{\varepsilon_0 m_e} \frac{1}{\omega_0^2 - \omega'^2 - i\Gamma\omega'} \tag{A5.4}$$

When $|\omega'|$ is quite large, $\chi(\omega')$ becomes almost zero. In the condition $|\omega'| \to \infty$, therefore, we get $I_B = 0$. For the integration of C in Fig. A5.1(d), we use the polar coordinates given by $\omega' = \omega + \delta \exp(i\theta)$. It follows that $d\omega' = i\delta \exp(i\theta)d\theta$. By applying these equations, we get

$$I_c = \int_C \frac{\chi(\omega')}{\omega' - \omega} d\omega' = \int_\pi^0 i\chi[\omega + \delta \exp(i\theta)]d\theta = -i\pi\chi(\omega) \tag{A5.5}$$

For the transformation of Eq. (A5.5), $\delta \to 0$ was used. Since $I_A = I_B = 0$, I in Eq. (A5.3) is given by

$$I = i\pi\chi(\omega) \tag{A5.6}$$

Using $\chi = \chi_1 + i\chi_2$ and $\varepsilon = 1 + \chi$, we obtain

$$\chi_1 = \varepsilon_1 - 1 \quad \chi_2 = \varepsilon_2 \tag{A5.7}$$

By substituting these into Eq. (A5.6), the following equation is obtained:

$$I = -\pi\varepsilon_2(\omega) + \mathrm{i}\{\pi[\varepsilon_1(\omega) - 1]\} \tag{A5.8}$$

From $\chi = \chi_1 + \mathrm{i}\chi_2$ and Eq. (A5.7), on the other hand, Eq. (A5.2) can be transformed as follows:

$$I = \left[P\int_{-\infty}^{+\infty} \frac{\varepsilon_1(\omega') - 1}{\omega' - \omega}\mathrm{d}\omega'\right] + \mathrm{i}\left[P\int_{-\infty}^{+\infty} \frac{\varepsilon_2(\omega')}{\omega' - \omega}\mathrm{d}\omega'\right] \tag{A5.9}$$

Comparison between Eqs. (A5.8) and (A5.9) yields

$$\varepsilon_1(\omega) - 1 = \frac{P}{\pi}\int_{-\infty}^{+\infty} \frac{\varepsilon_2(\omega')}{\omega' - \omega}\mathrm{d}\omega' \tag{A5.10a}$$

$$\varepsilon_2(\omega) = -\frac{P}{\pi}\int_{-\infty}^{+\infty} \frac{\varepsilon_1(\omega') - 1}{\omega' - \omega}\mathrm{d}\omega' \tag{A5.10b}$$

Notice that the range of the integration in Eq. (A5.10) is from $-\infty$ to $+\infty$. However, since ω' is a positive real number, the range should be from 0 to $+\infty$. This conversion can be performed as described below. With respect to $\varepsilon(\omega)$, it is known that there is a symmetry relation that satisfies the following equation [1,2]:

$$\varepsilon(-\omega) = \varepsilon(\omega)^* \tag{A5.11}$$

Accordingly,

$$\varepsilon_1(-\omega) = \varepsilon_1(\omega) \quad \text{and} \quad \varepsilon_2(-\omega) = -\varepsilon_2(\omega) \tag{A5.12}$$

In general, the following equation holds:

$$\begin{aligned} P\int_{-\infty}^{+\infty} \frac{f(x)}{x-a}\mathrm{d}x &= P\left(\int_{0}^{+\infty} \frac{f(x)}{x-a}\mathrm{d}x + \int_{-\infty}^{0} \frac{f(x)}{x-a}\mathrm{d}x\right) \\ &= P\int_0^{\infty}\left[\frac{f(x)}{x-a} + \frac{f(-x)}{-x-a}\right]\mathrm{d}x \\ &= P\int_0^{\infty} \frac{x[f(x) - f(-x)] + a[f(x) + f(-x)]}{x^2 - a^2}\mathrm{d}x \end{aligned} \tag{A5.13}$$

By applying Eqs. (A5.12) and (A5.13) to Eq. (A5.10), we finally obtain

$$\varepsilon_1(\omega) - 1 = \frac{P}{\pi}\int_{-\infty}^{+\infty} \frac{\varepsilon_2(\omega')}{\omega' - \omega}\mathrm{d}\omega' = \frac{2}{\pi}P\int_0^{\infty} \frac{\omega'\varepsilon_2(\omega')}{\omega'^2 - \omega^2}\mathrm{d}\omega' \tag{A5.14a}$$

$$\varepsilon_2(\omega) = -\frac{P}{\pi}\int_{-\infty}^{+\infty} \frac{\varepsilon_1(\omega') - 1}{\omega' - \omega}\mathrm{d}\omega' = -\frac{2\omega}{\pi}P\int_0^{\infty} \frac{\varepsilon_1(\omega') - 1}{\omega'^2 - \omega^2}\mathrm{d}\omega' \tag{A5.14b}$$

The above equations have been shown in Eq. (5.33).

REFERENCES

[1] F. Wooten, *Optical Properties of Solids*, Academic Press, New York (1972).
[2] J. D. Jackson, *Classical Electrodynamics*, 3rd edition, John Wiley & Sons, Inc., New York (1999).

Index

Italic page numbers refer to diagrams and tables.

Printed in the USA
CPSIA information can be obtained
at www.ICGtesting.com
LVHW042133191023
761416LV00066B/8